AF814980

Springer Series in

SOLID-STATE SCIENCES 130

Springer
Berlin
Heidelberg
New York
Barcelona
Hong Kong
London
Milan
Paris
Singapore
Tokyo

Springer Series in

SOLID-STATE SCIENCES

Series Editors:

M. Cardona P. Fulde K. von Klitzing R. Merlin H.-J. Queisser H. Störmer

126 **Physical Properties of Quasicrystals**
Editor: Z.M. Stadnik

127 **Positron Annihilation in Semiconductors**
Defect Studies
By R. Krause-Rehberg and H.S. Leipner

128 **Magneto-Optics**
Editors: S. Sugano and N. Kojima

129 **Computational Materials Science**
From Ab Initio to Monte Carlo Methods
By K. Ohno, K. Esfarjani, and Y. Kawazoe

130 **Contact, Adhesion and Rupture of Elastic Solids**
By D. Maugis

Volumes 1–125 are listed at the end of the book.

D. Maugis

Contact, Adhesion and Rupture of Elastic Solids

With 186 Figures

Springer

Dr. Daniel Maugis

Laboratoire des Matériaux et des Structures du Génie Civil
CNRS-LCPC, Cité Descartes – Parc club de la haute maison
2, allée Kepler, F77420 Champs sur Marne

Series Editors:

Professor Dr., Dres. h. c. Manuel Cardona
Professor Dr., Dres. h. c. Peter Fulde*
Professor Dr., Dres. h. c. Klaus von Klitzing
Professor Dr., Dres. h. c. Hans-Joachim Queisser

Max-Planck-Institut für Festkörperforschung, Heisenbergstrasse 1, D-70569 Stuttgart, Germany
* Max-Planck-Institut für Physik komplexer Systeme, Nöthnitzer Strasse 38
D-01187 Dresden, Germany

Professor Dr., Dres. h. c. Roberto Merlin

Department of Physics, 5000 East University, University of Michigan
Ann Arbor, MI 48109-1120, USA

Professor Dr. Horst Störmer

Dept. Phys. and Dept. Appl. Physics, Columbia University, New York, NY 10023 and
Bell Labs., Lucent Technologies, Murray Hill, NJ 07974, USA

ISSN 0171-1873

ISBN 3-540-66113-1 Springer-Verlag Berlin Heidelberg New York

Library of Congress Cataloging-in-Publication Data

Maugis, D. (Daniel), 1934-
Contact, adhesion, and rupture of elastic solids / D. Maugis.
p. cm. -- (Springer series in solid-state sciences, ISSN 0171-1873 ; 130)
Includes bibliographical references and index.
ISBN 3540661131 (alk. paper)
1. Adhesion. 2. Fracture mechanics. 3. Elastic solids. I. Title. II. Series.

QC183.M47 1999
530.4'17--dc21 99-050194

This work is subject to copyright. All rights reserved, wheter the whole or part of the material is concerned, specifically the right of translation, reprinting, reuse of illustrations, recitation, broadcasting, reproduction on microfilm or in any other way, and storage in data banks. Duplication of this publication or parts thereof is permitted only under the provisions of the German Copyright Law of September 9, 1965, in its current version, and permission for use must always be obtained from Springer-Verlag. Violations are liable for prosecution under the German Copyright Law.

© Springer-Verlag Berlin Heidelberg 2000
Printed in Germany

The use of descriptive names, registered names, trademarks, etc. in this publication does not imply, even in the absence of a specific statement, that such names are exempt from the relevant protective laws and regulations and therefore free for general use.

Production: Pro Edit GmbH, Heidelberg
Cover Concept: eStudio Calamar Steinen
Cover Production: *design & production* GmbH, Heidelberg
Typesetting: LE-TEX Jelonek, Schmidt & Vöckler GbR, Leipzig

Printed on acid-free paper SPIN 10732471 57/3144ML 5 4 3 2 1 0

Foreword

In 1970 an investigation into rubber friction, sponsored by a manufacturer of automobile windscreen wipers, was being carried out at the Cavendish Laboratory in Cambridge, England. When a smooth spherical slider of soft rubber was placed in contact with flat glass or perspex, the compliance of the rubber enabled it to conform to any slight roughness of the two surfaces such that perfect contact was obtained. In these circumstances the surfaces were found to adhere: under load the contact area exceeded that predicted by the Hertz theory of elastic contact, a contact area of finite size was seen at zero load and a tensile force was required to pull the surfaces apart. In an attempt to model these observations the JKR theory (Johnson, Kendall and Roberts, 1971) was born. At the same time, working in Moscow on adhesion of particles in colloidal suspension, Derjaguin, Muller and Toporov had developed a different ("DMT") theory of the adhesion of elastic spheres (DMT, 1975). At first it was thought that these theories were incompatible, until Tabor suggested that each applied to opposite ends of the spectrum of a non-dimensional parameter which expressed the ratio of the magnitude of the elastic deformation to the range of surface forces.

This work was followed by Maugis and Barquins in the CNRS Laboratory at Belle Vue, who recognised the analogy between adhesion and fracture. In a classic paper in 1978 they showed, both experimentally and theoretically, how a wide variety of geometries in adhesive contact could be brought together and treated in a unified and consistent way by the principles of linear elastic fracture mechanics. However, due to the destruction of adhesion by surface roughness, application of the theory was largely restricted to compliant materials or very small particles.

The past decade has seen new demands of surface finish by the recording and computer industry and also the development of micro-probe instruments such as the atomic force microscope (AFM) and micro-mechanical systems (MEM). With reduced roughness and decreasing contact size, adhesive forces assume greater importance. The regime of operation of the AFM falls into the transition zone between that JKR and DMT theories, for which Maugis has provided a neat analysis. All these developments have focused attention on the role of adhesive forces in mechanics.

A comprehensive treatment of any scientific topic by a leading player is to be welcomed. Here is one such; I wish it every success.

Cambridge, December 1998 *K.L. Johnson*

Preface

This book has its origin in a series of courses given at the University of Notre-Dame de la Paix in Namur (Belgium) in 1992. For that occasion I wrote some short chapters, which have since extended. My aim in writing this book has been to emphasize the role of surface energy in mechanics, i.e. fracture mechanics and mechanics of contact.

The first chapter is devoted to surface energy and surface tension of solids and to surface thermodynamics. This is followed by a chapter on elasticity where, in particular, the methods of Muskhelichvili and Hankel transforms for the resolution of plane and axisymmetric problems are studied. The essential concepts of fracture mechanics are presented in Chap. 3, with emphasis on the thermodynamic aspect of the problem. The reader will find most of the complete demonstrations of analytical results and the detailed calculations for cracks submitted to pressure distributions and the Dugdale model. Chapter 4, on contact mechanics, is more extensive, with theories of the concentrated force, of the flat punch, of Hertz, of JKR and the JKR/DMT transition, and of the conical punch; the contact and adherence of rough solids is also studied. One of the original aspects of this books is the systematic analytical study of the stress field around elliptical cavities, cracks or elastic contacts. Principal stresses, isopachics, isochromatic lines, von Mises criterion and displacement fields are all presented. Finally, the last chapter describes in detail the geometries generally used in fracture mechanics and adhesion testing: double cantilever beam, peeling, lap joint, double torsion and blister test.

I am particularly grateful to Christian Stefani, who has carefully corrected the manuscript and verified all the calculations, and to Prof. K.L. Johnson who has reread and corrected the English version, and has written the foreword. Without the encouragement of these two persons, this book would probably never have appeared.

Champs sur Marne, June 1999 *Daniel Maugis*

Contents

1. Elements of Surface Physics

1.1 Introduction

Let us recall that bonds between atoms can be divided into strong bonds (whose rupture corresponds to a large absorption of energy, around one hundred kcal/mole) and weak bonds (a few kcal/mole).

Among the strong bonds one distinguishes the following types.

- *The ionic bond* (or electrovalent bond), which results from the coulombic attraction between ions of opposite charge. It has no privileged direction in space, and it ensures the cohesion of ionic crystals such as NaCl.
- *The covalent bond* obtained by sharing one or more pairs of electrons to compose molecules such as H_2, Cl_2, N_2, or organic molecules or covalent crystals such as diamond, germanium, etc. Those bonds have preferential directions. Note that there exist an almost continuous range of intermediate cases between a pure ionic bond and a pure covalent bond (one shows, for example, that the HCl bond is 17% ionic and 83% covalent).
- *The metallic bond* due to free electrons running between positive ions, which ensures the cohesion of metals and their alloys in the solid or the liquid state.

Among the weak bonds, one has the following types.

- *The hydrogen bond*, linking molecules through a hydrogen atom belonging to one of them. That bond is essentially electrostatic: the hydrogen atom forms the positive pole of a molecule polarizes and attracts the nearby molecule. The smallness of the proton allows for a small approach (0.2 to 0.3 nm), and leads to a bonding energy of 3 – 7 kcal/mole. The hydrogen bond leads to the phenomenon of molecular association in the vapor, liquid (water) or solid (ice) phase.
- *The van der Waals forces*, due to electromagnetic interactions produced by continual fluctuations in the electronic distribution inside atoms or molecules. Attractive forces between molecules of a real gas provoke the departure from the law of perfect gases $PV = RT$, accounted for by van der Waals in his thesis in 1873 by his famous equation:

$$\boxed{\left(P + \frac{a}{V^2}\right)(V - b) = RT} \, .$$

The term b (the covolume) is due to the finite volume of molecules and to their impenetrability. The corrective term a/V^2 on the pressure arises from the fact that the pressure P on the walls is lower than the actual pressure inside the gas due to the mutual attraction between molecules (as if the bulk of the gas was slightly contracted under the action of attractive forces and thus produced less pressure on the walls). It is that bonding energy (less than 3 kcal/mole) which is responsible for the cohesion of liquid and solid rare gases. The van der Waals forces determine most of the properties of liquids: viscosity, heat of vaporization, surface tension, solubility in other liquids, etc. They also play a role in the stability and the coagulation of colloids, in the physisorption of a molecule onto a surface.
It can be noted that these van der Waals forces are oriented—that is, in addition to the force between two molecules there exists generally a torque, as in the interaction of a bar magnet and a bar of soft iron, where the attraction is accompanied by a torque which rotates the two bars until they point towards one another. In like manner molecules physisorbed onto a solid surface are oriented by these forces.

It should be emphasized that there are only four types of force in nature, which are, in decreasing order of intensity, strong interaction, electromagnetic force, weak interaction and gravitation. All the above-mentioned bonds belong to the second category.

The creation of a new surface in a solid or a liquid is obtained by breaking the bonds which ensure the cohesion of that condensed phase (and by slightly reorganizing atoms near that surface). That needs a given amount of energy (work + heat) per unit area of created surface. It is the *work* needed to create reversibly and isothermally a unit area of a new surface which is called the *surface energy* γ, expressed in J m^{-2} (1 mJ m^{-2} = 1 erg cm^{-2}).

Due to the dissymetry of the attraction forces undergone by atoms or molecules near a surface, a small variation in the interatomic or intermolecular distances perpendicular to the surface occurs, which produces, as we shall see below, a tensile force parallel to that surface, termed the *surface tension* σ and expressed in N/m (1 mN/m = 1 dyne/cm).

When the area of the surface is only increased by the climbing of atoms or molecules from the bulk (in liquids or in solids near their melting point), the work needed to increase that surface is the one which must be furnished to overcome the surface tension, and one has $\gamma \equiv \sigma$. We will see later that it is not generally the case for solids.

We will examine, at the microscopic level first, these two variables γ and σ, and then study them as thermodynamic intensive variables such as pressure or temperature.

1.2 Van der Waals Forces Between Molecules

To explain attractive forces between neutral molecules, and after elimination of the gravitation forces, which are too weak and too long range, forces between permanent dipoles (Keesom orientation forces) have been first proposed, followed by forces between permanent and induced dipoles (Debye induction forces). However, it was necessary to wait for quantum mechanics to explain attraction forces between nonpolar molecules (London dispersion forces).

A dipole is a combination of two opposite electric charges $+q$ and $-q$ set apart by a small distance l. It does not move in a uniform electric field but rotates. The quantity $\mu = ql$ is the *dipolar moment*. For example, if q is the charge of the electron (1.6×10^{-19} C) and if $l = 0.1$ nm, one obtains $\mu = 1.6\times 10^{-29}$ C m. The dipolar moments are often expressed in the system CGSUES where the unit is the *debye* with $1\,\mathrm{D} = 10^{-18}$ ues $= 3.336\times10^{-30}$ C m. As an example the dipolar moment of HCl is 1.08 D, that of NH_3 is 1.47 D, that of H_2O is 1.85 D, and that of KBr reaches 10.4 D. Let us recall that the distinction between polar and nonpolar materials is due to Lewis in 1916.

1.2.1 The Keesom Theory (1921)

Keesom's theory [1, 2] is modelled upon Langevin's theory of paramagnetism (1905). Two neighbouring dipoles tend to stand in a line and to attract one another, despite the thermal motion which tends to orient them at random. Let us consider (Fig. 1.1) two dipolar molecules situated at O and at M, and whose dipolar moments μ_1 and μ_2 are oriented at angles θ_1 and θ_2 to the line joining them. Let us compute their energy of interaction.

Field Due to a Dipole. The potential created at a point M by the dipole AB is

$$V = \frac{q}{4\pi\varepsilon_0}\left(\frac{1}{r_\mathrm{A}} - \frac{1}{r_\mathrm{B}}\right) .$$

If OM $= r$ is large compared to AB it becomes

$$V = \frac{\mu}{4\pi\varepsilon_0}\frac{\cos\theta_1}{r^2} .$$

The field E_1 at point M is in the plane ABM. Using the components of the gradient in cylindrical coordinates, the components of the field are

$$E_r = -\frac{\partial V}{\partial r} = \frac{2\mu_1}{4\pi\varepsilon_0}\frac{\cos\theta_1}{r^3} ,$$

$$E_\theta = -\frac{1}{r}\frac{\partial V}{\partial \theta} = \frac{\mu_1}{4\pi\varepsilon_0}\frac{\sin\theta_1}{r^3} ,$$

and hence the electric field of a dipole, in $1/r^3$ is given by

$$E_1 = \frac{1}{4\pi\varepsilon_0}\frac{\mu_1}{r^3}\sqrt{1+3\cos^2\theta_1}\ . \tag{1.1}$$

With r fixed, it has its maximum on the dipole axis. The differential equation of the lines of field is

$$\frac{\mathrm{d}r}{E_r} = \frac{r\,\mathrm{d}\theta}{E_\theta}$$

or

$$\frac{\mathrm{d}r}{r} = 2\frac{\cos\theta\,\mathrm{d}\theta}{\sin\theta}\ ,$$

which after integration gives

$$r = k\sin^2\theta\ .$$

Figure 1.1 shows one of these lines of field.

Due to the thermal motion, all orientations of the dipole AB in space are possible, with equal probability if it is isolated. However, the mean value of the field at point M is nonzero, for

$$\overline{\cos^2\theta} = \frac{\int\cos^2\theta\,\mathrm{d}\Omega}{\int\mathrm{d}\Omega} = \frac{\int_0^\pi\cos^2\theta\ 2\pi\sin\theta\,\mathrm{d}\theta}{4\pi} = \frac{1}{3}\ .$$

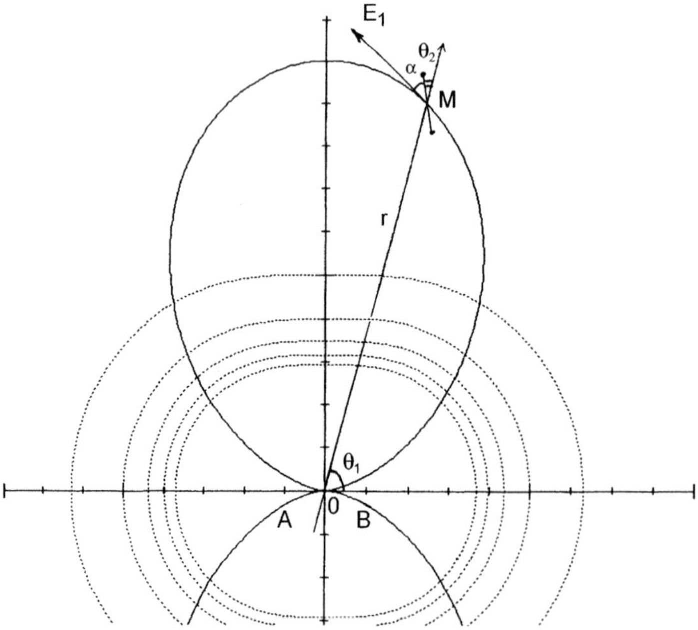

Fig. 1.1. Contours of the field (*dashed lines*) and lines of force of a dipole

The mean field at point M is thus

$$\overline{E_1} = \frac{\mu_1\sqrt{2}}{4\pi\varepsilon_0 r^3} \tag{1.2}$$

and is along OM by symmetry.

Orientation of Dipoles in an Electric Field. Let us consider a dipole of moment $\mu = 2aq$ perpendicular to an electric field E, and let it turn so as it makes an angle α with the field. So doing, each charge has received the work $-qEa\cos\alpha$, and hence the potential energy U of a dipole in a field E is given by

$$U = -\mu E\cos\alpha = -\boldsymbol{\mu}\cdot\boldsymbol{E}\,. \tag{1.3}$$

The thermal motion provokes a rotation of that dipole μ. If that rotation were continuous and regular, attractions and repulsions should be exactly compensated and interaction should be zero. In fact, in a gas in thermodynamic equilibrium, the number of dipoles μ which have that energy U is proportional to $\exp(-U/kT)$, which promotes orientations near those of the stable equilibrium. The continuation of the calculus is modelled upon that of Langevin for the paramagnetism. (A magnetic moment μ in an induction field B_0 has a potential energy $U = -B_0\mu\cos\alpha$.)

The number $\mathrm{d}N$ of dipoles per unit volume, whose axes are in a solid angle $\mathrm{d}\Omega = 2\pi\sin\alpha\,\mathrm{d}\alpha$ around the field E is

$$\begin{aligned}\mathrm{d}N &= A\exp(\mu E\cos\alpha/kT)\,\mathrm{d}\Omega\\ &= 2\pi A\mathrm{e}^{z\cos\alpha}\sin\alpha\,\mathrm{d}\alpha\,,\end{aligned}$$

letting $z = \mu E/kT$. The number A is found by writing that the total number of dipoles per unit volume is N, where

$$N = 2\pi A\int_0^\pi \mathrm{e}^{z\cos\alpha}\sin\alpha\,\mathrm{d}\alpha\,.$$

By reason of symmetry, only the polarizations parallel to E are additive, every dipole contributing by a quantity $\mu\cos\alpha$. These $\mathrm{d}N$ dipoles, making an angle α with the field E, lead to a dipolar moment

$$\mathrm{d}\mathcal{M} = \mu_2\cos\alpha\,\mathrm{d}N\,.$$

The total dipolar moment of these N dipoles is given by

$$\mathcal{M} = \int_0^\pi \mu_2\cos\alpha\,\mathrm{d}N\,.$$

The mean moment $\overline{\mu}$ of a dipole μ in the field E ($\overline{\mu}$ is oriented parallel to E by symmetry) is thus

$$\overline{\mu} = \frac{\mathcal{M}}{N} = \frac{\int_0^\pi \mu\cos\alpha\mathrm{e}^{z\cos\alpha}\,\mathrm{d}(\cos\alpha)}{\int_0^\pi \mathrm{e}^{z\cos\alpha}\,\mathrm{d}(\cos\alpha)}$$

or, with $x = \cos\alpha$,

$$\overline{\mu} = \mu \frac{\int_{-1}^{+1} x \mathrm{e}^{zx}\,\mathrm{d}x}{\int_{-1}^{+1} \mathrm{e}^{zx}\,\mathrm{d}x} = \mu \frac{1}{I}\frac{\mathrm{d}I}{\mathrm{d}z} \,,$$

where I is the integral in the denominator such that

$$I = \int_{-1}^{+1} \mathrm{e}^{zx}\,\mathrm{d}x = \frac{2\sinh z}{z} \,.$$

Hence:

$$\overline{\mu} = \mu \left(\coth z - \frac{1}{z} \right) \,.$$

The function $\mathcal{L}(z) = \coth z - 1/z$ is *Langevin's function.* When z is small, which is physically always the case, one has $\coth z \approx 1/z + z/3 + \cdots$, and it becomes

$$\boxed{\overline{\mu} = \frac{\mu^2 E}{3kT}} \,, \tag{1.4}$$

which is the mean moment of a dipole μ in a field E due to the thermal motion.

Interaction Between Two Dipoles. In the absence of thermal motion, the potential energy of a dipole μ_2, at point M, in the field E_1 of a dipole μ_1 is thus, from (1.3),

$$U = -\frac{1}{4\pi\varepsilon_0}\frac{\mu_1\mu_2}{r^3} \cos\alpha \sqrt{1 + 3\cos^2\theta_1} \,,$$

where α is the angle between the dipole μ_2 and the field E_1; that energy is minimum when the dipole μ_2 is aligned with the dipole μ_1 ($\theta_1 = \alpha = 0$, the position of stable equilibrium), and is equal to

$$U_{\mathrm{st}} = -\frac{1}{4\pi\varepsilon_0}\frac{2\mu_1\mu_2}{r^3} \,.$$

In the presence of thermal motion, the problem is more difficult. It is expedient to write that the mean interaction energy of two polar molecules at distance r and with dipolar moments μ_1 and μ_2 is equal to the potential energy of a mean dipole $\overline{\mu_2}$ in the mean field $\overline{E_1}$, i.e. $U = -\overline{\mu_2}\,\overline{E_1}$, or, from (1.2) and (1.4),

$$\boxed{U = -\frac{1}{(4\pi\varepsilon_0)^2}\frac{2\mu_1^2\mu_2^2}{3kT}\frac{1}{r^6}} \,. \tag{1.5}$$

The result is exact, but we have forgotten that the field of dipole 2 acts on dipole 1, whose orientations in space are no longer all equally probable. The correct calculation is slightly longer.

Equation (1.5) describes an attractive potential in $1/r^6$, corresponding to a force law in $1/r^7$, which is of the proper order of magnitude for the measured bonding energies. However, the numerical values computed from μ_1 and μ_2 and the variation with temperature do not agree with experimental results (the van der Waals forces being more or less independent of the temperature).

1.2.2 The Debye Theory (1920)

Since in an electric field E a molecule, by deformation of its electronic cloud, takes an induced dipolar moment $\mu' = \alpha_0 E$ (where α_0 is the electronic polarizability, which has the dimension of a volume multiplied by $4\pi\varepsilon_0$), Debye [3] remarks that polar molecules not only are oriented, but also are polarized by one another. Thus a polar molecule of moment μ_1 induces a dipolar moment μ'_2 on a polarizable neighbouring molecule; hence the mean energy of interaction $U = -\alpha_{02}\overline{E_1}^2/2$. The same result holds for molecule 2 on molecule 1, and the total energy of interaction of the two molecules (by the induction effect) is

$$U = -\frac{1}{(4\pi\varepsilon_0)^2}\frac{\alpha_{02}\mu_1^2 + \alpha_{01}\mu_2^2}{r^6} . \tag{1.6}$$

That energy of interaction, also in $1/r^6$, is independent of the temperature and is superimposed on that of Keesom. However, these induction forces are always weak.

Besides the electronic polarizability α_0, one can define an *orientational polarizability* $\mu^2/3kT$ by (1.4), and it can be seen that Keesom's equation is recovered by replacing α_0 in Debye's equation by $\mu^2/3kT$. Furthermore, the total polarizability α of a molecule is defined by the *Debye–Langevin equation*,

$$\alpha = \alpha_0 + \frac{\mu^2}{3kT} , \tag{1.7}$$

and is related to the static dielectric constant ε by the *Clausius–Mosotti equation*,

$$\alpha = 3\left(\frac{\varepsilon - 1}{\varepsilon + 2}\right)\varepsilon_0 v , \tag{1.8}$$

where $v = M/\varrho\mathcal{N}$ is the volume per molecule (M is the molecular mass, ϱ the specific mass and $\mathcal{N}$ Avogadro's number).

1.2.3 The London Theory (1930)

The above theories do not explain the cohesion of liquid or solid rare gases whose atoms are spherical and without permanent dipolar moment. We had to wait for quantum mechanics to understand, with London [4] in 1930, the physical origin of these forces. The effect is purely a quantum effect. Electrons in an atom or a molecule are never at rest, even at 0 K (residual energy). The centre of mass of these electrons moves continually with respect to the nucleus, whence the existence of a dipolar moment fluctuating rapidly (frequency ranging about that of electronic motions, i.e. 10^{15} to 10^{18} Hz) whose mean value is zero. In a neighbouring polarizable molecule, the oscillations of that instantaneous dipole create by induction another instantaneous dipole, which oscillates at the same frequency.

Let us consider a spherical atom and assume that at a given time the centre of mass of the Z electrons of charge q and mass m is at the distance $\varrho(x_1, y_1, z_1)$ from the nucleus, which is assumed to be fixed at the origin of coordinates. Let us compute the restoring force. The instantaneous dipolar moment is $\boldsymbol{\mu} = -Zq\boldsymbol{\varrho}$. By definition of the polarizability α, it can be obtained by submitting the atom to a field $\boldsymbol{E}$ such that $\boldsymbol{\mu} = \alpha \boldsymbol{E}$. The restoring force f is equal and opposite to that exerted by the field and is

$$\boldsymbol{f} = Zq\boldsymbol{E} = \frac{Zq}{\alpha}\boldsymbol{\mu} = -\frac{Z^2q^2}{\alpha}\boldsymbol{\varrho} \,.$$

As the force is central and proportional to the distance, the atom is an oscillator with a natural frequency

$$\nu_0 = \frac{1}{2\pi}\sqrt{\frac{Z^2q^2}{\alpha Z m}} = \frac{q}{2\pi}\sqrt{\frac{Z}{\alpha m}} \,.$$

Its total energy (kinetic and potential), given by

$$W = \frac{Zm}{2}\left(\dot{x}_1^2 + \dot{y}_1^2 + \dot{z}_1^2\right) + \frac{Z^2q^2}{2\alpha}\left(x_1^2 + y_1^2 + z_1^2\right) ,$$

is quantified and is

$$W_n = \left(n + \frac{3}{2}\right) h\nu_0 \,.$$

The minimum of the energy, in the fundamental state, is $W_0 = 3h\nu_0/2$ (being the residual energy which persists even at 0 K).

Let us now place a second identical oscillator on the z axis at a distance r from the first (charges at $(0, 0, r)$ and $(x_2, y_2, r + z_2)$). The interaction energy of the two instantaneous dipoles is

$$U = \frac{Z^2q^2}{4\pi\varepsilon_0 r} - \frac{Z^2q^2}{4\pi\varepsilon_0\sqrt{x_1^2 + y_1^2 + (r - z_1)^2}} - \frac{Z^2q^2}{4\pi\varepsilon_0\sqrt{x_2^2 + y_2^2 + (r + z_2)^2}}$$
$$+ \frac{Z^2q^2}{4\pi\varepsilon_0\sqrt{(x_1 - x_2)^2 + (y_1 - y_2)^2 + (r + z_2 - z_1)^2}} \,.$$

By developing in powers of $1/r$ and keeping only the leading terms, it becomes, with

$$(1+\varepsilon)^{-1/2} \approx 1 - \frac{\varepsilon}{2} + \frac{3\varepsilon^2}{8} ,$$

$$U = \frac{1}{4\pi\varepsilon_0} \frac{Z^2 q^2}{r^3} (x_1 x_2 + y_1 y_2 - 2 z_1 z_2) .$$

The total energy of the system (sum of kinetic, potential and interaction energies) is thus

$$\begin{aligned} W = {} & \frac{Zm}{2} (\dot{x}_1^2 + \dot{y}_1^2 + \dot{z}_1^2 + \dot{x}_2^2 + \dot{y}_2^2 + \dot{z}_2^2) \\ & + \frac{Z^2 q^2}{2\alpha} (x_1^2 + y_1^2 + z_1^2 + x_2^2 + y_2^2 + z_2^2) \\ & + \frac{Z^2 q^2}{4\pi\varepsilon_0 r^3} (x_1 x_2 + y_1 y_2 - 2 z_1 z_2) . \end{aligned}$$

By letting

$$\begin{aligned} \xi_1 &= \frac{1}{\sqrt{2}}(x_1 + x_2) , & \xi_2 &= \frac{1}{\sqrt{2}}(x_1 - x_2) , \\ \eta_1 &= \frac{1}{\sqrt{2}}(y_1 + y_2) , & \eta_2 &= \frac{1}{\sqrt{2}}(y_1 - y_2) , \\ \zeta_1 &= \frac{1}{\sqrt{2}}(z_1 + z_2) , & \zeta_2 &= \frac{1}{\sqrt{2}}(z_1 - z_2) , \end{aligned}$$

that energy can be written as a sum of squared terms:

$$\begin{aligned} W = {} & \left[\frac{Zm}{2} \dot{\xi}_1^2 + \frac{Z^2 q^2}{2} \left(\frac{1}{\alpha} + \frac{1}{4\pi\varepsilon_0 r^3} \right) \xi_1^2 \right] \\ & + \left[\frac{Zm}{2} \dot{\xi}_2^2 + \frac{Z^2 q^2}{2} \left(\frac{1}{\alpha} - \frac{1}{4\pi\varepsilon_0 r^3} \right) \xi_2^2 \right] \\ & + \left[\frac{Zm}{2} \dot{\eta}_1^2 + \frac{Z^2 q^2}{2} \left(\frac{1}{\alpha} + \frac{1}{4\pi\varepsilon_0 r^3} \right) \eta_1^2 \right] \\ & + \left[\frac{Zm}{2} \dot{\eta}_2^2 + \frac{Z^2 q^2}{2} \left(\frac{1}{\alpha} - \frac{1}{4\pi\varepsilon_0 r^3} \right) \eta_2^2 \right] \\ & + \left[\frac{Zm}{2} \dot{\zeta}_1^2 + \frac{Z^2 q^2}{2} \left(\frac{1}{\alpha} - \frac{2}{4\pi\varepsilon_0 r^3} \right) \zeta_1^2 \right] \\ & + \left[\frac{Zm}{2} \dot{\zeta}_2^2 + \frac{Z^2 q^2}{2} \left(\frac{1}{\alpha} + \frac{2}{4\pi\varepsilon_0 r^3} \right) \zeta_2^2 \right] . \end{aligned}$$

It is the sum of the energies of six independent oscillators, whose frequencies are

$$\nu_{\xi_1} = \nu_{\eta_1} = \frac{1}{2\pi}\sqrt{\frac{Z^2q^2}{Zm}\left(\frac{1}{\alpha} + \frac{1}{4\pi\varepsilon_0 r^3}\right)} \approx \nu_0\left(1 + \frac{\alpha}{8\pi\varepsilon_0 r^3} - \frac{\alpha^2}{128\pi^2\varepsilon_0^2 r^6}\right) ,$$

$$\nu_{\xi_2} = \nu_{\eta_2} = \frac{1}{2\pi}\sqrt{\frac{Z^2q^2}{Zm}\left(\frac{1}{\alpha} - \frac{1}{4\pi\varepsilon_0 r^3}\right)} \approx \nu_0\left(1 - \frac{\alpha}{8\pi\varepsilon_0 r^3} - \frac{\alpha^2}{128\pi^2\varepsilon_0^2 r^6}\right) ,$$

$$\nu_{\zeta_1} = \frac{1}{2\pi}\sqrt{\frac{Z^2q^2}{Zm}\left(\frac{1}{\alpha} - \frac{2}{4\pi\varepsilon_0 r^3}\right)} \approx \nu_0\left(1 - \frac{\alpha}{4\pi\varepsilon_0 r^3} - \frac{\alpha^2}{32\pi^2\varepsilon_0^2 r^6}\right) ,$$

$$\nu_{\zeta_2} = \frac{1}{2\pi}\sqrt{\frac{Z^2q^2}{Zm}\left(\frac{1}{\alpha} + \frac{2}{4\pi\varepsilon_0 r^3}\right)} \approx \nu_0\left(1 + \frac{\alpha}{4\pi\varepsilon_0 r^3} - \frac{\alpha^2}{32\pi^2\varepsilon_0^2 r^6}\right) .$$

(The power expansions are for $\alpha/2\pi\varepsilon_0 r^3 \ll 1$, which is always the case if r is larger than the atom diameter.)

The residual energy of the system is

$$\begin{aligned} W_0 &= \frac{1}{2}h\nu_{\xi_1} + \frac{1}{2}h\nu_{\eta_1} + \frac{1}{2}h\nu_{\zeta_1} + \frac{1}{2}h\nu_{\xi_2} + \frac{1}{2}h\nu_{\eta_2} + \frac{1}{2}h\nu_{\zeta_2} \\ &= 3h\nu_0 - \frac{3\alpha^2}{64\pi^2\varepsilon_0^2 r^6}h\nu_0 . \end{aligned} \tag{1.9}$$

It is thus lower than that of two isolated atoms, which is $3h\nu_0$. Between two atoms or two molecules separated by a distance r larger than their diameter, a potential energy and an attraction force exist given by

$$U = -\frac{C}{r^6} , \tag{1.10}$$

$$F = \frac{6C}{r^7} , \tag{1.11}$$

where

$$C = \frac{1}{(4\pi\varepsilon_0)^2}\frac{3\alpha^2 h\nu_0}{4} \tag{1.12}$$

is the *London constant* (about $10^{-79}\,\mathrm{J\,m^6}$).

It can be shown that these London attraction forces arise from the coupling of oscillations of two neighbouring molecules vibrating in resonance. The phase angle is near $\pi/2$ and the energy passes alternately from one oscillator to the other at the frequency $\alpha\nu_0/4\pi\varepsilon_0 r^3$. The polarizability α of the molecule and the frequency ν_0 corresponding to an absorption band in the ultraviolet can be deduced from the dispersion curve giving the variation of the index of refraction with frequency; hence the term *dispersion forces* given to London's forces. Furthermore, one can replace, without large error, $h\nu_0$ by the ionisation energy E_i of the molecule.

For two dissimilar molecules the *London constant* is written as

$$\boxed{C_{\mathrm{AB}} = \frac{1}{(4\pi\varepsilon_0)^2}\frac{3\alpha_{\mathrm{A}}\alpha_{\mathrm{B}}h\nu_{\mathrm{A}}\nu_{\mathrm{B}}}{2(\nu_{\mathrm{A}}+\nu_{\mathrm{B}})}}, \tag{1.13}$$

and it is easy to see that, in a mixture, attraction between similar molecules is energetically more favourable than between dissimilar molecules[1], and that the more the molecules are dissimilar the more the probability of miscibility is weak. It is the reason why the separation of two liquids by an interface into two phases can often be observed.

Those dispersion forces prevail over orientation forces or induction forces, except for very polarized molecules such as NH_3 or H_2O—for NH_3 the component of orientation is slightly more than half the component of dispersion, and for H_2O it reaches three times that component.

1.2.4 Lennard–Jones Potential

Of course, that energy of interaction does not tend towards infinity as C/r^6 when r tends towards zero, since a repulsion can be observed when the electronic clouds begin to overlap, and then molecules behave as if they were impenetrable. Quantum mechanics leads to an energy of repulsion related to $\exp(r_c/r)$, but for mathematical convenience, it is written as $1/r^n$ with $n > 10$. One of the more usual empirical equations for the energy of interaction is the *Lennard–Jones potential*:

$$\begin{aligned} U &= \frac{D}{r^{12}} - \frac{C}{r^6} \\ &= 4U_0\left[\left(\frac{a}{r}\right)^{12} - \left(\frac{a}{r}\right)^{6}\right]. \end{aligned} \tag{1.14}$$

That potential and the corresponding interaction force are drawn in Fig. 1.2. The potential energy displays a minimum $U_0 = -C/2r_0^6$ at the distance

$$r_0 = \left(\frac{2D}{C}\right)^{1/6} = 2^{1/6}a = 1.12a \tag{1.15}$$

---

[1] Let us consider two molecules A and two molecules B. For bonds of type A–A, B–B, the energy is proportional to $\alpha_{\mathrm{A}}^2\nu_{\mathrm{A}} + \alpha_{\mathrm{B}}^2\nu_{\mathrm{B}}$. For bonds of type A–B, A–B, it is proportional to $4\alpha_{\mathrm{A}}\nu_{\mathrm{A}}\alpha_{\mathrm{B}}\nu_{\mathrm{B}}/(\nu_{\mathrm{A}}+\nu_{\mathrm{B}})$. The difference in the energy can be written as

$$\Delta U = \frac{\alpha_{\mathrm{B}}^2\nu_{\mathrm{B}}^2}{\nu_{\mathrm{A}}+\nu_{\mathrm{B}}}\left[\left(\frac{\alpha_{\mathrm{A}}\nu_{\mathrm{A}}}{\alpha_{\mathrm{B}}\nu_{\mathrm{B}}} - 1\right)^2 + \frac{\nu_{\mathrm{A}}}{\nu_{\mathrm{B}}}\left(\frac{\alpha_{\mathrm{A}}}{\alpha_{\mathrm{B}}} - 1\right)^2\right] > 0,$$

which shows that bonds A–A, B–B are more probable than bonds A–B, A–B, and all the more so as molecules A, B are dissimilar.

which defines the radius of the van der Waals bond (about 0.1–0.2 nm). The attractive force displays a maximum $F = -2.69U_0/r_0$ at a distance $r_{\max} = 1.11r_0$.

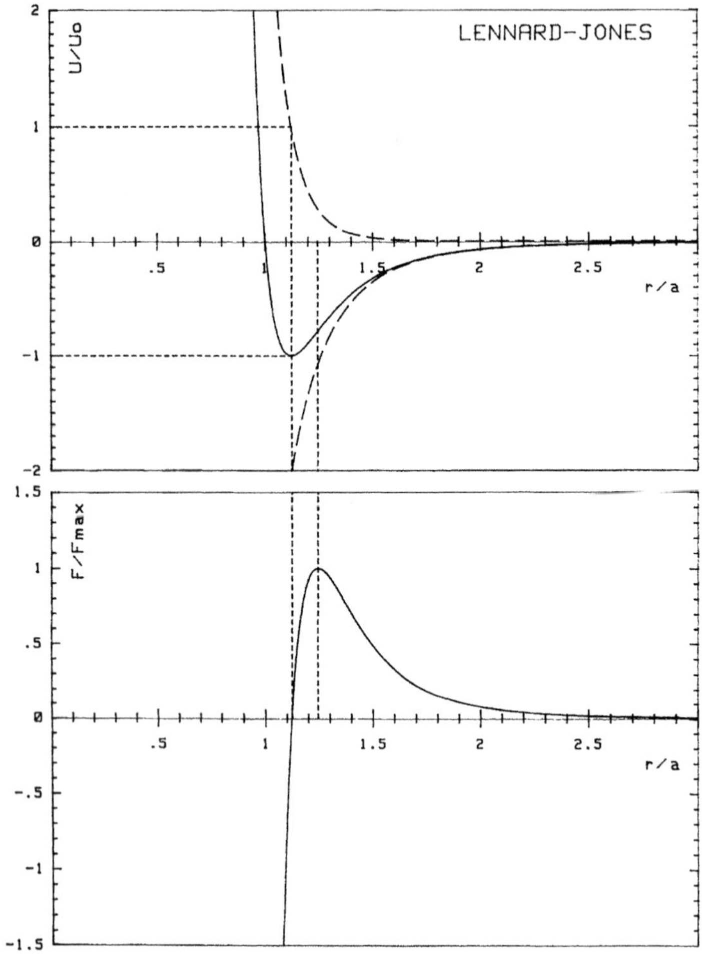

Fig. 1.2. Lennard–Jones potential and interaction force between two atoms

1.2.5 Retarded van der Waals Forces

As we shall see below, in 1937 Hamaker computed, by a simple integration, the attractive force between two spheres. Starting from those results, Verwey and Overbeek constructed in 1947 a theory of the stability of colloids, which did not agree with experiments in the case of suspensions of large particles. Overbeek concluded that, at large distance, the force between molecules decreases more rapidly than that according to London's equation in $1/r^7$, probably by an effect of "retardation" of those forces.

To compute his interaction potential $1/r^6$ between two atoms, London assumed that each atom responds instantaneously to the fluctuating electric field of the other. However, due to the finite speed of light, it takes a finite time for the electric field of the first atom to reach the second atom, and during this time the electronic distribution in the first atom has changed. The frequency at which the dipolar moment varies is related to the frequencies of radiation that the atom can emit or absorb. If λ is a characteristic wavelength of the absorption spectrum of the body (about 100 nm for most bodies), the sign of the dipole has changed after a time $\lambda/2c$, during which the field has propagated by $\lambda/2$. So, London's equations can be used only if $r \ll 50$ nm.

Casimir and Polder [5] computed this effect in 1948 but used the more general methods of quantum electrodynamics, for the above simple model is not suitable for computations. Their result was that the London formula must be multiplied by a corrective term, equal to 1 at small distances and proportional to $1/r$ at large distances (> 500 nm). At this distance, the interaction energy between two dissimilar molecules is

$$U = -\frac{C'_{12}}{r^7} \tag{1.16}$$

with

$$C'_{12} = \frac{1}{(4\pi\varepsilon_0)^2} \frac{23\hbar c}{4\pi} \alpha_1 \alpha_2 \,. \tag{1.17}$$

The method used by Casimir and Polder is rather complicated, but in 1949 Casimir [6] gave a simpler demonstration, valid only for large distances. The principle is as follows. It is known that a resonant electromagnetic cavity with perfectly conducting walls has an infinite number of natural oscillations of pulsation ω. According to quantum mechanics, the energy of every mode is quantified and its value at absolute zero is $\hbar\omega/2$. That leads to a total energy of the system $W = (\hbar/2)\sum \omega$, a divergent expression without physical significance. However, let us introduce into this cavity a particle of linear dimension $d \ll c/\omega$ and of polarizability α. It can be shown that the natural frequencies are modified, and the change in energy due to the particle is $\delta W = (\hbar/2)\sum \delta\omega$. It is still a divergent quantity, but if we study the variations of energy $\delta_r W$ and $\delta_\infty W$ for two positions of the particle, at a distance r from the wall and at large distance, the difference $\delta_r W - \delta_\infty W$ is finite and represents the interaction energy of the particle with the wall.

To study the interaction between two particles, it is necessary to study the influence of the second particle on the stationary waves of the system "cavity + first particle". The result obtained by Casimir and Polder is then repeated without too much difficulty.

1.3 Van der Waals Forces Between Solids

1.3.1 Nonretarded Forces

Assuming that van der Waals forces are additive, de Boer [7], in 1936, and Hamaker [8], in 1937, computed, by a simple integration, the energy and the interaction force between two parallel plates (infinite half-spaces), two spheres, or a sphere and a plane.

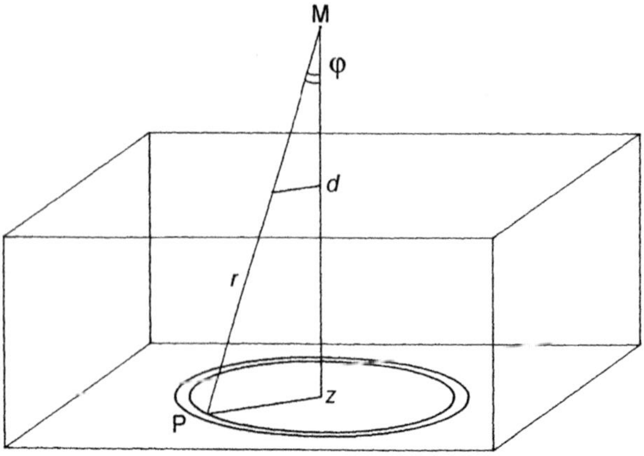

Fig. 1.3.

Let us consider a molecule M at a distance d from an infinite half-crystal having n atoms per unit volume (Fig. 1.3). The normal component of the force exerted on M by a molecule P situated at the distance r in a plane of ordinate z is, from (1.11), $6C\cos\varphi/r^7$. The contribution from the annulus of volume

$$\mathrm{d}V = 2\pi z^2 \frac{\sin\varphi}{\cos^3\varphi}\,\mathrm{d}\varphi\,\mathrm{d}z$$

is

$$\frac{6C\cos\varphi}{z^7}\cos^7\varphi\, n\,\mathrm{d}V = \frac{12\pi nC}{z^5}\cos^5\varphi\sin\varphi\,\mathrm{d}\varphi\,.$$

By integration, the interaction force between a molecule and a half-space is

$$F_{\mathrm{MP}}(d) = 12\pi nC\int_d^\infty \frac{\mathrm{d}z}{z^5}\int_0^{\pi/2}\cos^5\varphi\sin\varphi\,\mathrm{d}\varphi\,,$$

i.e.

$$F_{\mathrm{MP}}(d) = \frac{\pi nC}{2d^4}\,, \tag{1.18}$$

$$U_{\mathrm{MP}}(d) = -\frac{\pi nC}{6d^3}\,. \tag{1.19}$$

Therefore the interaction force between two half-spaces separated by a distance d is

$$F_{\mathrm{PP}}(d) = \int_d^{\infty} \frac{\pi n C}{2z^4} n \, \mathrm{d}z = \frac{\pi n^2 C}{6d^3} \, .$$

One generally introduces the *Hamaker constant* $A = \pi^2 n^2 C$ (for two dissimilar bodies $A_{12} = \pi^2 n_1 n_2 C_{12}$), which is about 10^{-19} J for interactions in a vacuum, and hence the force and interaction potential between two planes, per unit area, is given respectively by

$$\boxed{F_{\mathrm{PP}}(d) = \frac{A}{6\pi d^3} \, ,} \tag{1.20}$$

$$\boxed{U_{\mathrm{PP}}(d) = -\frac{A}{12\pi d^2}} \, . \tag{1.21}$$

The force between a sphere of radius R and an infinite half-crystal at a distance d can be computed in the same way: a point at a distance z from the surface undergoes a force $\pi n C/2z^4$, a layer of thickness $\mathrm{d}z$ at a distance z of area $S(z)$ undergoes a force $(\pi n C/2z^4) n S(z) \, \mathrm{d}z$, and the force between a sphere and a half-space is

$$F_{\mathrm{RP}}(d) = \frac{\pi n^2 C}{2} \int_d^{\infty} \frac{S(z) \, \mathrm{d}z}{z^4} \, .$$

With $S(z) = \pi[R^2 - (d + R - z)^2]$ and assuming $d \ll R$, we have

$$\boxed{F_{\mathrm{RP}}(d) = \frac{AR}{6d^2} \, ,} \tag{1.22}$$

$$\boxed{U_{\mathrm{RP}}(d) = -\frac{AR}{6d}} \, . \tag{1.23}$$

Likewise, the force and interaction potential between two spheres of radius R are

$$F_{\mathrm{RR}}(d) = \frac{AR}{12d^2} \, ,$$

$$U_{\mathrm{RR}}(d) = -\frac{AR}{12d} \, .$$

Eliminating A/d^2 between (1.22) and (1.21) we can write F_{RP} as a function of the interaction potential between two planes

$$F_{\mathrm{RP}}(d) = -2\pi R U_{\mathrm{PP}}(d) \, . \tag{1.22a}$$

It should be noted that this result, initially obtained by Bradley [9] in 1932, is independent of the force law, as we will see later with the Derjaguin approximation.

It must also be pointed out that the hypothesis of additivity, used above, is not always justified. In condensed media the electronic clouds of the atoms are very perturbed by the presence of neighbouring atoms, and the presence of any medium between two atoms modifies the electromagnetic field through which they interact. As we will see below, in connection with Lifshitz's theory, the Hamaker constants so calculated are very approximate.

1.3.2 Retarded Forces

If one replaces the London forces in $6C/r^7$ by forces in $7C'/r^8$ from Casimir and Polder, the above computations lead, for the retarded force per unit area between two planes, to

$$F_{\mathrm{PP}}(d) = \frac{B}{d^4} \tag{1.24}$$

and between a sphere and a plane:

$$F_{\mathrm{RP}}(d) = \frac{2\pi BR}{3d^3}\,, \tag{1.25}$$

where

$$B = \frac{\pi n_1 n_2 C'_{12}}{10} \tag{1.26}$$

is the *retarded Hamaker constant*, in J m.

Still, using the method of pertubation of the natural frequencies of an electromagnetic cavity, in 1948 Casimir [10] directly computed, without the assumption of additivity, the force and the interaction potential, per unit area, between two perfectly conducting parallel plates:

$$F_{\mathrm{PP}}(d) = \frac{\pi^2 \hbar c}{240 d^4}\,, \tag{1.27}$$

$$U_{\mathrm{PP}}(d) = -\frac{\pi^2 \hbar c}{720 d^3}\,. \tag{1.28}$$

For perfect conductors this force is independent of the nature of the materials.

1.3.3 Derjaguin Approximation

As the interaction forces $F(d)$ between a sphere and a plane or between crossed cylinders are more easily measurable, we can return readily to the interaction potential $U(d)$ between two plates by the Derjaguin approximation [11–13].

Let O_1 and O_2 be the points of smallest separation between two solids S_1 and S_2, and d this distance measured on the common normal to the two surfaces (Fig. 1.4) Let $\varepsilon_1, \varepsilon_1', \varepsilon_2$ and ε_2' be the principal curvatures of the two surfaces at these points. If d is small compared to the principal radii of curvature, we can compute by a simple integration the potential energy $V(d)$ between the two solids:

$$V(d) = \iint U(z)\,\mathrm{d}S$$

where $U(z)$ is the interaction potential per unit area between two parallel half-spaces separated by a distance z.

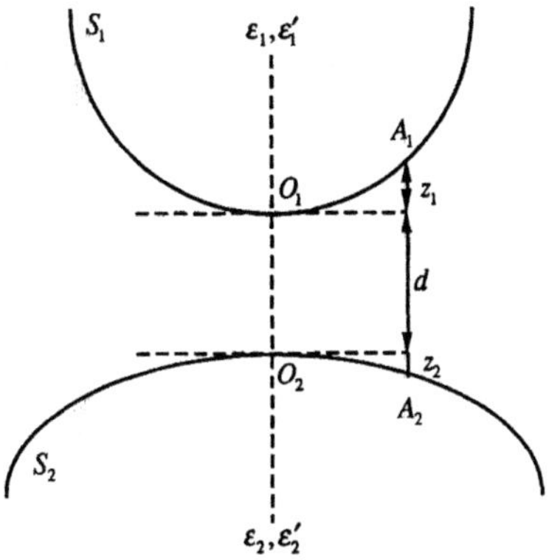

Fig. 1.4. Measuring the potential energy between two solids S_1 and S_2

Let us first compute the distance z between two opposite points A_1 and A_2, with coordinates x_1, y_1 and x_2, y_2 in the axes $O_1x_1y_1$ and $O_2x_2y_2$ parallel to the principal planes:

$$z = d + z_1 + z_2$$

with (to the second order)

$$z_1 = \frac{1}{2}\varepsilon_1 x_1^2 + \frac{1}{2}\varepsilon_1' y_1^2 ,$$
$$z_2 = \frac{1}{2}\varepsilon_2 x_2^2 + \frac{1}{2}\varepsilon_2' y_2^2 ,$$

where the principal curvatures ε and ε' are assumed to be positive. The axes O_1x_1 and O_2x_2 are not necessarily parallel, but make an angle $\omega = (O_1x_1, O_2x_2)$ so that we have

$$x_2 = x_1 \cos\omega + y_1 \sin\omega ,$$
$$y_2 = -x_1 \sin\omega + y_1 \cos\omega .$$

Putting these expressions in the above equations, we have

$$z = d + \frac{1}{2}[\varepsilon_1 + \varepsilon_2 \cos^2\omega + \varepsilon_2' \sin^2\omega]x_1^2 + \frac{1}{2}[\varepsilon_1' + \varepsilon_2 \sin^2\omega + \varepsilon_2' \cos^2\omega]y_1^2$$
$$+ [(\varepsilon_2 - \varepsilon_2') \sin\omega \cos\omega]x_1 y_1 .$$

The contours of equal separation z are curves of the type

$$Ax_1^2 + 2Bx_1y_1 + Cy_1^2 + D = 0\,,$$

with $D = 2(d - z)$, whose determinant

$$\begin{aligned}\Delta &= AC - B^2\\ &= [\varepsilon_1 + \varepsilon_2\cos^2\omega + \varepsilon_2'\sin^2\omega][\varepsilon_1' + \varepsilon_2\sin^2\omega + \varepsilon_2'\cos^2\omega]\\ &\quad - [(\varepsilon_2 - \varepsilon_2')^2\sin^2\omega\cos^2\omega]\\ &= (\varepsilon_1\varepsilon_1' + \varepsilon_2\varepsilon_2') + (\varepsilon_1\varepsilon_2 + \varepsilon_1'\varepsilon_2')\sin^2\omega + (\varepsilon_1\varepsilon_2' + \varepsilon_1'\varepsilon_2)\cos^2\omega\end{aligned}$$

is positive and invariant. They are thus ellipses. By a suitable rotation of the $O_1x_1y_1$ axes, we can eliminate the crossed terms in x_1y_1 and write

$$z = d + \frac{1}{2}\lambda x^2 + \frac{1}{2}\lambda' y^2$$

with $\lambda\lambda' = \Delta$.

Let us now return to the calculation of the integral, with $\mathrm{d}S \cong \mathrm{d}x\,\mathrm{d}y$:

$$V(d) = \int_{-\infty}^{+\infty}\mathrm{d}x\int_{-\infty}^{+\infty} U\left(d + \frac{1}{2}\lambda x^2 + \frac{1}{2}\lambda' y^2\right)\mathrm{d}y\,.$$

The limits may be taken at infinity, owing to the fast decrease of $U(z)$. Let $x = u\sqrt{2/\lambda}$ and $y = v\sqrt{2/\lambda'}$; then $u = \varrho\cos\theta$ and $v = \varrho\sin\theta$. Then

$$\begin{aligned}V(d) &= \frac{2}{\sqrt{\lambda\lambda'}}\int_{-\infty}^{+\infty}\int_{-\infty}^{+\infty} U(d + u^2 + v^2)\,\mathrm{d}u\,\mathrm{d}v\\ &= \frac{2}{\sqrt{\lambda\lambda'}}\int_0^{2\pi}\mathrm{d}\theta\int_{\infty}^{d} U(d + \varrho^2)\varrho\,\mathrm{d}\varrho\\ &= \frac{2\pi}{\sqrt{\lambda\lambda'}}\int_{\infty}^{d} U(l)\,\mathrm{d}l\,.\end{aligned}$$

The interaction force between the two solids being $F(d) = -\partial V/\partial d$, we obtain by differentiation under the integral

$$F(d) = -\frac{2\pi}{\sqrt{\Delta}}U_{\mathrm{PP}}(d)\,. \tag{1.29}$$

We can note that if the two solids are not spherical ($R_1 \neq R_1'$ and $R_2 \neq R_2'$), the interaction energy depends on the relative orientation ω and there exists a torque

$$-\frac{\partial V}{\partial\omega} = \frac{V(d)}{2\Delta}\sin 2\omega\left(\frac{1}{R_1} - \frac{1}{R_1'}\right)\left(\frac{1}{R_2} - \frac{1}{R_2'}\right),$$

which tends to align the principal axes of the two solids.

Let us consider the case of two spheres of dissimilar radii. With $\varepsilon_1 = \varepsilon_1' = 1/R_1$ and $\varepsilon_2 = \varepsilon_2' = 1/R_2$, we have

$$F_{\mathrm{RR}}(d) = -2\pi \frac{R_1 R_2}{R_1 + R_2} U_{\mathrm{PP}}(d) . \tag{1.30}$$

In the case of two cylinders of dissimilar radii, we have, with $\varepsilon_1 = 1/R_1$, $\varepsilon_1' = 0$, and $\varepsilon_2 = 1/R_2$, $\varepsilon_2' = 0$,

$$F(d) = -2\pi \frac{\sqrt{R_1 R_2}}{\sin\omega} U_{\mathrm{PP}}(d) . \tag{1.31}$$

In the case of a sphere of radius R and a plane, or in the case of two cylinders of radii R crossed at a right angle, these equations can be simplified as

$$\boxed{F_{\mathrm{RP}}(d) = -2\pi R U_{\mathrm{PP}}(d)} . \tag{1.32}$$

This *Derjaguin approximation*, valid for d small compared with the principal radii of curvature, is purely geometric in nature, independent of the interaction potential $U(d)$ and thus free of any reliance on additivity.

As we will see below, when d is reduced to the equilibrium interatomic distance Z_0, the interaction energy is equal, save the sign, to the adhesion energy w, i.e. 2γ if the solids are similar, or $\gamma_1 + \gamma_2 - \gamma_{12}$ if they are dissimilar (γ_{12} is the interfacial energy). The adherence force (the pull-off force) needed to separate a rigid sphere from a rigid plane is thus

$$\boxed{F = 2\pi w R} . \tag{1.32a}$$

A point must be emphasized: the force between a sphere and a plane (or between two spheres) does not vary with the distance as the force per unit area does between two planes! This is the same difference as between a function and its derivative! In effect, the force between two curved surfaces varies as $U(d)$, whereas the force per unit area between two plane surfaces varies as $-\,\mathrm{d}U(d)/\,\mathrm{d}d$.

1.3.4 Early Experiments

It was in 1950, i.e. two years after the publication of Casimir's and Polder's theory, that the first measurements of molecular attractive forces between macroscopic bodies were undertaken. However, the experiment presented a number of significant difficulties. The forces to be measured are about one tenth of a milligramm, which is not a problem in itself, but they only appear at a distance of a few tenths of micrometres and rapidly vary with the approach. As soon as the surfaces are approached they tend to adhere, and yet the gap must be held at a given value despite the force of attraction. Moreover, electrostatic charges lead to forces an order of magnitude larger and must be eliminated so as to avoid the phenomenon being completely masked.

The first to succeed experimentally were Derjaguin *et al.* [14] in 1954. The distance between a quartz plate and a quartz sphere (radius 10 cm) was measured by the method of Newton rings. The quartz plate was fastened to one arm of a balance actuated by an electronic control to maintain a fixed distance. This method enabled forces between $10^{-3}\,\mu$N and $200\,\mu$N and distances between 0.1 and $0.8\,\mu$m to be measured. By comparing their result with the Hamaker equation (non retarded), Derjaguin *et al.* obtained $A = 5\times10^{-21}$ J, whereas the Hamaker constant of quartz is approximately 10^{-19} J. The experimental results were thus 20 times too weak. Agreement was better in comparison with retarded forces, but this time the experimental results were still three times too large.

When Lifshitz heard of these results, he tackled the problem and gave the complete theory in 1956.

1.3.5 Lifschitz's Theory of van der Waals Forces

Lifshitz [15] followed a method completely different from the previous ones. He computed directly the attractive force between two parallel half-spaces, treating the phases as continuous media instead of using the individual properties of atoms. This method avoids the (incorrect) assumption of additivity of forces, i.e. the assumption that the force between two atoms is not affected by the presence of other atoms between them.

Lifshitz's theory is based upon the existence of a fluctuating electromagnetic field connected to thermodynamics fluctuations, extending out of the bodies partly as radiated progressive waves (thermal radiation) and partly as an evanescent stationary wave (which subsists at absolute zero and is connected to zero point oscillation of the electomagnetic field). Only the imaginary dielectric constant $\varepsilon''(\omega)$ of the bodies is introduced, which represents the electromagnetic energy absorbed at every frequency and is always positive.

Let us recall that the dielectric constant is a complex quantity

$$\varepsilon(\omega) = \varepsilon'(\omega) + \mathrm{i}\varepsilon''(\omega)$$

and that it is connected to the refractive index n and to the absortivity (or extinction index) κ by the equation

$$\varepsilon(\omega) = [n(\omega) + \mathrm{i}\kappa(\omega)]^2 \,. \tag{1.33}$$

By considering ε as a function of the complex variable $\boldsymbol{\omega} = \omega + \mathrm{i}\xi$ we obtain integral relations between $\varepsilon'(\boldsymbol{\omega})$ and $\varepsilon''(\boldsymbol{\omega})$ known as *Kramers–Krönig relations* (1927):

$$\varepsilon'(\boldsymbol{\omega}) - 1 = \frac{1}{\pi}\mathcal{P}\int_{-\infty}^{+\infty} \frac{\varepsilon''(\boldsymbol{\omega}')}{\boldsymbol{\omega}' - \boldsymbol{\omega}}\,\mathrm{d}\omega' \,, \tag{1.34}$$

$$\varepsilon''(\boldsymbol{\omega}) = -\frac{1}{\pi}\mathcal{P}\int_{-\infty}^{+\infty} \frac{\varepsilon'(\boldsymbol{\omega}') - 1}{\boldsymbol{\omega}' - \boldsymbol{\omega}}\,\mathrm{d}\omega' \,, \tag{1.35}$$

where $\mathcal{P}$marks the function "plus" defined in the upper half-plane of the complex ω.

The imaginary part $\varepsilon''(\omega)$, everywhere positive in the complex plane, cancels out on the imaginary axis (where pulsations are noted as $\mathrm{i}\xi$). On this axis the dielectric constant $\varepsilon(\mathrm{i}\xi)$ can be computed with the function of $\varepsilon''(\omega)$ for the real values ω by the equation

$$\varepsilon(\mathrm{i}\xi) = 1 + \frac{2}{\pi}\int_0^\infty \frac{\omega\varepsilon''(\omega)}{\omega^2+\xi^2}\,\mathrm{d}\omega\,, \tag{1.36}$$

which is a consequence of the Kramers–Krönig equations. As $\varepsilon''(\omega)$ is always positive, $\varepsilon(\mathrm{i}\xi)$ is a real and positive function which monotonically decreases from the value $\varepsilon(0)$ (static dielectric constant) for $\xi = 0$ to the value 1 for $\xi \to \infty$. It is this dielectric constant which contains all the information needed for the prediction of van der Waals forces. In principle, it suffices to measure indices and absorptivities $n(\omega)$ and $\kappa(\omega)$ in the whole range of frequencies to have $\varepsilon''(\omega) = 2n(\omega)\kappa(\omega)$ and then $\varepsilon(\mathrm{i}\xi)$.

The initial theory of Lifshitz [15] applied to two dielectric half-spaces separated by a vacuum. It was subsequently generalized by Dzyaloshinski *et al.* [16] in 1961 to treat, by the quantum theory of fields, the case of two dielectrics 1 and 2 separated by a medium 3 of thickness d. This theory joins up all the results obtained previously. The general equation is very complicated, but in the limiting cases of distances small compared to main wavelength of the optical absorption spectrum of the body and large compared with interatomic spacing, the interaction force per unit area is proportional to $1/d^3$, as in the London–Hamaker equation. The force and energy of interaction are given [16] by

$$F(d) = \frac{\hbar\overline{\omega}}{8\pi^2 d^3}\,,\quad \overline{\omega} = \int_0^\infty \left[\frac{\varepsilon_1(\mathrm{i}\xi)-\varepsilon_3(\mathrm{i}\xi)}{\varepsilon_1(\mathrm{i}\xi)+\varepsilon_3(\mathrm{i}\xi)}\right]\left[\frac{\varepsilon_2(\mathrm{i}\xi)-\varepsilon_3(\mathrm{i}\xi)}{\varepsilon_2(\mathrm{i}\xi)+\varepsilon_3(\mathrm{i}\xi)}\right]\mathrm{d}\xi \tag{1.37}$$

$$= \frac{A_{132}}{6\pi d^3}\,,$$

$$U(d) = -\frac{A_{132}}{12\pi d^2}\,, \tag{1.38}$$

where $\varepsilon_j(\mathrm{i}\xi)$ is the dielectric permittivity of the jth medium for the imaginary frequency $\mathrm{i}\xi$, $|\overline{\omega}|$ is a characteristic frequency of the absorption spectrum of the three media, and A_{132} is the Hamaker constant for the non-retarded forces between 1 and 2 across 3. (The last will be denoted by A_{12} if the medium 3 is a vacuum, and A if 1 and 2 are similar). By studying the interaction between two rarefied gases where forces are additive, one recovers exactly the London equation in $1/r^7$ for the force between two atoms.

In the case of large separations, the retardation effect is naturally introduced. The attraction force per unit area between two plates 1 and 2 separated by a medium 3 is proportional to $1/d^4$ and depends only on the static dielectric constants ε_{j0} of media j. It is given, cf [17], by:

$$F(d) = \frac{\pi^2 \hbar c}{240 d^4} \frac{1}{\sqrt{\varepsilon_{30}}} \left(\frac{\varepsilon_{10} - \varepsilon_{30}}{\varepsilon_{10} + \varepsilon_{30}} \right) \left(\frac{\varepsilon_{20} - \varepsilon_{30}}{\varepsilon_{20} + \varepsilon_{30}} \right) \varphi(\varepsilon_{10}, \varepsilon_{20}, \varepsilon_{30}) = \frac{B_{132}}{d^4} \tag{1.39}$$

where φ is a function varying from 1.00 down to 0.35 and B_{132} is the Hamaker constant for retarded forces. For two metals in a vacuum ($\varepsilon_{10} = \varepsilon_{20} = \infty, \varepsilon_{30} = 1$) the Casimir equation is obtained. The Casimir and Polder equation in $1/r^8$ for the force between two distant atoms is also exactly obtained.

When the media 1 and 2 are separated by a liquid metal (ε_3 infinite), (1.39) tends towards zero. In fact, cf [16], the decrease is in $1/d^5$ and the attraction force depends only on the electric conductivity σ of medium 3:

$$F(d) = 0.0034 \frac{\hbar c^2}{\sigma_3 d^5} . \tag{1.40}$$

1.3.6 Van der Waals Disjoining Pressure

It must be pointed out that the forces described above, which are always attractive in a vacuum where $\varepsilon_3 = 1$, can be repulsive or change sign with distance when $\varepsilon_3 \neq 1$. We can see from (1.37) that if $(\varepsilon_1 - \varepsilon_3)$ and $(\varepsilon_2 - \varepsilon_3)$ are of opposite sign on a large frequency range, then F is negative and solids 1 and 2 repel one another at small separations (the Hamaker A_{123} constant being negative). Likewise, at large separations, when the force depends on static dielectric constants, repulsion can arise if $(\varepsilon_{10} - \varepsilon_{30})$ and $(\varepsilon_{20} - \varepsilon_{30})$ are of opposite sign. As the relative magnitudes of $\varepsilon_1(\mathrm{i}\xi), \varepsilon_2(\mathrm{i}\xi), \varepsilon_3(\mathrm{i}\xi)$ in a given frequency range do not depend on the relative magnitudes of $\varepsilon_{10}, \varepsilon_{20}$ or ε_{30}, it is possible, in principle, to have cases where $F(d)$ changes its sign at a given separation d.

Israelachvili [18] has proposed a comparison with Newton's apple, which in air is attracted by the centre of the earth but which is repelled in water through Archimedes' principle. In fact, repulsive van der Waals forces can exist only if media 1 and 2 are dissimilar; two similar phases are always attracted by van der Waals forces, whatever the medium 3 separating them (for example two soap bubbles in a liquid attract one another).

The reason why colloidal particles, which should attract and coalesce because they are similar, do not do so is that these particles are surrounded by electric charges which repel (it is the DLVO theory of colloids); likewise, thin liquid films in foams are stabilized by ionized impurities. In a medium 3 separating two neighbouring phases 1 and 2, several types of pressure exist:

- $\Pi_A = -F(d)$, the electromagnetic pressure due to van der Waals interactions;
- Π_{el}, the electrostatic pressure due to the repulsion of double layers;
- Π_{s}, the pressure due to steric interactions between adsorbed molecules.

The sum of these pressures acting normally to the medium 3 is the *disjoining pressure* (so termed by Derjaguin in 1939) which, when positive, opposes the thinning down of the film (Fig. 1.5).

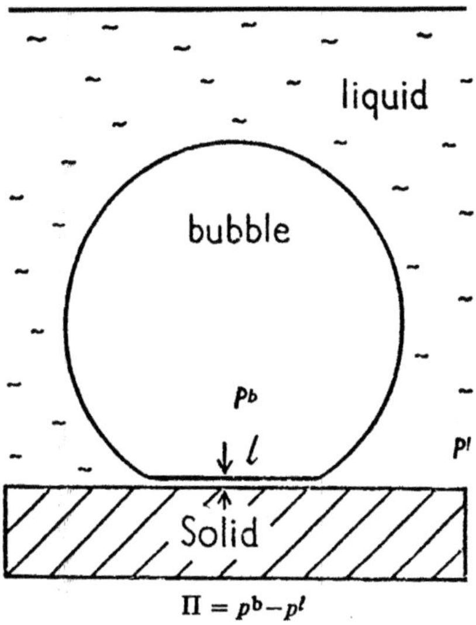

Fig. 1.5. Example of disjoining pressure between a solid and a vapour

An interesting example of an asymmetric film is that of wetting films [19]: in this case, medium 1 is a solid or a liquid on which is a film 3 of thickness d, and medium 2 is the vapour ($\varepsilon_2 = 1$). If $\Pi_A(d)$ is positive and decreasing, i.e. if $A_{132} < 0$, (Fig. 1.6a), the liquid completely wets the surface and forms a stable film of arbitrary thickness (depending on the vapour pressure).

If $\Pi_A(d)$ is a negative function, monotonically increasing with d (Fig. 1.6b), there is always attraction between the solid and the vapour: the liquid does not wet the surface and there is no wetting film. However, it is possible to have a monolayer (which in Lifshitz's theory cannot be treated as a distinct phase) or, on an horizontal surface, a non-wetting thick film stabilized by gravity (where if it is too thin, this metastable film will transform into small drops separated by dry zones).

If $\Pi_A(d)$ displays a minimum, the film cannot exceed the thickness d_{max} corresponding to point A (partial wetting); beyond this point the film is unstable (from A to B), then metastable. However, this thin film can coexist with a thick film, where $F(d) = 0$, or a drop of the liquid. We will return on these problems in Sect. 1.4.5.

It can be shown that $F(d)$ is the "van der Waals excess" of the chemical potential per unit volume, $\mu(d)$, in the film of thickness d. If μ_l is the chemical potential (per unit volume) in the bulk phase, that of the thin film will be:

$$\mu = \mu_l + F(d) . \tag{1.41}$$

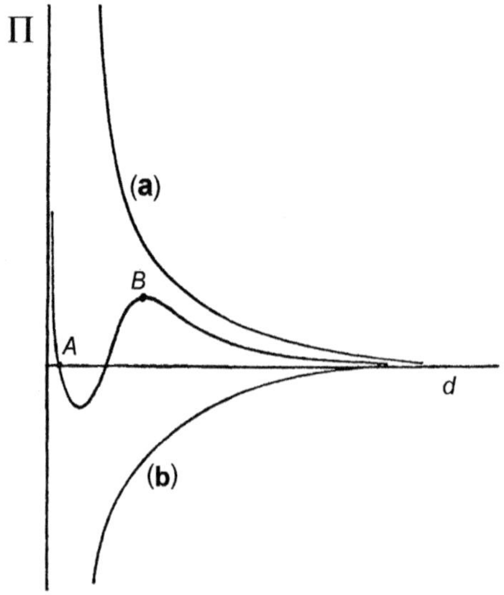

Fig. 1.6. Examples of isotherms of the disjoining pressure: **(a)** Π positive and decreasing; **(b)** Π negative

As chemical potentials vary with the pressure p as $\ln p$, we have:

$$\mu = \mu^0 + \frac{kT\varrho}{m} \ln p \,,$$
$$\mu_l = \mu^0 + \frac{kT\varrho}{m} \ln p_{\mathrm{sat}} \,,$$

where m is the mass of the molecule, ϱ the density of the liquid, and p_{sat} is the saturated vapour pressure (for which the gas condenses in bulk liquid phase). From this we deduce

$$\mu - \mu_l = \frac{kT\varrho}{m} \ln \frac{p}{p_{\mathrm{sat}}} = F(d) = -\Pi_A = \begin{cases} \dfrac{A_{132}}{6\pi} \dfrac{1}{d^3} \\ \dfrac{B_{132}}{d^4} \end{cases} , \tag{1.42}$$

an equation which gives the thickness of the wetting film as a function of the pressure p.

In the case of a wetting film on a vertical wall, one must take into account the barometric variation of the vapour pressure with height (known as the *barometric Boltzmann formula*):

$$p(z) = p(0)\mathrm{e}^{-mgz/kT} \,. \tag{1.43}$$

In general, $p(0) = p_{\mathrm{sat}}$ (except in a dynamical regime where the pressure at the liquid level is lowered, when one then speaks of “unsaturated” film), and (1.42) leads to:

$$F(d) + \varrho g z = 0 \,, \tag{1.44}$$

which gives the profile of the liquid. This equation represents a balance between gravity forces which tend to thin down the film and the disjoining pressure which tends to thicken it. For van der Waals interactions alone, the thickness will vary as $1/z^{1/4}$ in the lower part (retarded forces), and as $1/z^{1/3}$ for nonretarded forces. In reality, this film does not rise indefinitely, but stops at a given height H, which cannot be computed by (1.44) because a curvature term is missing in the equilibrium of pressures.

1.3.7 Experimental Verifications of Lifshitz's Theory

Lifshitz's theory of attractive forces between solids has been perfectly verified, both for retarded forces [20–22] and the nonretarded forces [23, 24]. Figure 1.7 displays the attraction force between a silica plate and a silica sphere after Rouweler and Overbeek [22].

Figure 1.8 shows, after Israelachvili and Tabor [24], the transition between retarded and nonretarded forces for two crossed mica cylinders (whose geometry is similar to the geometry of sphere–plane) where the separation varies from 2 nm to 130 nm. The attraction forces vary between R/d^2 and R/d^3.

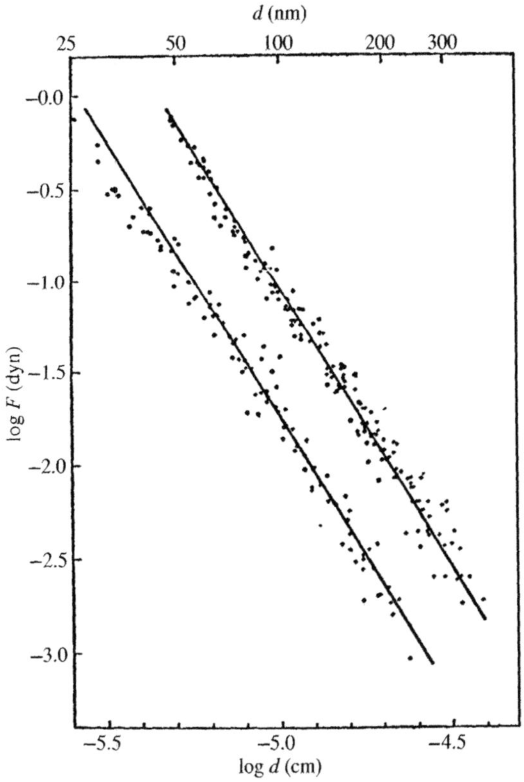

Fig. 1.7. Attraction between a flat plate and a lens of fused silica, of radius $R = 413.5$ cm (*upper line*) or $R = 83.75$ cm (*lower line*). The slope is −3, in agreement with (1.25) for retarded forces, with $B = 1.05 \times 10^{-28}$ Jm

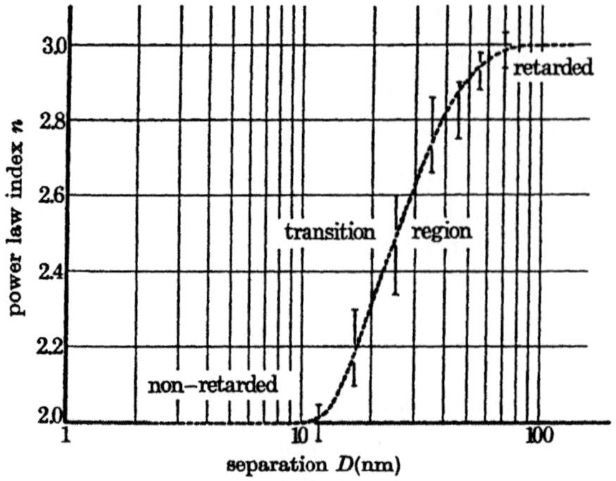

Fig. 1.8. van der Waals law of force between crossed mica cylinders, showing a variation of the power law index with distance

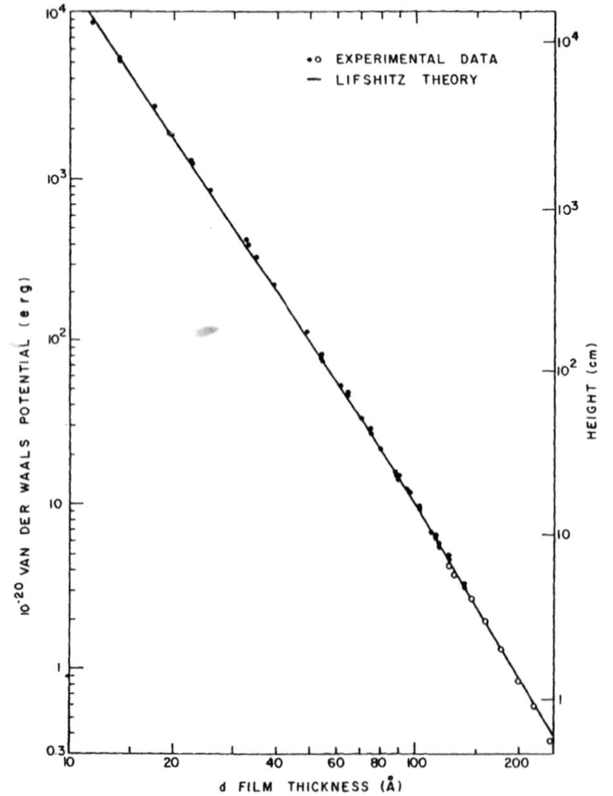

Fig. 1.9. Chemical potential per helium atom at the surface of the helium film which is adsorbed on a SrF_2 clived surface as a function of the film thickness at 1.38 K. The data represented by *closed* and *open points* are for saturated and unsaturated films respectively

Israelachvili and Tabor [24] have also studied the effect of a monolayer of stearic acid on van der Waals forces between mica sheets: the Hamaker constant tends towards that of mica for separations larger than 5 nm, whereas for separations less than 2.5 nm it tends towards that of stearic acid. This result is in agreement with Langbein's predictions [25, 26]: the effect of adsorbed layers predominates over that of the substrate only for separations less than the thickness of the layer.

Lifshitz's theory of thin films was verified within 1% by Sabisky and Anderson [27] in 1973 for helium films on a vertical wall (Fig. 1.9). Film heights of 100 m are of course not observable, but the corresponding thicknesses are obtained for unsaturated films ($p_0 < p_{\mathrm{sat}}$), for which the following equation is easily proved:

$$F(d) + \varrho g \left(z + \frac{kT}{mg} \ln \frac{p_{\mathrm{sat}}}{p_0} \right) = 0 \, .$$

This theory has been verified with the same precision by Blake [28] for the disjoining pressure of alkane films (Fig. 1.10). Likewise Richmond *et al.* [29] have predicted, from the values of dielectric constants, that pentane, hexane and heptane spread over water, whereas octane and dodecane do not, a result which has been experimentally verified.

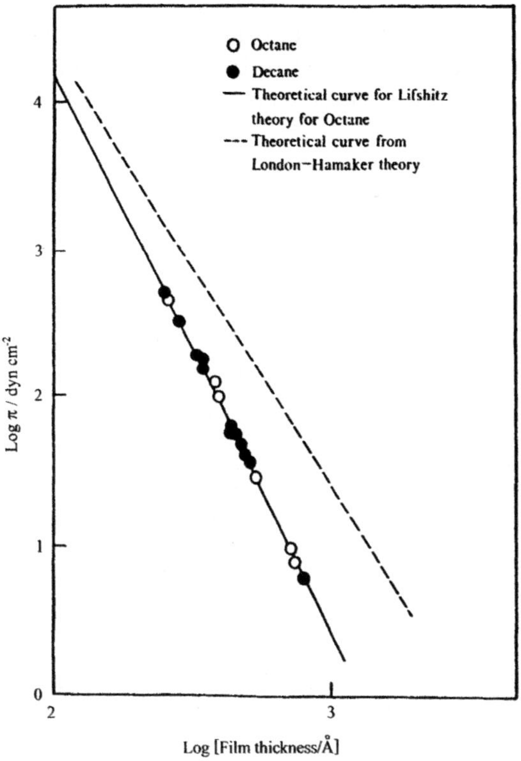

Fig. 1.10. Octane and decane films on sapphire, showing disjoining pressure as a function of the film thickness

1.4 Surface Energy

The surface energy γ of a solid (or a liquid) in the presence of a gaseous phase is defined as the work $\gamma\,dA$ needed to create reversibly and isothermally an elemental area dA of new surface in equilibrium with the medium. It is given in J/m^2 ($1\,mJ\,m^{-2} = 1\,erg\,cm^{-2}$).

This definition needs some comments.

1. The work needed to create the surface only is all that is relevant, any work in the bulk being either nonexistent or taken into account and eliminated. The energy of a system is thus the sum of a bulk term and a surface term: $U = U_0 + \gamma A$.
2. The issue is to create a new surface (on which the interatomic distances are the equilibrium ones) and not to stretch a formed one. We will return on this point in Sect. 1.5 concerning surface tension.
3. The surface remains in equilibrium with the gaseous phase, so that any creation of a new surface is generally accompanied by adsorption.
4. During this reversible and isothermal process, the system exchanges heat with the outer medium—heat whose origins are the entropy difference between bulk atoms and surface atoms, and the entropy difference between atoms of the gas and adsorbed atoms. However, this calorific energy is not taken into account in the surface energy. It is the reason why the term "specific surface work" is sometimes preferred.
5. The reversibility is needed only for its ease to define the surface energy by the work which the system has furnished to the outer medium, but the surface energy of a surface does not depend on the way followed to obtain it. It is an intensive quantity like pressure or temperature.[2]
6. Unlike bulk atomic planes, atomic planes in the vicinity of the surface undergo asymmetrical interactions which provoke a modification of few percentage points in the reticular distances. This phenomenon, termed *surface relaxation*, minimizes the energy of the system and occurs in any creation of a new surface.

Let us return to the microscopic study of the surface energy. If we imagine a solid cut along any plane and attempt to separate the two surfaces in a vacuum, we must do work against the attractive forces which ensure the

[2] In fact the creation of a new surface by cleavage or brittle fracture is accompanied by irreversible processes (work hardening, viscous losses). Even if the work of elastic deformation in the bulk has been correctly evaluated and eliminated, the remaining work involves an important part irreversibly dissipated as heat or spent to create defects (work hardening). This remaining work is generally written as $\gamma_f\,dA$ and γ_f is termed *specific energy of fracture*. Unfortunately, this energy, which can reach $10{,}000\gamma$. is not a characteristic of the surface and varies with experimental conditions. For viscoelastic solids $\gamma_f \to \gamma$ when the velocity of deformation tends towards zero and the reversibility is approached. It is not the case for the plastic deformation of metals where, even at low velocity of deformation, the motion of dislocations dissipates energy.

cohesion of the solid. The work so spent to create a unit area of new surface is the surface energy. For a molecular solid, one has to work only against van der Waals forces and the surface energy is weak; but if the solids are ionic, covalent or metallic, most of the work is done against the strong bonds which are predominant at short distance, and the contribution of van der Waals forces is weak.

So the surface energy of a solid reflects the nature of the bonds between the atoms which constitute it. Metals and covalents have high surface energies ($1000-3000\,\mathrm{mJ/m^2}$). The surface energies of ionic crystals are weaker ($100-500\,\mathrm{mJ/m^2}$), and those of molecular crystals still weaker (less than $100\,\mathrm{mJ/m^2}$). By adsorption, the short range forces are more or less screened and the surface energy decreases. This is why a metallic surface in air, ill degreased, will have a very low surface energy, corresponding to van der Waals interaction energies (Fig. 1.11).

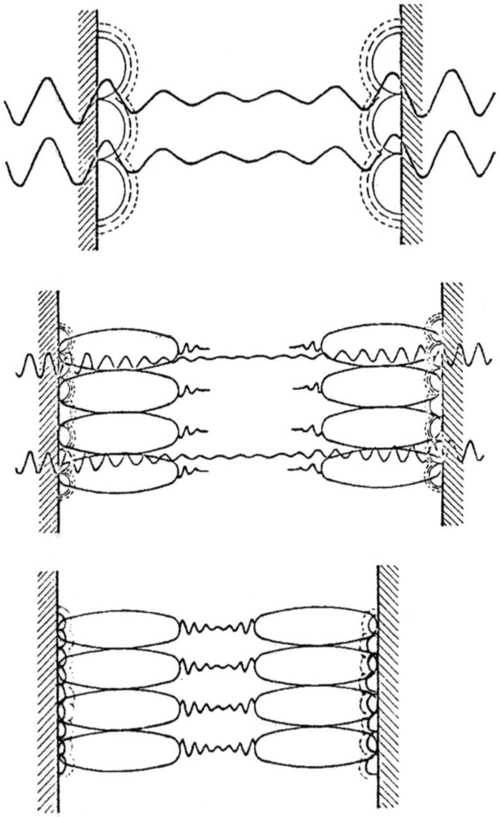

Fig. 1.11. The damped sinusoids schematize the electromagnetic interactions, and the circle arcs the short-range forces. Adsorption of molecules, like stearic acid for example, screens these short-range forces. However, the effect of an adsorbed layer predominates on that of the substrate in the computation of van der Waals forces only for separations less than the thickness of the layer

Israelachvili and Tabor [24] have shown that for two clean surfaces of mica (recovered nevertheless by their monomolecular layer of water), the attraction forces are of the van der Waals type from infinity to 2 nm. By extrapolating these forces down to 0.2 nm, they have calculated that the surface energy should be 40 mJ/m^2, instead of the experimental value found of 300 mJ/m^2. This means that short-range forces (less than 2 nm) are dominant in the calculation of the surface energy of the mica. On the other hand, if the two mica surfaces are covered with a monolayer of stearic acid, whose thickness is 2.5 nm, the short-range forces are screened and the above calculation makes sense. For mica covered by stearic acid, it leads to a surface energy of 30 mJ/m^2, close to the experimental value. (The difference from the 40 mJ/m^2 computed above arises because the Hamaker constant of the stearic acid has been taken into account for distances less than 2.5 nm, as said above.)

When two solids 1 and 2 are in contact over an area A, their energy is $U = U_0 + \gamma_{12}A$, where γ_{12} is their *interfacial energy*. When they are separated, their energy is $U = U_0 + (\gamma_1 + \gamma_2)A$. To separate them reversibly and isothermally, one has to furnish the work

$$w = \gamma_1 + \gamma_2 - \gamma_{12} , \tag{1.45}$$

termed *Dupré's energy of adhesion* (and dating from 1869). When the two solids 1 and 2 form a single crystal, the interfacial energy γ_{12} is zero, and the energy $w = 2\gamma$ represents the theoretical cohesive energy.[3] When they are two grains of a polycrystal, the interface energy γ_G (the grain boundary energy) is a function of the misorientation of the grains. For a grain boundary of arbitrary orientation, γ_G is about one-third of the surface energy γ_S, and for a twin boundary about one-fiftieth.

1.4.1 Surface Energy of a Lennard–Jones Solid

Integration of the Lennard–Jones Potential. We have computed in Sect. 1.3.1 the attraction force per unit area between two parallel plates, by integration of the London forces. As this force tends towards infinity when the plates are approached, let us perform again the calculation with a Lennard–Jones potential, using (1.14), to introduce an equilibrium distance Z_0. We obtain

$$\sigma(z) = \frac{A}{6\pi z^3} - \frac{B}{z^9} ,$$

where $A \equiv \pi^2 n^2 C$ is the Hamaker constant of the solid, and n the number of atoms per unit volume. The constant B is deduced from $\sigma = 0$ at the equilibrium distance Z_0, and it gives:

[3] These terms do not have the same significance for physicists of surfaces and physicists of solids. For the latter, it is the energy gained when individual atoms are approached from infinity to form a crystal.

$$\boxed{\begin{aligned}\sigma(z) &= \frac{A}{6\pi Z_0^3}\left[\left(\frac{Z_0}{z}\right)^3 - \left(\frac{Z_0}{z}\right)^9\right] \\ &= \frac{3\sqrt{3}}{2}\sigma_{\mathrm{th}}\left[\left(\frac{Z_0}{z}\right)^3 - \left(\frac{Z_0}{z}\right)^9\right]\end{aligned}}\,, \tag{1.46}$$

$$\boxed{\begin{aligned}U(z) &= -\frac{A}{12\pi Z_0^2}\left[\left(\frac{Z_0}{z}\right)^2 - \frac{1}{4}\left(\frac{Z_0}{z}\right)^8\right] \\ &= -\frac{8\gamma}{3}\left[\left(\frac{Z_0}{z}\right)^2 - \frac{1}{4}\left(\frac{Z_0}{z}\right)^8\right]\end{aligned}}\,, \tag{1.47}$$

where

$$\boxed{\sigma_{\mathrm{th}} = \frac{A}{9\pi\sqrt{3}Z_0^3}} \tag{1.48}$$

is the maximum stress (termed *theoretical stress*) at $Z_{\mathrm{m}} = 3^{1/6}Z_0 = 1.201Z_0$, and where

$$\boxed{U_0 = 2\gamma = \frac{A}{16\pi Z_0^2}} \tag{1.49}$$

is the work to separate the two plates from the equilibrium distance to infinity. With $A \cong 10^{-19}$ J and $Z_0 \cong 2\times10^{-10}$ m, (1.49) gives $\gamma \cong 25\,\mathrm{mJ/m^2}$, which is of the correct order of magnitude.

Figures 1.12a and 1.12b represent graphically (1.47) and (1.46), respectively. The curve showing stress-displacement has two branches. The first, with a positive slope, represents the *restoring elastic forces*. Its slope at $z = Z_0$, allows us to define Young's modulus

$$E = Z_0\left(\frac{\mathrm{d}\sigma}{\mathrm{d}z}\right)_{z=Z_0} = \frac{A}{\pi Z_0^3}\,, \tag{1.50}$$

from which one deduces that

$$\boxed{\sigma_{\mathrm{th}} = 0.06E = 2.05\frac{\gamma}{Z_0} = 0.36\sqrt{\frac{E\gamma}{Z_0}}}\,. \tag{1.51}$$

This theoretical stress is enormous compared with the elastic limit. (The elastic limit at 0.5% of deformation—an optimistic value—is displayed as the dashed lines in Fig. 1.12b). The second branch, with a negative slope,

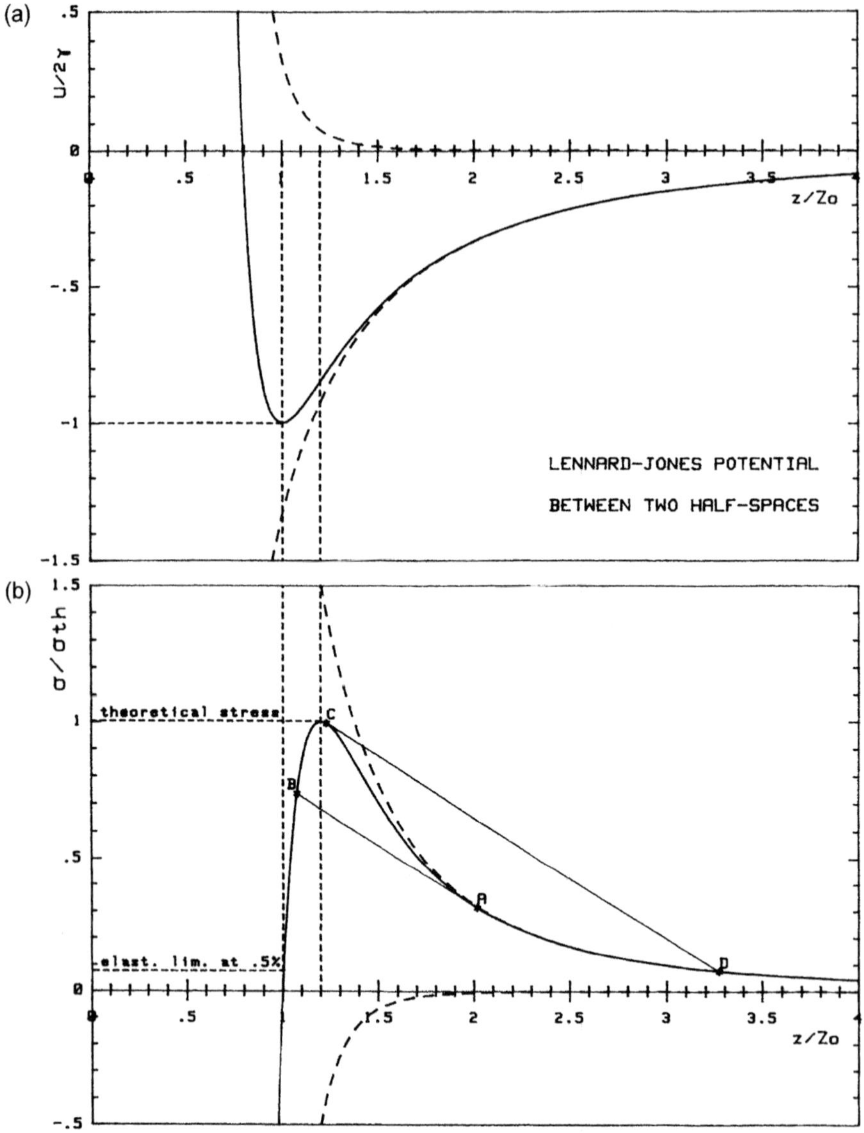

Fig. 1.12. Lennard–Jones potential, showing energy (**a**) and interaction stress (**b**) between two half-spaces as a function of the distance. If the measuring apparatus has a finite stiffness, jumps are observed from A to B, and from C to D

represents the adhesion forces and is also enormous compared with the elastic limit since one must go up to $z = 7Z_0$ before the forces return to the tenth of this value. (Let us note, in passing, that interactions between two plates clearly extend to larger distance than those between two atoms.)

As we will see, it is this theoretical stress which prevails at a crack tip (or on the edge of a contact). If one considers the plane of symmetry of a crack, all the stresses corresponding to interatomic distances $z < Z_m$ on both sides of this plane belong to the elasticity of the solid (strongly nonlinear near the tip). All those corresponding to $z > Z_m$ are regarded as attraction forces between the crack faces (termed also adhesion or cohesion forces between the surfaces). There is here a semantic problem. As can be seen in Fig. 1.12b, the elastic traction stresses are essentially due to attractive van der Waals forces. It would be not necessary to make the distinction between elastic forces and adhesion forces if the continuum theory of elasticity did not require an univocal relation between stress and deformation. Furthermore, the definition of surface energy avoids this distinction since 2γ is merely the area under the (equilibrium) stress-displacement curve. However, let us note that the contribution of the "adhesion forces" to the surface energy is 85% for a Lennard–Jones solid, since the interaction energy at $z = Z_m$ is $U = 0.85U_0$. The van der Waals forces having a range δ_t of about 10 nm, there is at a crack tip a zone of length d along which surfaces are in interaction and beyond which they are free. It must be borne in mind that integration along this zone d, i.e. from Z_m to δ_t, does not correspond to the surface energy, this latter being defined by an integration from Z_0 to δ_t.

We can see that it is physically unacceptable to admit elastic traction stresses in an area of contact and to refuse adhesion forces outside it, or, reciprocally, to accept adhesion forces between neighbouring surfaces and to have only compression forces in the contact. We will return on these points, which relate to the Barenblatt and Dugdale models of cracks and to the theories of JKR and DMT for the contact of spheres.

Summation of the Lennard–Jones Potential. The previous method treats the solid as a continuum, which can give correct results only when the plates are at a distance large compared with Z_0. When the distances z become small, the integrations must be replaced by summations, and the structure of the solid and its surface must be taken into account.

Let us examine, as an example, the case of rare gas crystals, following Shuttleworth [30] in 1949. Imagine such a crystal cut along a crystallographic plane, and move away the two parts I and II to infinity. In a first stage, we will assume that atoms in each semi-crystal are held in their relative positions by constraints and that the work to make the separation is $2U_0'$ per unit area. Then we will relax the constraints: surface atoms are no longer in equilibrium and move to a position of lower energy, which reduces the energy of each surface by a quantity U_0'' per unit area. The surface energy at a temperature of 0 K is:

$$\gamma = U_0 = U_0' - U_0'' ,$$

where U_0'' is a small correction.

The lattice of inert-gas crystal is face-centred cubic; this may be regarded as a simple cubic lattice in which only alternate sites are occupied. Let us calculate the interaction energy through a plane such as that displayed

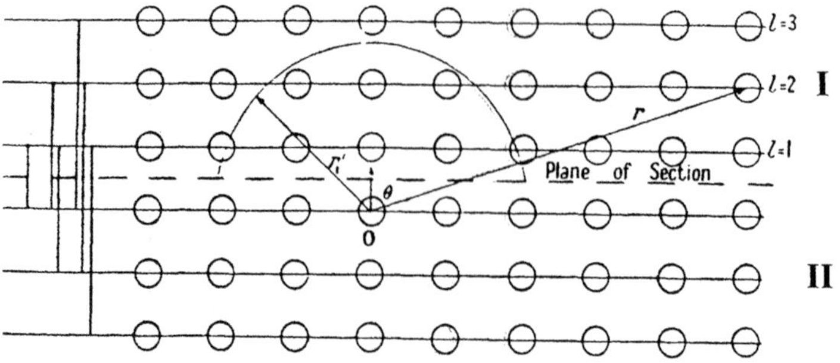

Fig. 1.13. Summation of the Lennard–Jones potential (○ atoms, — planes)

as a dashed line on Fig. 1.13: it is the sum of the potentials between all planes in part I and all planes in part II. Let a be the equilibrium distance between planes parallel to the section; it is seen from the Fig. 1.13 that the contribution of a pair of planes at separation a occurs once, at separation $2a$ twice, at separation $3a$ three times, and so on. If the planes in part I are defined via $l = 1, 2, 3, \ldots$, then the mutual potential (per atom in the surface plane) is

$$-2u_0' = \sum_{l \geq 1} lU(r) , \tag{1.52}$$

where $U(r)$ is the Lennard–Jones potential between atoms (see (1.14)) and r is the distance between an atom in the surface plane ($l = 0$) of part II and an atom in the plane l of part I; the summation is over all atoms in part II. The distance from the origin O to any point in the lattice is:

$$r = \sqrt{x^2 + y^2 + z^2}$$
$$= a\sqrt{m_1^2 + m_2^2 + m_3^2} ,$$

where x, y, z are given by $m_1 a, m_2 a, m_3 a$ with m_1, m_2, m_3 integers. Then (1.52) becomes

$$-2u_0' = \frac{D}{a^{12}} \sum_{\substack{(m_1+m_2+m_3)\ \text{even} \\ l \geq 1}} \frac{l}{(m_1^2 + m_2^2 + m_3^2)^6}$$
$$- \frac{C}{a^6} \sum_{\substack{(m_1+m_2+m_3)\ \text{even} \\ l \geq 1}} \frac{l}{(m_1^2 + m_2^2 + m_3^2)^3} ,$$
$$-2u_0' = B_{12}\frac{D}{a^{12}} - B_6\frac{C}{a^6} , \tag{1.53}$$

where only the even values of $m_1 + m_2 + m_3$ are used since one site in two is unoccupied. Variables B_6 and B_{12} are the lattice sums computed by Shuttleworth for planes (111), (110) and (100).

From the constants C and D one can compute u_0', but Shuttleworth preferred to eliminate them by introducing the heat of sublimation per atom at 0 K, ε_0, which can also be expressed in terms of lattice sums:

$$-2\varepsilon_0 = A_{12}\frac{D}{a^{12}} - A_6\frac{C}{a^6} \, . \tag{1.54}$$

At 0 K the lattice potential energy ε_0 is a minimum with respect to variations in a, hence

$$2A_{12}\frac{D}{a^{12}} = A_6\frac{C}{a^6} \, . \tag{1.55}$$

Eliminating C and D from (1.53), (1.54), and (1.55), we obtain

$$u_0' = \left(\frac{2B_6}{A_6} - \frac{B_{12}}{A_{12}}\right)\varepsilon_0 \, . \tag{1.56}$$

Multiplying by the number of atoms per unit area of the considered plane, we further obtain U_0' for this plane as a function of the heat of sublimation of the crystal.

Let us assume that the outer plane only is relaxed after the cut. The potential energy per atom of this surface plane is:

$$\begin{aligned}\varphi &= \frac{D}{a^{12}}\sum_{l\geq 2}\left(\frac{a}{r}\right)^{12} - \frac{C}{a^6}\sum_{l\geq 2}\left(\frac{a}{r}\right)^{6} \\ &= C_{12}(c)\frac{D}{a^{12}} - C_6(c)\frac{C}{a^6} \, ,\end{aligned} \tag{1.57}$$

where ca is the distance of this surface plane, $l = 1$, from the next plane, $l = 2$; and r is the distance between a particular atom in the surface plane and any atom in the remainder of the crystal. The equilibrium position of the surface plane occurs at the value of c for which φ is a minimum, namely

$$\frac{\mathrm{d}C_{12}}{\mathrm{d}c} = \frac{2A_{12}}{A_6}\frac{\mathrm{d}C_6}{\mathrm{d}c} \, ,$$

where C and D have been eliminated by (1.55). The solution is found by numerical integration. For a (100) plane, for example, we find that at equilibrium $c = 1.025$, i.e. a distance 2.5% greater than the distance at the interior of the crystal.

The energy variation of a surface atom is thus:

$$u_0'' = \Delta\varphi = \Delta C_{12}\frac{D}{a^{12}} - \Delta C_6\frac{C}{a^6} \, ,$$

where c varies from unity to its equilibrium value. The calculation of lattice sums for these values of c shows that the correction to the surface energy is less than 1%.

Since the time of Shuttleworth's work in the late forties, more precise calculations have been made on computers, relaxing more atomic planes. Table 1.1 after Benson and Claxton [31] gives the surface energy of inert gases computed at 0 K, and Fig. 1.14 after Allen and de Wette [32] displays the relaxation of the first (111), (100) and (110) planes. The displacements for all three surfaces fall off approximately as the inverse cube of the distance from the surface.

Table 1.1. Surface energy of inert gases at 0 K

Crystal	*Face*	γ (mJ/m^2)
Ne	(111)	19.701
	(100)	20.339
	(110)	21.339
Ar	(111)	43.168
	(100)	44.568
	(110)	46.759
Kr	(111)	52.793
	(100)	54.504
	(110)	57.186
Xe	(111)	62.116
	(100)	64.130
	(110)	67.284

1.4.2 Surface Energy of Metals

In solid-state physics it is possible to compute the interaction potential $U(z)$ between two surfaces in the range of metallic bonds. This potential is essentially due to the exchange energy associated with the antisymmetry of wave functions. In 1981, Rose *et al.* [33] discovered that it is possible to obtain a master curve for all metals by plotting U/U_0 as a function of a scaled separation

$$z^* = \frac{z - Z_0}{\lambda} , \tag{1.58}$$

where λ is the *Thomas–Fermi screening length* (analogous to the Debye length in an electrolyte). The length λ defines the "size" of an atom in a metallic environment:

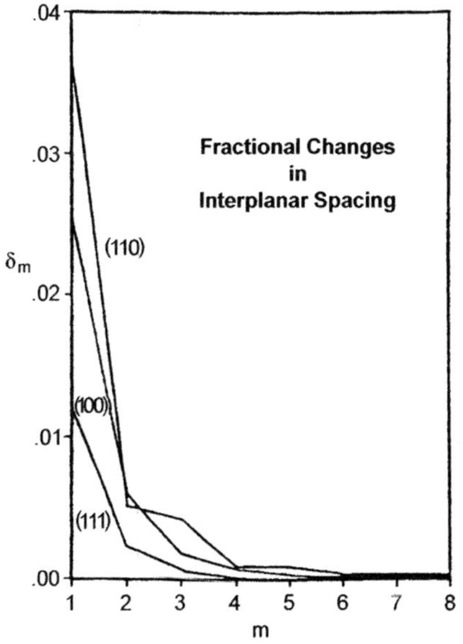

Fig. 1.14. Relaxation of first atomic planes in an inert gas crystal

$$\lambda = \frac{1}{3}\left(\frac{243\pi}{64}\right)^{1/6} n^{-1/6}(Z_0)\,, \tag{1.59}$$

where $n(Z_0)$ is the electron density at equilibrium position. In (1.58) and (1.59), z, Z_0 and λ are expressed in atomic units (1 a.u. corresponds to the Bohr radius, $a_B = \hbar/me^2$ in ESU, and is 5.286×10^{-9} cm). Figure 1.15 after Ferrante *et al.* [34] displays this scaled potential as a function of the scaled separation for various metals. The results fit well with the curve

$$U^*(z^*) = \frac{U}{U_0} = -(1 + \beta z^*)e^{-\beta z^*} \tag{1.60}$$

with $U_0 = 2\gamma$ and $\beta = 0.9$. Hence by differentiation:

$$\frac{\sigma}{\sigma_{th}} = \beta z^* e^{1-\beta z^*}\,, \tag{1.61}$$

with

$$\sigma_{th} = \frac{2\gamma\beta}{\lambda e} = \frac{\sqrt{2}}{e}\sqrt{\frac{E\gamma}{Z_0}} \cong 0.5\sqrt{\frac{E\gamma}{Z_0}}\,. \tag{1.62}$$

1.4.3 Theoretical Stress: Orowan's Model

The theoretical stress σ_{th} of a solid at rupture was estimated by Orowan [35] in 1945, replacing the stress–strain relation by a semi-sinusoid:

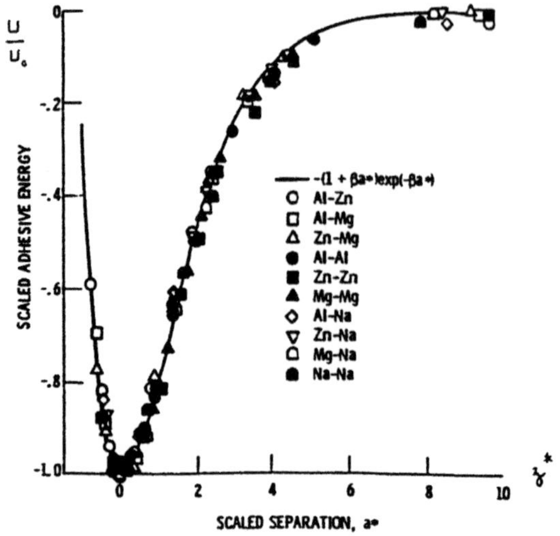

Fig. 1.15. Scaled potential as a function of scaled separation, for various metals

$$\sigma = \sigma_{\text{th}} \sin \frac{2\pi x}{\lambda} \,, \tag{1.63}$$

with $x = z - Z_0$, where Z_0 is the equilibrium separation between two planes. When x tends towards zero, $\sigma \to \sigma_{\text{th}} 2\pi x/\lambda$, an expression which must be identified with Hooke's law. So we can write

$$\sigma = \varepsilon E = \frac{x}{Z_0} E \,,$$

and hence $\lambda/2\pi = Z_0\sigma_{\text{th}}/E$. The semi-sinusoid has the same slope at the origin as the actual curve (i.e. same Young's modulus), and is thus

$$\sigma = \sigma_{\text{th}} \sin \frac{Ex}{Z_0\sigma_{\text{th}}} \,.$$

The value of σ_{th} must be such that the area under the curve is the work 2γ to separate the two surfaces:

$$\frac{2Z_0\sigma_{\text{th}}^2}{E} = 2\gamma \,;$$

hence

$$\sigma_{\text{th}} = \sqrt{\frac{E\gamma}{Z_0}} \tag{1.64}$$

Comparison with (1.51) and (1.62) shows that (1.64) overestimates the theoretical stress. It is generally agreed that the theoretical stress for rupture is of the order of $E/10$ to $E/5$.

1.4.4 Adhesive Avalanche

Let us return to Fig. 1.12b and to the practical problem of measuring the interaction force between two parallel plates. If one of the plates is fixed and the other is hanging from a spring of stiffness k (or more precisely k per unit area), whose elongation allows computation of the attraction force, equilibrium is reached when the total force per unit area is zero:

$$F = \sigma(z) - k(x - x_0) = 0 , \tag{1.65}$$

where $x - x_0$ is the spring elongation. This equilibrium is stable if the total force is a restoring force, i.e. if $\mathrm{d}F/\mathrm{d}z > 0$. With $\mathrm{d}z = -\mathrm{d}x$, this condition is written

$$\frac{\mathrm{d}\sigma}{\mathrm{d}z} + k > 0 . \tag{1.66}$$

During the approach, the slope of the curve becomes larger (in magnitude) than the stiffness of the spring at point A and we can observe a jump from A to B, as shown in Fig. 1.11b. At separation, the jump will be observed from C to D. (The analogy with the case of a magnet suspended by a rubber band, and which is approached to or removed from a steel plate, is very close.) From (1.46) the minimum slope of the curve is

$$\left(\frac{\mathrm{d}\sigma}{\mathrm{d}z}\right)_{\mathrm{max}} = -1.22\frac{\sigma_{\mathrm{th}}}{Z_0}$$

at point $z/Z_0 = 1.40$. For the whole curve to be followed, the stiffness of the spring must satisfy the condition

$$k > 1.22\frac{\sigma_{\mathrm{th}}}{Z_0} .$$

It was by observing this jump that Tabor and Winterton [23] were able to measure the retarded and non-retarded van der Waals forces between crossed cylinders of mica, a method widely used later. The stiffness of the spring is varied and the critical separation is measured at which the jump is observed.

However, Pethica and Sutton [36] discovered in 1988, from molecular dynamics computations, that instabilities of the same type could appear whatever the stiffness of the measuring apparatus, owing to the bulk elasticity of the two solids. These authors simulated the approach of two f.c.c. plates, each half-crystal being divided into a inner region of 11 atomic planes free to move under the action of van der Waals forces exerted by other planes, and a outer rigid shoulder which is not relaxed. Calculations show steep variations in the force and energy of interaction (Fig. 1.16) and in the distances between the plate surfaces when a regular displacement is imposed to the shoulders (Fig. 1.17). The same phenomenon has been observed with other interaction

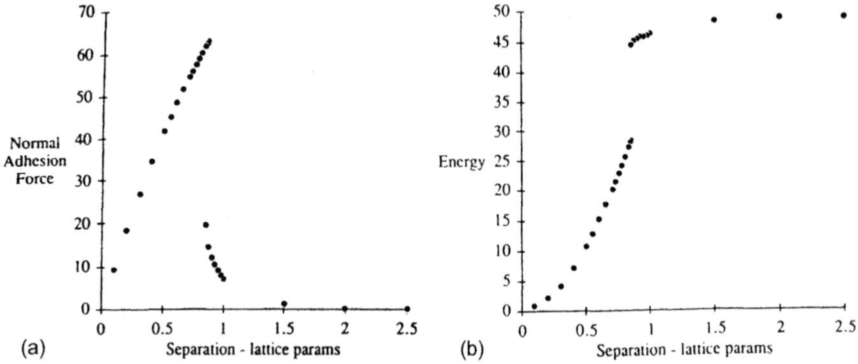

Fig. 1.16. Adhesive force **(a)** and interaction energy **(b)** between atom blocks as a function of rigid shoulders displacement (each block has 11 atomic planes free to relax)

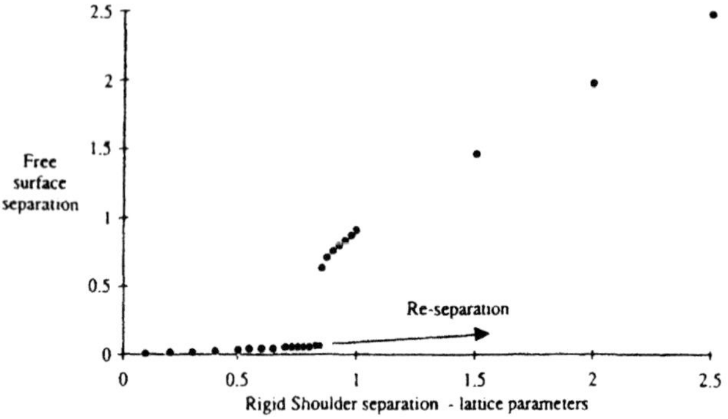

Fig. 1.17. Separation of the free surface of blocks as a function of displacement of rigid shoulders from a bulk lattice position, for approaching blocks (*points*) and separation (*trend*)

potentials [37–40] and has been termed an *adhesive avalanche* by Smith *et al.* [37].

It is instructive to obtain this behaviour by a simple calculation [41] based on (1.46) and (1.47). Let us consider two half-crystals rigidly fixed on one edge and having $N/2$ atomic planes free to move (Fig. 1.18). Let D be the rigid interfacial separation, when the reticular distance is maintained equal to the equilibrium separation Z_0, and let d be the interfacial separation when the two half-crystals are elastically deformed by forces acting between them. Clearly we have $d \cong D$ when $D \gg Z_0$. We will make the following simplifying hypotheses:

1. The reticular distance z is constant in each half-crystal, i.e. we will neglect the surface relaxation.
2. The interaction stress σ between blocks depends only on the interfacial distance d and is independent of the elastic deformation of the crystals,

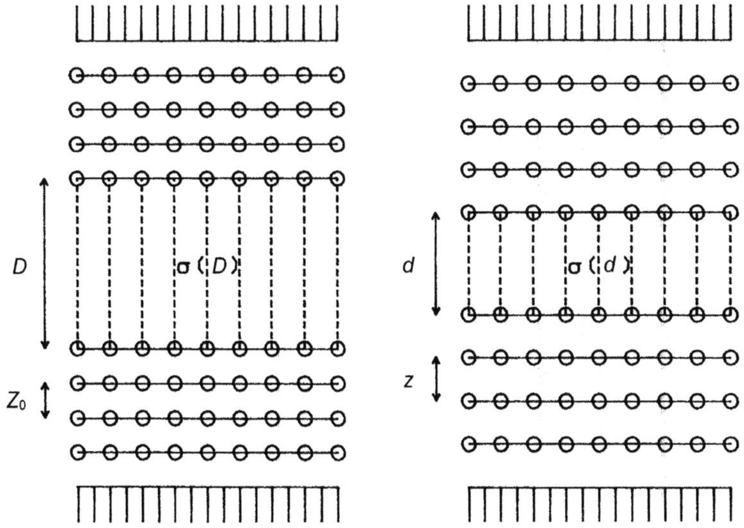

Fig. 1.18. Two half-crystals having each $N/2$ atomic planes free to relax

i.e. the number n of atoms per unit volume is kept constant in the Hamaker constant A.

3. The elastic stress in each crystal, when the reticular distance is z, is given by (1.46), i.e. it is computed as an interaction stress between two rigid blocks separated by an interfacial distance z.
4. The stress $\sigma(z)$ in each half-crystal is equal to the interaction stress $\sigma(d)$ acting between them.

We have seen in Sect. 1.4.1 that the maximum stress (the theoretical stress) was situated where $z = Z_{\mathrm{m}} = 1.20Z_0$. If $d < Z_{\mathrm{m}}$, we deal with one stretched or compressed crystal with $z = d$. If $d > Z_{\mathrm{m}}$, we deal with two half-crystals with interaction, and the solution of $\sigma(z) = \sigma(d)$ is

$$\left(\frac{z}{Z_0}\right)^3 = \frac{2\,(d/Z_0)^3}{-1+\sqrt{4\,(d/Z_0)^6 - 3}} \,. \tag{1.67}$$

Examination of Fig. 1.18 shows that the rigid interfacial distance D is related to the actual interfacial distance d by

$$D = d + N(z - Z_0) \quad \text{if} \quad d > Z_{\mathrm{m}} \,, \tag{1.68a}$$

$$D = d + N(d - Z_0) \quad \text{if} \quad d < Z_{\mathrm{m}} \,, \tag{1.68b}$$

with z given by (1.67). This relation is displayed graphically in Fig. 1.19 for $N = 20$. Figures 1.20a and 1.20b show the potential and the interaction stress between surfaces as a function of the rigid interfacial distance D (the dotted curves are for $d = D$, i.e. $N = 0$, and are thus identical to those of Fig. 1.12).

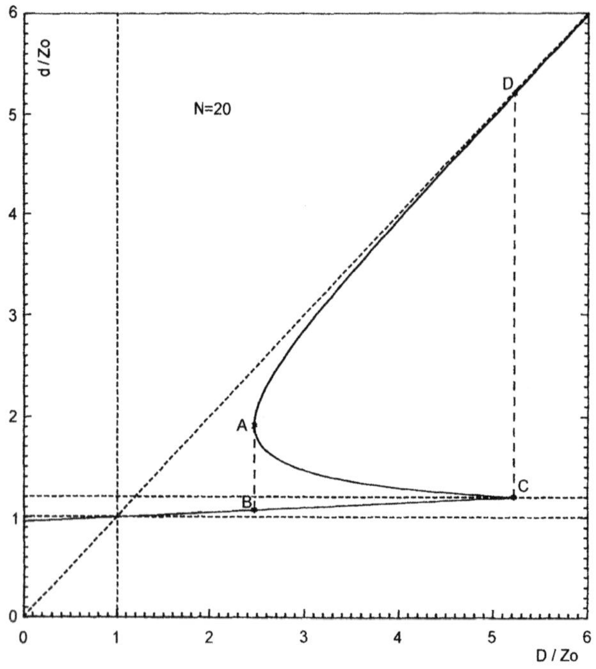

Fig. 1.19. Separation of the free surfaces as a function of the rigid interfacial distance imposed. Compare this simple model to the numerical simulation shown in Fig. 1.17.

The experimenter can vary only the rigid distance D—with a micrometer screw for example—and d adjusts itself. Starting with two half-crystals without interaction, at large distance, let us uniformly decrease D: the separation d decreases continuously until a jump occurs from A to B, where in B we have one stretched crystal. This is the phenomenon of adhesive avalanche. By continuing to decrease D, the stress in the single crystal decreases and cancels when $D = Z_0$. Point A corresponds to the point where the gradient of surface stresses

$$\frac{\mathrm{d}\sigma}{\mathrm{d}d} = \frac{16\gamma}{Z_0^2}\left[-\left(\frac{Z_0}{d}\right)^4 + 3\left(\frac{Z_0}{d}\right)^{10}\right] \tag{1.69a}$$

is equal to the gradient of the elastic restoring force at fixed D

$$\frac{\mathrm{d}\sigma}{\mathrm{d}d} = -\frac{1}{N}\frac{\mathrm{d}\sigma}{\mathrm{d}z} = -\frac{1}{N}\frac{16\gamma}{Z_0^2}\left[-\left(\frac{Z_0}{z}\right)^4 + 3\left(\frac{Z_0}{z}\right)^{10}\right], \tag{1.69b}$$

with z given by (1.67). The numerical solution gives d_c as a function of N, hence the rigid critical separation D_c by (1.68a), and the reticular distance of the single crystal at point B by substituting the value of D_c in (1.68b).

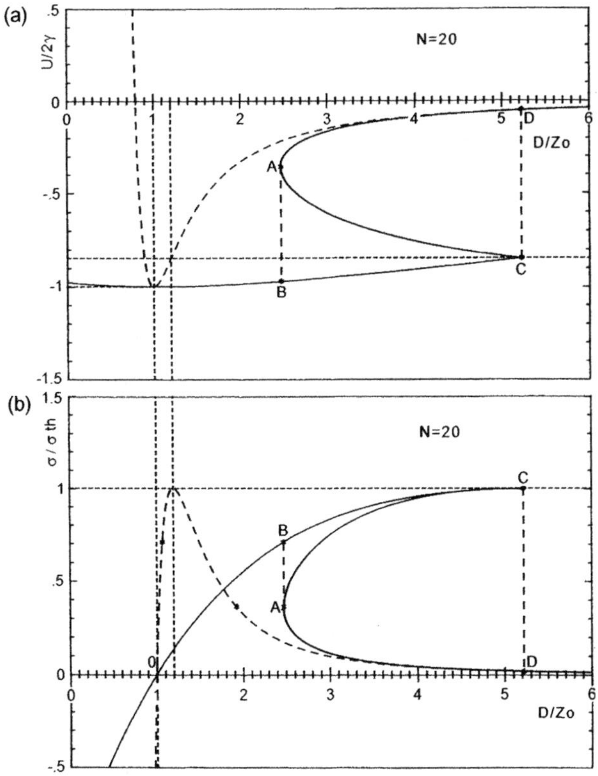

Fig. 1.20. Potential **(a)** and interaction stress **(b)** as a function of the rigid interfacial distance D. *Dotted lines* show where $d = D$

Let us now increase D from Z_0 to infinity. The stress in the single crystal increases up to the theoretical stress σ_{th} at point C where the separation occurs abruptly. Point C ($z = Z_{\text{m}}$) corresponds also to equivalence of (1.69a) and (1.69b) with a zero gradient.

When the number N of planes capable of moving increases, the critical distance for a jump to occur, d_{c}, increases slowly. When $d_{\text{c}} \to \infty$, we have $z/Z_0 \to 1$ and

$$\frac{d_{\text{c}}}{Z_0} \cong \left(\frac{N}{2}\right)^{1/4} . \tag{1.70}$$

The potential and the interaction stress as a function of the rigid separation distance D are displayed in Figs. 1.21a and 1.21b for various values of N. These results, obtained in a very simple manner, resemble very much the results given by molecular dynamics [36, 37, 40]. They have the advantage of displaying the complete curves with their unstable branches. Examination of Fig. 1.21a shows that all curves pass through the point $(1, -1)$, which means that areas under the curves $\sigma(d)$ are all equal to 2γ. So, in Fig. 1.20b, if

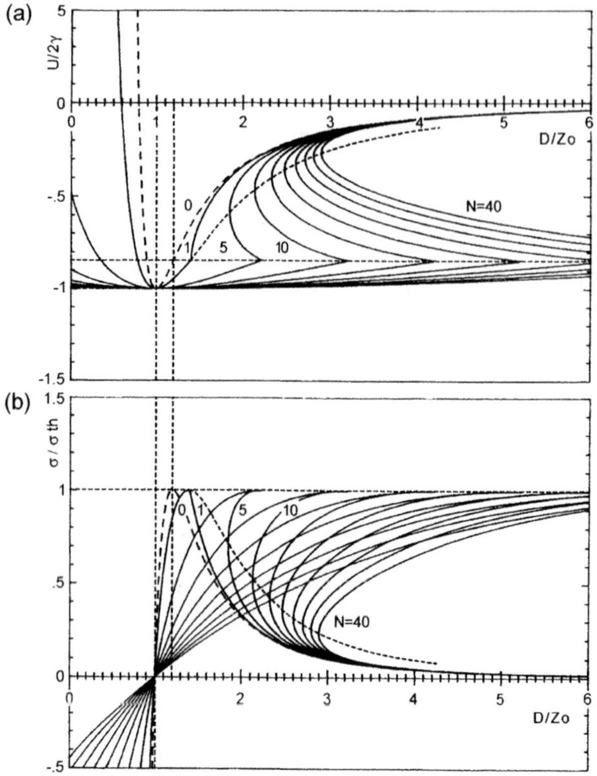

Fig. 1.21. Potential **(a)** and interaction stress **(b)** as a function of the rigid seperation distance D, for various values of N

the area under the curve OBCAD... is 2γ, the work done in the approach experiment (...DABO) is less, and the work done in the pull-off experiment (OBCD...) is larger than 2γ.

This result is interesting. The surface energy of a given surface is a state function which does not depend on the way the surface has been obtained. However, it can be computed from the work done only if this transformation is reversible and isothermal: in this case the heat received by the solid is well-defined, and compensates exactly the entropy difference between surface atoms and bulk atoms.

Remark. Finally, it should be noted that this method to create a new surface, by imagining a crystal cut along a plane and removing progressively the two halves until they are out of the range of interaction forces, is not a good one and leads to misinterpretations. By this method it is not possible to define an increase of the surface area, say of 10%. This difficulty arises from the unrealistic symmetry used for the calculation. The least angular error breaks this symmetry and leads to partial contact during the approach, with separation occuring on a part of the surface. The separation of two solids never occurs as a whole as in Fig. 1.18, but instead by propagation of a crack whose increment of displacement creates an increment of new surface. The theoretical stress, instead of invading the whole solid, is localized at the tip of this crack, which thus acts as a stress transformer, breaking the bonds one by one as the crack advances, rather like a zip fastener. Elastic forces act ahead of the crack tip and adhesion (or cohesion) forces act over a length termed the *cohesion zone*, beyond which the surfaces are without interaction. If it is possible to compute the variation $G\,\mathrm{d}A$ of elastic and potential energies when the crack area varies by $\mathrm{d}A$ (G is termed the *energy release rate* in fracture mechanics), then its value gives the energy needed to create an element $\mathrm{d}A$ of surface area. This energy varies with crack speed due to dissipative phenomena. If all dissipation mechanisms (viscoelastic or plastic) are absent, $G_0 = 2\gamma$. We will return in detail on these problems in Chapt. 3.

Let us add that, for a crack, the problem of the adhesive avalanche due to the elasticity of the solid does not arise. The elasticity of the solid which led to a jump when the parallel surfaces were close enough leads here to a continuous deformation of the surfaces near the crack tip. As shown by Barenblatt [42], the crack tip takes a cusp shape, which will be considered later.

1.4.5 Wetting Films

We have seen in Sect. 1.3.5 that, depending on the sign of the disjoining pressure and its derivative, we could have or have not a wetting film on a surface. Let us examine more closely this idea of wetting, and the definitions of "dry" and "wetted" surfaces (cf. [43]).

Let us consider a vessel made of a solid 1, of surface energy γ_{S} in a vacuum and closed by a piston. Let us introduce a vapour of matter 2, and increase progressively the pressure at constant temperature. The surface is first covered by a number of matter-2 atoms, the coverage being dependent on the vapour pressure. This is the phenomenon of *adsorption* (physisorption or chemisorption). It cannot be treated by Lifschitz's theory, which only applies to phases of thickness larger than molecular dimensions, but thermodynamics indicates (Gibbs' adsorption equation) that the surface energy of the solid can only decrease. The quantity

$$\pi_{\mathrm{S}} = \gamma_{\mathrm{S}} - \gamma_{\mathrm{SV}} \tag{1.71}$$

is the *surface pressure* of the adsorbed layer, not to be confused with the disjoining pressure Π. (In the case of adsorption on a liquid this surface pressure, in N/m, can easily be measured with a floating barrier, used as a "surface piston".) For surfaces of high surface energy, like metals or covalents, this surface pressure can be enormous due to the screening of short-range forces. We shall say that such a surface covered by a monolayer or a fraction of a monolayer of adsorbed molecules, is a "dry" surface.

When the pressure is increased, a film can be formed on the surface. In the case of pure van der Waals interactions, the problem is simple. When the disjoining pressure is positive ($A_{132} < 0$), the thickness of the film (in the nonretarded regime) on a horizontal surface is given by (1.42), i.e.

$$h^3 = \frac{A_{\mathrm{SLV}}\overline{v}}{6\pi kT \ln(p/p_{\mathrm{sat}})}\,, \tag{1.72}$$

where $\overline{v} = m/\varrho$ is the molecular volume. It regularly increases when the pressure increases, and then diverges when $p = p_{\mathrm{sat}}$ (the vapour condenses on the wall). We shall say that where the vapour and the macroscopic liquid coexist, the surface is "completely wetted". If a drop is deposited onto the surface, it will spread completely.

However, the disjoining pressure is not always a monotonic and decreasing positive function of thickness. A thin film can be formed which then stops increasing (cf Fig. 1.6). At the saturation pressure, the vapour does not condense on the wall, which remains apparently dry, but instead in bulk from condensation sites. If a macroscopic drop is deposited onto the surface, it remains in equilibrium with the thin film, forming a contact angle. Depending on this angle, we say that the liquid "does not wet" or "partially wets" the surface, at least at the temperature of the experiment, since complete wetting can appear at higher temperature; this is the *wetting transition*. Figure 1.22 after [43] displays the adsorption isotherms of methane on gold (111): at 87.1 K methane does not wet gold, at 89.8 K it wets it completely.

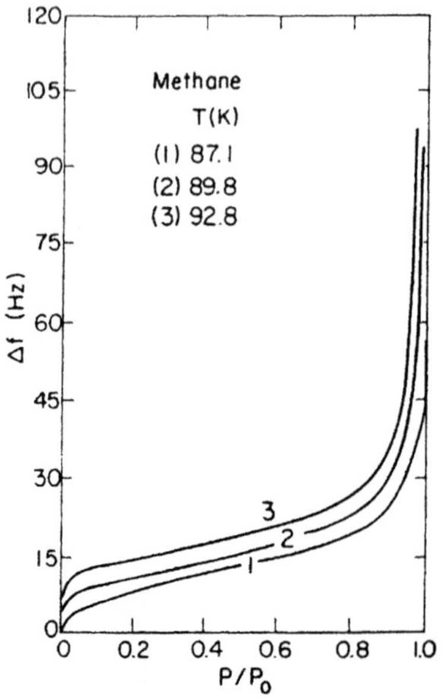

Fig. 1.22. Adsorption isotherms of methane on gold (111), as measured by the frequency shift Δf of a quartz microbalance. The wetting transition appears between 87.1 K and 89.8 K

Let us consider a solid covered with a thick macroscopic liquid film. There is no interaction between the solid and the vapour, and the disjoining pressure is zero. The work γ_{SV}^{m} needed to create a new unit area of this wetted surface is the sum of the work γ_{SL} (the interfacial energy) to create the solid–liquid interface and of the work γ_L to create the liquid surface:

$$\gamma_{SV}^{m} = \gamma_{SL} + \gamma_L \, .$$

When the solid is covered with a thin film of thickness h, the work $\gamma_{SV}(h)$ needed to create a new unit area of this surface is the sum of the work γ_{SL} (the interfacial energy) to create the solid–liquid interface and of the work γ_L to create the liquid surface increased by the interaction energy (in J/m^2)

$$U(h) = -\int_{\infty}^{h} \Pi(h)\,dh = -\frac{A_{132}}{12\pi h^2} \tag{1.73}$$

between the solid and the vapour (see (1.38)). In total, we have

$$\gamma_{SV}(h) = \gamma_{SL} + \gamma_L + U(h) \, , \tag{1.74}$$

which relation is represented on Fig. 1.23, after [44].

In wetting problems, the quantity termed *spreading parameter* was introduced by Marangoni in 1865 and by Cooper and Nuttal in 1915 (see [45]) as

$$S = \gamma_{SV} - (\gamma_{SL} + \gamma_L) \, . \tag{1.75}$$

It is the difference between the surface energy of a dry solid and the surface energy of the solid wetted by a liquid thick film ($U(h) \to 0$). If the Dupré energy of adhesion is introduced, (1.45) becomes

$$w_{SL} = \gamma_S + \gamma_L - \gamma_{SL} \tag{1.76}$$

and we get

$$S = w_{SL} - 2\gamma_L \, , \tag{1.77}$$

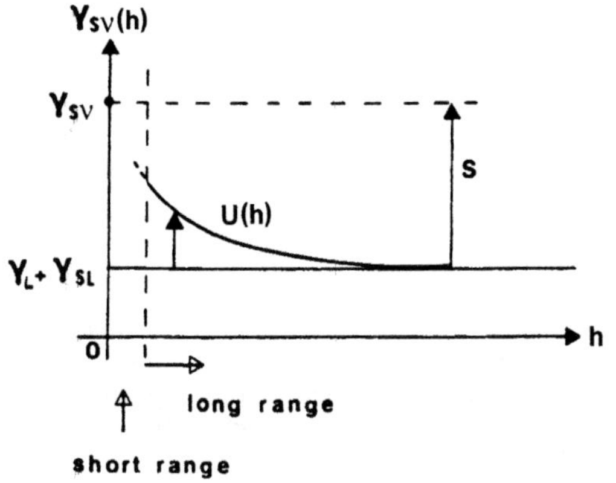

Fig. 1.23. Surface energy of a wetted surface as a function of the thickness h of the film created

which shows that a liquid spreads on a dry surface if its energy of adhesion to the solid is larger than its energy of cohesion. When S is negative, the dry surface has the lowest free energy and the liquid does not spread (no wetting or partial wetting), and when S is positive or zero the wetting is complete.[4]

The type of wetting is defined by the observed value of the *equilibrium contact angle* θ_e of a drop on the surface. If this angle is larger than 90°, as in the case of mercury on glass, no wetting occurs. If the angle is between 90° and 0°, we have partial wetting. If this angle is zero, i.e. if the drop completely spreads on the surface, we have complete wetting. The lateral equilibrium of a drop is determined by *Young's equation* (1805):

$$\boxed{\gamma_{SV} - \gamma_{SL} = \gamma_L \cos\theta_e} . \tag{1.78a}$$

Expressed in the form

$$S = -\gamma_L(1 - \cos\theta_e) , \tag{1.78b}$$

it follows that lateral equilibrium can be maintained only if S is negative.

1.4.6 Kinetics of Spreading and Precursor Films

In the spreading of drops of wetting liquids, we must distinguish between very volatile liquids and nonvolatile liquids. In the first case there is a liquid film brought by the vapour and the drop will then spread on its own film; in the second case, the drop will spread on a dry surface and the equilibrium figure will be a *pancake*.

When a droplet of volume Ω of a nonvolatile wetting liquid is deposited on a dry surface (such as silicon oil on glass), this droplet spreads by the flow of a *precursor film* of a few tens of nanometres thick through a purely hydrodynamic process (Fig. 1.24), as shown by Joanny and de Gennes [46–48]. The apparent contact angle θ_a and the visible radius of the droplet vary with time, first as *Tanner's law*:

$$\theta_a(t) \propto \left(\frac{\gamma}{\eta}t\right)^{-3/10} , \tag{1.79}$$

$$R(t) \propto \Omega^{3/10}\left(\frac{\gamma}{\eta}t\right)^{1/10} ; \tag{1.80}$$

---

[4] When the surface pressure π_S is very high, the phenomenon of *autophobicity* can be observed. A drop containing impurities begins to spread on a dry surface of high surface energy ($S > 0$); then, after adsorption of the impurities, γ_{SV} becomes so weak that the surface is no longer wettable ($S < 0$); the drop then retracts, leaving behind it a dry surface with a low surface energy.

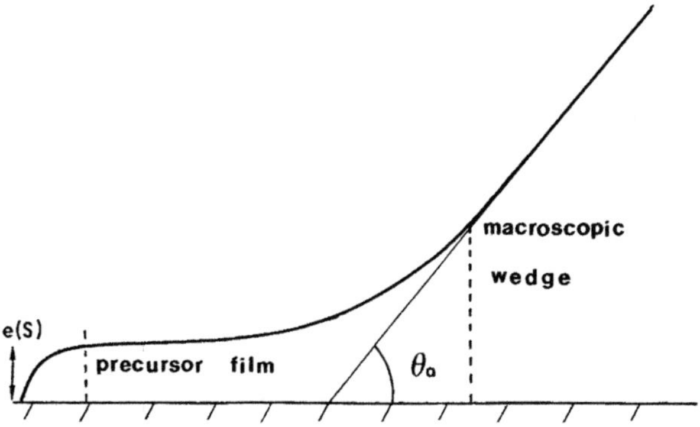

Fig. 1.24. Spreading of a droplet and precursor film

and then, as gravity effects become predominant [49], as

$$R(t) \propto \Omega^{3/8} \left(\frac{\varrho g}{\eta} t \right)^{1/8} , \tag{1.81}$$

where η is the viscosity of the liquid. These kinetics, independent of S, are well verified experimentally [50, 51]. Ultimately, competition is set up between the tendency to spread due to S and the tendency to thicken due to the disjoining pressure, and at equilibrium the droplet takes the form of a pancake (Fig. 1.25). The free energy of a film of thickness h and area A is:

$$F = F_0 - SA + U(h)A .$$

At equilibrium it must be at a minimum, with the constraint that the volume $\Omega = Ah =$ const.. Thus

$$S = U(h_0) + h_0 \Pi(h_0) . \tag{1.82}$$

Hence, with

$$\Pi(h) = -\frac{A_{\mathrm{SLV}}}{6\pi h^3} > 0 ,$$

the thickness of the pancake [46] is

$$h_0 = \sqrt{\frac{|A_{\mathrm{SLV}}|}{4\pi S}} . \tag{1.83}$$

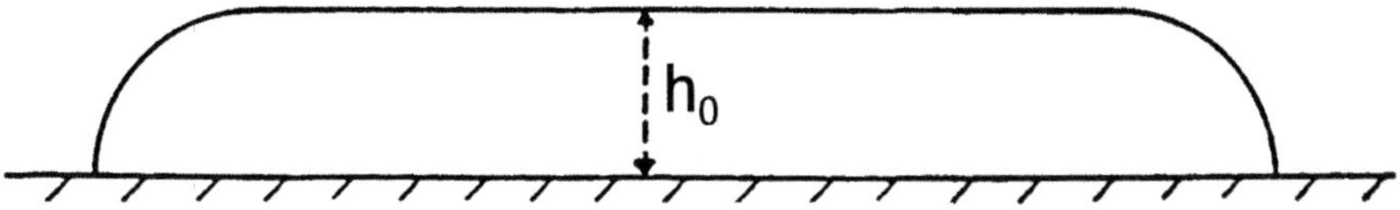

Fig. 1.25. Droplet in the form of a pancake

The higher the spreading parameter is, the thinner the film and the larger the occupied surface.

When the droplet is volatile, or when the surface is in equilibrium with the vapour, the problem is slightly different since the droplet spreads on its own wetting film, but (1.79–81) are still valid.

1.4.7 De-wetting of Films

When a liquid does not wet a surface ($S < 0$), it is still possible to deposit a film on a horizontal surface. Above a critical thickness [52]

$$h_c = \left(\frac{2|S|}{\varrho g}\right)^{1/2} = 2\left(\frac{\gamma_L}{\varrho g}\right)^{1/2} \sin\frac{\theta_e}{2}\,, \tag{1.84}$$

of the order of a millimeter, the liquid is stabilized by gravitational forces. Let us remark that this critical thickness is that of a large drop deposited on a surface. For smaller thicknesses, yet higher than the van der Waals range, the film is metastable and can de-wet by the nucleation and growth of large dry zones. If it is very thin, with a thickness less than

$$h_i = \left(\frac{|A_{SLV}|}{6\pi\varrho g}\right)^{1/4}, \tag{1.85}$$

the film is unstable and breaks into a number of droplets, each making a contact angle θ_e with the dry surface. These results have been obtained recently by Brochard *et al.* [53–57].

1.5 Surface Tension

Besides the surface energy γ (expressed in $\mathrm{J\,m^{-2}}$) another quantity can be defined in purely mechanical terms, namely the *surface tension* σ (expressed in $\mathrm{N\,m^{-1}}$). This quantity, which plays a role in problems of the mechanical equilibrium of surfaces, is more delicate to introduce. In the case of liquids, surface energy and surface tension are identical; but, as shown by Gibbs in 1876, the surface tension of a solid can greatly differ from its surface energy.

Through the observation of soap bubbles and lamellae, the fact that the surfaces of liquids have a tendency to contract led very early to the concept of surface tension. As recalled by Shuttleworth [58], this concept was introduced by Cabeo (1629) and established more explicitly by Seguer (1751). Theories explaining surface tension in terms of molecular forces have been proposed by Young (1805), Laplace (1806), Poisson (1830), Worlington (1884), Bakker (1928), Brown (1947) and Prandtl (1947).

The concept of surface energy is more recent, of course, since the concept of energy itself has only been elaborated at the beginning of the 19th century (the word "energy" being coined by Young in 1807). Surface energy was introduced by Gauss in 1830. It had such a success that for a long time surface tension was considered no more than a mathematical abstraction, as recalled by Brown [59] in 1947. The physical origin of this tension was explained as follows. A molecule in the bulk of a liquid undergoes from neighbouring molecules attractive forces in all directions, whereas a molecule at or near the surface undergoes only forces directed towards the interior of the liquid, acting normally to the surface. Work must therefore be spent to bring molecules from the bulk to the surface, i.e. surface molecules have higher potential energy than bulk molecules. Now, to bring molecules at the surface is equivalent to increasing its area, and we can say that we must undertake work to increase the area of the surface. As every system has a tendency to reduce its potential energy, the surface of a liquid tends to minimize, and thus to contract. It is as if the surface of a liquid was stretched, like a rubber membrane. However, this argument does not explain how molecular forces acting normally to the surface lead to tangential forces stretching the surface.

There are shortcomings in the above argument. Molecules near the surface (of course we are speaking of time mean values, due to molecular motion) cannot be subjected to a force directed to the interior of the liquid without moving towards it. Surface layers will have a tendency to decrease their density, until the density gradient is sufficient to cancel this tendency. At equilibrium, molecules near the surface are subjected to a system of forces equivalent to zero, the pressure gradient equilibrating exactly the dissymmetry of attractive forces. The problem is similar to the Boltzmann barometric distribution: molecules are distributed so that the density is lower in the regions of high potential energy, but in this case the gradient extends to a few molecular layers (two or three) only. Thus there exists near to the surface (in the condensed phase as well in the vapour phase) a transition zone in which the density varies. The credit for having shown that a density gradient normal to a surface leads to surface tension is generally attributed to Bakker (1911).

1.5.1 Mechanical Definition of the Surface Tension of a Liquid

This definition, which makes use of the internal pressure, requires us to return briefly to the notion of pressure inside a fluid.

The pressure in any fluid phase is an intensive parameter defined by the work which the fluid can provide to the external world by a volume variation: $\mathrm{d}W = -p\,\mathrm{d}V$. In thermodynamics it is defined as the the derivative of the free energy of the system:

$$p = -\left(\frac{\partial F}{\partial V}\right)_{\mathrm{T}} \tag{1.86}$$

This free energy involves a term of kinetic energy of the molecules, but also a term of interaction which corresponds to a binding or cohesive energy of the system. Thus, the pressure inside a fluid containing n molecules per unit volume can be written as

$$p = nkT - p_{\mathrm{i}} \,, \tag{1.87}$$

where p_{i} is the "internal" or "interparticular" pressure. It is this corrective term $p_{\mathrm{i}} = a/V^2$ which appears in the van der Waals equation, and statistical mechanics shows that it can be written as

$$p_{\mathrm{i}} = \frac{2\pi}{3} n^2 \int_0^\infty \frac{\mathrm{d}\varphi(r)}{\mathrm{d}r} g(r) r^3 \,\mathrm{d}r \,, \tag{1.88}$$

where $\varphi(r)$ is the interaction potential for two molecules, and the *radial distribution function*, $g(r)$, is the density probability of finding two molecules at distance r from one another. It describes the local order around a molecule.

Another way to conceive the internal pressure is as follows. Consider a perfect gas on which a piston exerts the pressure $p = n_0 kT$, and let the molecular attraction forces act. The gas will contract and receive work (and it will also furnish heat since entropy decreases by organization of the fluid). The density of the gas will increase from n_0 to n, but the pressure exerted by the piston will remain equal to p; so we must now write $p = nkT - p_{\mathrm{i}}$. The pressure p_{i} is the excess pressure which should be exerted on the piston to maintain the gas perfect at this density, but it is not measurable by external means since molecular attractive forces decrease the pressure exerted on any wall.

In a gas at normal conditions of temperature and pressure, the density is $n = 2.7 \times 10^{19}$ molecules per cm^3; in a liquid the density is more than 1000 times higher, and almost independent of the applied pressure. The internal pressure is thus enormous (more than 1000 atmospheres, at atmospheric pressure) and varies steeply with the mean interparticular distance. It is what explains the cohesion of liquids and the fact that they can sustain tension.

Near a surface, molecular interactions decrease and become anisotropic, being weaker in a direction perpendicular to the surface than parallel to it (Fig. 1.26). The density thus varies and the pressure becomes an axially symmetric tensor

$$\mathbf{p} = \begin{vmatrix} p_{\mathrm{N}} & 0 & 0 \\ 0 & p_{\mathrm{T}} & 0 \\ 0 & 0 & p_{\mathrm{T}} \end{vmatrix} \,,$$

with

$$p_{\mathrm{N}}(z) = n(z)kT - p_{\mathrm{iN}}(z) \,, \tag{1.89}$$

$$p_{\mathrm{T}}(z) = n(z)kT - p_{\mathrm{iT}}(z) \,. \tag{1.90}$$

Since $p_{\mathrm{iN}}(z) < p_{\mathrm{iT}}(z)$, it follows that $p_{\mathrm{T}}(z) < p_{\mathrm{N}}(z)$. We shall see that this transverse component is strongly negative and is the origin of surface tension.

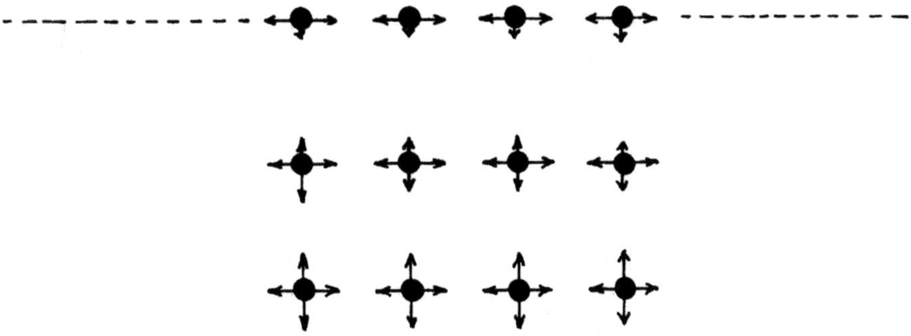

Fig. 1.26. Modification of molecular interactions near the surface of a liquid

Let two homogeneous fluids, α (liquid) and β (vapour), be separated by a thin layer of thickness h, anisotropic and inhomogeneous, in which the pressure is defined by the above tensor, and let us consider (Fig. 1.27) a slab of unit width in the y direction. The total force exerted in the x direction by this slab is

$$F_{\mathrm{T}} = \int_{-h/2}^{+h/2} p_{\mathrm{T}}(z)\,\mathrm{d}z\;. \tag{1.91}$$

However, if the fluid had been isotropic, the three principal pressures would have been equal to $p_{\mathrm{N}}(z)$ and the force would be:

$$\int_{-h/2}^{+h/2} p_{\mathrm{N}}(z)\,\mathrm{d}z\;.$$

The anisotropy thus introduces a deficiency in the horizontal force of

$$\int_{-h/2}^{+h/2} (p_{\mathrm{N}} - p_{\mathrm{T}})\,\mathrm{d}z\;.$$

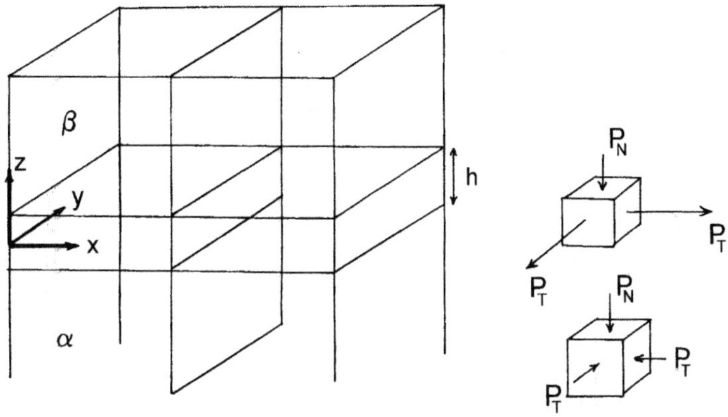

Fig. 1.27. Two homogeneous fluids α and β separated by a thin slab of height h

This deficiency is localized in the layer, since in the adjacent phases we have $p_N(z) = p_T(z)$, and we may define the surface tension by Bakker's formula:

$$\boxed{\sigma = \int_{-\infty}^{+\infty} (p_N - p_T)\,dz} . \tag{1.92}$$

For a flat transition layer, we have $dp_N/dz = 0$ (since the layer is very thin, hydrostatic effects can be neglected) and the continuity condition of the stresses can be written as

$$p_N^\alpha = p_N(z) = p_N^\beta = p .$$

(We can see, in this case, that the steep variation of the density $n(z)$ is due to the variation of internal pressure $p_{iN}(z)$, i.e. to the variation of molecular forces.) Bakker's formula becomes

$$\sigma = ph - F_T . \tag{1.93}$$

Although we can conceive of the existence of surface tension without any tension in the transition layer (it suffices to have $F_T < ph$), it happens that the quantity ph is generally negligible, so that the surface tension σ represents roughly the tension $-F_T$ exerted by the transition layer.

Let us evaluate the mean pressure $\overline{p}_T = F_T/h$ in this layer. Take $p = 1\,\text{atm}, \sigma = 30\,\text{mN/m}$ and $h = 15\times10^{-10}\,\text{m}$ (i.e. about 3 molecular diameters). In accordance with (1.93) we obtain $\overline{p}_T = p - (\sigma/h) = -199\,\text{atm}$. Thus, the component p_T of the pressure, equal to $p_N = 1\,\text{atm}$ in the two homogeneous phases, passes through a sharp minimum in the transition layer (Fig. 1.28).

Furthermore, it is σ and not F_T which is measured experimentally. Let us consider an elementary transformation of the volume and the surface area.

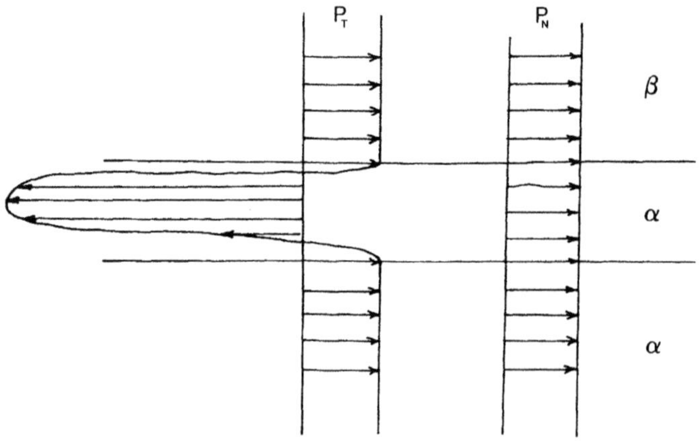

Fig. 1.28. The minimization of p_T in the transition layer

The work received arises from the pressure forces $-p\left(\mathrm{d}V^{\alpha}+\mathrm{d}V^{\beta}\right)$ in the homogeneous phases, to which must be added the work $\mathrm{d}W^{\mathrm{s}}$ received from the pressure forces $p_{\mathrm{T}}(z)$ exerted in the slab of thickness h (assumed constant) and whose resultant is F_{T} per unit width. The thickness h being constant, the work done by $p_{\mathrm{N}}(z)$ is zero and we have

$$\begin{aligned}\mathrm{d}W^{\mathrm{s}} &= -F_{\mathrm{T}}\,\mathrm{d}A\\ &= -ph\,\mathrm{d}A+\sigma\,\mathrm{d}A\\ &= -p\,\mathrm{d}V^{\mathrm{s}}+\sigma\,\mathrm{d}A\,.\end{aligned}$$

The total work received by the system can be written as

$$\mathrm{d}W=-p\left(\mathrm{d}V^{\alpha}+\mathrm{d}V^{\beta}+\mathrm{d}V^{\mathrm{s}}\right)+\sigma\,\mathrm{d}A\,,$$

i.e.

$$\boxed{\mathrm{d}W=-p\,\mathrm{d}V+\sigma\,\mathrm{d}A}\,. \tag{1.94}$$

Thus, the surface tension appears as an excess parameter, which must be introduced when we do not want to study the transition layer as a distinct volumetric phase. As suggested by Young in 1805, it is as if phases α and β were homogeneous up to a surface Σ, termed a *surface of tension*, similar to a membrane subjected to a tension σ. However, contrary to the case of a rubber membrane, this tension σ remains constant whatever the variation of the surface area. That arises from the fact that, if the variation of area is not too fast, molecules move from the bulk towards the surface (in less than a millisecond) to maintain the situation in the transition layer in which σ is constant. There is an analogy of plasticity in this aspect of the behaviour of this pseudo-membrane. The mechanical work $\sigma\,\mathrm{d}A$ done to increase by $\mathrm{d}A$ an area A, is in fact used to create a new surface $\mathrm{d}A$ of surface energy γ and thus to provide the energy $\gamma\,\mathrm{d}A$. For a liquid we thus have $\sigma=\gamma$.

In the case of a curved surface, the mechanical equilibrium requires that p^{α} is different from p^{β}, so that the normal pressure $p_{\mathrm{N}}(z)$ varies in the transition layer. The surface tension, given by Bakker's equation (1.92), in fact varies in a negligible manner with the radius of curvature R, as long as R is large compared with the thickness h of the transition layer. In this case, it is easy to compute the pressure difference $(p^{\alpha}-p^{\beta})$ between the two phases. Let us consider a sphere of radius R, cut it in two halves, and study the equilibrium of forces in one half: the surface tension σ acting on the diameter is counterbalanced by the overpressure $p^{\alpha}-p^{\beta}$ acting on the area of the diameter

$$\sigma 2\pi R=\left(p^{\alpha}-p^{\beta}\right)\pi R^{2}\;;$$

hence the overpressure in the sphere

$$\boxed{p^{\alpha}-p^{\beta}=\frac{2\sigma}{R}}\,. \tag{1.95}$$

This is the *Laplace equation* (1806) (called the Kelvin equation (1858) by the English!). It is this excess pressure which explains why small bubbles empty into large ones, unlike India rubber balls. Likewise droplets of mist are compressed, and their saturated vapour pressure is higher for small ones than for large ones. They thus distil onto a plane surface of water: which is why mist appears to float some ten centimetres above ponds.

The location of the surface of tension force is perfectly well defined. The two systems, transition layer of thickness h and surface of tension at height z_σ above the origin of the layer (taken, for example, at the origin of phase α), are equivalent if they have the same resultant force

$$F_{\mathrm{T}} = \int_0^h p_{\mathrm{T}}(z)\,\mathrm{d}z = p^\alpha z_\sigma + p^\beta\,(h - z_\sigma) - \sigma$$

and the same momentum (as measured from the origin O, for example):

$$\int_0^h z p_{\mathrm{T}}(z)\,\mathrm{d}z = -\sigma z_\sigma + p^\alpha z_\sigma \left(\frac{z_\sigma}{2}\right) + p^\beta\,(h - z_\sigma)\left(\frac{h + z_\sigma}{2}\right) .$$

We thus have two equations,

$$\sigma = \int_0^{z_\sigma} \left[p^\alpha - p_{\mathrm{T}}(z)\right]\mathrm{d}z + \int_{z_\sigma}^h \left[p^\beta - p_{\mathrm{T}}(z)\right]\mathrm{d}z$$

and

$$\sigma z_\sigma = \int_0^{z_\sigma} \left[p^\alpha - p_{\mathrm{T}}(z)\right] z\,\mathrm{d}z + \int_{z_\sigma}^h \left[p^\beta - p_{\mathrm{T}}(z)\right] z\,\mathrm{d}z ,$$

which determine completely the value of σ and the position of the surface of tension once the microscopic structure of the transition layer is known.

1.5.2 Surface Tension of Thin Films

Let us consider, as per Fig. 1.29, a thin film of thickness h (a soap film, for example) in equilibrium with its vapour at pressure p^β. The pressure in the film is p^α, and the difference $\Pi = p^\beta - p^\alpha$ is the disjoining pressure which acts against the thinning of the film (and which can be considered as due to the interaction between the two faces of the film). This film exerts a tension σ_{f} on the frame on which it is placed, but this tension is not equal to twice the surface tension σ^α of the liquid, for two reasons.

The first reason is that the surface tension $\sigma(h)$ of a face of a thin film is not equal to the surface tension σ^α of the liquid, since this face is in the force field of the second face. The difference between the two corresponds to the work which must be done against the disjoining pressure to thin the thick film down to the thickness h:

$$2\sigma(h) - 2\sigma^\alpha = -\int_\infty^h \Pi\,\mathrm{d}h .$$

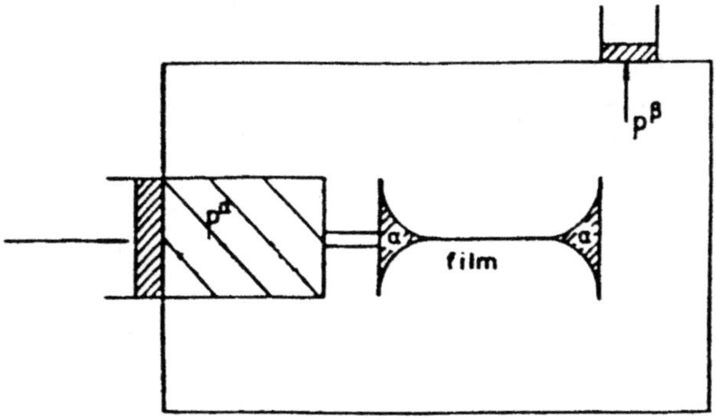

Fig. 1.29. The disjoining pressure $\Pi(h)$ being positive, a pressure $p^\beta = p^\alpha + \Pi(h)$ must be applied to maintain a film of thickness h

Secondly, the disjoining pressure must be taken into account in the mechanical equilibrium of the film (Fig. 1.30):

$$\begin{aligned}\sigma_{\mathrm{f}} &= 2\sigma(h) + \Pi h \\ &= 2\sigma^\alpha - \int_\infty^h \Pi \,\mathrm{d}h + \Pi h \,.\end{aligned}$$

Hence

$$\sigma_{\mathrm{f}} = 2\sigma^\alpha + \int_0^\Pi h \,\mathrm{d}\Pi \,. \tag{1.96}$$

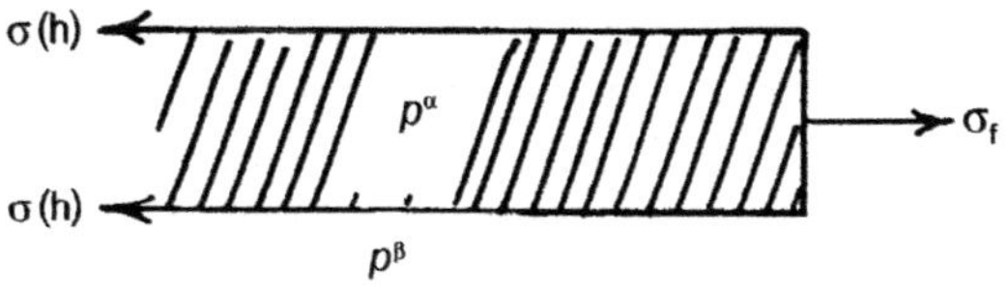

Fig. 1.30. Mechanical equilibrium of a thin film

1.5.3 Surface Tension of Solids

As seen in Sect. 1.4.1, and as for liquids, the dissymmetry of interactions at the surface of a solid causes a modification of the lattice parameters of the first atomic planes of the order of 1 – 2%.[5] This variation of interatomic distances can occur perpendicularly to the surface (normal relaxation) or parallel to it (tangential relaxation). A form of residual stress is set up at the surface, and leads to surface tension.

[5] These values are far from negligible. Let us recall that the elastic limit of a number of metals corresponds to a variation of interatomic distances of about 0.1%.

The phenomenon can be pictured as follows. Let us assume, as a hypothesis, that lattice parameters increase isotropically near the surface, in the same manner as for an increase of temperature. We can apply in this zone the equation of thermoelasticity:

$$\varepsilon_x - \alpha T = \frac{1}{E}\left[\sigma_x - \nu\left(\sigma_y + \sigma_z\right)\right] ,$$
$$\varepsilon_y - \alpha T = \frac{1}{E}\left[\sigma_y - \nu\left(\sigma_z + \sigma_x\right)\right] ,$$
$$\varepsilon_z - \alpha T = \frac{1}{E}\left[\sigma_z - \nu\left(\sigma_x + \sigma_y\right)\right] ,$$

where α is the coefficient of thermal expansion. If z is the axis perpendicular to the surface, we must have $\sigma_z = 0$ and $\varepsilon_x = \varepsilon_y = 0$, which gives

$$\varepsilon_z = \alpha T \frac{1+\nu}{1-\nu} ,$$
$$\sigma_x = \sigma_y = -\frac{E}{1+\nu}\varepsilon_z .$$

Let us take $E = 1.3 \times 10^{11}$ Pa, $\nu = 0.3$ and assume that the normal relaxation $\varepsilon = 2.5\%$. We find $\sigma_x = \sigma_y = 2.5$ GPa. Assume that these stresses act on a thickness $h = 0.5$ nm, the integrated stress will be

$$\sigma = \sigma_x h = -1.25\,\mathrm{N/m} .$$

This method of computation is that used to compute the surface compression due to ionic implantation [61], where α is the dilation due to implanted ions. We can see that this computation, based on the hypothesis of a positive and isotropic relaxation, leads to surface compression and not to surface tension.

In fact, calculations based on the hypothesis of central forces between first and second neighbours[6] always lead to a positive normal relaxation (expansion), which corresponds to the relaxation of the compressive force of the first neighbours but also to the possibility of a tangential relaxation of variable sign according to crystal faces [62]. This tangential relaxation can only be observed on the edge of the crystals or on monoatomic ledges. At a distance from any edge, the tangential relaxation is impossible ($\varepsilon_x = \varepsilon_y = 0$) and surface stresses appear which can be compressions or tensions.

Instead of a positive relaxation, experimental results for clean metallic surfaces (Fig. 1.31, after van Hove [63]) show a tendency to a reduction of in-

[6] Let us recall that the wrong hypothesis of central forces between first neighbours led Navier (1821), Cauchy (1827) and Poisson (1828) to the theory of rarefied constants of elasticity involving only 15 constants of elasticity (instead of 21) in the general case, and only one instead of two (E and ν) in the case of isotropic solids (where ν took the value 1/4). The controversy lasted more than 50 years. Note that the introduction of an attractive force of the second neighbours causes the first neighbours to compress the central atom.

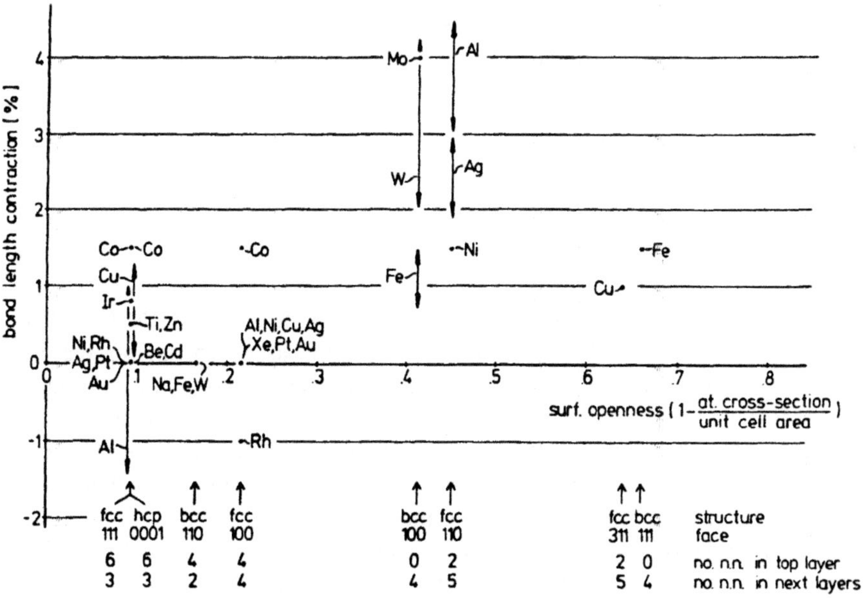

Fig. 1.31. Plot of observed bond-length contractions between surface atoms and underlying atoms versus surface openness (lack of close packing) for metals. The number of nearest neighbours in the surface plane and in the next layers of a surface atom is indicated for each type of surface

teratomic distances perpendicularly to the surface, especially for faces which are not very dense. Moreover, a negative relaxation is generally transformed into positive relaxation after adsorption of foreign atoms. The explanation generally given is that on the surface of metals there is a redistribution of ions and electrons, with the formation of a dipole layer. On undense faces, this electronic cloud has the effect of contracting the first atomic plane, and this effect cannot be analysed in term of central forces.

1.5.4 Mechanical Effects Due to the Surface Tension of Solids

Generally speaking, any residual stress gives rise to elastic deformation, and the measurement of this deformation is a method of determination of the residual stresses. In the X-ray method, the variation of the lattice cell is measured in subsuperficial layers; in Sachs' method the dimensional variations during the removal of surface layers are measured, i.e. during the removal of surface stresses. Mechanical effects of the same type can be attributed to the influence of surface tension.

Dimensional Effects. Let us consider a ribbon of thickness h, width b and length L. Assume that the surface tension increases by $\Delta\sigma$. A force

$F = 2(b+h)\Delta\sigma$ will then apply along the length. With E = Young's modulus this length will vary by

$$\frac{\Delta L}{L} = -\frac{2(b+h)\Delta\sigma}{Ebh} .$$

If the ribbon is thin enough, we can write

$$\frac{\Delta L}{L} = -\frac{2\Delta\sigma}{Eh} . \tag{1.97}$$

As a numerical example take $\Delta\sigma = -100\,\mathrm{mN/m}, E = 10\,\mathrm{GPa}, h = 10\,\mu\mathrm{m}$, which gives $\Delta L/L = 2 \times 10^{-6}$. This method of the extensometer has been used by Beck *et al.* [64, 65] to study the electrocapillary maximum of gold. Beck fastened a gold ribbon ($E = 8\,\mathrm{GPa}$) of thickness of around $15\,\mu\mathrm{m}$ to a spring of weak stiffness and measured the variation in length of the ribbon immersed in an electrolyte as a function of the applied potential. The spring, being of negligible stiffness, plays no part in the calculation.

This analysis can be applied to dimensional effects observed during adsorption on very thin sheets (solids with large specific area). Consider the unit mass of a porous medium made of a set of N very thin sheets, of thickness h, width b, length L. Its total area (its specific area) is $\Sigma = 2NbL$, and its volume $1/\varrho = NbLh$ where ϱ is the volumetric mass of the porous medium. One can deduce the mean thickness of the sheets: $h = 2/\varrho\Sigma$. If all the sheets are in parallel planes, (1.97) gives:

$$\frac{\Delta L}{L} = -\frac{\varrho\Sigma}{E}\Delta\sigma ,$$

an equation presented by Bangham and Maggs [66] in 1943.

Let us take the example of an exfoliated graphite (Grafoil or Papyex). It is made of graphite lamellae ($E \cong 10\,\mathrm{GPa}$) whose side is around $3\,\mu\mathrm{m}$, with their basal planes parallel to each other; the volumetric mass is $\varrho \cong 1\,\mathrm{g/cm^3}$, and the specific area is $\Sigma \cong 20\,\mathrm{m^2/g}$. The equation gives $\Delta L/L = 2 \times 10^{-3}\Delta\sigma$ (with $\Delta\sigma$ in N/m), which is the correct order of magnitude of the 1976 experimental results [67].

Let us assume, on the contrary, that these sheets have no preferential orientation. The expansion (or contraction) of the solid is thus isotropic. Every platelet exerts a force $2b\Delta\sigma$ of random orientation. If $n \equiv N\varrho$ is the number of platelets by unit volume, a calculation modelled upon the kinetic theory of gases shows that the pressure exerted is

$$\Delta p = \frac{2}{3}nL2b\Delta\sigma = \frac{2}{3}\varrho\Sigma\Delta\sigma .$$

Introducing the compressibility $\chi = -1/V(\mathrm{d}V/\mathrm{d}p)$ gives

$$\frac{\Delta L}{L} = \frac{1}{3}\frac{\Delta V}{V} = -\frac{1}{3}\chi\Delta p ,$$

i.e.

$$\frac{\Delta L}{L} = -\frac{2}{9}\chi\varrho\Sigma\Delta\sigma\ .$$

This is the *Yates equation* [68, 69] which applies to microporous media of isotropic structure.

Thus, in general, we can write

$$\boxed{\frac{\Delta L}{L} = -\lambda\Delta\sigma}\ ,$$

which is the expression of the *Bangham's law* [70] (see also the review by Sereda and Feldman [71]). In fact, Bangham employed $\Delta\gamma$ instead of $\Delta\sigma$. As the surface energy can only decrease by adsorption, Bangham's law was unable to explain the contractions (termed "anomalous contractions") often observed at the first stages of adsorption (see Fig. 1.32 after Yates [72]). We will see below that σ and γ can vary in a different way by adsorption.

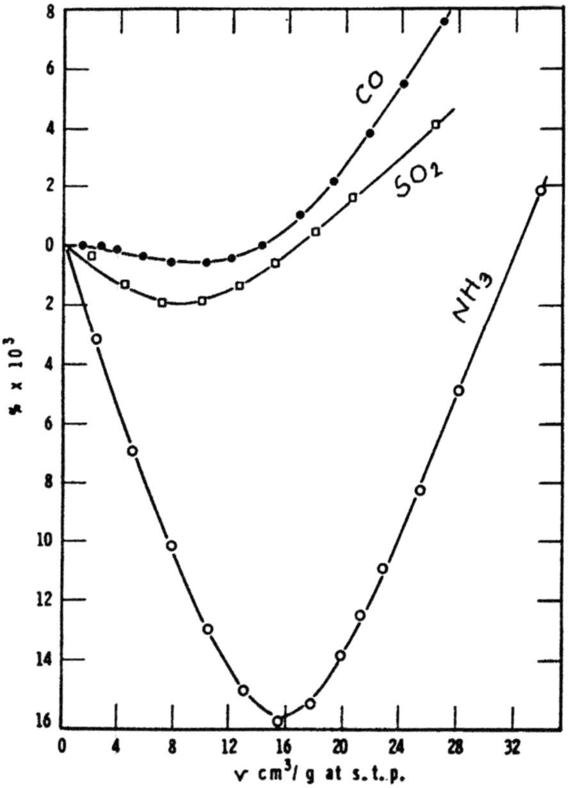

Fig. 1.32. Length change of a porous glass, on the adsorption of gases

Curvature of Thin Sheets. Another mechanical effect is the bending which appears when the surface tension of one side only of a thin sheet is modified. This effect has been used since Stoney [73] in 1909 to compute the stress in a film deposited on one side of a thin sheet [74]. It is also used in ionic implantation [61, 75] and in electrochemistry [76, 77]. In the latter case, a metallic film of about 100 nm is deposited under vacuum on a thin sheet of glass (about a hundred micrometres thick). Immersed in a liquid, this electrode displays a deflection which varies with the degree of adsorption on the metallic film.

Let h, b, L be the thickness, width and length, respectively, of this sheet. The force exerted on one side is $b\Delta\sigma$ (and in the case of a deposited film $\Delta\sigma$ is the product of the stress in the film by its thickness). The bending moment is constant and has the value

$$\mathcal{M}_{\mathrm{f}} = \frac{1}{2} b h \Delta\sigma \,.$$

The curvature is then given by

$$\frac{1}{R} = \frac{\mathrm{d}\theta}{\mathrm{d}x} = \frac{\mathcal{M}_{\mathrm{f}}}{EI} = \frac{6\Delta\sigma}{Eh^2} \,,$$

which is the Stoney equation [73]. This formula has been modified to take into account the fact that $\Delta\sigma$ is isotropic and has a transverse component

$$\frac{1}{R} = \frac{1-\nu}{E}\frac{6\Delta\sigma}{h^2} \,,$$

where ν is Poisson's ratio. The rotation of the end and its deflection are respectively

$$\alpha = \frac{1-\nu}{E}\frac{6L}{h^2}\Delta\sigma \,,$$

$$\delta = \frac{1-\nu}{E}\frac{3L^2}{h^2}\Delta\sigma \,.$$

As a numerical example, take a glass sheet ($E = 74.5\,\mathrm{GPa}, \nu = 0.25$), $L = 12.5\,\mathrm{cm}, b = 1.5\,\mathrm{cm}, h = 85\,\mu\mathrm{m}$ and $\Delta\sigma = 100\,\mathrm{mN/m}$. These values give $\delta = 6\,\mu\mathrm{m}$.

Lattice Contraction. Another effect, forecast by Lennard, Jones and Dent in 1928, is the variation of pressure, and hence the variation of lattice parameter, under the action of surface tension in small crystals of about ten nanometres. This effect is easy to compute from Laplace's law, where $\Delta p = 2\sigma/R$. Introducing the compressibility χ gives

$$\sigma = \frac{R}{2}\Delta p = -\frac{R}{2\chi}\frac{\Delta V}{V} = -\frac{3R}{2\chi}\frac{\Delta a}{a} \,.$$

Hence

$$\frac{\Delta a}{a} = -\frac{2}{3}\frac{\chi\sigma}{R},$$

where Δa is the variation of the lattice cell due to surface tension. This is the basis of the method of determination of residual stresses by X-rays.

The experimental study of cell contraction by electronic diffraction has been performed by a number of authors since Boswell [78] in 1951. The experimental difficulties are such that it took until 1968 to obtain reliable experimental proof of this variation of lattice size. Boswell [78] observed a lattice contraction for LiF, NaCl, NaBr, and KCl particles, but his results have been criticized by Berry (1952), Libowitz *et al.* (1955) and Rymer (1957). Nicholson (1955) has found a contraction for MgO and NaCl, whereas no effect was observed by Rymer *et al.* (1952) for NaCl, KCl and CsI, by Brunauer *et al.* (1956) for CaO, by Harrison *et al.* (1957) for NaCl. Halliday *et al.* (1954) observed a dilation with LiF.

In 1964, de Planta *et al.* [79] obtained an important relative variation $\Delta a/a$ for Ag (a value near to that of Berry in 1952: 2.7% for particles of 31 Å). This variation, linear in $1/R$, leads to $\sigma \cong 6000\,\mathrm{mN/m}$. Wasserman and Vermaak [80] showed that this value, which is too high, arose from the choice of the diffraction peak (111) whose position is mistaken by the existence of the continuous base line. More precise results (Table 1.2) were obtained by Vermaak *et al.* for gold [81, 82], for silver [80], and for copper and platinum [83]. For gold, surface tension decreases linearly with increasing temperatures, from 1175 mN/m at 50 °C to 410 mN/m at 985 °C [82].

Table 1.2. Relative variation results for various metals

	T (°C)	σ (mN/m)	γ (mJ/m^2)
Au	50	1175 ± 93	1370
Ag	55	1415 ± 300	1520
Cu	48	0 ± 450	1720
Pt	65	2574 ± 400	2100

Other more physical methods, such as ferromagnetic resonance, allow for the determination of internal stresses due to surface tension in thin ferromagnetic films (2 – 100 nm). Here also residual stresses vary considerably with adsorption: adsorption of a monolayer of oxygen almost completely relaxes the stresses in a film of nickel [84].

1.6 The Relation Between γ and σ

The distinction for solids between the work $\gamma\,\mathrm{d}A$ to create an elementary area of new surface, and the work $\sigma\,\mathrm{d}A$ to stretch a surface elastically, was clearly established by Gibbs in 1876. In the first case the surface retains its equilibrium structure; in the second case the area per atom increases. The relation between the two quantities σ and γ was established by Shuttleworth [58] in 1950.

1.6.1 Shuttleworth's Equation

Let us consider (Fig. 1.33) a crystal which has been cleaved along a plane of area A and surface energy γ, and which is then stretched elastically in its two halves by a quantity $\mathrm{d}A_{\mathrm{el}}$. The (excess) work provided is $2\gamma A + 2\sigma\,\mathrm{d}A_{\mathrm{el}}$. However, the crystal can first be stretched and then cleaved. If $\gamma + \mathrm{d}\gamma$ is the surface energy of the face so stretched, the work provided is $2(\gamma + \mathrm{d}\gamma)(A + \mathrm{d}A)$. (Only the excess work relating to the cleaved face is taken into account.) We thus have

$$2\gamma A + 2\sigma\,\mathrm{d}A_{\mathrm{el}} = 2(\gamma + \mathrm{d}\gamma)(A + \mathrm{d}A_{\mathrm{el}})\ ,$$

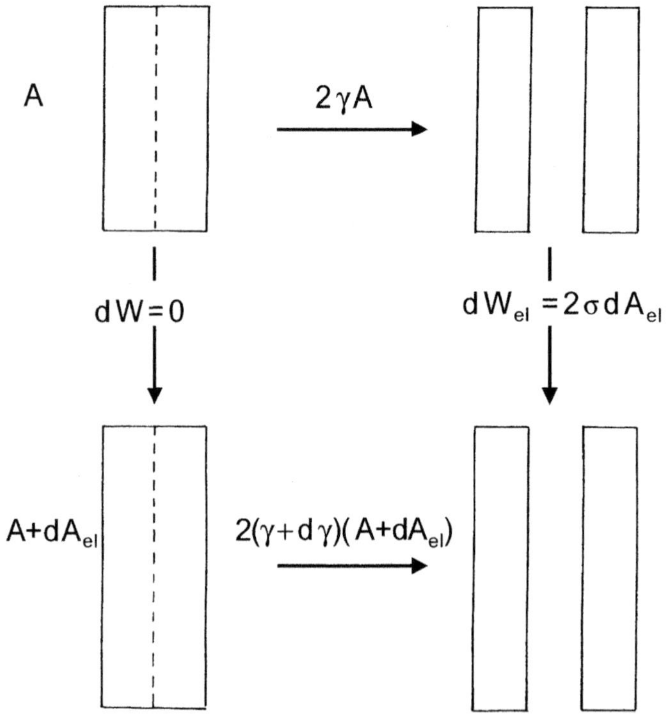

Fig. 1.33. Stretched and cleaved crystal

i.e.

$$\sigma = \gamma + \frac{\mathrm{d}\gamma}{(\mathrm{d}A_{\mathrm{el}}/A)} ,$$

or, taking a to be the area per atom:

$$\boxed{\sigma = \gamma + a\frac{\mathrm{d}\gamma}{\mathrm{d}a}} , \tag{1.98}$$

which is the Shuttleworth's formula.

1.6.2 Surface Stress g in Any Reversible Stretching

Let us now consider (Fig. 1.34), an area A of a surface containing N atoms. If a is the area per atom, we have

$$A = Na ,$$
$$\mathrm{d}A = a\,\mathrm{d}N + N\,\mathrm{d}a .$$

In the above formulae, the term $a\,\mathrm{d}N$ describes an increase of area during which the area per atom remains constant, the number of atoms increasing to create a surface of rigorously the same aspect as the initial surface A. During this increase in area, the surface energy remains constant. It is the case of liquid surfaces for not too fast deformations: molecules are so mobile that in less than one millisecond they move from the bulk to the surface. It is also the case of solids at high temperature for slow deformations: atomic mobility is high enough for the area per atom to be the initial area at any time. These deformations are said to be *purely plastic.*

The term $N\,\mathrm{d}a$ describes an increase of area during which surface atoms have the same number but move apart from one another. The surface so stretched has no longer the same microscopic aspect as the initial one, and the surface energy is modified ($\mathrm{d}\gamma/\mathrm{d}a \neq 0$). We will say that this deformation is *purely elastic.*

In the general case, where the deformation is partly elastic and partly plastic, we can write

$$\frac{\mathrm{d}A}{A} = \varepsilon = \varepsilon_{\mathrm{pl}} + \varepsilon_{\mathrm{el}} ,$$

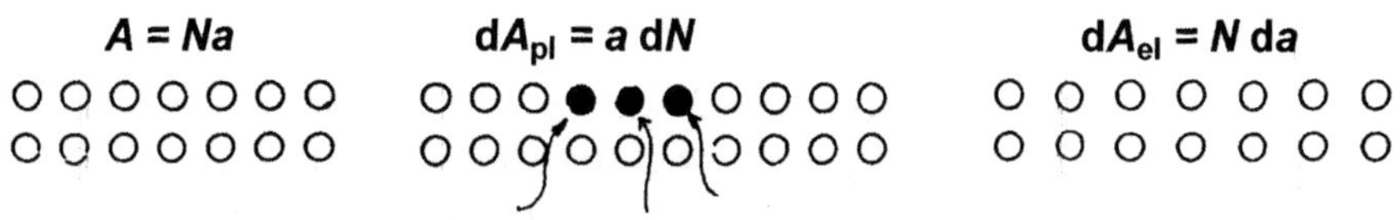

Fig. 1.34. Surface deformation

with

$$\varepsilon_{\mathrm{pl}} = \frac{a\,\mathrm{d}N}{Na}\,,$$
$$\varepsilon_{\mathrm{el}} = \frac{N\,\mathrm{d}a}{Na}\,.$$

The various contributions to the work will be defined by

$$\begin{aligned}\mathrm{d}W_{\mathrm{pl}} &= \gamma\,\mathrm{d}A_{\mathrm{pl}}\\ \mathrm{d}W_{\mathrm{el}} &= \sigma\,\mathrm{d}A_{\mathrm{el}}\\ \mathrm{d}W &= g\,\mathrm{d}A\,.\end{aligned}$$

The element of work written as $2\sigma\,\mathrm{d}A_{\mathrm{el}}$ in Fig. 1.33 becomes, in the general case $2g\,\mathrm{d}A$. We then have

$$\gamma A + g\,\mathrm{d}A = (\gamma + \mathrm{d}\gamma)(A + \mathrm{d}A)\,,$$

i.e.

$$g\,\mathrm{d}A = \gamma\,\mathrm{d}A + A\,\mathrm{d}\gamma$$
$$\begin{aligned}g\,\mathrm{d}A &= \gamma a\,\mathrm{d}N + \gamma N\,\mathrm{d}a + Na\frac{\mathrm{d}\gamma}{\mathrm{d}a}\,\mathrm{d}a\\ &= \gamma a\,\mathrm{d}N + \left(\gamma + a\frac{\mathrm{d}\gamma}{\mathrm{d}a}\right)N\,\mathrm{d}a\,.\end{aligned}$$

We can recognize that the term in parentheses is the surface tension σ, and hence

$$g\,\mathrm{d}A = \gamma a\,\mathrm{d}N + \sigma N\,\mathrm{d}a\,,$$

i.e. after division of the two members by A we have

$$\boxed{g = \frac{\varepsilon_{\mathrm{pl}}}{\varepsilon}\gamma + \frac{\varepsilon_{\mathrm{el}}}{\varepsilon}\sigma}\,. \tag{1.99}$$

This formula was given in 1972 by Couchman *et al.* [85]. The measured surface stress is, according to the case, either γ or σ:

$$g_{\mathrm{pl}} = \gamma\,,$$
$$g_{\mathrm{el}} = \sigma\,.$$

We can show graphically in a qualitative way (Fig. 1.35) the variation of the measured stress g with the velocity of the deformation [86]. For very slow velocities $g = g_{\mathrm{pl}} = \gamma$, whereas for very fast velocities no rearrangement from the bulk to the surface occurs (even for liquids) and $g = g_{\mathrm{el}} = \sigma$.

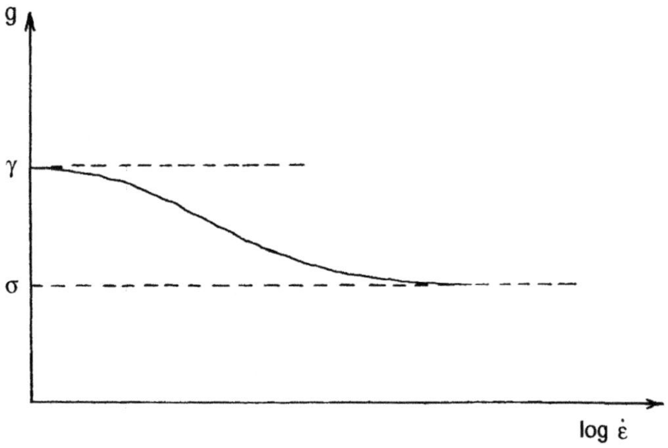

Fig. 1.35. Variation of surface stress with the velocity of deformation

1.7 Surface Thermodynamics

1.7.1 Dividing Surface and Adsorption

Let us consider the system made of a condensed phase α (solid or liquid) and a vapour phase β, in contact with one another (Fig. 1.36). The concentrations of the components vary from c^α to c^β by crossing the transition zone, as shown in the figure.

Let us also consider a *dividing surface* Σ placed somewhere in this transition zone. All extensive thermodynamic quantities of the system, such as the free energy F, can be written as

$$F = F^\alpha + F^\beta + F^\sigma \ , \tag{1.100}$$

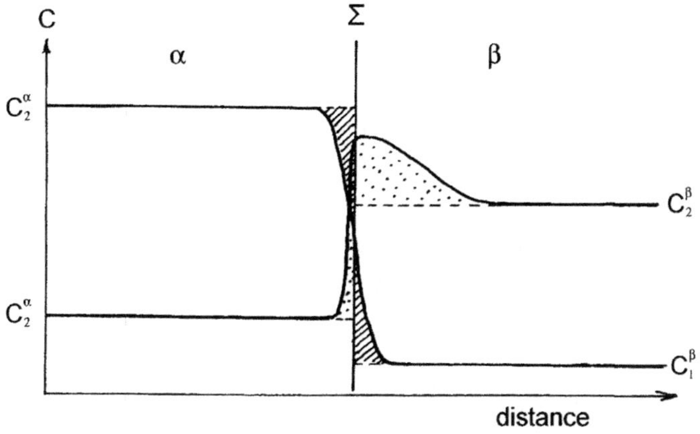

Fig. 1.36. Solid/liquid and vapour in contact, with a transition zone

where F^α and F^β are the free energies of volumes V^α and V^β of phases α and β assumed to be homogeneous up to the surface of division Σ. F^σ is thus the excess free energy of the real system compared with the ideal system in which the two phases α and β would present a sharp discontinuity at Σ.

Similarly, the total number of molecules i of the system can be written as

$$N_i = N_i^\alpha + N_i^\beta + N_i^\sigma \,, \tag{1.101}$$

where N_i^σ is the number of molecules i which represents the difference (positive or negative) between the number of molecules i in the ideal system and in the real system.

For reversible transformations, we can write, for the phases α and β,

$$\begin{aligned} \mathrm{d}F^\alpha &= \mathrm{d}W^\alpha - S^\alpha\,\mathrm{d}T + \sum_i \mu_i\,\mathrm{d}N_i^\alpha \,, \\ \mathrm{d}F^\beta &= \mathrm{d}W^\beta - S^\beta\,\mathrm{d}T + \sum_i \mu_i\,\mathrm{d}N_i^\beta \,. \end{aligned}$$

By analogy, let us write in a formal way

$$\mathrm{d}F^\sigma = \mathrm{d}W^\sigma - S^\sigma\,\mathrm{d}T + \sum_i \mu_i\,\mathrm{d}N_i^\sigma \,. \tag{1.102}$$

Let us increase by $\mathrm{d}A$ the area of the dividing surface (Fig. 1.37) by a reversible and isothermal ($\mathrm{d}T = 0$) transformation, at constant volume ($\mathrm{d}W^\alpha = \mathrm{d}W^\beta = 0$), the number N_i of molecules i remaining constant.

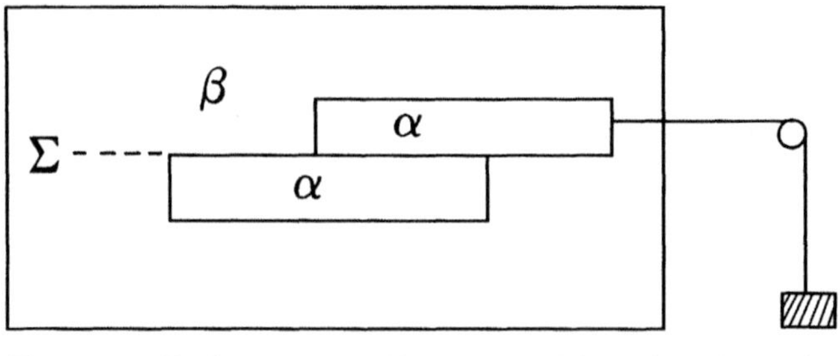

Fig. 1.37. Surface increased by a reversible and isothermal transformation

The work received by the system in this transformation is $g\,\mathrm{d}A$ for any stretching ($\gamma\,\mathrm{d}A$ for a plastic stretching, $\sigma\,\mathrm{d}A$ for an elastic stretching). It is equal to the variation $\mathrm{d}F$ of the free energy of the system, so that

$$\begin{aligned} g\,\mathrm{d}A = \mathrm{d}F &= \mathrm{d}F^\alpha + \mathrm{d}F^\beta + \mathrm{d}F^\sigma \\ &= \mathrm{d}W^\sigma + \sum_i \mu_i \left(\mathrm{d}N_i^\alpha + \mathrm{d}N_i^\beta + \mathrm{d}N_i^\sigma\right) \\ &= \mathrm{d}W^\sigma \,. \end{aligned}$$

The term $\mathrm{d}W^\sigma$, introduced in a formal manner, is thus $g\,\mathrm{d}A$, and we can write for the surface phase

$$\mathrm{d}F^\sigma = g\,\mathrm{d}A - S^\sigma\,\mathrm{d}T + \sum_i \mu_i\,\mathrm{d}N_i^\sigma\ . \tag{1.103}$$

If A is the area of the dividing surface, we can put:

$$\Gamma_i = \frac{N_i^\sigma}{A}\ , \tag{1.104}$$

where Γ_i represents the adsorption of the component i at the dividing surface, expressed in atoms/m^2 or in moles/m^2 and with Γ_i positive or negative. Note immediately that Γ_i can vary not only at A = constant by variation of N_i, but also at N_i = constant by variation of the area A. We can then write

$$\left(\frac{\partial N_i^\sigma}{\partial A}\right)_{\mu_i} = \Gamma_i + A\left(\frac{\partial \Gamma_i}{\partial A}\right)_{\mu_i}\ .$$

In the case of an adsorption on a nondeformable surface, or of a plastic increase of area (or cleavage), or also of an elastic increase of area when adsorption is not localized on well-defined sites of the surface, we have $(\partial\Gamma_i/\partial A)_{\mu_i} = 0$, i.e.

$$\left(\frac{\partial N_i^\sigma}{\partial A}\right)_{\mu_i} = \Gamma_i\ .$$

In the case of an elastic increase of area, at constant number of adsorbed atoms (*localized adsorption*), we have

$$\left(\frac{\partial N_i^\sigma}{\partial A}\right)_{\mu_i} = \Gamma_i + A\left(\frac{\partial \Gamma_i}{\partial A}\right)_{\mu_i} = 0\ .$$

1.7.2 Plastic Stretching: Gibbs Adsorption Formula

For a plastic stretching we have $g = \gamma$, which gives

$$\boxed{\mathrm{d}F^\sigma = \gamma\,\mathrm{d}A - S^\sigma\,\mathrm{d}T + \sum_i \mu_i\,\mathrm{d}N_i^\sigma}\ . \tag{1.105}$$

When the dividing surface increases by $\mathrm{d}A = A\,\mathrm{d}\xi$ at constant temperature, the number of molecules i adsorbed increases by $\mathrm{d}N_i^\sigma = N_i^\sigma\,\mathrm{d}\xi$, the structure of the new surface remaining identical to the old. Since only the quantity of matter is modified, we have $\mathrm{d}F^\sigma = F^\sigma\,\mathrm{d}\xi$ (F^σ is a homogeneous function of degree one of the extensive variables). Hence

$$F^\sigma \, \mathrm{d}\xi = \gamma A \, \mathrm{d}\xi + \sum_i \mu_i N_i^\sigma \, \mathrm{d}\xi \,,$$

i.e.

$$F^\sigma = \gamma A + \sum_i \mu_i N_i^\sigma \,, \tag{1.106}$$

or, dividing by A,

$$\gamma = f^\sigma - \sum_i \mu_i \Gamma_i \,, \tag{1.107}$$

where $f^\sigma = F^\sigma / A$ is the specific surface free energy.[7]

For a system with one component, it is always possible to choose the dividing surface to have $\Gamma = 0$, so that the surface energy γ and the specific surface free energy f^σ are identical. However, for a multicomponent system $\sum_i \mu_i \Gamma_i \neq 0$ and the surface energy cannot be identified with the specific suface free energy. By convention, it is assumed that the dividing surface is so placed that the adsorption of the major component on it is zero.

By differentiating (1.106) we get

$$\mathrm{d}F^\sigma = \gamma \, \mathrm{d}A + A \, \mathrm{d}\gamma + \sum_i \mu_i \, \mathrm{d}N_i^\sigma + \sum_i N_i^\sigma \, \mathrm{d}\mu_i \,,$$

which, compared with

$$\mathrm{d}F^\sigma = \gamma \, \mathrm{d}A - S^\sigma \, \mathrm{d}T + \sum_i \mu_i \, \mathrm{d}N_i^\sigma \,,$$

leads to a Gibbs–Duhem type of equation:

$$\boxed{\mathrm{d}\gamma = -s^\sigma \, \mathrm{d}T - \sum_i \Gamma_i \, \mathrm{d}\mu_i} \,. \tag{1.108}$$

This is the famous *Gibbs adsorption formula*, which links the reduction of surface energy to the adsorption; s^σ is the specific surface entropy.

By making $\mathrm{d}T = 0$ we obtain the *Gibbs adsorption isotherm*

$$\boxed{\Gamma_i = -\left(\frac{\partial \gamma}{\partial \mu_i}\right)_{T, \mu_{j \neq i}}} \,. \tag{1.109}$$

This equation applies to the adsorption of a gas on a solid surface, as well as to the segregation of an impurity in a solid from the bulk towards the surface.

---

[7] The expression $\mathrm{d}F = -P \, \mathrm{d}V^\alpha - P \, \mathrm{d}V^\beta + g \, \mathrm{d}A - S \, \mathrm{d}T + \sum_i \mu_i \, \mathrm{d}N_i$ leads, for an increase $\mathrm{d}A$ of area at constant volumes and temperature, in a closed system ($\mathrm{d}N_i = 0$), to $(\partial F / \partial A)_{V^\alpha, V^\beta, T} = \gamma$. The expressions $\left[(\partial F / \partial A)_{V,T} = \gamma\right]$ and $\left[(\partial F^\sigma / \partial A)_{V,T} = F^\sigma / A = f^\sigma\right]$ must not be confused.

1.7.3 The Case of Reversible Stretching: the Gibbs–Eriksson Equation

Let us now consider a system in equilibrium whose area of the dividing surface is A. This surface can be considered as having be created by cleavage, and its free energy is

$$F^\sigma = \gamma A + \sum_i \mu_i N_i^\sigma .$$

Let us increase this area by $\mathrm{d}A$ with any reversible stretching. We have seen that the variation of the free energy of the surface phase is

$$\mathrm{d}F^\sigma = g\,\mathrm{d}A - S^\sigma\,\mathrm{d}T + \sum_i \mu_i\,\mathrm{d}N_i^\sigma .$$

Differentiating the first equation and comparing the result with the second, we obtain a Gibbs–Duhem type of equation:

$$A\,\mathrm{d}\gamma = -S^\sigma\,\mathrm{d}T - \sum_i N_i^\sigma\,\mathrm{d}\mu_i + (g-\gamma)\,\mathrm{d}A ;$$

or, dividing by A,

$$\boxed{\mathrm{d}\gamma = -s^\sigma\,\mathrm{d}T - \sum_i \Gamma_i\,\mathrm{d}\mu_i + (g-\gamma)\frac{\mathrm{d}A}{A}} , \tag{1.110}$$

which is the *Gibbs–Eriksson equation* [87,88].

This equation reduces to that of Gibbs for $\mathrm{d}A = 0$ (adsorption on non-deformable surfaces) or for $g = \gamma$ (purely plastic stretching). In the case of a purely elastic stretching ($g = \sigma$), isothermal ($\mathrm{d}T = 0$) and at constant chemical potential ($\mathrm{d}\mu_i = 0$), we obtain once more the Shuttleworth equation:

$$\sigma = \gamma + A\left(\frac{\partial \gamma}{\partial A_{\mathrm{el}}}\right)_{T,\mu_i} = \left[\frac{\partial(\gamma A)}{\partial A_{\mathrm{el}}}\right]_{T,\mu_i} .$$

An additional variable $\mathrm{d}\varepsilon \equiv \mathrm{d}A/A$ appears in the Gibbs–Eriksson equation which, contrarily to the Gibbs equation, permits the study of adsorption on surfaces capable of dilation (or contraction) by adsorption or variation of temperature.

1.7.4 Thermodynamics of the Energy of Adhesion

When a solid containing impurities is heated, a segregation of impurities occurs at the surface and at the grain boundaries reducing the surface energy γ_{S}

and the interfacial energy (grain boundary energy γ_G). As for surface energy, this grain boundary energy is measurable: it is deduced from the dihedral angle formed at high temperature (by thermal attack) at the junction of a grain boundary and the surface. This angle is such that the grain boundary tension equilibrates the two surface tensions. (We have seen that at high temperature both surface tension and surface energy are identical.) From the variation of surface energies and grain boundary energies with the concentration of impurities in a solid, we can deduce the segregations Γ_S and Γ_G by the Gibbs adsorption formula (1.108). These segregations can also be measured by examination of surfaces and grain boundaries with Auger electrons after intergranular rupture in a vacuum.

Most of the studies in this field have been carried on metals and on their embrittlement by segregation of impurities. Figure 1.38, for example, after Hondros and McLean [89], compares the variations of rupture stress and grain boundary energy with the bismuth content in copper. These problems of intergranular fracture are very similar to the problems of adherence (we deal here with adherence of grain boundaries), and are thus directly related to the Dupré energy of adhesion

$$w = 2\gamma_S - \gamma_G ,$$

as suggested for the first time by Inman and Tipler [90] in 1963.

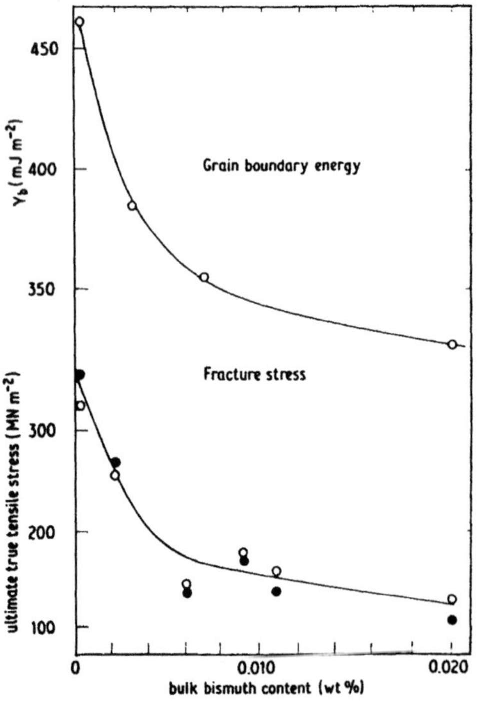

Fig. 1.38. Variation of the grain boundary energy and the ultimate tensile stress of copper as a function of its bismuth content

However, two difficulties appear in the thermodynamics of the energy of adhesion. First, the Gibbs equation implies that both γ_S and γ_G decrease by segregation, but it says nothing about w. If γ_G decreases more rapidly than γ_S, then w can eventually increase by segregation. Second, as pointed out by Inman and Tipler [90], the surface energy of the newly created surface of rupture can be different from the equilibrium surface energy γ_S.

Let us assume that the impurity makes an ideal solid solution. Its chemical potential varies with the concentration in accordance with

$$\mu = \mu_0 + RT \ln c ,$$

and the Gibbs adsorption formula can be written, for a constant temperature, as

$$\mathrm{d}\gamma = -\Gamma RT \frac{\mathrm{d}c}{c} . \tag{1.111}$$

Let us assume, moreover, that the segregation Γ varies with the concentration c according to a Langmuir-type law:

$$\Gamma = \Gamma_{\max} \frac{ac}{1+ac} , \tag{1.112}$$

with $a = \exp(-\Delta G/RT)$ and where ΔG is the free energy of segregation.

Integration of (1.111) yields

$$\Delta\gamma = -RT\Gamma_{\max} \ln(1+ac) , \tag{1.113}$$

which represents the decrease of γ for a concentration c. For a very dilute solution, $(ac \ll 1)$, it reduces to

$$\Delta\gamma = -\Gamma RT .$$

We can thus write the energy of adhesion as

$$\begin{aligned} w &= 2\gamma_S^0 - 2\Gamma_S RT - \gamma_G^0 + \Gamma_G RT \\ &= w_0 - RT(2\Gamma_S - \Gamma_G) . \end{aligned} \tag{1.114}$$

For a very slow crack propagation at high temperature, Γ_S and Γ_G can be the thermodynamic equilibrium values, and w can vary by segregation. But for a fast-crack propagation at low temperature, impurities segregated at the grain boundary meet again on the surface of rupture without being in equilibrium with the bulk. We then have $\Gamma_S = \Gamma_G/2$, so that, by (1.114), there is no variation of the energy of adhesion even when large embrittlement by segregation is observed. This paradox, due to Seah [91, 92], was resolved by Hirth and Rice [93] in 1980.

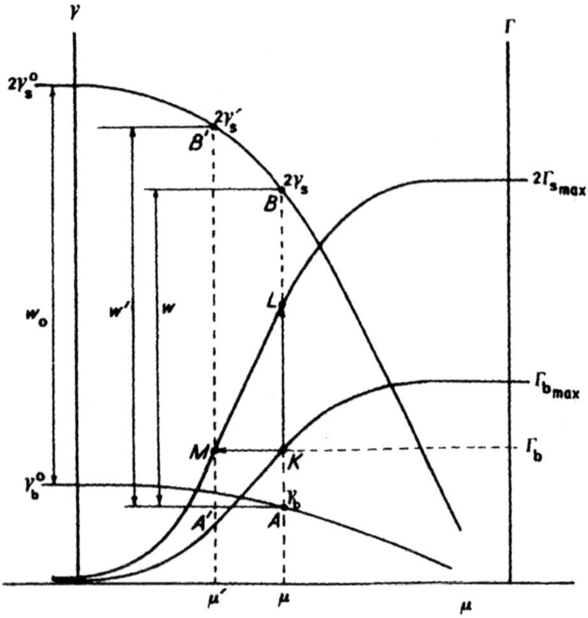

Fig. 1.39. Variation of twofold surface energy, grain boundary energy and adsorption with the chemical potential of the impurity, assuming a segregation of the Langmuir type

Figure 1.39 displays the variation of $2\gamma_S$ and γ_G with the chemical potential μ of the impurity, for an isotherm of the Langmuir type. The slope of these curves gives $2\Gamma_S$ and Γ_G after (1.109). For a quasistatic crack propagation, at constant chemical potential μ, surface and grain boundary segregation is represented by points K and L, the surface and intergranular energies by A and B, and the Dupré energy of adhesion by the segment AB. For a fast crack, we have $\Gamma_S = \Gamma_G/2$ (point M) and, assuming a local thermodynamic equilibrium, the surface energy $2\gamma'_S$ corresponds to the point B′, as for a chemical potential μ'. The Dupré energy of adhesion is $w' = $ A′B′. Hirth and Rice have shown this new Dupré energy of adhesion to be

$$w' = w_0 - (\mu - \mu')\Gamma_G \,. \tag{1.115}$$

Using (1.112) for Γ_G (at the concentration c) and $\Gamma_S = \Gamma_G/2$ (at the concentration c'), and assuming $\Gamma_{G_{max}} = \Gamma_{S_{max}}$, we have [94]

$$w' = w_0 + \Gamma_G RT \left(\frac{U}{RT} - \ln 2 \right) , \tag{1.116}$$

with

$$\begin{aligned} U &= \Delta G_S - \Delta G_G \\ &= \Delta H_S - \Delta H_G - T(\Delta S_S - \Delta S_G) \,. \end{aligned} \tag{1.117}$$

A similar treatment of (1.114) yields

$$w = w_0 + \Gamma_G RT \left(1 - e^{-U/RT}\right) . \tag{1.118}$$

In both cases, the Dupré energy of adhesion varies linearly with the segregation at the grain boundary. It decreases with segregation if $\Delta G_S - \Delta G_G < RT \ln 2$, but more slowly for fast cracks (Fig. 1.40). Seah [94], using the method of pair bonding, has directly obtained (1.116), but by neglecting the entropic terms of (1.117), and has shown that $w_0 - w'$ is proportional to the difference $H_A^{\text{sub}^*} - H_B^{\text{sub}^*}$ in the enthalpy of sublimation per unit area of the pure matrix and the impurity, so that the embrittlement (or, in contrast, the beneficent effect) of atoms segregated at grain boundaries can be predicted from tables of thermodynamic data. Figure 1.41 after Seah [94], displays these enthalpies of sublimation as a function of the ion size. Considering an element in Fig. 1.41 as a matrix element, those lower in the figure, considered as impurities, will provoke an embrittlement by segregation at grain boundaries, whereas elements higher in the figure will have a reinforcing effect. Roughly speaking, atoms with small radius increase the grain boundary cohesion, whereas atoms with large radius decrease it.

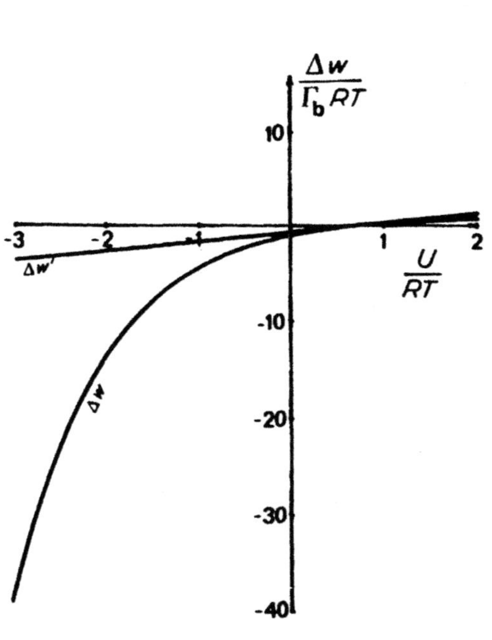

Fig. 1.40. Variation of the Dupré energy of adhesion with $U = \Delta G_S - \Delta G_G$ for slow or fast cracks

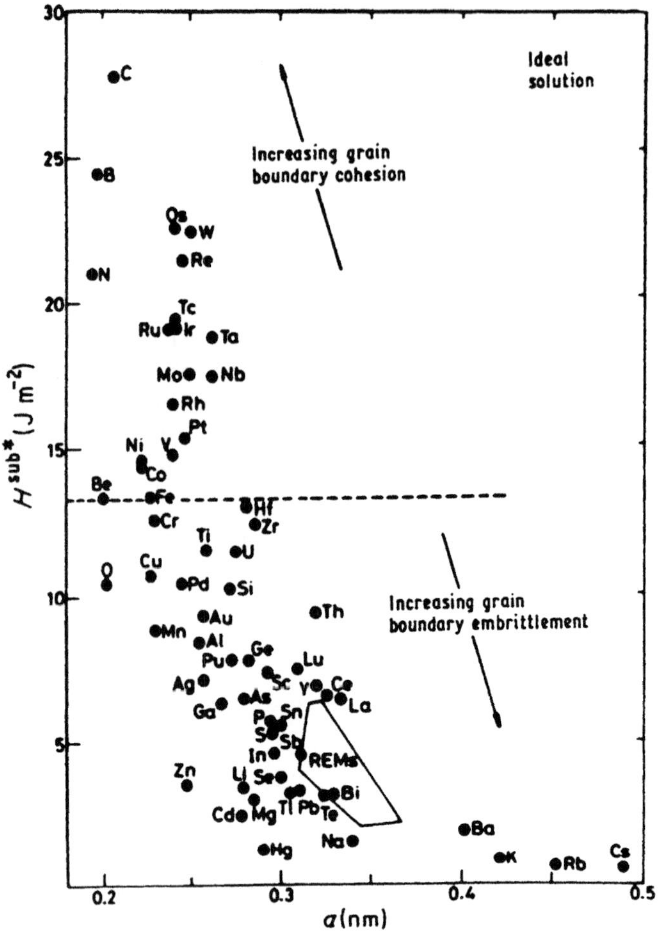

Fig. 1.41. Enthalpies of sublimation as a function of the ion size (after [94])

References

1. W.H. Keesom: Koninkl. Ned. Akad. Wetenschap. Proc. **24**, 162 (1921)
2. W.H. Keesom: Physik Z. **22**, 129 (1921)
3. P. Debye: Physik Z. **21**, 178 (1920)
4. F. London: Z. Physik **63**, 245 (1930); Z. Physik. Chem. B **11**, 222 (1930)
5. H.B.G. Casimir, D. Polder: Phys. Rev. **73**, 360 (1948)
6. H.B.G. Casimir: J. Chim. Phys. **46**, 407 (1949)
7. J.H. de Boer: Trans. Faraday Soc. **32**, 10 (1936)
8. H.C. Hamaker: Physica **4**, 1058 (1937)
9. R.S. Bradley: Phil. Mag. **13**, 853 (1932)
10. H.B.G. Casimir: Proc. Koninkl. Ned. Akad. Wetenschap. B **51**, 793 (1948)
11. B.V. Deryagin: Kolloid Z. **69**, 155 (1934)
12. B.V. Deryagin, N.A. Krotova, V.P. Smilga: *Adhesion of solids* (Consultant Bureau, New York 1978)
13. L.R. White: J. Colloid Interface Sci. **95**, 286 (1983) (cf also B.V. Derjaguin: J. Colloid Interface Sci. **107**, 579 (1985))
14. B.V. Derjaguin, A.S. Titijevskaia, I.I. Abrikossova, A.D. Malkina: Discuss. Faraday Soc. **18**, 24 (1954)
15. E.M. Lifshitz: Sov. Phys. JETP **2**, 73 (1956)
16. I.E. Dzyaloshinski, E.M. Lifshitz, L.P. Pitaevski: Adv. Phys. **10**, 165 (1961)
17. J.N. Israelachvili, D. Tabor: Prog. Surf. Membrane Sci. **7**, 1 (1973)
18. J.N. Israelachvili: Contemp. Phys. **15**, 159 (1974)
19. I.E. Dzyaloshinski, E.M. Lifshitz, L.P. Pitaevski: Sov. Phys. JETP **37**, 161 (1960)
20. B.V. Derjaguin, I.I. Abrikossova, E.M. Lifshitz: Quat. Rev. Chem. Soc. **10**, 295 (1956)
21. W. Black, J.G.V. de Jongh, J.Th.G. Overbeek, M.J. Sparnaay: Trans. Faraday Soc. **56**, 1597 (1960)
22. G.C.J. Rouweler, J.Th.G. Overbeek: Trans. Faraday Soc. **67**, 2117 (1971)
23. D. Tabor, R.H.S. Winterton: Proc. Roy. Soc. A **312**, 435 (1969)
24. J.N. Israelachvili, D. Tabor: Proc. Roy. Soc. A **331**, 19 (1972)
25. D. Langbein: J. Phys. Chem. Solids **32**, 1657 (1971)
26. D. Langbein: J. Adhesion **6**, 1 (1974)
27. E.S. Sabisky, C.H. Anderson: Phys. Rev. **7**, 790 (1973)
28. T.D. Blake: J. C. S. Faraday Trans. I **2**, 192 (1975)
29. P. Richmond, B.W. Ninham, R.H. Ottewill: J. Colloid. Interface Sci. **45**, 69 (1973)
30. R. Shuttleworth: Proc. Phys. Soc. A **62**, 167 (1949)
31. G.C. Benson, T.A. Claxton: J. Phys. Chem. Solids **25**, 367 (1964)
32. R.E. Allen, F.W. de Wette: Phys. Rev. **179**, 873 (1969)
33. J.H. Rose, J. Ferrante, J.R. Smith: Phys. Rev. Lett. **47**, 675 (1981)
34. J. Ferrante, J.R. Smith, J.H. Rose: 'Universal binding energy relations in metallic adhesion'. In: *Microscopic aspects of adhesion and lubrication*, ed. by J.M. Georges (Elsevier, 1982) pp. 19–30
35. E. Orowan: Trans. Inst. Eng. Shipbuilders Scotland **89**, 165 (1945)
36. J.B. Pethica, A.P. Sutton: J. Vac. Sci. Technol. A **6**, 2490 (1988)
37. J.R. Smith, G. Bozzolo, A. Banerjea, J. Ferrante: Phys. Rev. Lett. **63**, 1269 (1989)
38. U. Landman, W.D. Luedtke, N.A. Burnham, R.J. Colton: Science **24**, 454 (1990)
39. U. Landman, W.D. Luedtke: J. Vac. Sci. Technol. B **9**, 414 (1991)
40. P.A. Taylor, J.S. Nelson, B.W. Dodson: Phys. Rev. B **44**, 5834 (1991)

41. D. Maugis: On the adhesive avalanche (to appear)
42. G.I. Barenblatt: Adv. Appl. Mech. **7**, 55 (1962)
43. M. Schick: 'Introduction to wetting phenomena'. In: *Liquides aux interfaces / Liquids at interfaces*, ed. by J. Charvolin, J.F. Joanny, J. Zinn-Justin (Elsevier, 1990) pp. 415–497
44. A.M. Cazabat: Contemp. Phys. **28**, 347 (1987)
45. S. Ross, P. Becher: J. Colloid Interface Sci. **149**, 575 (1992)
46. J.F. Joanny, P.G. de Gennes: C. R. Acad. Sci. II **299**, 279 (1984)
47. P.G. de Gennes: Rev. Mod. Phys. **57**, 827 (1985)
48. J.F. Joanny, P.G. de Gennes: J. Phys. **47**, 121 (1986)
49. J. Lopez, C.A. Miller, E. Rückenstein: J. Colloid Interface Sci. **56**, 460 (1976)
50. L.H. Tanner: J. Phys. D: Appl. Phys. **12**, 1473 (1979)
51. A.M. Cazabat, M.A. Cohen Stuart: C. R. Acad. Sci. II **301**, 1337 (1985)
52. G.I. Taylor, D.H. Michael: J. Fluid Mech. **58**, 625 (1973)
53. F. Brochard-Wyart, J.M. di Meglio, D. Quere: C. R. Acad. Sci. II **304**, 553 (1987)
54. F. Brochard, C. Redon, F. Rondelez: C. R. Acad. Sci. II **306**, 1143 (1988)
55. F. Brochard-Wyart, J. Daillant: Can. J. Phys. **68**, 1084 (1990)
56. C. Redon, F. Brochard-Wyart, F. Rondelez: Phys. Rev. Lett. **66**, 715 (1991)
57. F. Brochard-Wyart, C. Redon, C. Sykes: C. R. Acad. Sci. II **314**, 19 (1992)
58. R. Shuttleworth: Proc. Phys. Soc. A **63**, 444 (1950)
59. R.C. Brown: Proc. Phys. Soc. **59**, 429 (1947)
60. J.A. de Feijter, J.B. Rijnbout, A. Vrij: J. Colloid Interface Sci. **64**, 258 (1978)
61. E.P. eer Nisse: Appl. Phys. Lett. **18**, 581 (1971)
62. J. Friedel: Ann. Phys. **1**, 257 (1976)
63. M.A. van Hove: Surf. Sci. **81**, 1 (1979)
64. T.R. Beck: J. Phys. Chem. **73**, 466 (1969)
65. K.F. Lin, T.R. Beck: J. Electrochem. Soc. **123**, 1145 (1976)
66. D.H. Bangham, F.A.P. Maggs: *Conf. on the ultra-fine structure of coals and cokes* (British Coal Utilization Association Committe, London 1943) p. 118
67. J.G. Dash, J. Suzanne, H. Shechter, P.E. Peierls: Surf. Sci. **60**, 411 (1976)
68. D.J.C. Yates: Proc. Phys. Soc. B **65**, 80 (1952)
69. D.J.C. Yates: Proc. Roy. Soc. A **224**, 526 (1954)
70. D.H. Bangham, N. Fakhoury, A.F. Mohamed: Proc. Roy. Soc. A **147**, 152 (1934)
71. P.J. Sereda, R.F. Feldman: 'Mechanical properties and the solid-gas interface'. In: *The solid-gas interface*, vol. 2, ed. by E.A. Flood (Dekker, New York 1967) pp. 729–764
72. D.J.C. Yates: In: *Advances in catalysis*, vol. 9, ed. by E.A. Flood (Dekker, New York 1956) p. 481
73. G.G. Stoney: Proc. Roy. Soc. A **82**, 172 (1909)
74. J.D. Wilcock, D.S. Campbell, J.C. Anderson: Thin Solid Films **3**, 13 (1969)
75. N.E.W. Hartley: J. Vac. Sci. Technol. **12**, 485 (1975)
76. R.A. Fredlein, A. Damjanovic, J.O'M. Bockris: Surf. Sci. **25**, 261 (1971)
77. R.A. Fredlein, J.O'M. Bockris: Surf. Sci. **46**, 641 (1974)
78. F.W.C. Boswell: Proc. Phys. Soc. A **64**, 465 (1951)
79. T. de Planta, R. Ghez, F. Piuz: Helv. Phys. Acta **37**, 74 (1964)
80. H.J. Wasserman, J.S. Vermaak: Surf. Sci. **22**, 164 (1970)
81. C.W. Mays, J.S. Vermaak, D. Kuhlmann-Wilsdorf: Surf. Sci. **12**, 134 (1968)
82. J.S. Vermaak, D. Kuhlmann-Wilsdorf: J. Phys. Chem. **72**, 4150 (1968)
83. H.J. Wasserman, J.S. Vermaak: Surf. Sci. **32**, 168 (1972)
84. M.M.P. Janssen: J. Appl. Phys. **41**, 384 (1970)
85. P.R. Couchman, W.A. Jesser, D. Kuhlmann-Wilsdorf, J.P. Hirth: Surf. Sci. **33**, 429 (1972)

86. J.S. Vermaak, C.W. Mays, D. Kuhlmann-Wilsdorf: Surf. Sci. **12**, 128 (1968)
87. J.C. Eriksson: Surf. Sci. **14**, 221 (1969)
88. P.R. Couchman, D.H. Everett, W.A. Jesser: J. Colloid Interface Sci. **52**, 410 (1975)
89. E.D. Hondros, D. McLean: Phil. Mag. **29**, 771 (1974)
90. M.C. Inman, H.R. Tipler: Metall. Rev. **8**, 105 (1963)
91. M.P. Seah: Surf. Sci. **53**, 168 (1975)
92. M.P. Seah: Proc. Roy. Soc. Lond. A **349**, 535 (1976)
93. J.P. Hirth, J.R. Rice: Metall. Trans. **11A**, 1501 (1980)
94. M.P. Seah: Acta Metall. **28**, 955 (1980)

2. Elements of Elasticity

One can date the development of the linear elasticity from 1676, the year when Robert Hooke proposed his anagram *ceiiinosssttuu* which he explained two years later as *ut tensio sic vis* (force is proportional to extension). Since then, several centuries of experiments have been needed to distinguish the various fields of linear and nonlinear elasticity, plasticity, viscoelasticity, viscoplasticity, and almost two centuries of thinking to introduce concepts which appear familiar today. If we think that the usual concept of pressure in a fluid, defined as the force per unit area that an element exerts on a neighbouring element, appeared only with Jean Bernoulli and Euler around 1750, that Coulomb in 1773 was the first to consider the stresses acting on various planes passing by the same point and the first to introduce the concept of shear stress on a plane, we can understand that the foundations of elasticity do not appear self-evident for anyone who approaches elasticity for the first time.

In fact, the foundations of the mechanics of continuous media and elasticity were established definitely only by Cauchy in 1822. It is he who introduced the notion of stress at a point by forces per unit area acting on any plane passing through this point and who showed that this stress was expressed by six components. He also defined the state of strain around a point by six components defined as partial derivatives of three components of the displacement at that point. These six numbers are the components of a quantity termed "tensor" by Voigt in 1910.

2.1 Stress Tensor at a Point

2.1.1 Definitions

To define the state of stress at a point $\mathrm{P}(x, y, z)$ of a continuous medium, let us consider a surface S passing through P and cutting the body into two parts (1) and (2); and let $\boldsymbol{n}$, the vector normal to S, oriented outside (1) at P. Let ΔS be an element of S containing P. The forces exerted by (2) on (1) through ΔS reduce at P to a resultant force $\Delta \boldsymbol{T}$ and to a resultant momentum $\Delta \boldsymbol{G}$. We have

$$\lim_{\Delta S \to 0} \frac{\Delta \boldsymbol{T}}{\Delta S} = \boldsymbol{T}^{(n)} \qquad \text{with } \boldsymbol{T}^{(n)} \text{ depending on } S \text{ only by } \boldsymbol{n},$$

$$\lim_{\Delta S \to 0} \frac{\Delta \boldsymbol{G}}{\Delta S} = 0 ,$$

where $\boldsymbol{T}^{(n)}$ is the *stress traction vector* acting at P on plane with a normal vector $\boldsymbol{n}$. In any orthonormed coordinates $(\boldsymbol{e}_x, \boldsymbol{e}_y, \boldsymbol{e}_z)$ we will write

$$\boldsymbol{T}^{(n)} = X\boldsymbol{e}_x + Y\boldsymbol{e}_y + Z\boldsymbol{e}_z . \tag{2.1}$$

In general, this vector is not collinear with $\boldsymbol{n}$. On the three faces perpendicular to axes (x, y, z) and passing through P we can write the three stress traction vectors in the particular form:

$$\begin{aligned} \boldsymbol{T}^{(x)} &= \tau_{xx}\boldsymbol{e}_x + \tau_{xy}\boldsymbol{e}_y + \tau_{xz}\boldsymbol{e}_z , \\ \boldsymbol{T}^{(y)} &= \tau_{yx}\boldsymbol{e}_x + \tau_{yy}\boldsymbol{e}_y + \tau_{yz}\boldsymbol{e}_z , \\ \boldsymbol{T}^{(z)} &= \tau_{zx}\boldsymbol{e}_x + \tau_{zy}\boldsymbol{e}_y + \tau_{zz}\boldsymbol{e}_z . \end{aligned} \tag{2.2}$$

The first index indicates the face by the direction of its normal, the second indicates the component of the traction vector acting on that face. The τ_{ii} are the *normal stresses*, the τ_{ij} are the tangential stresses or *shear stresses.* In the following we will write simply σ_i for τ_{ii}. We shall see that these nine quantities, homogeneous to pressures, are the components of a second-order tensor, the *stress tensor*, and that they allow for the complete determination of the state of stress at any point.

Besides these surface forces, body forces can exist which act directly on a volume element ΔV around P—as, for example, gravitational forces, magnetic forces or inertial forces. They reduce at P to a resultant force $\Delta \boldsymbol{R}$ and to a resultant momentum $\Delta \boldsymbol{M}$. We will admit that

$$\lim_{\Delta V \to 0} \frac{\Delta \boldsymbol{R}}{\Delta V} = \boldsymbol{F} ,$$

$$\lim_{\Delta V \to 0} \frac{\Delta \boldsymbol{M}}{\Delta V} = \mathcal{M} .$$

$\mathcal{M}$ represents a distribution of body moments as found in dielectric media submitted to an electric field. We shall assume in the following that $\mathcal{M} = 0$.

When the nine components σ_i and τ_{ij} are known, the traction vector $\boldsymbol{T}$ (components X, Y, Z) acting on any face passing through that point can be computed. Let us consider the tetrahedron PABC in Fig. 2.1, in which three faces of whose are normal to axes and the fourth has its normal $\boldsymbol{n}$ defined by the three direction cosines p, q, r. The signs of the nine components are defined with reference to the direction of the outward normal of the faces. In accordance with (2.2), if the outward normal of a face is in the same direction as one of the axes (x, y, z), a component (normal or tangential) on that face is positive if it is directed in the direction of its corresponding axis,

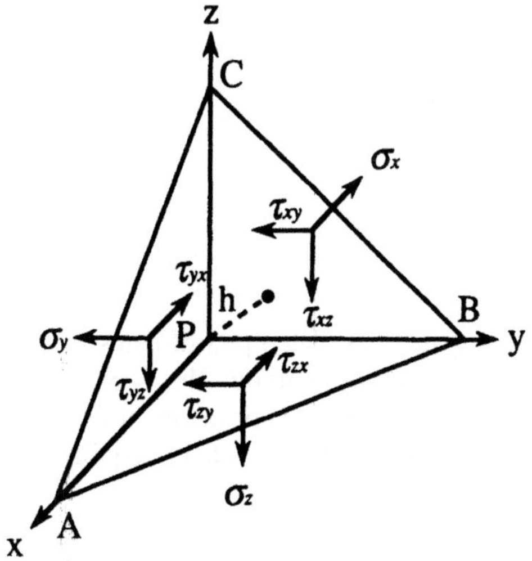

Fig. 2.1. Example tetrahedron

and negative in the contrary case. As per the law of action and reaction, if the outward normal is in a direction opposite to one of the axes (x, y, z), a component on this face is positive if it is directed in a direction opposite to its corresponding axis, negative in the opposite case. For example, the nine components in the Fig. 2.1 are all positive stresses.

(So, tensile stresses acting on two opposite faces of a small volume element will both be positive, and compressive stresses will be negative.) If the area of ABC is ΔS, the areas PBC, PCA, PAB are $p\Delta S$, $q\Delta S$, $r\Delta S$ and the volume of the tetrahedron is $\Delta V = \frac{1}{3}h\Delta S$. Let us write the equilibrium of the forces acting on the tetrahedron. In the direction y for example, we have

$$\Delta R_y + \Delta S(Y - p\tau_{xy} - q\sigma_y - r\tau_{zy}) = 0 \, .$$

(The stresses being generally not constant on faces, the stresses τ_{xy}, σ_y, τ_{zy} are to be taken each on a point of the face defined by the formula of the mean.) Let us divide by ΔS and let h tend to zero with

$$\lim_{\Delta V \to 0} \frac{\Delta R_y}{\Delta S} = \lim_{\Delta V \to 0} \frac{\Delta R_y}{\Delta V}\frac{\Delta V}{\Delta S} = 0 \, .$$

We thus obtain the three equations

$$\begin{aligned} X &= p\sigma_x + q\tau_{yx} + r\tau_{zx} \, , \\ Y &= p\tau_{xy} + q\sigma_y + r\tau_{zy} \, , \\ Z &= p\tau_{xz} + q\tau_{yz} + r\sigma_z \, , \end{aligned} \tag{2.3a}$$

where this time the stresses are taken at point P. They can be rewritten in matrix notation thus:

$$\boxed{\begin{pmatrix} X \\ Y \\ Z \end{pmatrix} = \begin{pmatrix} \sigma_x & \tau_{yx} & \tau_{zx} \\ \tau_{xy} & \sigma_y & \tau_{zy} \\ \tau_{xz} & \tau_{yz} & \sigma_z \end{pmatrix} \begin{pmatrix} p \\ q \\ r \end{pmatrix}} , \tag{2.3b}$$

or also:

$$\boldsymbol{T} = \overline{\overline{\sigma}} \cdot \boldsymbol{n} .$$

Let us consider now a small rectangular parallelepiped of sides a, b, c, parallel to the axes and centred on P. The moment about the z-axis of the forces acting on this volume element involves the moment of the body forces $\Delta \boldsymbol{M}_z$ and the moment of the surface forces drawn in Fig. 2.2. The equilibrium is thus given by

$$\Delta \boldsymbol{M}_z - b(ac\tau_{yx}) + a(bc\tau_{xy}) = 0 ,$$

and hence, dividing by the volume $V = abc$, we have

$$\frac{\Delta \boldsymbol{M}_z}{V} - \tau_{yx} + \tau_{xy} = 0$$

and by making V tending towards zero we obtain

$$\mathcal{M}_z - \tau_{yx} + \tau_{xy} = 0 .$$

In the following we shall assume that these body moments are zero, so that we shall have

$$\boxed{\tau_{ij} = \tau_{ji}} .$$

The stress tensor will be always symmetrical and its eigenvalues (its principal stresses) will always be real.

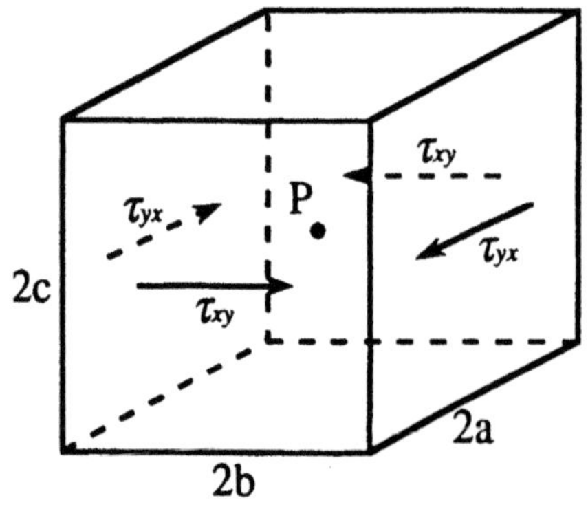

Fig. 2.2. Example parallelipiped

2.1.2 Principal Stresses and Principal Axes

At every point P of a solid, three orthogonal planes always exist, and these are termed *principal planes*, so that the traction vector $\boldsymbol{T}$ is directed along the normal $\boldsymbol{n}$ (shear stresses beeing thus zero on these principal planes). The normal to such a plane is termed the *principal axis*. On one of these principal planes we thus have

$$\begin{aligned} X &= \sigma p \,, \\ Y &= \sigma q \,, \\ Z &= \sigma r \,, \end{aligned} \tag{2.4}$$

where p, q, r are the direction cosines of the normal. Comparing with (2.3) we get

$$\begin{aligned} p(\sigma_x - \sigma) + q\tau_{yx} + r\tau_{zx} &= 0 \,, \\ p\tau_{xy} + q(\sigma_y - \sigma) + r\tau_{zy} &= 0 \,, \\ p\tau_{xz} + q\tau_{yz} + r(\sigma_z - \sigma) &= 0 \,, \end{aligned} \tag{2.5}$$

with

$$p^2 + q^2 + r^2 = 1 \,.$$

These equations have solutions for p, q, r only if the determinant is zero:

$$\begin{vmatrix} \sigma_x - \sigma & \tau_{yx} & \tau_{zx} \\ \tau_{xy} & \sigma_y - \sigma & \tau_{zy} \\ \tau_{xz} & \tau_{yz} & \sigma_z - \sigma \end{vmatrix} = -\sigma^3 + I_1\sigma^2 - I_2\sigma + I_3 = 0 \,, \tag{2.6}$$

i.e. if the three principal stresses σ_1, σ_2, σ_3 (eigenvalues of the stress tensor) are roots of a cubic equation (characteristic equation) where I_1, I_2, I_3 are scalar quantities termed *tensor invariants* since it is possible to show that they are independent of the coordinate system used to express the stresses. Thus

$$\begin{aligned} I_1 &= \sigma_x + \sigma_y + \sigma_z \,, \\ I_2 &= (\sigma_x\sigma_y + \sigma_y\sigma_z + \sigma_z\sigma_x) - \tau_{xy}^2 - \tau_{yz}^2 - \tau_{zx}^2 \,, \\ I_3 &= \begin{vmatrix} \sigma_x & \tau_{xy} & \tau_{xz} \\ \tau_{yx} & \sigma_y & \tau_{yz} \\ \tau_{zx} & \tau_{zy} & \sigma_z \end{vmatrix} \,. \end{aligned}$$

I_1 is the trace of the tensor, I_2 is the sum of the cofactors of the main diagonal terms and I_3 is its determinant. In this principal coordinate system (attached to the point P), the stress tensor is thus diagonal:

$$\overline{\overline{\sigma}} = \begin{pmatrix} \sigma_1 & 0 & 0 \\ 0 & \sigma_2 & 0 \\ 0 & 0 & \sigma_3 \end{pmatrix} \,,$$

and the tensor invariants become

$$\begin{aligned} I_1 &= \sigma_1 + \sigma_2 + \sigma_3 \,, \\ I_2 &= \sigma_1\sigma_2 + \sigma_2\sigma_3 + \sigma_3\sigma_1 \,, \\ I_3 &= \sigma_1\sigma_2\sigma_3 \,. \end{aligned}$$

If at point P all three principal stresses are nonzero, the stress state is said to be "triaxial". If one of the principal stresses *is* zero ($I_3 = 0$, $I_2 \neq 0$), the state is termed "biaxial", and it is "uniaxial" if two principal stresses are zero. If the three principal stresses are equal, the tensor is termed "spherical" and any face is a principal face (as in the case of a fluid at rest). If two principal stresses are equal, any axis orthogonal to the third principal axis is principal.

Let us consider the particular case where the z-axis is a principal axis. We have thus $\tau_{zx} = 0$, $\tau_{zy} = 0$ and (2.6) is written, with $\sigma = \sigma_z$ as an obvious solution, as

$$\begin{vmatrix} \sigma_x - \sigma & \tau_{yx} & 0 \\ \tau_{xy} & \sigma_y - \sigma & 0 \\ 0 & 0 & \sigma_z - \sigma \end{vmatrix} = (\sigma_z - \sigma)\left[\sigma^2 - (\sigma_x + \sigma_y)\sigma + \sigma_x\sigma_y - \tau_{xy}^2\right] = 0 \,.$$

The two principal stresses in the plane xy will be:

$$\boxed{\begin{aligned} \sigma_1 &= \frac{\sigma_x + \sigma_y}{2} + \sqrt{\left(\frac{\sigma_x - \sigma_y}{2}\right)^2 + \tau_{xy}^2} \,, \\ \sigma_2 &= \frac{\sigma_x + \sigma_y}{2} - \sqrt{\left(\frac{\sigma_x - \sigma_y}{2}\right)^2 + \tau_{xy}^2} \,, \end{aligned}} \qquad (2.7)$$

taking $\sigma_1 > \sigma_2$.

Let us compute the eigenvector, i.e. the principal axis, corresponding to the principal stress σ_1, and let p_1, q_1, r_1 be its direction cosines. Equations (2.5) become:

$$\begin{aligned} p_1(\sigma_x - \sigma_1) + q_1\tau_{yx} &= 0 \,, \\ p_1\tau_{xy} + q_1(\sigma_y - \sigma_1) &= 0 \,, \\ r_1 &= 0 \,. \end{aligned}$$

This vector is in the xy plane and makes with the x-axis an angle α_1 defined by

$$\tan\alpha_1 = \frac{q_1}{p_1} = \frac{\sigma_1 - \sigma_x}{\tau_{xy}} = \frac{\tau_{xy}}{\sigma_1 - \sigma_y} \,,$$

i.e.:

$$\tan\alpha_1 = \frac{\sigma_y - \sigma_x}{2\tau_{yx}} + \sqrt{\left(\frac{\sigma_y - \sigma_x}{2\tau_{yx}}\right)^2 + 1}\,. \tag{2.8a}$$

For the other principal axis we have

$$\tan\alpha_2 = \frac{\sigma_y - \sigma_x}{2\tau_{yx}} - \sqrt{\left(\frac{\sigma_y - \sigma_x}{2\tau_{yx}}\right)^2 + 1} = \frac{-1}{\tan\alpha_1}\,, \tag{2.8b}$$

both equations reducing to

$$\boxed{\tan 2\alpha = \frac{2\tau_{xy}}{\sigma_x - \sigma_y}}\,. \tag{2.9}$$

Note that (2.7) and (2.9) can be obtained directly by the method of the Mohr circle, which we will not discuss here.

2.1.3 Deviatoric Stress Tensor and von Mises Criterion

It is often useful to decompose the stress tensor into a *spherical stress tensor* and a *deviatoric stress tensor*, thus:

$$\overline{\overline{\sigma}} = \begin{pmatrix} \sigma_x & \tau_{yx} & \tau_{zx} \\ \tau_{xy} & \sigma_y & \tau_{zy} \\ \tau_{xz} & \tau_{yz} & \sigma_z \end{pmatrix} = \begin{pmatrix} \sigma_{\mathrm{m}} & 0 & 0 \\ 0 & \sigma_{\mathrm{m}} & 0 \\ 0 & 0 & \sigma_{\mathrm{m}} \end{pmatrix} + \begin{pmatrix} \sigma_x - \sigma_{\mathrm{m}} & \tau_{yx} & \tau_{zx} \\ \tau_{xy} & \sigma_y - \sigma_{\mathrm{m}} & \tau_{zy} \\ \tau_{xz} & \tau_{yz} & \sigma_z - \sigma_{\mathrm{m}} \end{pmatrix}$$

with $\sigma_{\mathrm{m}} = \frac{1}{3}(\sigma_x + \sigma_y + \sigma_z) = \frac{1}{3}(\sigma_1 + \sigma_2 + \sigma_3)$ for the spherical tensor to have the same trace as the original. σ_{m} is the *mean stress*. The trace of the deviatoric stress tensor is zero by definition, and its three invariants are:

$$\begin{aligned} J_1 &= 0 \\ J_2' &= (\sigma_x - \sigma_{\mathrm{m}})(\sigma_y - \sigma_{\mathrm{m}}) + (\sigma_y - \sigma_{\mathrm{m}})(\sigma_z - \sigma_{\mathrm{m}}) + (\sigma_z - \sigma_{\mathrm{m}})(\sigma_x - \sigma_{\mathrm{m}}) \\ &\quad - \tau_{xy}^2 - \tau_{yz}^2 - \tau_{zx}^2 \\ J_3 &= \begin{vmatrix} \sigma_x - \sigma_{\mathrm{m}} & \tau_{yx} & \tau_{zx} \\ \tau_{xy} & \sigma_y - \sigma_{\mathrm{m}} & \tau_{zy} \\ \tau_{xz} & \tau_{yz} & \sigma_z - \sigma_{\mathrm{m}} \end{vmatrix}\,. \end{aligned}$$

It is generally preferred to use instead of J_2', which is negative, its opposite J_2, which can be written as

$$\boxed{\begin{aligned} J_2 &= \frac{1}{6}\left[(\sigma_x - \sigma_y)^2 + (\sigma_y - \sigma_z)^2 + (\sigma_z - \sigma_x)^2\right] \\ &\quad + \tau_{xy}^2 + \tau_{yz}^2 + \tau_{zx}^2 \\ &= \frac{1}{6}\left[(\sigma_1 - \sigma_2)^2 + (\sigma_2 - \sigma_3)^2 + (\sigma_3 - \sigma_1)^2\right] \end{aligned}}\,. \tag{2.10}$$

When an experiment in uniaxial tension is performed, on a metal for example, one observes that the stress–strain relation remains linear up to a value σ_0 of stress, termed the *elastic limit* or *flow stress*, after which the deformation is plastic. The task is to write a criterion for the initiation of the plastic deformation under a multiaxial state of stress, as a function of σ_0. As experiments show that the elastic limit is not affected by the hydrostatic pressure, this criterion therefore use only of the deviatoric stress tensor and therefore be a function of the invariants J_2 and J_3 only since it must be independent of the choice of the coordinate system. The most simple criterion (von Mises, 1913) consists of neglecting the influence of J_3 and writing $J_2 = k^2$. Since $J_2 = \sigma_1^2/3$ in a uniaxial stress state, we find $k = \sigma_0/\sqrt{3}$. The criterion of plastic deformation will thus be

$$\sqrt{J_2} = \frac{\sigma_0}{\sqrt{3}}\,, \tag{2.11}$$

or else $\sigma_{\mathrm{eq}} = \sigma_0$, with σ_{eq} defined by

$$\sigma_{\mathrm{eq}}^2 = \frac{1}{2}\left[(\sigma_1-\sigma_2)^2 + (\sigma_2-\sigma_3)^2 + (\sigma_3-\sigma_1)^2\right] = 3J_2\,. \tag{2.12}$$

2.1.4 Description of a Stress Field

At any point in a solid is an associated stress tensor with its principal axes. We can draw in symmetry planes the contours of the principal stresses σ_1, σ_2, σ_3, as well as the contours of $\sqrt{J_2}$ or σ_{eq} to find the locus of first plastic deformations.

If the chosen plane is a principal plane for all its points, we can display the principal axes by their isoclines or their isostatics (stress trajectories). *Isoclines* are the locus of the points where the principal axes make a given angle with an axis. If x and y are the axes of the plane, (2.9) shows that isoclines 0° (principal stresses parallel to axes) and 45° are respectively given by $\tau_{xy} = 0$ and $(\sigma_x - \sigma_y) = 0$. If there exist in the plane singular points where $\sigma_1 = \sigma_2$ (isotropic points), all isoclines pass through these points since, in accordance with (2.9), we have both $\tau_{xy} = 0$ and $(\sigma_x - \sigma_y) = 0$. *Stress trajectories* are the "trajectories" of the principal axes. In any point of the principal plane xOy we have two stress trajectories; they cut at right angles since the principal axes are tangential to these lines. The slope $\mathrm{d}y/\,\mathrm{d}x$ of a stress trajectory being defined by (2.8a), a first network of stress trajectories will be given by the differential equation

$$\frac{\mathrm{d}y}{\mathrm{d}x} = \frac{\sigma_y - \sigma_x}{2\tau_{yx}} + \mathrm{sign}(\tau_{yx})\sqrt{\left(\frac{\sigma_y - \sigma_x}{2\tau_{yx}}\right)^2 + 1} \tag{2.13}$$

where $\mathrm{sign}(\tau_{yx})$ insure the continuity of the slope through the isocline 0°. The second network of stress trajectories corresponds to the minus sign in front of the radical. If an isotropic point exists, an infinite number of stress trajectories pass through it.

We can complete this description by drawing isochromatic lines and isopachics. *Isochromatic lines* are the locus of points where the difference $(\sigma_1 - \sigma_2)$ of the principal stress has a given value. We can obtain such a network of isochromatic lines by observing between crossed nicols a birefringent plate subjected to stresses. *Isopachics* are the locus of points where $(\sigma_1 + \sigma_2) = \text{const.}$ They are the locus of points of equal thickness of a plate submitted to plane stresses $(\sigma_3 = 0)$. They can be observed by holography.

2.2 Basic Equations

Given a homogeneous and isotropic solid subjected on its surface to displacements or forces of surface density F_S, and eventually to body forces $F(x, y, z)$, we can attempt to determine the stress tensor and the strain tensor at any point. For that purpose, we have at our disposal several equations elaborated during the 19th century.

2.2.1 Equations of Equilibrium

These give the equilibrium of a volume element (*Cauchy's equations*, 1822):

$$\boxed{\begin{aligned} &\frac{\partial \sigma_x}{\partial x} + \frac{\partial \tau_{yx}}{\partial y} + \frac{\partial \tau_{zx}}{\partial z} + F_x = 0\,, \\ &\frac{\partial \tau_{xy}}{\partial x} + \frac{\partial \sigma_y}{\partial y} + \frac{\partial \tau_{zy}}{\partial z} + F_y = 0\,, \\ &\frac{\partial \tau_{xz}}{\partial x} + \frac{\partial \tau_{yz}}{\partial y} + \frac{\partial \sigma_z}{\partial z} + F_z = 0\,; \end{aligned}} \tag{2.14}$$

or, in condensed form,

$$\frac{\partial \tau_{ij}}{\partial x_j} + F_i = 0\,.$$

The boundary conditions on the surface, can be given in terms either of stresses (direction cosines p, q, r)

$$\boxed{\begin{aligned} &\sigma_x p + \tau_{yx} q + \tau_{zx} r = F_{sz}\,, \\ &\tau_{xy} p + \sigma_y q + \tau_{zy} r = F_{sy}\,, \\ &\tau_{xz} p + \tau_{yz} q + \sigma_z r = F_{sz}\,, \end{aligned}} \tag{2.15a}$$

or of displacements

$$\begin{aligned} u_x &= U , \\ u_y &= V , \\ u_z &= W , \end{aligned} \tag{2.15b}$$

or both (mixed boundary conditions), where at some points of the surface displacements are imposed and at others the stresses are imposed.

2.2.2 Compatibility Equations

The strain in the neighbouring of a point $\mathrm{M}(x, y, z)$ is defined as a function of its displacement (u_x, u_y, u_z):

$$\boxed{\begin{aligned} \varepsilon_x &= \frac{\partial u_x}{\partial x} ; & \varepsilon_{yz} &= \frac{1}{2}\left(\frac{\partial u_z}{\partial y} + \frac{\partial u_y}{\partial z}\right) ; \\ \varepsilon_y &= \frac{\partial u_y}{\partial y} ; & \varepsilon_{zx} &= \frac{1}{2}\left(\frac{\partial u_x}{\partial z} + \frac{\partial u_z}{\partial x}\right) ; \\ \varepsilon_z &= \frac{\partial u_z}{\partial z} ; & \varepsilon_{xy} &= \frac{1}{2}\left(\frac{\partial u_y}{\partial x} + \frac{\partial u_x}{\partial y}\right) . \end{aligned}} \tag{2.16}$$

There exist equations insuring the compatibility of these deformations (the *Saint Venant equations*, 1864), due to the fact that the ε are not whatever but defined as partial derivative of the displacements by (2.16):

$$\boxed{\begin{aligned} \frac{\partial^2 \varepsilon_x}{\partial y \partial z} &= \frac{\partial}{\partial x}\left(-\frac{\partial \varepsilon_{yz}}{\partial x} + \frac{\partial \varepsilon_{zx}}{\partial y} + \frac{\partial \varepsilon_{xy}}{\partial z}\right) ; \\ \frac{\partial^2 \varepsilon_y}{\partial z \partial x} &= \frac{\partial}{\partial y}\left(-\frac{\partial \varepsilon_{zx}}{\partial y} + \frac{\partial \varepsilon_{xy}}{\partial z} + \frac{\partial \varepsilon_{yz}}{\partial x}\right) ; \\ \frac{\partial^2 \varepsilon_z}{\partial x \partial y} &= \frac{\partial}{\partial z}\left(-\frac{\partial \varepsilon_{xy}}{\partial z} + \frac{\partial \varepsilon_{yz}}{\partial x} + \frac{\partial \varepsilon_{zx}}{\partial y}\right) ; \\ 2\frac{\partial^2 \varepsilon_{yz}}{\partial y \partial z} &= \frac{\partial^2 \varepsilon_z}{\partial y^2} + \frac{\partial^2 \varepsilon_y}{\partial z^2} ; \\ 2\frac{\partial^2 \varepsilon_{zx}}{\partial z \partial x} &= \frac{\partial^2 \varepsilon_x}{\partial z^2} + \frac{\partial^2 \varepsilon_z}{\partial x^2} ; \\ 2\frac{\partial^2 \varepsilon_{xy}}{\partial x \partial y} &= \frac{\partial^2 \varepsilon_y}{\partial x^2} + \frac{\partial^2 \varepsilon_x}{\partial y^2} . \end{aligned}} \tag{2.17}$$

2.2.3 Constitutive Equations Expressing the Linear Elastic Relationship Between Stresses and Strains

These equations are given either with the *Lamé coefficients* λ and μ (μ is the *shear modulus* or *Coulomb modulus*)

$$\begin{array}{|ll|}\hline \sigma_x = \lambda\theta + 2\mu\varepsilon_x \,, & \tau_{yz} = 2\mu\varepsilon_{yz} \,, \\ \sigma_y = \lambda\theta + 2\mu\varepsilon_y \,, & \tau_{zx} = 2\mu\varepsilon_{zx} \,, \\ \sigma_z = \lambda\theta + 2\mu\varepsilon_z \,, & \tau_{xy} = 2\mu\varepsilon_{xy} \,, \\ \hline \end{array} \tag{2.18}$$

where θ is the cubic dilation given by

$$\theta = \varepsilon_x + \varepsilon_y + \varepsilon_z \,;$$

or with the *Young's modulus* E and the *Poisson's ratio* ν (via Hooke's law), with

$$\boxed{\begin{aligned} \varepsilon_x &= \frac{1}{E}[\sigma_x - \nu(\sigma_y + \sigma_z)] \,, \\ \varepsilon_y &= \frac{1}{E}[\sigma_y - \nu(\sigma_z + \sigma_x)] \,, \\ \varepsilon_z &= \frac{1}{E}[\sigma_z - \nu(\sigma_x + \sigma_y)] \,. \end{aligned}} \tag{2.19}$$

The Lamé coefficients are related to E and ν by

$$\lambda = \frac{\nu E}{(1+\nu)(1-2\nu)} \,, \tag{2.20}$$

$$\mu = \frac{E}{2(1+\nu)} \,. \tag{2.21}$$

By using (2.19) and (2.14), the Saint Venant equation of compatibility (2.17) can be rewritten as a function of the stresses. They bear the name of *Beltrami–Michell* equations (Beltrami, 1892, for no or constant body forces; and Michell, 1900, for the general case). The Beltrami equations of compatibility are:

$$\begin{aligned} &(1+\nu)\Delta\sigma_x + \frac{\partial^2}{\partial x^2}(\sigma_x + \sigma_y + \sigma_z) = 0 \,; \\ &(1+\nu)\Delta\sigma_y + \frac{\partial^2}{\partial y^2}(\sigma_x + \sigma_y + \sigma_z) = 0 \,; \\ &(1+\nu)\Delta\sigma_z + \frac{\partial^2}{\partial z^2}(\sigma_x + \sigma_y + \sigma_z) = 0 \,; \\ &(1+\nu)\Delta\tau_{yz} + \frac{\partial^2}{\partial y \partial z}(\sigma_x + \sigma_y + \sigma_z) = 0 \,; \\ &(1+\nu)\Delta\tau_{zx} + \frac{\partial^2}{\partial z \partial x}(\sigma_x + \sigma_y + \sigma_z) = 0 \,; \\ &(1+\nu)\Delta\tau_{xy} + \frac{\partial^2}{\partial x \partial y}(\sigma_x + \sigma_y + \sigma_z) = 0 \,. \end{aligned} \tag{2.22}$$

All these complicated equations are more easily solved in states of plane elasticity (plane strain, plane stress) and in the axisymmetric case. In these states one of the axes (Oz for example) is a principal axis. *In the following we will assume that the body forces F are zero.*

2.3 Plane State Problems

2.3.1 Plane Strain

Let us consider the case where the displacements u_z (in the z direction) are zero, and where the displacements u_x and u_y are independent of z. We have from (2.16):

$$\varepsilon_z = \varepsilon_{xz} = \varepsilon_{yz} = 0\,, \tag{2.23a}$$

$$\varepsilon_x, \varepsilon_y, \varepsilon_{xy} \quad \text{independent of } z\,, \tag{2.23b}$$

which defines a state of plane strain. It is the case of a plate held between two rigid walls preventing any deformation along z. There are also the conditions which prevail in the bulk of a plate which is very thick in the z-direction. We can see, from (2.18), that a stress $\sigma_z = \nu(\sigma_x + \sigma_y)$ arises because the material is not free to deform along the z-axis. The stress tensors and strain tensors reduce to

$$\begin{bmatrix} \varepsilon_x & \varepsilon_{xy} & 0 \\ \varepsilon_{xy} & \varepsilon_y & 0 \\ 0 & 0 & 0 \end{bmatrix}\,, \qquad \begin{bmatrix} \sigma_x & \tau_{xy} & 0 \\ \tau_{xy} & \sigma_y & 0 \\ 0 & 0 & \nu(\sigma_x + \sigma_y) \end{bmatrix}\,.$$

It is a state of biaxial strain (plane), and a state of triaxial stress with three principal stresses: σ_1, σ_2, σ_3 (cf (2.7)).

Replacing σ_z by its value $\nu(\sigma_x + \sigma_y)$, Hooke's equations can be rewritten as

$$\boxed{\begin{array}{ll} \varepsilon_x = \dfrac{1}{E^*}(\sigma_x - \nu^* \sigma_y)\,; & \varepsilon_{xy} = \tau_{xy}/2\mu\,; \\ \varepsilon_y = \dfrac{1}{E^*}(\sigma_y - \nu^* \sigma_x)\,; & \varepsilon_{yz} = 0\,; \\ \varepsilon_z = 0\,; & \varepsilon_{zx} = 0\,, \end{array}} \tag{2.24}$$

with

$$E^* = \frac{E}{1-\nu^2}\,,$$

$$\nu^* = \frac{\nu}{1-\nu}\,.$$

With $\partial/\partial z = 0$ and with the body forces being neglected as aforementioned, Cauchy's equilibrium equations (2.14) reduce to:

$$\boxed{\begin{aligned} \frac{\partial \sigma_x}{\partial x} + \frac{\partial \tau_{xy}}{\partial y} = 0 \,, \\ \frac{\partial \tau_{xy}}{\partial x} + \frac{\partial \sigma_y}{\partial y} = 0 \,. \end{aligned}} \tag{2.25}$$

Airy showed in 1862 that if one introduces a stress function $\Phi(x, y)$ such that

$$\boxed{\begin{aligned} \sigma_x &= \frac{\partial^2 \Phi}{\partial y^2} \,, \\ \sigma_y &= \frac{\partial^2 \Phi}{\partial x^2} \,, \\ \tau_{xy} &= -\frac{\partial^2 \Phi}{\partial x \partial y} \,, \end{aligned}} \tag{2.26}$$

the equilibrium equations (2.25) are automatically and reciprocally satisfied.

As regards the compatibility equations (2.17), with $\partial/\partial z = 0$ and $\varepsilon_z = \varepsilon_{zx} = \varepsilon_{zy} = 0$ they reduce to only one:

$$2\frac{\partial^2 \varepsilon_{xy}}{\partial x \partial y} = \frac{\partial^2 \varepsilon_y}{\partial x^2} + \frac{\partial^2 \varepsilon_x}{\partial y^2} \,. \tag{2.27}$$

Let us write ε_x, ε_y, ε_z as a function of σ_x, σ_y, σ_z with the help of Hooke's equations (2.24), equation (2.27) then becomes

$$2\frac{\partial^2 \tau_{xy}}{\partial x \partial y} = (1-\nu)\frac{\partial^2 \sigma_y}{\partial x^2} - \nu\frac{\partial^2 \sigma_x}{\partial x^2} + (1-\nu)\frac{\partial^2 \sigma_x}{\partial y^2} - \nu\frac{\partial^2 \sigma_y}{\partial y^2} \,.$$

(We can observe that Young's modulus E^* no longer appears.) By differentiating Cauchy's equations (2.25), we can write

$$2\frac{\partial^2 \tau_{xy}}{\partial x \partial y} = -\frac{\partial^2 \sigma_x}{\partial x^2} - \frac{\partial^2 \sigma_y}{\partial y^2} \,,$$

and the compatibility equation becomes

$$\boxed{\Delta(\sigma_x + \sigma_y) = 0} \,, \tag{2.28}$$

where the Laplacian operator reduces to:

$$\Delta = \frac{\partial^2}{\partial x^2} + \frac{\partial^2}{\partial y^2} \,,$$

a result given by J. C. Maxwell in 1868. We observe that Poisson's ratio ν^*, similarly to E^*, no longer appears.

By introducing Airy's stress function, (2.28) becomes

$$\boxed{\Delta\Delta\Phi(x,y) = 0}\,, \tag{2.29}$$

so that to solve a problem of elasticity in plane strain it suffices to find biharmonic functions (i.e. solutions of (2.29)) satisfying to the boundary conditions of the problem. The equilibrium equations (2.25) and the compatibility equation (2.27) will then be automatically satisfied.

2.3.2 Plane Stress

The question of the plane stress is a delicate one, and a demonstration of the asymptotic character of most of the solutions is rarely given. However, such an analysis is very instructive. We will follow that of Timoshenko and Goodier [1].

By analogy with plane strain, let us define the state of plane stress by the following conditions:

$$\sigma_z = \tau_{xz} = \tau_{yz} = 0\,, \tag{2.30a}$$

$$\sigma_x\,,\ \sigma_y\,,\ \sigma_{xy} \qquad \text{independent of } z\,. \tag{2.30b}$$

Let us first examine the implications of (2.30a). Hooke's laws, given in (2.19), become

$$\boxed{\begin{aligned} \varepsilon_x &= \frac{1}{E}(\sigma_x - \nu\sigma_y)\,; & \varepsilon_{xy} &= \tau_{xy}/2\mu\,; \\ \varepsilon_y &= \frac{1}{E}(\sigma_y - \nu\sigma_x)\,; & \varepsilon_{yz} &= 0\,; \\ \varepsilon_z &= -\frac{\nu}{1-\nu}(\varepsilon_x + \varepsilon_y)\,; & \varepsilon_{zx} &= 0\,. \end{aligned}} \tag{2.31}$$

The stress tensor and the strain tensor reduce to

$$\begin{bmatrix} \sigma_x & \tau_{xy} & 0 \\ \tau_{xy} & \sigma_y & 0 \\ 0 & 0 & 0 \end{bmatrix}\,, \qquad \begin{bmatrix} \varepsilon_x & \varepsilon_{xy} & 0 \\ \varepsilon_{xy} & \varepsilon_y & 0 \\ 0 & 0 & -\frac{\nu}{1-\nu}(\varepsilon_x + \varepsilon_y) \end{bmatrix}\,.$$

It is a state of biaxial stress (with two principal stresses σ_1 and σ_2) and a state of triaxial strain.

Furthermore, Cauchy's equilibrium equations (2.14) reduce to

$$\boxed{\begin{aligned} \frac{\partial\sigma_x}{\partial x} + \frac{\partial\tau_{xy}}{\partial y} &= 0\,, \\ \frac{\partial\tau_{xy}}{\partial x} + \frac{\partial\sigma_y}{\partial y} &= 0\,, \end{aligned}}$$

as in plane strain, and they will be automatically satisfied by writing the stresses in the following form:

$$\sigma_x = \frac{\partial^2 \Phi}{\partial y^2} ,$$
$$\sigma_y = \frac{\partial^2 \Phi}{\partial x^2} ,$$
$$\tau_{xy} = -\frac{\partial^2 \Phi}{\partial x \partial y} ,$$

where Φ is an Airy's stress function. (Equation (2.30a) alone does not imply that Φ is independent of z.)

Let us now examine the implications of (2.30b). From (2.31) this hypothesis implies that ε_x, ε_y, ε_{xy} do not vary with z; the variable z can be eliminated as in plane strain. The compatibility equations (2.17) reduce to four equations (instead of one in plane strain):

$$\frac{\partial^2 \varepsilon_z}{\partial x \partial y} = \frac{\partial^2 \varepsilon_z}{\partial y^2} = \frac{\partial^2 \varepsilon_z}{\partial x^2} = 0 ,$$
$$2\frac{\partial^2 \varepsilon_{xy}}{\partial x \partial y} = \frac{\partial^2 \varepsilon_y}{\partial x^2} + \frac{\partial^2 \varepsilon_x}{\partial y^2} .$$

The first three imply that ε_z is a polynomial of the first degree in x and y (and thus also for $\sigma_x + \sigma_y$ from Hooke's law):

$$\varepsilon_z = -\frac{\nu}{E}(\sigma_x + \sigma_y) = -\frac{\nu}{E}\Delta\Phi = a + bx + cy . \tag{2.32}$$

The fourth, which is the same as in plane deformation, implies

$$\Delta(\sigma_x + \sigma_y) = \Delta\Delta\Phi = 0 , \tag{2.33}$$

which is automatically satisfied if $\Delta\Phi$ is a polynomial of the first degree in x and y. Unfortunately, this last condition is very restrictive and allows only for a small number of cases to be treated.

Let us instead renounce the hypothesis of (2.30b). The stresses σ_x, σ_y, τ_{xy} now being considered dependent upon z, Airy's stress function will be a function $\Phi(x, y, z)$ and we will write its Laplacian in the form

$$\Delta\Phi = \Delta_\perp \Phi + \frac{\partial^2 \Phi}{\partial z^2} .$$

By using the stress function Φ, and after some simple transformations, Beltrami's equations of compatibility, (2.22), can be written as

$$\frac{\partial^2}{\partial y^2}\left[\nu\Delta_\perp\Phi+(1+\nu)\frac{\partial^2\Phi}{\partial z^2}\right]+\Delta_\perp\Delta_\perp\Phi=0\,,$$
$$\frac{\partial^2}{\partial x^2}\left[\nu\Delta_\perp\Phi+(1+\nu)\frac{\partial^2\Phi}{\partial z^2}\right]+\Delta_\perp\Delta_\perp\Phi=0\,,$$
$$\frac{\partial^2}{\partial z^2}\Delta_\perp\Phi=0\,,$$
$$\frac{\partial^2}{\partial z\partial y}\Delta_\perp\Phi=0\,,$$
$$\frac{\partial^2}{\partial z\partial x}\Delta_\perp\Phi=0\,,$$
$$\frac{\partial^2}{\partial x\partial y}\left[\nu\Delta_\perp\Phi+(1+\nu)\frac{\partial^2\Phi}{\partial z^2}\right]=0\,.$$

The third, fourth and fifth equations show that $\Delta_\perp\Phi$ is a function in the form

$$\Delta_\perp\Phi=kz+F(x,y)\,. \tag{2.34}$$

Adding together the two first equations, we get:

$$(2+\nu)\Delta_\perp\Delta_\perp\Phi+(1+\nu)\frac{\partial^2}{\partial z^2}\Delta_\perp\Phi=0\,,$$

which, using (2.34), reduces to

$$\Delta_\perp\Delta_\perp\Phi=0\,; \tag{2.35}$$

hence

$$\Delta F(x,y)=0\,. \tag{2.36}$$

Finally, using (2.34), Beltrami's equations of compatibility reduce to

$$\frac{\partial^2}{\partial y^2}\left[\nu F+(1+\nu)\frac{\partial^2\Phi}{\partial z^2}\right]=0\,,$$
$$\frac{\partial^2}{\partial x^2}\left[\nu F+(1+\nu)\frac{\partial^2\Phi}{\partial z^2}\right]=0\,,$$
$$\frac{\partial^2}{\partial x\partial y}\left[\nu F+(1+\nu)\frac{\partial^2\Phi}{\partial z^2}\right]=0\,,$$

which show that the bracketed term is a polynomial of the first degree in x and y and

$$\frac{\partial^2\Phi}{\partial z^2}+\frac{\nu}{1+\nu}F(x,y)=a(z)+b(z)x+c(z)y\,.$$

Integrating this equation twice, we get

$$\begin{aligned}\Phi(x,y,z) = &-\frac{\nu}{2(1+\nu)}F(x,y)z^2 + A(z) + B(z)x + C(z)y \\ &+ \Phi_1(x,y)z + \Phi_0(x,y)\,.\end{aligned}$$

The terms $A(z)$, $B(z)x$, $C(z)y$ can be taken as zero since they disappear in the calculation of stresses. Furthermore, if one limits oneself to the case of plates in which the stress distribution is symmetrical about the median plane $z = 0$, the term $\Phi_1 z$ can also be taken as zero. Then we get

$$\Phi(x,y,z) = \Phi_0(x,y) - \frac{\nu}{2(1+\nu)}F(x,y)z^2\,.$$

If we take the perpendicular Laplacian of it, using (2.36) we obtain

$$\Delta_\perp \Phi(x,y,z) = \Delta\Phi_0(x,y)\,,$$

which, compared with (2.34), shows that $k = 0$ and that

$$F(x,y) = \Delta\Phi_0(x,y)\,. \tag{2.37}$$

We thus obtain the following result:

$$\boxed{\Phi(x,y,z) = \Phi_0(x,y) - \frac{\nu}{2(1+\nu)}z^2\Delta\Phi_0(x,y)}\,, \tag{2.38}$$

with, using (2.37) and (2.36):

$$\boxed{\Delta\Delta\Phi_0(x,y) = 0}\,. \tag{2.39}$$

From Airy's function, (2.38), expressions of the stresses will have two parts:

$$\begin{aligned}\sigma_x &= \sigma_x^0 - \frac{\nu}{2(1+\nu)}z^2\frac{\partial^2}{\partial y^2}\Delta\Phi_0\,,\\ \sigma_y &= \sigma_y^0 - \frac{\nu}{2(1+\nu)}z^2\frac{\partial^2}{\partial x^2}\Delta\Phi_0\,,\\ \tau_{xy} &= \tau_{xy}^0 + \frac{\nu}{2(1+\nu)}z^2\frac{\partial^2}{\partial x\partial y}\Delta\Phi_0\,.\end{aligned}$$

The first arises, as in plane strain, from the biharmonic function $\Phi_0(x,y)$; the second is proportional to z^2 and can thus be made as small as desired by reducing the thickness of the plate. These stresses will be rigorously independent of z only if $\Delta\Phi_0$ is a linear function of x and y, which is indeed the condition of (2.32).

As in plane stress, we can search for biharmonic functions (solutions of (2.39)) satisfying the boundary conditions of the problem. These solutions σ_x^0, σ_y^0, τ_{xy}^0 will have in general an asymptotic character: the thinner the plate the closer the exact solution. The only difference between plane strain and plane stress will be the use of E^* and ν^* instead of E and ν: when a problem is solved in plane stress, the corresponding solution in plane strain may be obtained by replacing E and ν by E^* and ν^*.

2.3.3 Remark

If the boundary conditions comprise only stresses, the internal stresses σ_x, σ_y, τ_{xy} are independent of the elastic properties of the solid, at least for a simply connected solid. (For a multiply connected solid (with holes), there exist implicit boundary conditions in terms of displacements.) The values of σ_x, σ_y, τ_{xy} will be the same in plane stress and plane strain, but the stress field will be different due to the presence of σ_z in plane deformation—in particular the contours of the von Mises criterion for plastic deformation will be different in plane stress and plane strain.

2.4 Axisymmetric Problems

We will take (Fig. 2.3) Oz as the axis of symmetry, and we will use the cylindrical coordinates (r, θ, z). The components of the displacement of a point M are $(u_r, 0, u_z)$ with u_r and u_z independent of θ. The strains are given by

$$\begin{aligned} \varepsilon_r &= \frac{\partial u_r}{\partial r}\,; & \varepsilon_{r\theta} &= 0\,; \\ \varepsilon_\theta &= \frac{u_r}{r}\,; & \varepsilon_{rz} &= \frac{1}{2}\left(\frac{\partial u_r}{\partial z} + \frac{\partial u_z}{\partial r}\right)\,; \\ \varepsilon_z &= \frac{\partial u_z}{\partial z}\,; & \varepsilon_{\theta z} &= 0\,. \end{aligned}$$

The volumetric dilation is

$$\theta = \frac{\partial u_r}{\partial r} + \frac{u_r}{r} + \frac{\partial u_z}{\partial z}\,,$$

and hence the stress tensor is given by

$$\boxed{\begin{aligned} \sigma_r &= \lambda\theta + 2\mu\frac{\partial u_r}{\partial r}\,; & \tau_{rz} &= \mu\left(\frac{\partial u_r}{\partial z} + \frac{\partial u_z}{\partial r}\right)\,; \\ \sigma_\theta &= \lambda\theta + 2\mu\frac{u_r}{r}\,; & \tau_{z\theta} &= 0\,; \\ \sigma_z &= \lambda\theta + 2\mu\frac{\partial u_z}{\partial z}\,; & \tau_{\theta r} &= 0\,. \end{aligned}} \tag{2.40}$$

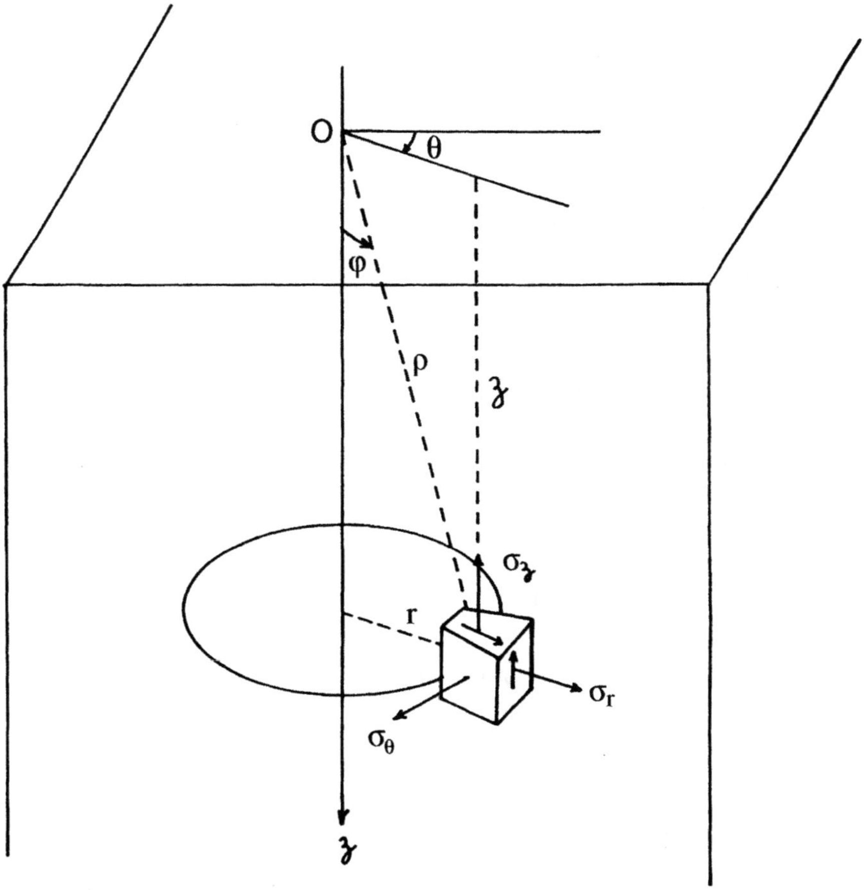

Fig. 2.3. Cylindrical coordinates

The equilibrium equations of a small volume element ($\mathrm{d}r$, $r\,\mathrm{d}\theta$, $\mathrm{d}z$) are:

$$\frac{\partial \sigma_r}{\partial r} + \frac{\partial \tau_{rz}}{\partial z} + \frac{1}{r}(\sigma_r - \sigma_\theta) = 0 \,, \qquad (2.41a)$$

$$\frac{\partial \tau_{rz}}{\partial r} + \frac{\partial \sigma_z}{\partial z} + \frac{\tau_{rz}}{r} = 0 \,. \qquad (2.41b)$$

The terms in $1/r$ result from the fact that the stress σ_θ turns by $\mathrm{d}\theta$ between the faces θ and $\theta + \mathrm{d}\theta$, and that the stresses $\tau_{rz} + (\partial \tau_{rz}/\partial r)\,\mathrm{d}r$ and $\sigma_r + (\partial \sigma_r/\partial r)\,\mathrm{d}r$ act on a face larger than the face where σ_r and σ_θ act. By symmetry, the stress σ_θ is a principal stress, known as the *hoop stress.* On the z-axis we will have $\varepsilon_r = \varepsilon_\theta$, $\sigma_r = \sigma_\theta$, $\tau_{rz} = 0$; this axis is a principal axis.

The equations of elasticity for which a solution is required are the two equilibrium equations from (2.41), and Beltrami's equations of compatibility, (2.22), which in axial symmetry are written:

$$\boxed{\begin{aligned}
&\Delta\sigma_r - \frac{2}{r^2}(\sigma_r - \sigma_\theta) + \frac{1}{1+\nu}\frac{\partial^2}{\partial r^2}(\sigma_r + \sigma_\theta + \sigma_z) = 0\,, \\
&\Delta\sigma_\theta + \frac{2}{r^2}(\sigma_r - \sigma_\theta) + \frac{1}{1+\nu}\frac{\partial}{r\partial r}(\sigma_r + \sigma_\theta + \sigma_z) = 0\,, \\
&\Delta\sigma_z + \frac{1}{1+\nu}\frac{\partial^2}{\partial z^2}(\sigma_r + \sigma_\theta + \sigma_z) = 0\,, \\
&\Delta\tau_{rz} - \frac{1}{r^2}\tau_{rz} + \frac{1}{1+\nu}\frac{\partial^2}{\partial r\partial z}(\sigma_r + \sigma_\theta + \sigma_z) = 0\,,
\end{aligned}} \tag{2.42}$$

and where the Laplacian operator reduces to

$$\Delta = \frac{\partial^2}{\partial r^2} + \frac{1}{r}\frac{\partial}{\partial r} + \frac{\partial^2}{\partial z^2}\,.$$

By substitution, it can be verified that with

$$\boxed{\begin{aligned}
\sigma_r &= \frac{\partial}{\partial z}\left(\nu\Delta\Phi - \frac{\partial^2\Phi}{\partial r^2}\right)\,, \\
\sigma_\theta &= \frac{\partial}{\partial z}\left(\nu\Delta\Phi - \frac{1}{r}\frac{\partial\Phi}{\partial r}\right)\,, \\
\sigma_z &= \frac{\partial}{\partial z}\left[(2-\nu)\Delta\Phi - \frac{\partial^2\Phi}{\partial z^2}\right]\,, \\
\tau_{rz} &= \frac{\partial}{\partial r}\left[(1-\nu)\Delta\Phi - \frac{\partial^2\Phi}{\partial z^2}\right]\,,
\end{aligned}} \tag{2.43}$$

where $\Phi(r,z)$ is a stress function (the *Love function*), the first equilibrium equation, (2.41a) is identically satisfied, and the second (2.41b) is also satisfied if Φ is a solution of the biharmonic equation

$$\boxed{\Delta\Delta\Phi(r,z) = 0}\,. \tag{2.44}$$

It can be shown that this condition also satisfies the compatibility equations (2.42), so that the problem is to find a biharmonic stress function Φ which is consistent with the imposed boundary conditions of stress or displacement (the semi-inverse method of Saint Venant).

The displacements are easily written as a function of Φ by noting that $u_r = r\varepsilon_\theta$ and that $u_z = \int \varepsilon_z\,\mathrm{d}z$, and by using Hooke's law. We get

$$\boxed{\begin{aligned}
2\mu u_r &= -\frac{\partial^2\Phi}{\partial r\partial z}\,, \\
2\mu u_z &= 2(1-\nu)\Delta\Phi - \frac{\partial^2\Phi}{\partial z^2}\,.
\end{aligned}} \tag{2.45}$$

Let us note that in axisymmetric problems the stress field always depends on Poisson's ratio.

2.5 Resolution of $\Delta\Delta\Phi = 0$ in Plane Elasticity

2.5.1 Analytic Functions

A function $\varphi(z)$ of the complex variable $z = x + \mathrm{i}y$ is said to be "analytic" if the expression

$$\lim_{\Delta z \to 0} \frac{\varphi(z + \Delta z) - \varphi(z)}{\Delta z}$$

is independent of how Δz tends towards zero; in this case the above limit is the derivative.[1]

Let us consider the analytic function

$$\varphi(z) = p(x, y) + \mathrm{i}q(x, y) \, . \tag{2.46}$$

Its partial derivatives

$$\begin{aligned} \frac{\partial}{\partial x}\varphi(z) &= \frac{\mathrm{d}}{\mathrm{d}z}\varphi(z)\frac{\partial z}{\partial x} = \varphi'(z) \, , \\ \frac{\partial}{\partial y}\varphi(z) &= \frac{\mathrm{d}}{\mathrm{d}z}\varphi(z)\frac{\partial z}{\partial y} = \mathrm{i}\varphi'(z) \end{aligned} \tag{2.47a}$$

are written

$$\begin{aligned} \frac{\partial}{\partial x}\varphi(z) &= \frac{\partial p}{\partial x} + \mathrm{i}\frac{\partial q}{\partial x} \, , \\ \frac{\partial}{\partial y}\varphi(z) &= \frac{\partial p}{\partial y} + \mathrm{i}\frac{\partial q}{\partial y} \, . \end{aligned} \tag{2.47b}$$

By comparison of the two forms in (2.47a) and (2.47b), we get

$$\frac{\partial p}{\partial x} = \frac{\partial q}{\partial y} \, , \tag{2.48}$$

$$\frac{\partial p}{\partial y} = -\frac{\partial q}{\partial x} \, . \tag{2.49}$$

These are the Cauchy–Riemann conditions, which are necessary and sufficient conditions for a function $\varphi(z) = p + \mathrm{i}q$ to be analytic. We deduce that $\Delta p = \Delta q = 0$, i.e. that the real part $p(x, y)$ and the imaginary part $q(x, y)$ of an analytic function are harmonic. The functions p and q are said to be *conjugate harmonics*.

[1] Note that $\overline{z} = x - \mathrm{i}y$ is not an analytic function of z, for its derivative depends on the direction of the displacement. It is defined by

$$\lim \frac{\Delta x - \mathrm{i}\Delta y}{\Delta x + \mathrm{i}\Delta y} \, , \qquad \Delta x \to 0 \, , \qquad \Delta y \to 0 \, .$$

For $\Delta y = 0$ this limit is 1; for $\Delta x = 0$ its value is -1.

Let us show that any biharmonic function $\Phi(x, y)$ can be written in the form

$$\boxed{\Phi(x, y) = p_1(x, y) + xp(x, y) + yq(x, y)}\,, \tag{2.50}$$

where p_1 is harmonic and p and q are conjugate harmonics (and hence $p + \mathrm{i}q$ is analytic). Let

$$P = \Delta\Phi\,.$$

By definition P is harmonic, and a conjugate harmonic function Q can be defined so that the function

$$F(z) = P + \mathrm{i}Q$$

is analytic. Let us call $4\varphi(z)$ its integral. Then

$$\varphi(z) = p + \mathrm{i}q = \frac{1}{4}\int F(z)\,\mathrm{d}z\,.$$

If we differentiate with respect to x, we obtain

$$\begin{aligned}\frac{\partial p}{\partial x} + \mathrm{i}\frac{\partial q}{\partial x} &= \frac{\partial}{\partial x}\varphi(z) = \varphi'(z)\frac{\partial z}{\partial x} = \varphi'(z)\\ &= \frac{1}{4}F(z) = \frac{1}{4}(P + \mathrm{i}Q)\,.\end{aligned}$$

Identifying the real parts and using (2.48), we get

$$\Delta\Phi = 4\frac{\partial p}{\partial x} = 4\frac{\partial q}{\partial y}\,. \tag{2.51}$$

Let us consider now the function

$$p_1 = \Phi - xp - yq$$

and show that it is harmonic. In effect we have

$$\begin{aligned}\Delta(xp) &= 2\frac{\partial p}{\partial x} + x\left(\frac{\partial^2 p}{\partial x^2} + \frac{\partial^2 p}{\partial y^2}\right) = 2\frac{\partial p}{\partial x}\,,\\ \Delta(yq) &= 2\frac{\partial q}{\partial y} + y\left(\frac{\partial^2 q}{\partial x^2} + \frac{\partial^2 q}{\partial y^2}\right) = 2\frac{\partial q}{\partial y}\,.\end{aligned}$$

Using (2.51), we can verify that $\Delta p_1 = 0$. Thus any biharmonic function $\Phi(x, y)$ can be written as (2.50). Reciprocally, we can verify easily that xp and yq are biharmonic.

2.5.2 The Method of Kolossov and Muskhelishvili

Let us return to the analytic function $\varphi(z) = p + \mathrm{i}q$, and introduce a second analytic function

$$\chi(z) = p_1 + \mathrm{i}q_1 ,$$

whose real part is p_1. We can see that the real part of

$$(x - \mathrm{i}y)\varphi(z) + \chi(z)$$

is identical to that in (2.50). We can thus express any biharmonic function in the form (due to Goursat in 1898)

$$\boxed{\Phi(x,y) = \mathrm{Re}[\bar{z}\varphi(z) + \chi(z)]} \; . \tag{2.52}$$

Then, we can deduce by differentiation the stresses and the displacements.

First, let us remark that, as any function $f(z)$ can be written $f(z) = \alpha + \mathrm{i}\beta$ with α and β real, we have, with $\overline{f(z)} = \alpha - \mathrm{i}\beta$,

$$f(z) + \overline{f(z)} = 2\alpha = 2\,\mathrm{Re}\, f(z) \, .$$

We can thus write

$$2\Phi(x,y) = \bar{z}\varphi(z) + \chi(z) + z\overline{\varphi(z)} + \overline{\chi(z)} \, .$$

By differentiation we get

$$2\frac{\partial \Phi}{\partial x} = \bar{z}\varphi'(z) + \varphi(z) + \chi'(z) + z\overline{\varphi'(z)} + \overline{\varphi(z)} + \overline{\chi'(z)} \, ,$$
$$2\frac{\partial \Phi}{\partial y} = \mathrm{i}\big[\bar{z}\varphi'(z) - \varphi(z) + \chi'(z) - z\overline{\varphi'(z)} + \overline{\varphi(z)} - \overline{\chi'(z)}\big] \, .$$

(To get $\overline{\varphi'(z)}$ we must first compute $\varphi'(z) = \mathrm{d}\varphi/\,\mathrm{d}z$ and then take its complex conjugate.)

We can deduce that

$$\frac{\partial \Phi}{\partial x} + \mathrm{i}\frac{\partial \Phi}{\partial y} = \varphi(z) + z\overline{\varphi'(z)} + \overline{\chi'(z)} \, . \tag{2.53}$$

Taking the partial derivatives with respect to x and y we get

$$\frac{\partial^2 \Phi}{\partial x^2} + \mathrm{i}\frac{\partial^2 \Phi}{\partial x \partial y} = \varphi'(z) + z\overline{\varphi\prime\prime(z)} + \overline{\varphi'(z)} + \overline{\chi''(z)} \, ,$$
$$\mathrm{i}\frac{\partial^2 \Phi}{\partial x \partial y} - \frac{\partial^2 \Phi}{\partial y^2} = -\varphi'(z) + z\overline{\varphi''(z)} - \overline{\varphi'(z)} + \overline{\chi''(z)} \, .$$

Combining them and introducing the stresses, as per (2.26), we get

$$\begin{aligned} \sigma_x + \sigma_y &= 2\varphi'(z) + 2\overline{\varphi'(z)} \,, \\ \sigma_y - \sigma_x - 2\mathrm{i}\tau_{xy} &= 2\big[\overline{z\varphi''(z)} + \overline{\chi''(z)}\big] \,. \end{aligned} \tag{2.54}$$

Taking the conjugate, this last equation becomes

$$\sigma_y - \sigma_x + 2\mathrm{i}\tau_{xy} = 2\big[\overline{z}\varphi''(z) + \chi''(z)\big] \,. \tag{2.55}$$

A similar calculation gives u_x and u_y. Let us consider the case of plane stress. The first two (left-hand) equations in (2.31) can be written as

$$\begin{aligned} E\varepsilon_x &= E\frac{\partial u_x}{\partial x} = \sigma_x + \sigma_y - (1+\nu)\sigma_y \,, \\ E\varepsilon_y &= E\frac{\partial u_y}{\partial y} = \sigma_x + \sigma_y - (1+\nu)\sigma_x \,, \end{aligned}$$

or, introducing the stress function Φ

$$\begin{aligned} 2\mu\frac{\partial u_x}{\partial x} &= -\frac{\partial^2 \Phi}{\partial x^2} + \frac{1}{1+\nu}\Delta\Phi \,, \\ 2\mu\frac{\partial u_y}{\partial y} &= -\frac{\partial^2 \Phi}{\partial y^2} + \frac{1}{1+\nu}\Delta\Phi \,. \end{aligned}$$

Using (2.51), we then have

$$\begin{aligned} 2\mu\frac{\partial u_x}{\partial x} &= -\frac{\partial^2 \Phi}{\partial x^2} + \frac{4}{1+\nu}\frac{\partial p}{\partial x} \,, \\ 2\mu\frac{\partial u_y}{\partial y} &= -\frac{\partial^2 \Phi}{\partial y^2} + \frac{4}{1+\nu}\frac{\partial q}{\partial y} \,, \end{aligned}$$

whose integration gives

$$\begin{aligned} 2\mu u_x &= -\frac{\partial \Phi}{\partial x} + \frac{4}{1+\nu}p + f(y) \,, \\ 2\mu u_y &= -\frac{\partial \Phi}{\partial y} + \frac{4}{1+\nu}q + g(x) \,. \end{aligned}$$

The arbitrary functions $f(y)$ and $g(x)$ are governed by the relation

$$\tau_{xy} = 2\mu\varepsilon_{xy} = \mu\left(\frac{\partial u_y}{\partial x} + \frac{\partial u_x}{\partial y}\right) \,,$$

which can be written as

$$-2\frac{\partial^2 \Phi}{\partial x \partial y} = -\frac{\partial^2 \Phi}{\partial x \partial y} + \frac{4}{1+\nu}\frac{\partial q}{\partial x} + g'(x) - \frac{\partial^2 \Phi}{\partial x \partial y} + \frac{4}{1+\nu}\frac{\partial p}{\partial y} + f'(y) \,.$$

Using the Cauchy–Rieman equation (2.49) this last equation reduces to

$$f'(y) + g'(x) = 0 ,$$

i.e.

$$f(y) = \alpha y + \beta ,$$
$$g(x) = -\alpha x + \gamma ,$$

which represent a rigid body displacement and can be discarded. We have thus:

$$2\mu u_x = -\frac{\partial \Phi}{\partial x} + \frac{4}{1+\nu} p ,$$
$$2\mu u_y = -\frac{\partial \Phi}{\partial y} + \frac{4}{1+\nu} q ,$$

i.e.

$$2\mu(u_x + \mathrm{i} u_y) = -\left(\frac{\partial \Phi}{\partial x} + \mathrm{i}\frac{\partial \Phi}{\partial y}\right) + \frac{4}{1+\nu}(p + \mathrm{i}q) .$$

Then, using (2.46) and (2.53) we get

$$2\mu(u_x + \mathrm{i} u_y) = \kappa\varphi(z) - z\overline{\varphi'(z)} - \overline{\chi'(z)} , \tag{2.56}$$

with, for plane stress,

$$\boxed{\kappa = \frac{3-\nu}{1+\nu} \qquad \text{(plain stress)}} .$$

For plane deformation, we get the same equation, but with

$$\boxed{\kappa = 3 - 4\nu \qquad \text{(plain strain)}} ,$$

since we must replace ν by $\nu/(1-\nu)$.

The set of equations (2.54), (2.55) and (2.56) constitutes the famous *equations of Kolossov and Muskhelishvili:*

$$\sigma_x + \sigma_y = 2\left[\varphi'(z) + \overline{\varphi'(z)}\right] = 4\,\mathrm{Re}\,\varphi'(z) , \tag{2.54}$$
$$\sigma_y - \sigma_x + 2\mathrm{i}\tau_{xy} = 2\left[\bar{z}\varphi''(z) + \chi''(z)\right] , \tag{2.55}$$
$$2\mu(u_x + \mathrm{i} u_y) = \kappa\varphi(z) - z\overline{\varphi'(z)} - \overline{\chi'(z)} . \tag{2.56}$$

These equations, initially obtained by Kolossov in 1909, remained unknown to the West for a long time, since Muskhelishvili's book written in 1933, was translated into English [2] only in 1953. During the intervening period, several authors independently achieved the same result, for example Stevenson [3] in 1945.

Let us note that we have some degrees of freedom to define the complex potentials $\varphi(z)$ and $\chi(z)$. From (2.54) and (2.55), $\varphi_0(z)$ and $\varphi(z)$ will give the same stresses if

$$\operatorname{Re}\varphi_0'(z) = \operatorname{Re}\varphi'(z) ,$$
$$\varphi_0''(z) = \varphi''(z) ,$$

i.e. if

$$\varphi_0(z) = \varphi(z) + \mathrm{i}cz + \alpha ,$$

where c is a real constant and α an arbitrary complex number. Similarly, in accordance with (2.55), $\chi_0(z)$ and $\chi(z)$ will give the same stresses if

$$\chi_0'(z) = \chi'(z) + \beta ,$$

where β is an arbitrary complex constant. If, moreover, we require that the displacements are the same, we must have

$$\kappa\varphi_0(z) - z\overline{\varphi_0'(z)} - \overline{\chi_0'(z)} = \kappa\varphi(z) - z\overline{\varphi'(z)} - \overline{\chi'(z)} ,$$

i.e. $c = 0$ and $\overline{\beta} = \kappa\alpha$. Finally, $\varphi(z)$ is defined except for a complex constant α.

2.5.3 The Old Method of Westergaard

We have seen that any biharmonic function could be written as

$$\Phi = p_1 + xp + yq ,$$

where p_1, p and q are harmonic. In the case of symmetry about Ox we will have:

$$\Phi = p_1 + yq .$$

Westergaard [4] in 1939 chose the functions p_1, p and q as real and imaginary parts of an analytic function of $z = x + \mathrm{i}y$. (As aforementioned, these real or imaginary parts are harmonic.) Let Z be an analytic function and $Z' = \mathrm{d}Z/\mathrm{d}z$ be its derivative. Let $\tilde{Z}$ and $\tilde{\tilde{Z}}$ be the integrals of Z ($Z = \mathrm{d}\tilde{Z}/\mathrm{d}z$, $\tilde{Z} = \mathrm{d}\tilde{\tilde{Z}}/\mathrm{d}z$). The Cauchy–Rieman conditions, given in (2.48) and (2.49), allow us to write

$$\frac{\partial \operatorname{Re} Z}{\partial x} = \frac{\partial \operatorname{Im} Z}{\partial y} = \operatorname{Re} Z' ,$$
$$\frac{\partial \operatorname{Im} Z}{\partial x} = -\frac{\partial \operatorname{Re} Z}{\partial y} = \operatorname{Im} Z' ,$$

and similarly for $\tilde{Z}$ and $\tilde{\tilde{Z}}$.

In the case of symmetry about Ox, Westergaard takes

$$\Phi = \operatorname{Re} \tilde{\tilde{Z}} + y \operatorname{Im} \tilde{Z} ,$$

from which we can deduce

$$\frac{\partial \Phi}{\partial x} = \operatorname{Re} \tilde{Z} + y \operatorname{Im} Z ,$$
$$\frac{\partial \Phi}{\partial y} = y \operatorname{Re} Z ,$$

and

$$\boxed{\begin{aligned} \sigma_x &= \operatorname{Re} Z - y \operatorname{Im} Z' , \\ \sigma_y &= \operatorname{Re} Z + y \operatorname{Im} Z' , \\ \tau_{xy} &= -y \operatorname{Re} Z' \end{aligned}} \; . \qquad \begin{aligned} &(2.57) \\ &(2.58) \\ &(2.59) \end{aligned}$$

Then

$$\boxed{\begin{aligned} 2\mu u_x &= \frac{\kappa - 1}{2} \operatorname{Re} \tilde{Z} - y \operatorname{Im} Z , \\ 2\mu u_y &= \frac{\kappa + 1}{2} \operatorname{Im} \tilde{Z} - y \operatorname{Re} Z \end{aligned}} \; . \qquad \begin{aligned} &(2.60) \\ &(2.61) \end{aligned}$$

These equations have been widely used for some thirty years, and have been the basis for the development of fracture mechanics. However, as shown by Sih [5] in 1966, Eftis and Liebowitz [6] in 1972 and Sun and Farris [7] in 1989, they are in error: some terms are missing!

In the case of a symmetry about Ox, we must have $\tau_{xy} = 0$ for $y = 0$, i.e. using (2.55) of Kolossov and Muskhelishvili we can derive

$$\operatorname{Im} \left[\bar{z} \varphi''(z) + \chi''(z) \right] = 0 \qquad \text{for} \quad y = 0 .$$

This condition is fulfilled if and only if

$$z\varphi''(z) + \chi''(z) + A = 0 ,$$

where A is a real constant. By putting this value of $\chi''(z)$ in the equations of Kolossov and Muskhelishvili, we get

$$\begin{aligned} \sigma_x + \sigma_y &= 4 \operatorname{Re} \varphi'(z) , \\ \sigma_y - \sigma_x + 2\mathrm{i}\tau_{xy} &= -4\mathrm{i}y\varphi''(z) - 2A , \end{aligned}$$

and, with $\chi'(z) = \varphi(z) - z\varphi'(z) - Az$, we can write

$$2\mu(u_x + \mathrm{i}u_y) = (\kappa - 1)\varphi(z) + 2\mathrm{i} \operatorname{Im} \varphi(z) - 2\mathrm{i}y\overline{\varphi'(z)} + A\bar{z} .$$

Hence we have

$$\begin{aligned}
\sigma_x &= 2\,\mathrm{Re}\,\varphi'(z) - 2y\,\mathrm{Im}\,\varphi''(z) + A\,,\\
\sigma_y &= 2\,\mathrm{Re}\,\varphi'(z) + 2y\,\mathrm{Im}\,\varphi''(z) - A\,,\\
\tau_{xy} &= -2y\,\mathrm{Re}\,\varphi''(z)\,,\\
2\mu u_x &= (\kappa-1)\,\mathrm{Re}\,\varphi(z) - 2y\,\mathrm{Im}\,\varphi'(z) + Ax\,,\\
2\mu u_y &= (\kappa+1)\,\mathrm{Im}\,\varphi(z) - 2y\,\mathrm{Re}\,\varphi'(z) - Ay\,.
\end{aligned}$$

Now let $Z = 2\varphi'(z) - A$. Then we get

$$\sigma_x = \mathrm{Re}\,Z - y\,\mathrm{Im}\,Z' + 2A\,, \tag{2.62}$$

$$\sigma_y = \mathrm{Re}\,Z + y\,\mathrm{Im}\,Z'\,, \tag{2.63}$$

$$\tau_{xy} = -y\,\mathrm{Re}\,Z'\,, \tag{2.64}$$

$$2\mu u_x = \frac{\kappa-1}{2}\,\mathrm{Re}\,\tilde{Z} - y\,\mathrm{Im}\,Z + \frac{\kappa+1}{2}Ax\,, \tag{2.65}$$

$$2\mu u_y = \frac{\kappa+1}{2}\,\mathrm{Im}\,\tilde{Z} - y\,\mathrm{Re}\,Z + \frac{\kappa-3}{2}Ay\,. \tag{2.66}$$

A uniaxial constant stress $\sigma_x = 2A$ acting everywhere parallel to Ox is thus missing in Westergaard's equations, which modifies considerably the distribution of principal stresses in the solid. Furthermore, the displacements are wrong. This omission of the constant A by Westergaard has been the cause of some obscurities and confusions in fracture mechanics, particularly in the study of nonsingular terms in the power expansion of stresses.

2.6 Resolution of $\Delta\Delta\Phi = 0$ in Axisymmetry

Harding and Sneddon [8] showed in 1945 that by using the Hankel transforms, the biharmonic equation $\Delta\Delta\Phi(r,z) = 0$ of the two independent variables r and z can be reduced to an ordinary differential equation of only the variable z. The reader is also referred to Sneddon [9] and to Little [10].

The Hankel transform of order n of a function $f(r)$ is defined by

$$f_n^{\mathrm{H}}(\xi) = \mathcal{H}_n[f(r);\xi] = \int_0^\infty r f(r) J_n(\xi r)\,\mathrm{d}r\,, \tag{2.67}$$

and its inverse by

$$f(r) = \mathcal{H}_n[f_n^{\mathrm{H}}(\xi);r] = \int_0^\infty \xi f_n^{\mathrm{H}}(\xi) J_n(\xi r)\,\mathrm{d}\xi\,, \tag{2.68}$$

where J_n denotes the Bessel function of the first kind, of order n, a solution of differential equations of the type

$$\frac{\mathrm{d}^2 y}{\mathrm{d}x^2} + \frac{1}{x}\frac{\mathrm{d}y}{\mathrm{d}x} + \left(1 - \frac{n^2}{x^2}\right) y = 0 . \tag{2.69}$$

Let us show that if Φ is a solution of the biharmonic equation $\Delta\Delta\Phi(r,z) = 0$, its Hankel transform of zero-order $\Phi_0^{\mathrm{H}}(\xi, z)$ is a solution of the ordinary differential equation

$$\left(\frac{\mathrm{d}^2}{\mathrm{d}z^2} - \xi^2\right)^2 \Phi_0^{\mathrm{H}} = 0 , \tag{2.70}$$

where ξ is a parameter and whose general solution is

$$\Phi_0^{\mathrm{H}}(\xi, z) = (A + B\xi z)\mathrm{e}^{-\xi z} + (C + D\xi z)\mathrm{e}^{\xi z} . \tag{2.71}$$

Consider the Hankel transform of $\Delta\varphi$,

$$I = \int_0^\infty r \Delta\varphi J_0(\xi r)\, \mathrm{d}r ,$$

where

$$\begin{aligned} \Delta &= \frac{\partial^2}{\partial r} + \frac{1}{r}\frac{\partial}{\partial r} + \frac{\partial^2}{\partial z^2} \\ &= \frac{1}{r}\frac{\partial}{\partial r}\left(r\frac{\partial}{\partial r}\right) + \frac{\partial^2}{\partial z^2} . \end{aligned}$$

It becomes

$$I = \int_0^\infty \frac{\partial}{\partial r}\left(r\frac{\partial\varphi}{\partial r}\right) J_0(\xi r)\, \mathrm{d}r + \frac{\partial^2}{\partial z^2}\int_0^\infty r\varphi J_0(\xi r)\, \mathrm{d}r .$$

Integrating by parts yields

$$\begin{aligned} I &= -\xi \int_0^\infty r\frac{\partial\varphi}{\partial r} J_0'(\xi r)\, \mathrm{d}r + \frac{\partial^2}{\partial z^2}\int_0^\infty r\varphi J_0(\xi r)\, \mathrm{d}r \\ &= \xi \int_0^\infty \left[J_0'(\xi r) + \xi r J_0''(\xi r)\right] \varphi\, \mathrm{d}r + \frac{\partial^2}{\partial z^2}\int_0^\infty r\varphi J_0(\xi r)\, \mathrm{d}r . \end{aligned}$$

Taking into account the fact that J_0 is a solution of (2.69), i.e. that

$$J_0''(\xi r) + \frac{1}{\xi r} J_0'(\xi r) + J_0(\xi r) = 0 ,$$

we get finally

$$\begin{aligned}\int_0^\infty r\Delta\varphi J_0(\xi r)\,\mathrm{d}r &= \left(\frac{\mathrm{d}^2}{\mathrm{d}z^2}-\xi^2\right)\int_0^\infty r\varphi J_0(\xi r)\,\mathrm{d}r \\ &= \left(\frac{\mathrm{d}^2}{\mathrm{d}z^2}-\xi^2\right)\varphi_0^{\mathrm{H}}\,.\end{aligned} \tag{2.72}$$

Let us replace φ by $\Delta\Phi$. The above result applied twice shows that

$$\begin{aligned}\int_0^\infty r\Delta\Delta\Phi J_0(\xi r)\,\mathrm{d}r &= \left(\frac{\partial^2}{\partial z^2}-\xi^2\right)^2\int_0^\infty r\Phi(r,z)J_0(\xi r)\,\mathrm{d}r \\ &= \left(\frac{\partial^2}{\partial z^2}-\xi^2\right)^2\Phi_0^{\mathrm{H}}\,.\end{aligned}$$

Thus, if Φ is biharmonic, its Hankel transform Φ_0^{H} is indeed a solution of (2.70). We will thus write the stresses and displacements, (2.43) and (2.45), and their Hankel transforms, as functions of Φ_0^{H}. The arbitrary constants A, B, C, D, generally functions of ξ, will be determined by the boundary conditions.

Let us consider the displacement u_z in (2.45). Its Hankel transform is by definition

$$2\mu(u_z)_0^{\mathrm{H}} = 2(1-\nu)\int_0^\infty r\Delta\Phi J_0(\xi r)\,\mathrm{d}r - \frac{\partial^2}{\partial z^2}\int_0^\infty r\Phi J_0(\xi r)\,\mathrm{d}r\,,$$

i.e., using (2.72) for the Hankel transform of a Laplacian,

$$2\mu(u_z)_0^{\mathrm{H}} = (1-2\nu)\frac{\mathrm{d}^2\Phi_0^{\mathrm{H}}}{\mathrm{d}z^2} - 2(1-\nu)\xi^2\Phi_0^{\mathrm{H}}\,,$$

and, by inversion,

$$\boxed{2\mu u_z = \int_0^\infty \xi\left[(1-2\nu)\frac{\mathrm{d}^2\Phi_0^{\mathrm{H}}}{\mathrm{d}z^2} - 2(1-\nu)\xi^2\Phi_0^{\mathrm{H}}\right]J_0(\xi r)\,\mathrm{d}\xi}\,. \tag{2.73}$$

In the displacement u_r, a derivative with respect to r appears and we use the fact that the Hankel transform of the first order of the derivative of a function is expressed as a function of the Hankel transform of the zero-order of that function:

$$\mathcal{H}_1\left[\frac{\mathrm{d}f}{\mathrm{d}r};\xi\right] = -\xi\mathcal{H}_0[f;\xi]\,,$$

i.e.

$$\mathcal{H}_1[2\mu u_r;\xi] = \mathcal{H}_1\left[-\frac{\partial}{\partial r}\left(\frac{\partial\Phi}{\partial z}\right);\xi\right] = \xi\mathcal{H}_0\left[\frac{\partial\Phi}{\partial z};\xi\right]\,,$$

or else

$$2\mu(u_r)_1^{\mathrm{H}} = \xi \frac{\mathrm{d}\Phi_0^{\mathrm{H}}}{\mathrm{d}z} \,,$$

and, by inversion,

$$\boxed{2\mu u_r = \int_0^\infty \xi^2 \frac{\mathrm{d}\Phi_0^{\mathrm{H}}}{\mathrm{d}z} J_1(\xi r)\,\mathrm{d}\xi} \,. \tag{2.74}$$

For the stresses σ_z and τ_{rz}, the calculation is similar and yields

$$(\sigma_z)_0^{\mathrm{H}} = (1-\nu)\frac{\mathrm{d}^3\Phi_0^{\mathrm{H}}}{\mathrm{d}z^3} - (2-\nu)\xi^2 \frac{\mathrm{d}\Phi_0^{\mathrm{H}}}{\mathrm{d}z} \,,$$

$$(\tau_{rz})_1^{\mathrm{H}} = \nu\xi \frac{\mathrm{d}^2\Phi_0^{\mathrm{H}}}{\mathrm{d}z^2} + (1-\nu)\xi^3\Phi_0^{\mathrm{H}} \,,$$

and, by inversion,

$$\sigma_z = \int_0^\infty \xi \left[(1-\nu)\frac{\mathrm{d}^3\Phi_0^{\mathrm{H}}}{\mathrm{d}z^3} - (2-\nu)\xi^2 \frac{\mathrm{d}\Phi_0^{\mathrm{H}}}{\mathrm{d}z}\right] J_0(\xi r)\,\mathrm{d}\xi \,, \tag{2.75}$$

$$\tau_{rz} = \int_0^\infty \xi^2 \left[\nu \frac{\mathrm{d}^2\Phi_0^{\mathrm{H}}}{\mathrm{d}z^2} + (1-\nu)\xi^2\Phi_0^{\mathrm{H}}\right] J_1(\xi r)\,\mathrm{d}\xi \,. \tag{2.76}$$

The stresses σ_r and σ_θ are difficult to transform separately, in contrast with their sum. In effect, we have

$$\begin{aligned}\sigma_r + \sigma_\theta &= \frac{\partial}{\partial z}\left[2\nu\Delta\Phi - \left(\frac{\partial^2\Phi}{\partial r^2} + \frac{1}{r}\frac{\partial\Phi}{\partial r}\right)\right] \\ &= \frac{\partial}{\partial z}\left[\frac{\partial^2\Phi}{\partial z^2} - (1-2\nu)\Delta\Phi\right] .\end{aligned}$$

Hence

$$(\sigma_r + \sigma_\theta)_0^{\mathrm{H}} = \frac{\partial^3}{\partial z^3}\int_0^\infty r\Phi J_0(\xi r)\,\mathrm{d}r - (1-2\nu)\frac{\partial}{\partial z}\int_0^\infty r\Delta\Phi J_0(\xi r)\,\mathrm{d}r$$

and

$$(\sigma_r + \sigma_\theta)_0^{\mathrm{H}} = 2\nu \frac{\mathrm{d}^3\Phi_0^{\mathrm{H}}}{\mathrm{d}z^3} + (1-2\nu)\xi^2 \frac{\mathrm{d}\Phi_0^{\mathrm{H}}}{\mathrm{d}z} \,,$$

and then, by inversion,

$$\boxed{\sigma_r + \sigma_\theta = \int_0^\infty \xi \left[2\nu \frac{\mathrm{d}^3\Phi_0^{\mathrm{H}}}{\mathrm{d}z^3} + (1-2\nu)\xi^2 \frac{\mathrm{d}\Phi_0^{\mathrm{H}}}{\mathrm{d}z}\right] J_0(\xi r)\,\mathrm{d}\xi} \,. \tag{2.77}$$

The equilibrium equation (2.41a) can be written as

$$\frac{\partial}{\partial r}(r^2\sigma_r) = r(\sigma_r + \sigma_\theta) - r^2\frac{\partial \tau_{rz}}{\partial z}\,.$$

Using (2.76) and (2.77) we get

$$\begin{aligned}\frac{\partial}{\partial r}(r^2\sigma_r) = r&\int_0^\infty \xi\left[2\nu\frac{\mathrm{d}^3\Phi_0^{\mathrm{H}}}{\mathrm{d}z^3} + (1-2\nu)\xi^2\frac{\mathrm{d}\Phi_0^{\mathrm{H}}}{\mathrm{d}z}\right] J_0(\xi r)\,\mathrm{d}\xi\\ &- r^2\int_0^\infty \xi^2\left[\nu\frac{\mathrm{d}^3\Phi_0^{\mathrm{H}}}{\mathrm{d}z^3} + (1-\nu)\xi^2\frac{\mathrm{d}\Phi_0^{\mathrm{H}}}{\mathrm{d}z}\right] J_1(\xi r)\,\mathrm{d}\xi\,.\end{aligned}$$

Let us integrate with respect to r, noting that only J_0 and J_1 depend upon r, and that (cf tables of integrals):

$$\int rJ_0(\xi r)\,\mathrm{d}r = \frac{r}{\xi}J_1(\xi r)\,,$$

$$\int r^2 J_1(\xi r)\,\mathrm{d}r = \frac{2r}{\xi^2}J_1(\xi r) - \frac{r^2}{\xi}J_0(\xi r)\,.$$

We get[2]

$$\boxed{\begin{aligned}\sigma_r = &\int_0^\infty \xi\left[\nu\frac{\mathrm{d}^3\Phi_0^{\mathrm{H}}}{\mathrm{d}z^3} + (1-\nu)\xi^2\frac{\mathrm{d}\Phi_0^{\mathrm{H}}}{\mathrm{d}z}\right] J_0(\xi r)\,\mathrm{d}\xi\\ &-\frac{1}{r}\int_0^\infty \xi^2\frac{\mathrm{d}\Phi_0^{\mathrm{H}}}{\mathrm{d}z}J_1(\xi r)\,\mathrm{d}\xi\end{aligned}}\,. \tag{2.78}$$

Then, putting this expression in (2.77), we obtain

$$\boxed{\begin{aligned}\sigma_\theta = \nu&\int_0^\infty \xi\left[\frac{\mathrm{d}^3\Phi_0^{\mathrm{H}}}{\mathrm{d}z^3} - \xi^2\frac{\mathrm{d}\Phi_0^{\mathrm{H}}}{\mathrm{d}z}\right] J_0(\xi r)\,\mathrm{d}\xi\\ &+\frac{1}{r}\int_0^\infty \xi^2\frac{\mathrm{d}\Phi_0^{\mathrm{H}}}{\mathrm{d}z}J_1(\xi r)\,\mathrm{d}\xi\end{aligned}}\,. \tag{2.79}$$

We will make use of these results in the study of elastic contacts (see chap. 4).

2.7 The Inglis Elliptical Cavity: Stress Concentration Factors

Inglis [11] computed in 1913 the stress distribution around an elliptical cavity (semi-axes $a > b$) traversing a plate subjected, at infinity, either to a uni-axial stress σ or to an isotropic stress σ. However, his calculation is rather complicated and we will follow a more modern approach.

[2] It is necessary that $r^2\sigma_r \to 0$ when $r \to 0$, which is always the case.

2.7.1 Elliptical Coordinates

Let us consider the general case of a plate submitted to two principal stresses: σ along Oy' and $k\sigma$ along Ox'. The ellipse, centred on O, has its major axis along Ox, which makes an angle β with Oy' (Fig. 2.4). Coordinates x and y refer to the axes of the ellipse.

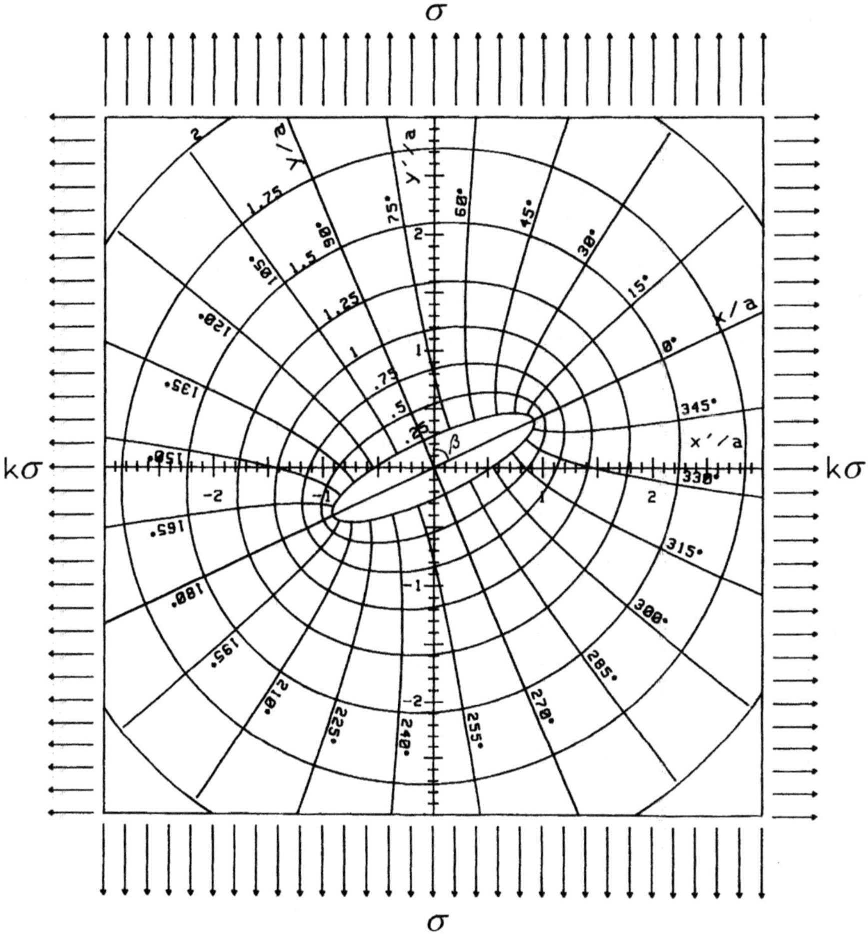

Fig. 2.4. Elliptical coordinates

We will make use of elliptical coordinates, defined by

$$x = c \cosh \xi \cos \eta \,, \tag{2.80}$$

$$y = c \sinh \xi \sin \eta \,, \tag{2.81}$$

with $c > 0$, $\xi > 0$, $0 \leq \eta \leq 2\pi$. Let us note that the Ox axis is described by $\xi = 0$ for $|x| < c$, and by $\sin \eta = 0$ for $|x| > c$. Eliminating ξ or η, we get

$$\frac{x^2}{c^2\cosh^2\xi}+\frac{y^2}{c^2\sinh^2\xi}=1\,, \tag{2.82}$$

$$\frac{x^2}{c^2\cos^2\eta}-\frac{y^2}{c^2\sin^2\eta}=1\,. \tag{2.83}$$

At constant ξ or η, these two equations respectively describe homofocal ellipses or hyperbolas. They are shown in Fig. 2.4 with their value of ξ or η. The elliptical cavity is defined by $\xi=\xi_0$, with $a=c\cosh\xi_0$ and $b=c\sinh\xi_0$, i.e.

$$c^2=a^2-b^2$$

(so that the direct study of the circular cylindrical cavity is prevented).

2.7.2 Complex Potentials

It can be shown (cf [1] or [2]) that in this system of coordinates the normal stresses σ_ξ (faces perpendicular to hyperbolas), σ_η (faces perpendicular to ellipses) and shear stresses $\tau_{\xi\eta}$ are related to the stresses σ_x, σ_y, τ_{xy} by

$$\sigma_\eta+\sigma_\xi=\sigma_y+\sigma_x\,, \tag{2.84}$$

$$\sigma_\eta-\sigma_\xi+2\mathrm{i}\tau_{\xi\eta}=\mathrm{e}^{2\mathrm{i}\alpha}(\sigma_y-\sigma_x+2\mathrm{i}\tau_{xy})\,, \tag{2.85}$$

with

$$\mathrm{e}^{2\mathrm{i}\alpha}=\frac{\sinh\zeta}{\sinh\overline{\zeta}}\,, \tag{2.86}$$

$$z=x+\mathrm{i}y=c\cosh(\xi+\mathrm{i}\eta)=c\cosh\zeta\,. \tag{2.87}$$

The equations of Kolossov–Muskhelishvili, (2.54) and (2.55), will then allow the obtaining of stresses σ_ξ, σ_η, $\tau_{\xi\eta}$. The complex potentials $\varphi(z)$ and $\chi(z)$ for an uniaxial loading ($k=0$) have been computed by Stevenson [3] with the boundary conditions

$$\sigma_{y'}=\sigma\,,$$

$$\sigma_{x'}=\tau_{x'y'}=0$$

at infinity, and

$$\sigma_\xi=\tau_{\xi\eta}=0$$

on the cavity itself ($\xi=\xi_0$). Let us note carefully that the cavity walls are assumed to be stress-free: when the minor axis b tends towards zero, the appearance of attractive molecular forces of attraction between the walls will be excluded. Stevenson has shown that these boundary conditions are satisfied[3] by taking:

[3] In fact, Stevenson wrote $\cos(2\beta)$ instead of $\exp(2\mathrm{i}\beta)$ in the term with $\cosh\zeta$ in (2.88), so that the condition of no rotation at infinity [1], $\operatorname{Im}\varphi'(z)\to 0$ when

$$4\varphi(z) = c\sigma\left[e^{2\xi_0}e^{2i\beta}\cosh\zeta + (1 - e^{2\xi_0}e^{2i\beta})\sinh\zeta\right], \tag{2.88}$$

$$4\chi'(z) = -\frac{c\sigma}{\sinh\zeta}\left[\cosh 2\xi_0 - \cos 2\beta + e^{2\xi_0}\sinh 2(\zeta - \xi_0 - i\beta)\right]. \tag{2.89}$$

By superposition, we can obtain the complex potentials for a biaxial loading. It suffices to add to each expression the same expression with σ replaced by $k\sigma$ and β replaced by $\beta + \pi/2$. Letting

$$m = 1 + k, \tag{2.90}$$

$$n = 1 - k, \tag{2.91}$$

we get

$$4\varphi(z) = c\sigma\left[ne^{2\xi_0}e^{2i\beta}\cosh\zeta + (m - ne^{2\xi_0}e^{2i\beta})\sinh\zeta\right], \tag{2.92}$$

$$\begin{aligned} 4\chi'(z) = -\frac{c\sigma}{\sinh\zeta}\Big[&m\cosh 2\xi_0 - n\cos 2\beta \\ &+ ne^{2\xi_0}\sinh 2(\zeta - \xi_0 - i\beta)\Big]. \end{aligned} \tag{2.93}$$

The computation[4] of stresses and displacements is rather tedious. Differentiating with respect to z, with $\mathrm{d}/\mathrm{d}z = (1/c\sinh\zeta)(\mathrm{d}/\mathrm{d}\zeta)$, and introducing

$$\frac{1}{\lambda} = 2\sinh\zeta\sinh\bar{\zeta} = \cosh 2\xi - \cos 2\eta, \tag{2.94}$$

we get

$$\varphi'(z) = \frac{\sigma}{4}\left[ne^{2\xi_0}e^{2i\beta} + (m - ne^{2\xi_0}e^{2i\beta})\frac{\cosh\zeta}{\sinh\zeta}\right],$$

$$\bar{z}\varphi''(z) = -2\sigma\lambda^3(m - ne^{2\xi_0}e^{2i\beta})\sinh^3\bar{\zeta}\cosh\bar{\zeta},$$

$$\begin{aligned} \chi''(z) = &-2\sigma\lambda^2\left[ne^{2\xi_0}\cosh 2(\zeta - \xi_0 - i\beta)\right]\sinh^2\bar{\zeta} \\ &+ 2\sigma\lambda^3\Big[m\cosh 2\xi_0 - n\cos 2\beta \\ &+ ne^{2\xi_0}\sinh 2(\zeta - \xi_0 - i\beta)\Big]\sinh^3\bar{\zeta}\cosh\zeta. \end{aligned}$$

$|z| \to \infty$, was not satisfied. The modification brought by [12], which corresponds to a rigid-body rotation, does not change the stress tensor (where only $\mathrm{Re}\,\varphi'(z)$ and $\varphi''(z)$ appear), but allow for the correct computation of displacements.

[4] Let us recall some classical results:

$$\sinh(\xi \pm i\eta) = \sinh\xi\cos\eta \pm i\cosh\xi\sin\eta;$$

$$\cosh(\xi \pm i\eta) = \cosh\xi\cos\eta \pm i\sinh\xi\sin\eta;$$

$$\tanh(\xi \pm i\eta) = \frac{\sinh 2\xi \pm i\sin 2\eta}{\cosh 2\xi + \cos 2\eta};$$

$$\coth(\xi \pm i\eta) = \frac{\sinh 2\xi \mp i\sin 2\eta}{\cosh 2\xi - \cos 2\eta}$$

Then, with

$$A = m - ne^{2\xi_0} \cos 2\beta \,, \tag{2.95}$$

$$B = ne^{2\xi_0} \cos 2\beta \,, \tag{2.96}$$

$$C = ne^{2\xi_0} \sin 2\beta \,, \tag{2.97}$$

we get

$$\frac{\sigma_\eta + \sigma_\xi}{\sigma} = B + \lambda(A \sinh 2\xi - C \sin 2\eta) \,, \tag{2.98}$$

$$\begin{aligned}\frac{\sigma_\eta - \sigma_\xi}{\sigma} &= -\lambda\big[\cosh 2(\xi - \xi_0) - 1\big](B \cos 2\eta + C \sin 2\eta) \\ &\quad + \lambda(\cosh 2\xi_0 - \cos 2\eta) \\ &\quad \times \big[B + \lambda(A \sinh 2\xi - C \sin 2\eta)\big] \,,\end{aligned} \tag{2.99}$$

$$\begin{aligned}\frac{2\tau_{\xi\eta}}{\sigma} &= -(C + \lambda B \sin 2\eta) \sinh 2(\xi - \xi_0) \\ &\quad + \lambda C \sinh 2\xi\big[\cosh 2(\xi - \xi_0) - 1\big] \\ &\quad + \lambda^2(\cosh 2\xi - \cosh 2\xi_0)(A \sin 2\eta + C \sinh 2\xi) \,,\end{aligned} \tag{2.100}$$

$$\begin{aligned}u_x = \frac{c\sigma}{8\mu}\big[&(\kappa + 1)(A \sinh\xi \cos\eta + B \cosh\xi \cos\eta + C e^{-\xi} \sin\eta) \\ &- 4 \sinh(\xi - \xi_0) \sinh\xi_0 (B \cos\eta + C \sin\eta) \\ &- 2\lambda(\cosh 2\xi - \cosh 2\xi_0)(A \sinh\xi \cos\eta - C \cosh\xi \sin\eta)\big]\end{aligned} \tag{2.101}$$

$$\begin{aligned}u_y = \frac{c\sigma}{8\mu}\big[&(\kappa + 1)(A \cosh\xi \sin\eta + B \sinh\xi \sin\eta + C e^{-\xi} \cos\eta) \\ &- 4 \sinh(\xi - \xi_0) \cosh\xi_0 (B \sin\eta - C \cos\eta) \\ &- 2\lambda(\cosh 2\xi - \cosh 2\xi_0)(A \cosh\xi \sin\eta + C \sinh\xi \cos\eta)\big] \,.\end{aligned} \tag{2.102}$$

We can verify that, on the cavity ($\xi = \xi_0$), we have $\sigma_\xi = \tau_{\xi\eta} = 0$. This stress tensor, as defined in (2.98–100), was given by Chang and Wu [13] in 1980.

2.7.3 Isotropic Loading

When $k = 1$, ($m = 2$, $n = 0$), (2.98–102) are considerably simplified:

$$\frac{\sigma_\xi}{\sigma} = \lambda^2 \sinh 2\xi(\cosh 2\xi - \cosh 2\xi_0) \,, \tag{2.103}$$

$$\frac{\sigma_\eta}{\sigma} = \lambda \sinh 2\xi \left[2 - \lambda(\cosh 2\xi - \cosh 2\xi_0)\right] , \tag{2.104}$$

$$\frac{\tau_{\xi\eta}}{\sigma} = \lambda^2 \sin 2\eta (\cosh 2\xi - \cosh 2\xi_0) , \tag{2.105}$$

$$u_x = \frac{c\sigma}{4\mu}\left[\kappa + 1 - 2\lambda(\cosh 2\xi - \cosh 2\xi_0)\right] \sinh \xi \cos \eta , \tag{2.106}$$

$$u_y = \frac{c\sigma}{4\mu}\left[\kappa + 1 - 2\lambda(\cosh 2\xi - \cosh 2\xi_0)\right] \cosh \xi \sin \eta . \tag{2.107}$$

The first three equations were given by Inglis [11]. On the cavity itself, we have $\sigma_\xi = \tau_{\xi\eta} = 0$ and

$$\frac{\sigma_\eta}{\sigma} = \frac{2 \sinh 2\xi_0}{\cosh 2\xi_0 - \cos 2\eta} , \tag{2.108}$$

which takes the value $2b/a$ on the minor axis and is at its maximum at the tip of the ellipse ($\eta = 0$):

$$\sigma_y = (\sigma_\eta)_{\eta=0} = 2\frac{a}{b}\sigma . \tag{2.109}$$

The factor:

$$\boxed{K_t = 2\frac{a}{b} = 2\sqrt{\frac{a}{\varrho}}} , \tag{2.110}$$

where $\varrho = b^2/a$ is the radius of curvature at the tip of the major axis of the elliptical cavity, is the *stress concentration factor* (a non-dimensional number), which depends only on the ellipticity of the cavity and not on its size. For very flat ellipses, the local stresses can reach the theoretical stress of the material needed to break the bonds.

On the Ox and Oy axes, the stresses can be written in Cartesian coordinates. On the Ox-axis ($\eta = 0$) where $\sigma_y \equiv \sigma_\eta$, $\sigma_x \equiv \sigma_\xi$ and $\tau_{xy} = 0$, we have

$$\left(\frac{\sigma_x}{\sigma}\right)_{y=0} = \frac{x(x^2 - a^2)}{(x^2 - a^2 + b^2)^{3/2}} , \tag{2.111}$$

$$\left(\frac{\sigma_y}{\sigma}\right)_{y=0} = \frac{x}{(x^2 - a^2 + b^2)^{1/2}}\left(1 + \frac{b^2}{x^2 - a^2 + b^2}\right) ; \tag{2.112}$$

and on the Oy axis ($\eta = \pi/2$), where $\sigma_y \equiv \sigma_\xi$ and $\sigma_x \equiv \sigma_\eta$, we have

$$\left(\frac{\sigma_x}{\sigma}\right)_{x=0} = \frac{y}{(y^2 + a^2 - b^2)^{1/2}}\left(1 + \frac{a^2}{y^2 + a^2 - b^2}\right) , \tag{2.113}$$

$$\left(\frac{\sigma_y}{\sigma}\right)_{x=0} = \frac{y(y^2 - b^2)}{(y^2 + a^2 - b^2)^{3/2}} . \tag{2.114}$$

(a)
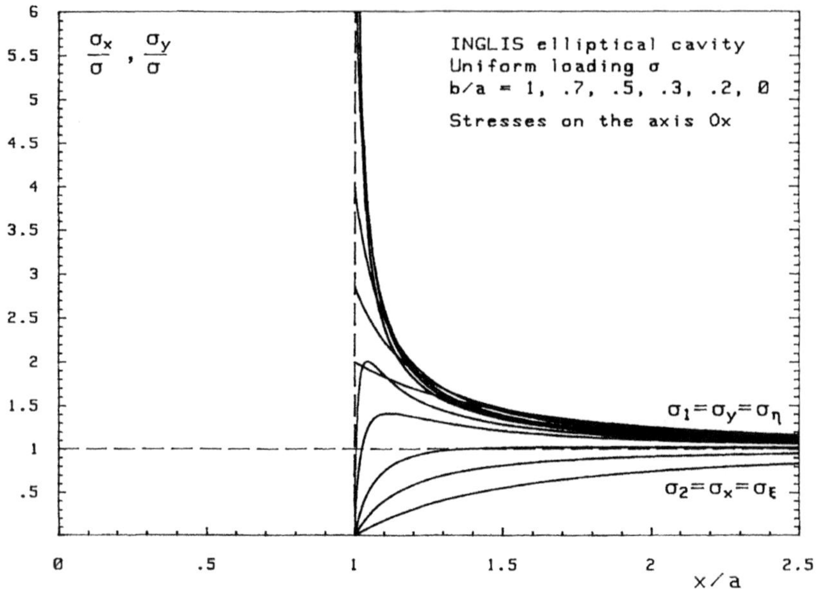

(b)
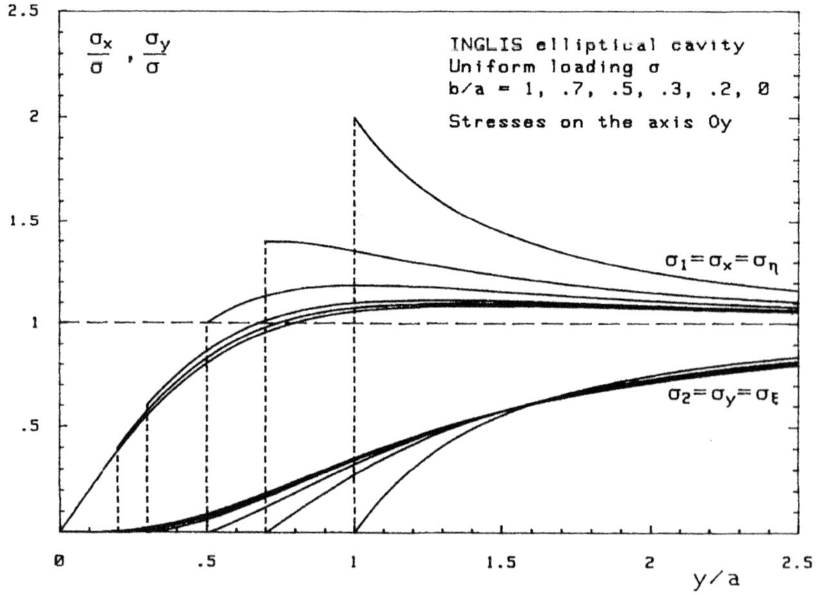

Fig. 2.5. Elliptical cavity under isotropic loading. Stresses on the Ox (**a**) and Oy (**b**) axes, for various values of b/a

These stresses are displayed in Fig. 2.5 for various values of b/a. On the Ox-axis the stress σ_y starts with the higher value the smaller is b/a, and then it decreases to approach the value σ imposed at large distances.

The displacements of the cavity points are simply

$$u_x = 2bS\cos\eta\,, \tag{2.115}$$
$$u_y = 2aS\sin\eta\,, \tag{2.116}$$

with

$$\begin{aligned} S &= \frac{\kappa+1}{8\mu}\sigma\,, \\ &= \frac{1-\nu^2}{E}\sigma \qquad \text{(plane strain)}\,, \\ &= \frac{\sigma}{E} \qquad \text{(plane stress)}\,. \end{aligned} \tag{2.117}$$

One sees easily that the semi-axes of the deformed ellipse are

$$a' = a + 2bS\,, \tag{2.118}$$
$$b' = b + 2aS\,. \tag{2.119}$$

In plane strain the smaller is Poisson's ratio, the more the ellipse is deformed.

Displayed in Fig. 2.6, and for $b/a = 1/2$, are the contours of the principal stresses, the isopachics, the isochromatic lines, the contours of von Mises criterion, as also the stress trajectories and the displacements field (with $\sigma/\mu = 0.5$ so as to have visible displacements).

2.7.4 Uniaxial Loading with $\beta = \pi/2$

For $k = 0$ ($m = n = 1$, $C = 0$) the modifications are small. On the cavity itself we have

$$\left(\frac{\sigma_\eta}{\sigma}\right)_{\xi=\xi_0} = \frac{\sinh 2\xi_0 - 1 + e^{2\xi_0}\cos 2\eta}{\cosh 2\xi_0 - \cos 2\eta}\,, \tag{2.120}$$

which is at its maximum at the tip of the major axis of the ellipse:

$$\sigma_y = (\sigma_\eta)_{\eta=0} = K_t\sigma\,, \tag{2.121a}$$

with

$$\boxed{K_t = 1 + 2\frac{a}{b}} \tag{2.121b}$$

(which is the result given by Inglis), and zero for

$$\cos\eta = \left(1 + \frac{b}{a}\right)^{-1}\,. \tag{2.122}$$

In these last points the two principal stresses are thus equal (they are *isotropic points*) and zero in value, and the slope of the stress trajectories is not defined.

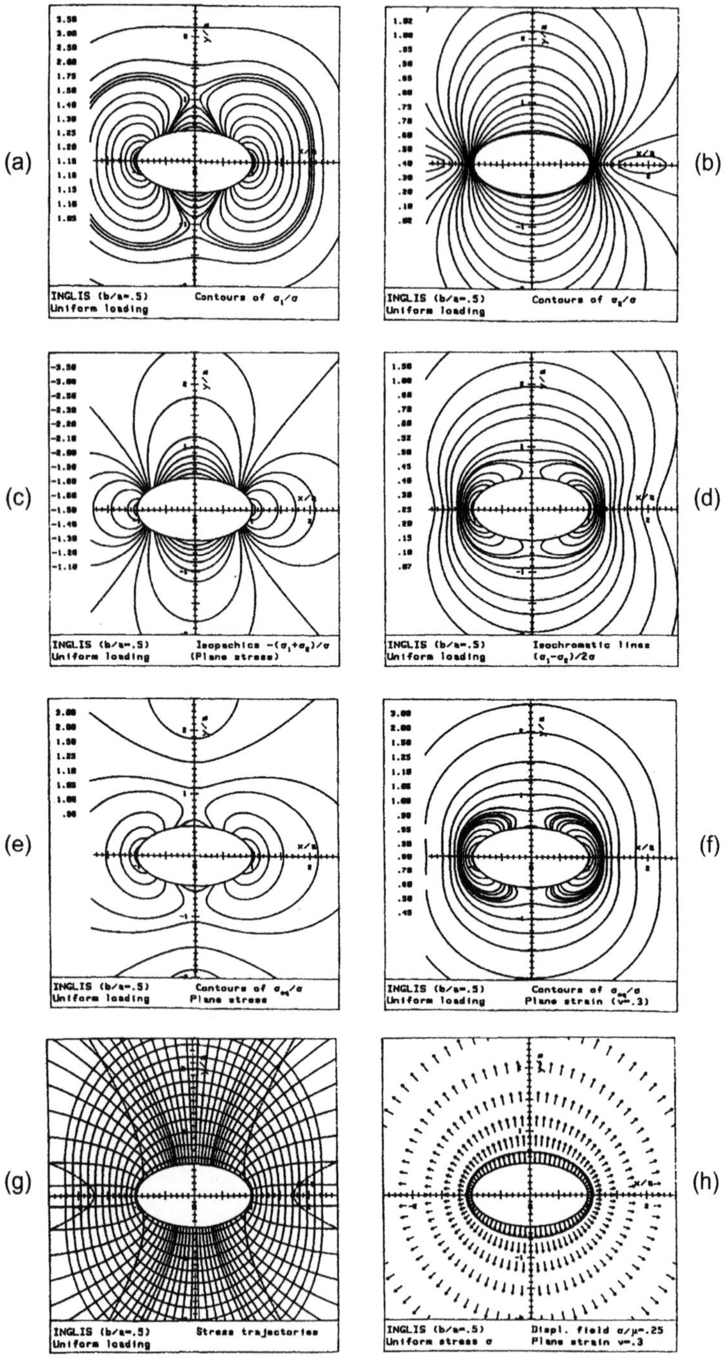

Fig. 2.6. Contours, for $b/a = 0.5$ and under isotropic loading, of: (**a**) and (**b**) the two principal stresses; (**c**) the isopachics; (**d**) the isochromatic lines; (**e**) von Mises criterion under plain stress; (**f**) von Mises criterion under plain strain; (**g**) stress trajectories; and (**h**) the displacement field

Here also the stresses on the axes can be easily written in Cartesian coordinates. On the Ox-axis we have

$$(a-b)^2\left(\frac{\sigma_x}{\sigma}\right)_{y=0} = -a^2 + \frac{a\,|x|}{(x^2-a^2+b^2)^{1/2}}\left[a - \frac{b^2(a-b)}{x^2-a^2+b^2}\right] , \tag{2.123}$$

$$(a-b)^2\left(\frac{\sigma_y}{\sigma}\right)_{y=0} = +b^2 + \frac{a\,|x|}{(x^2-a^2+b^2)^{1/2}}\left[a - 2b + \frac{b^2(a-b)}{x^2-a^2+b^2}\right] , \tag{2.124}$$

and on the Oy-axis we have

$$(a-b)^2\left(\frac{\sigma_x}{\sigma}\right)_{x=0} = -a^2 + \frac{a\,|y|}{(y^2+a^2-b^2)^{1/2}}\left[a + \frac{a^2(a-b)}{y^2+a^2-b^2}\right] , \tag{2.125}$$

$$(a-b)^2\left(\frac{\sigma_y}{\sigma}\right)_{x=0} = +b^2 + \frac{a\,|y|}{(y^2+a^2-b^2)^{1/2}}\left[a - 2b - \frac{a^2(a-b)}{y^2+a^2-b^2}\right] . \tag{2.126}$$

These stresses are displayed in Fig. 2.7 for various values of b/a. Let us note that on the minor axis we have $\sigma_x = -\sigma$ instead of $(2b/a)\sigma$ for an isotropic loading.

The displacements of the cavity surface are

$$u_x = -aS\cos\eta , \tag{2.127}$$

$$u_y = (2a+b)S\sin\eta , \tag{2.128}$$

and the new semi axes are

$$a' = a - aS , \tag{2.129}$$

$$b' = b + (b+2a)S . \tag{2.130}$$

The stress fields and the displacement field are shown in Fig. 2.8 for $b/a = 1/2$. The existence of compressive stresses (for $\sigma > 0$), independent of the Poisson's ratio on σ_1 and σ_2 must be noted. (The Poisson's ratio does not appear in σ_1 and σ_2; it only affects $\sigma_z = \nu(\sigma_1+\sigma_2)$ in plane strain.)

2.7.5 General Case

Let us study the case of an inclined elliptical cavity under biaxial loading. On the cavity itself we have $\sigma_\xi = \tau_{\xi\eta} = 0$ and

$$\left(\frac{\sigma_\eta}{\sigma}\right)_{\xi=\xi_0} = B + \frac{A\sinh 2\xi_0 - C\sin 2\eta}{\cosh 2\xi_0 - \cos 2\eta} . \tag{2.131}$$

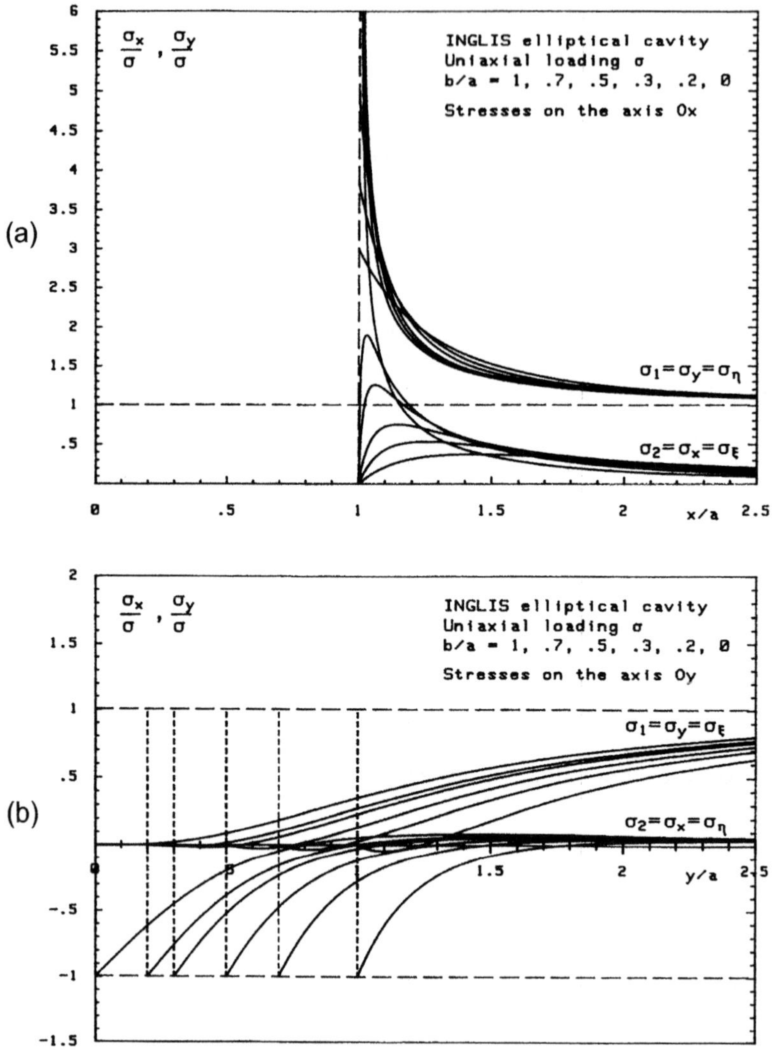

Fig. 2.7. Elliptical cavity under uniaxial loading. Stresses on the Ox (**a**) and Oy (**b**) axes, for various values of b/a

For $m = n = 1$ (uniaxial case), it is the result given by Inglis. The cavity points where $\sigma_\eta = 0$ are the isotropic points, since the two principal stresses are equal. Let us note that, on the cavity, σ_η passes through a maximum (the calculation needs $\xi_0 \neq 0$) at points defined by

$$\tan \eta = \frac{A \pm \sqrt{A^2 + C^2}}{C} \tanh \xi_0 \ , \tag{2.132}$$

and these points lie on the axes of the ellipse only for $\beta = 0$, $\beta = \pi/2$ or $n = 0$.

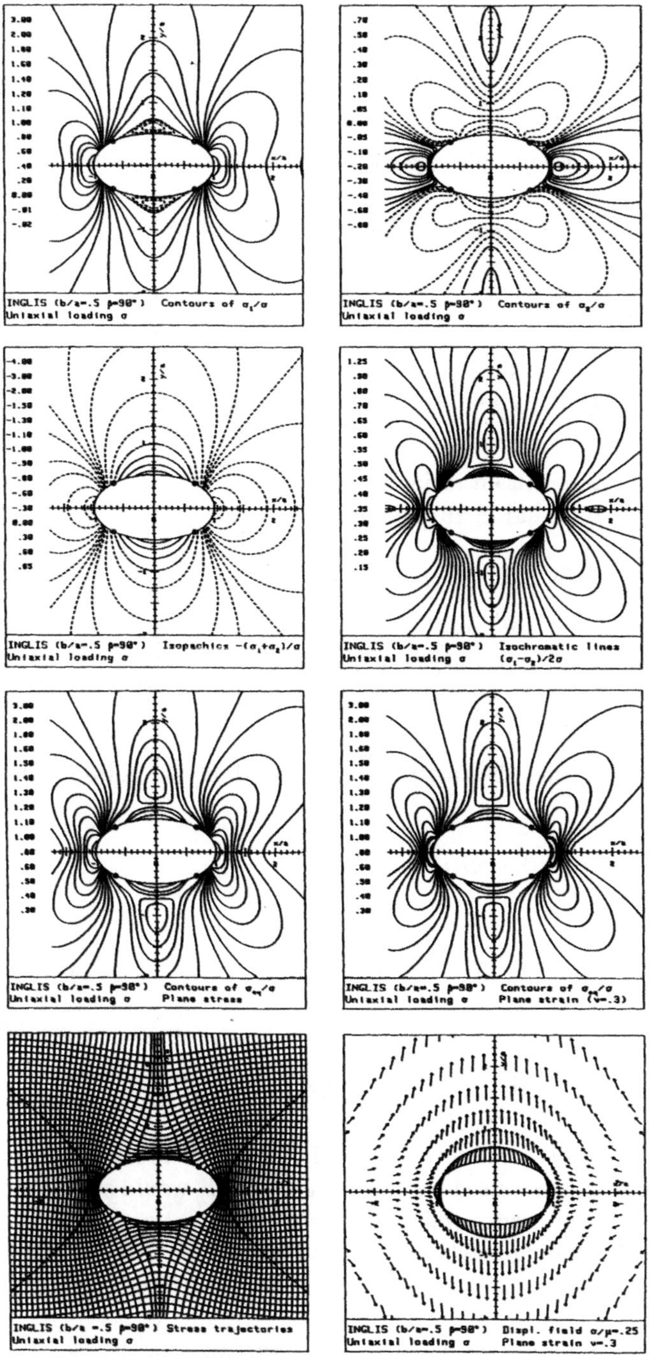

Fig. 2.8. Contours, for $b/a = 0.5$ and under uniaxial loading, of: (**a**) and (**b**) the two principal stresses; (**c**) isopachics; (**d**) the isochromatic lines; (**e**) von Mises criterion under plain stress; (**f**) von Mises criterion under plain strain; (**g**) stress trajectories; and (**h**) the displacement field

On the minor axis ($\eta = \pi/2$) we have

$$\left(\frac{\sigma_\eta}{\sigma}\right)_{\xi=\xi_0} = \frac{\sigma_x}{\sigma} = n\cos 2\beta + (m + n\cos 2\beta)\frac{b}{a} , \tag{2.133}$$

and on the major axis ($\eta = 0$) we have

$$\left(\frac{\sigma_\eta}{\sigma}\right)_{\xi=\xi_0} = \frac{\sigma_y}{\sigma} = -n\cos 2\beta + (m - n\cos 2\beta)\frac{a}{b} , \tag{2.134}$$

with, for $\beta = \pi/2$

$$\boxed{\sigma_y = K_t\sigma = \left(n + 2\frac{a}{b}\right)\sigma} , \tag{2.135}$$

where K_t is the stress concentration factor.

For $\xi = \xi_0$, (2.100) and (2.101) reduce to

$$\begin{aligned} u_x = Sc\big[&(A\sinh\xi_0 + B\cosh\xi_0)\cos\eta \\ &+ C(\cosh\xi_0 - \sinh\xi_0)\sin\eta\big] , \end{aligned} \tag{2.136}$$

$$\begin{aligned} u_y = Sc\big[&(A\cosh\xi_0 + B\sinh\xi_0)\sin\eta \\ &+ C(\cosh\xi_0 - \sinh\xi_0)\cos\eta\big] . \end{aligned} \tag{2.137}$$

It can be shown [12] that the cavity is transformed into an ellipse with semi-axes

$$a' = a(1 + Sn\cos 2\beta) + bS(m + n\cos 2\beta) , \tag{2.138}$$

$$b' = b(1 - Sn\cos 2\beta) + aS(m - n\cos 2\beta) , \tag{2.139}$$

rotated through an angle α defined by

$$\sin 2\alpha = 2Sne^{2\xi_0}\sin 2\beta . \tag{2.140}$$

The stress fields and the displacement field are displayed in Figs. 2.9 and 2.10 for $b/a = 1/2$ and for various values of β and of the factor of biaxiality k. For $\beta \neq \pi/2$, we have traced on the maps of σ_1 and σ_2 the stress trajectories starting from the more stressed points of the cavity. It would seem that they are the trajectories of cracks which will appear in traction or compression when the external loading reach a critical value [12].

2.8 Elliptical Cavity Subjected to Hydrostatic Pressure

Let us consider again an elliptical cavity (semi-axes $a > b$) traversing a plate subjected to an inner hydrostatic pressure p_0 (> 0), the stresses being zero at

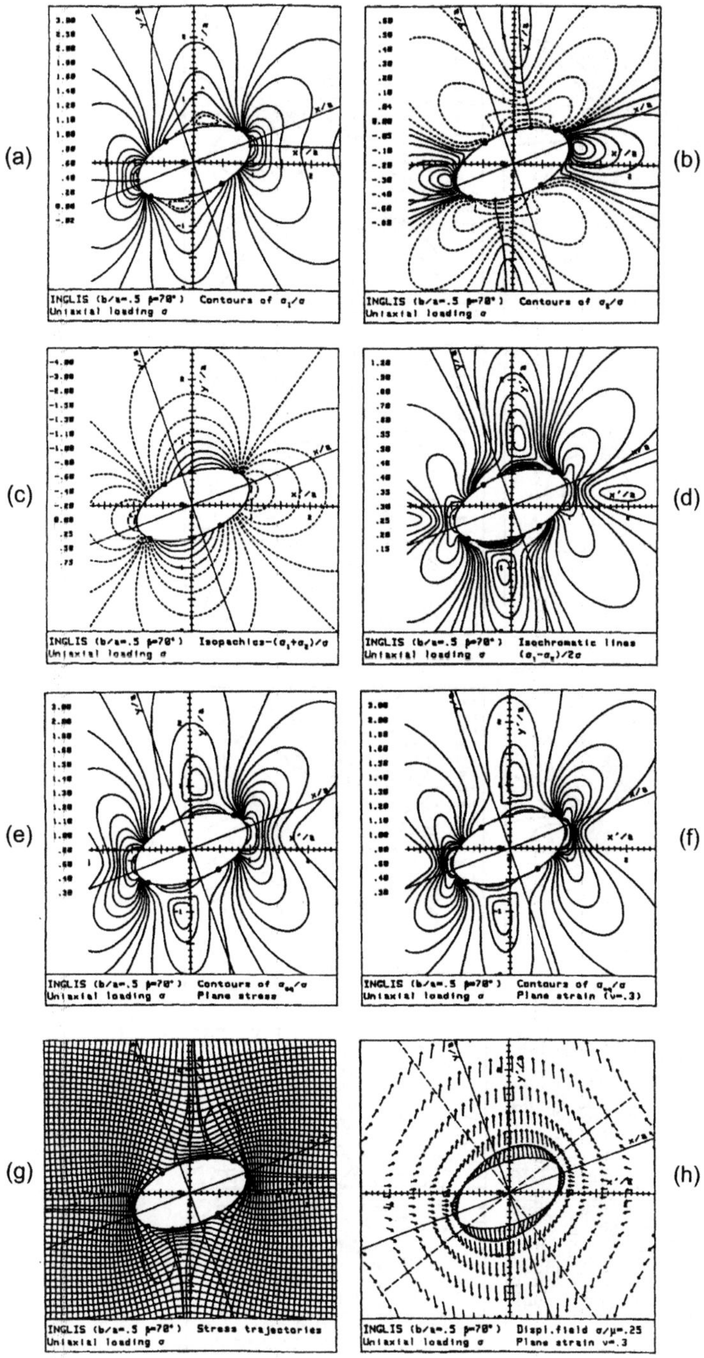

Fig. 2.9. Contours, for $b/a = 0.5$ and under inclined uniaxial loading, of: (**a**) and (**b**) the two principal stresses; (**c**) the isopachics; (**d**) the isochromatic lines; (**e**) von Mises criterion under plain stress; (**f**) von Mises criterion under plain strain; (**g**) stress trajectories; and (**h**) the displacement field

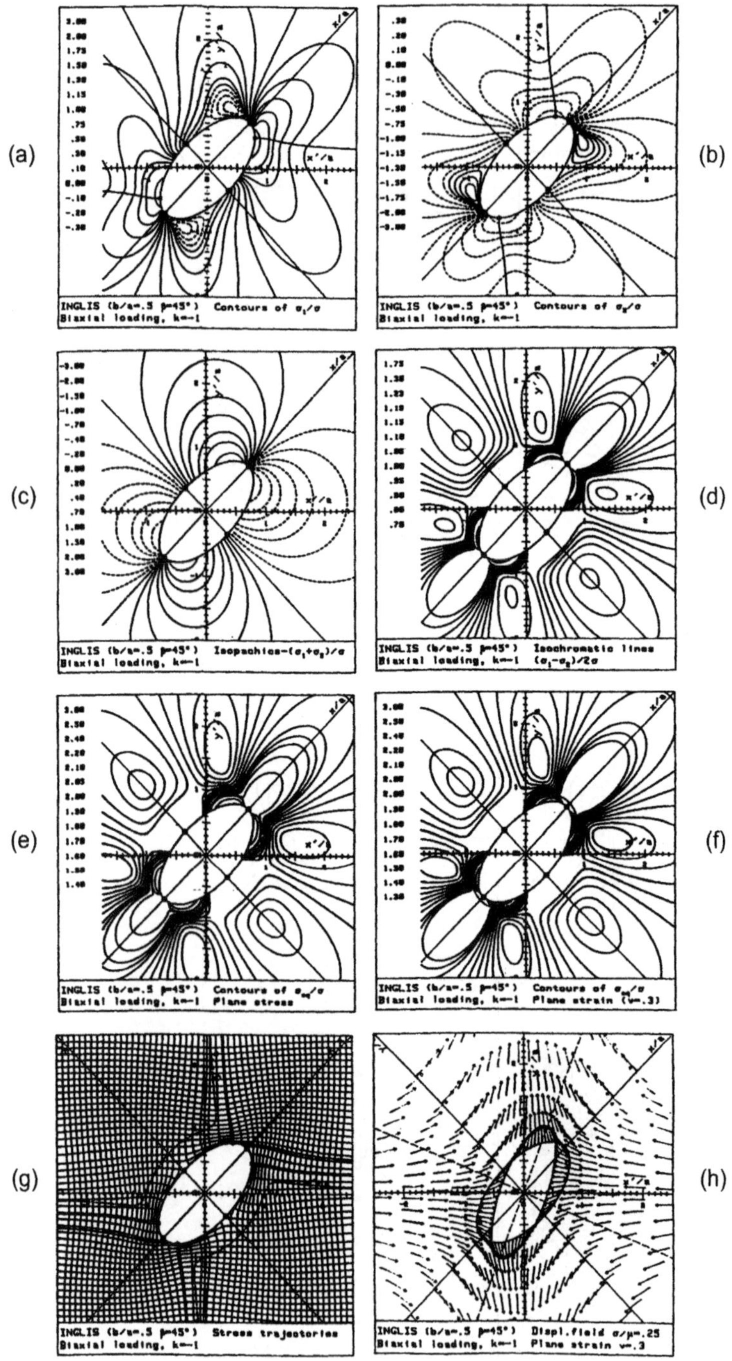

Fig. 2.10. Contours, for $b/a = 0.5$ and under biaxial loading, of: (**a**) and (**b**) the two principal stresses; (**c**) the isopachics; (**d**) the isochromatic lines; (**e**) von Mises criterion under plain stress; (**f**) von Mises criterion under plain strain; (**g**) stress trajectories; and (**h**) the displacement field

infinity. The complex potentials have been given by Muskhelishvili [2] using the conformal transformation:

$$z = R\left(\zeta_1 + \frac{m}{\zeta_1}\right), \qquad R \geq 0\,, \quad 0 \leq m \leq 1\,,$$

with $a = R(1+m)$ and $b = R(1-m)$,

$$\varphi(\zeta_1) = -\frac{p_0 R m}{\zeta_1}\,, \tag{2.141}$$

$$\psi(\zeta_1) = \chi'(\zeta_1) = -(1+m^2) p_0 R \frac{\zeta_1}{\zeta_1^2 - m}\,. \tag{2.142}$$

If we return to elliptical coordinates, with

$$R = \frac{c}{2}\mathrm{e}^{\xi_0}\,,$$
$$m = \mathrm{e}^{-2\xi_0}\,,$$
$$\zeta_1 = \mathrm{e}^{-\xi_0}\mathrm{e}^{\zeta}\,,$$

(2.141) and (2.142) are then written as

$$\varphi(z) = -\frac{1}{2}p_0 c \mathrm{e}^{-\zeta} = \frac{1}{2}p_0 c \sinh\zeta - \frac{1}{2}p_0 z\,, \tag{2.143}$$

$$\chi'(z) = -\frac{1}{2}p_0 c \cosh 2\xi_0 \frac{1}{\sinh\zeta}\,. \tag{2.144}$$

Using (2.85) to change the coordinates, the equations of Kolossov and Muskhelishvili, (2.54–56), allow the computation of the stresses and the displacements.

Instead, let us illustrate the principle of superposition. Consider first the state (1) of a plate traversed by an elliptical cavity and subjected at infinity to an isotropic loading $\sigma = p_0$. The stresses in the plate are given by (2.103–105). Consider now as state (2) the same plate subjected at infinity to an isotropic compression $\sigma = -p_0$ and where the cavity is subjected to an internal pressure p_0. The stresses in any point of the plate are given by

$$\sigma_\xi = -p_0\,,$$
$$\sigma_\eta = -p_0\,,$$
$$\tau_{\xi\eta} = 0\,.$$

Superpose now these two states. The stresses imposed at infinity cancel out, and an isotropic pressure p_0 prevails inside the cavity, which corresponds correctly to the problem studied. By addition, the stresses in the plate are thus:

$$\frac{\sigma_\xi}{p_0} = \lambda^2 \sinh 2\xi(\cosh 2\xi - \cosh 2\xi_0) - 1\,, \tag{2.145}$$

$$\frac{\sigma_\eta}{p_0} = \lambda \sinh 2\xi\big[2 - \lambda(\cosh 2\xi - \cosh 2\xi_0)\big] - 1\,, \tag{2.146}$$

$$\frac{\tau_{\xi\eta}}{p_0} = \lambda^2 \sin 2\eta(\cosh 2\xi - \cosh 2\xi_0)\,. \tag{2.147}$$

The displacements are computed in the same manner. To the displacements in state (1) given by (2.106) and (2.107), we must add the displacements of state (2). Hooke's laws are given by (2.23) in plane strain and by (2.28) in plane stress. With $\sigma_x = \sigma_y = -p_0$, $\tau_{xy} = 0$ and introducing κ (see Sect. 2.5.1), they reduce to

$$\varepsilon_x = -\frac{\kappa - 1}{4\mu}p_0\,,$$

$$\varepsilon_y = -\frac{\kappa - 1}{4\mu}p_0\,.$$

By integration, and by cancelling the constant terms to have displacements zero at the origin, we get:

$$u_x = -\frac{\kappa - 1}{4\mu}p_0 x\,,$$

$$u_y = -\frac{\kappa - 1}{4\mu}p_0 y\,.$$

Adding these displacements to those of state (1), we then obtain:

$$u_x = \frac{p_0 c}{4\mu}\Big\{\big[\kappa + 1 - 2\lambda(\cosh 2\xi - \cosh 2\xi_0)\big] \times \sinh\xi - (\kappa - 1)\cosh\xi\Big\}\cos\eta\,, \tag{2.148}$$

$$u_y = \frac{p_0 c}{4\mu}\Big\{\big[\kappa + 1 - 2\lambda(\cosh 2\xi - \cosh 2\xi_0)\big] \times \cosh\xi - (\kappa - 1)\sinh\xi\Big\}\sin\eta\,. \tag{2.149}$$

The displacements of the lips of the cavity ($\xi = \xi_0$) are given by:

$$u_x = \frac{p_0}{4\mu}\big[(\kappa + 1)b - (\kappa - 1)a\big]\cos\eta\,, \tag{2.150}$$

$$u_y = \frac{p_0}{4\mu}\big[(\kappa + 1)a - (\kappa - 1)b\big]\sin\eta\,. \tag{2.151}$$

One finds easily that the cavity remains elliptical, with semi-axes given by

$$a' = a + \frac{p_0}{4\mu}\left[(\kappa+1)b - (\kappa-1)a\right] , \tag{2.152}$$

$$b' = b + \frac{p_0}{4\mu}\left[(\kappa+1)a - (\kappa-1)b\right] . \tag{2.153}$$

On the Ox-axis, the stresses are just written as

$$\left(\frac{\sigma_x}{p_0}\right)_{y=0} = \frac{x(x^2-a^2)}{(x^2-a^2+b^2)^{3/2}} - 1 , \tag{2.154}$$

$$\left(\frac{\sigma_y}{p_0}\right)_{y=0} = \frac{x}{(x^2-a^2+b^2)^{1/2}}\left(1+\frac{b^2}{x^2-a^2+b^2}\right) - 1 . \tag{2.155}$$

2.9 Stresses Crossing an Interface

Let us finally in this chapter consider two solids (1) and (2) with different elastic properties (E_1, ν_1, E_2, ν_2), separated by an interface situated on the axis $y = 0$. The stress σ_y is continuous, and if any sliding at the interface is forbidden, the component ε_x of the strain must be continuous. Thus:

$$\varepsilon_x^{(1)} = \varepsilon_x^{(2)} .$$

But the stress σ_x generally displays a discontinuity. Assume, for simplicity, plane stress and apply Hooke's law (2.28). We get

$$\sigma_x^{(2)} = \frac{E_2}{E_1}\sigma_x^{(1)} + \left[\nu_2 - \nu_1\frac{E_2}{E_1}\right]\sigma_y . \tag{2.156}$$

The corresponding equation in plane strain is obtained by replacing E and ν by E^* and ν^*:

$$\sigma_x^{(2)} = \frac{E_2}{E_1}\frac{1-\nu_1^2}{1-\nu_2^2}\sigma_x^{(1)} + \left[\frac{\nu_2}{1-\nu_2} - \nu_1\frac{1+\nu_1}{1-\nu_2^2}\frac{E_2}{E_1}\right]\sigma_y . \tag{2.157}$$

A common formulation can be obtained by introducing the parameter κ and the shear modulus μ:

$$\sigma_x^{(2)} = \frac{1+\kappa_1}{1+\kappa_2}\frac{\mu_2}{\mu_1}\sigma_x^{(1)} + \left[\frac{3-\kappa_2}{1+\kappa_2} - \frac{3-\kappa_1}{1+\kappa_2}\frac{\mu_2}{\mu_1}\right]\sigma_y . \tag{2.158}$$

Dundurs [14] has shown that it was instructive to introduce the following interaction parameters (termed the *Dundurs parameters*):

$$\boxed{\alpha = \frac{\frac{\kappa_1+1}{\mu_1} - \frac{\kappa_2+1}{\mu_2}}{\frac{\kappa_1+1}{\mu_1} + \frac{\kappa_2+1}{\mu_2}}}\ , \tag{2.159}$$

$$\boxed{\beta = \frac{\frac{\kappa_1-1}{\mu_1} - \frac{\kappa_2-1}{\mu_2}}{\frac{\kappa_1+1}{\mu_1} + \frac{\kappa_2+1}{\mu_2}}}\ . \tag{2.160}$$

These parameters are zero if the two solids have the same elastic properties. The parameter β arises in contact mechanics and for interfacial fractures; in plane strain it vanishes if one of the solids is rigid and the second is incompressible. In the general case, and with $0 \leq \nu \leq \frac{1}{2}$, we have $-1 \leq \alpha \leq 1$ and $-\frac{1}{2} \leq \beta \leq \frac{1}{2}$.

Introducing these mismatch parameters, the stress discontinuity at the interface $[\sigma_x] = \sigma_x^{(2)} - \sigma_x^{(1)}$ is written as

$$[\sigma_x] = \frac{2\alpha}{1-\alpha}\sigma_x^{(1)} - \frac{2(\alpha - 2\beta)}{1-\alpha}\sigma_y\ . \tag{2.161}$$

If we consider a composite solid stretched parallel to the interface (thus along the x-axis), there will be no discontinuity if we have $\sigma_y = 0$ and $\alpha = 0$. If this solid is stretched perpendicularly to the interface and if $\sigma_x^{(1)} = 0$, we will have also $\sigma_x^{(2)} = 0$ if $\alpha - 2\beta = 0$.

References

1. S. Timoshenko, J.N. Goodier: *Théorie de l'élasticité* (Béranger, Paris 1961)
2. N.I. Muskhelishvili: *Some basic problems of the mathematical theory of elasticity* (Noordhoff, Leyden 1975)
3. A.C. Stevenson: Proc. Roy. Soc. A **184**, 129 (1945)
4. H.W. Westergaard: Trans. ASME **61**, A49 (1936)
5. G.C. Sih: Int. J. Fracture **2**, 628 (1966)
6. J. Eftis, H. Liebowitz: Int. J. Fracture **8**, 383 (1972)
7. C.T. Sun, T.N. Farris: Int. J. Fracture **40**, 73 (1989)
8. J.W. Harding, I.N. Sneddon: Proc. Cambridge Phil. Soc. **41**, 16 (1945)
9. I.N. Sneddon: *The use of integral transforms* (Mc Graw-Hill, New-York 1972)
10. R.W. Little: *Elasticity* (Prentice-Hall, 1973)
11. C.E. Inglis: Trans. Inst. Nav. Arch. **55**, 219 (1913)
12. D. Maugis: Eng. Fracture Mech. **43**, 217 (1992)
13. K.J. Chang, H.C. Wu: J. Appl. Mech. **47**, 57 (1980)
14. J. Dundurs: ASME J. Appl. Mech. **36**, 650 (1969)

3. Rupture and Adherence of Elastic Solids

3.1 Introduction

If two solids were to be separated as a whole, all the bonds being broken simultaneously, the force needed would be proportional to the contact area, and the stress (corresponding to the peak of intermolecular forces) would be enormous, reaching the theoretical stress of the material. This is never the case. Separation starts at a defect, an interfacial crack then propagates, breaking in its displacement all the bonds one after another, like a zip fastener does. If one keeps in mind this comparison, one can see that the displacement of the point of application of the force is greatly amplified, and that force is less since the work done must be the same.

It is important to understand where the energy to break the bonds is taken. Let us consider the more simple case of the peeling of an inextensible strip of width b under a load P applied at 90° ($\pi/2$ peeling). When the area of contact varies by $\mathrm{d}A = -b\,\mathrm{d}L$, the weight goes down by $\mathrm{d}L$ and the work $w\,\mathrm{d}A$ to break the bonds is provided by the potential energy of the weight. Consider now the case of a double cantilever of width b into which is inserted a wedge keeping the opening at a constant value δ so that the work of the external forces is zero. An elastic energy U_{E} is stored in the bent arms of the double cantilever. If the contact area could vary by $\mathrm{d}A$ (< 0), the arms of the double cantilever could extend, and the stored elastic energy could decrease ($(\partial U_{\mathrm{E}}/\partial A)_\delta\,\mathrm{d}A < 0$); but in order to do that, bonds must be broken on an area $|\,\mathrm{d}A|$ and a work $w\,|\,\mathrm{d}A|$ must be provided. The arms of the cantilever can extend, i.e. the crack can advance, only if $(\partial U_{\mathrm{E}}/\partial A)_\delta > w$. (Let us note that it was Griffith [1] who, for the first time, introduced in 1920 the concept of surface energy into a problem of the mechanics of solids.) This differential is the *strain energy release rate*, a quantity introduced and denoted G (initial letter of Griffith) by Irwin [2,3] in 1952. In this example (at an imposed displacement δ) the work is only taken to the elastic energy stored in the system, but in general G is the mechanical (elastic and potential) energy release rate, i.e. the mechanical energy which can be recovered in a virtual elemental decrease of the contact area, and which will be used to break the bonds if $G > w$. If $G < w$, there is not enough mechanical energy available to break the bonds, i.e. the decrease $G\,\mathrm{d}A$ of the mechanical energy of the system is not sufficient to compensate for the increase $w\,\mathrm{d}A$ in surface energy.

In this case the crack recedes, and the energy recovered in forming new bonds is used to increase the mechanical energy of the system until an equilibrium is reached—defined, indeed, not by the equality of mechanical energy and surface energy but instead by the point where the variations of mechanical energy compensate exactly for the variations of surface energy, i.e. where $G = w$. Such an equilibrium can be stable, unstable or neutral, and this point is, as we will see later, of great importance in the definition of an adherence force.

Let us note, moreover, that the work needed to break the bonds can also be provided by the chemical energy of the system. Molecules, being adsorbed at the very tip of the crack where bonds are just opening, can decrease the interaction forces between surfaces, and thus the work required for complete separation.

3.2 The Thermodynamic Aspect

All what was said in the Introduction to this chapter above can be couched more rigorously in term of thermodynamics. Let us consider a system made of two elastic solids in contact over an area A. A force P (compression or traction) can be applied on the two solids, either with a weight (Fig. 3.1a) or with a spring of stiffness k_{m} (Fig. 3.1b). The area of contact can vary at fixed load P, or at fixed displacement if k_{m} is infinite, or more generally at constant displacement Δ, so that the state of the system depends generally upon two independent parameters (P, A) or (δ, A), or even (Δ, A). The edge of the contact can be seen as an interfacial crack that recedes or advances

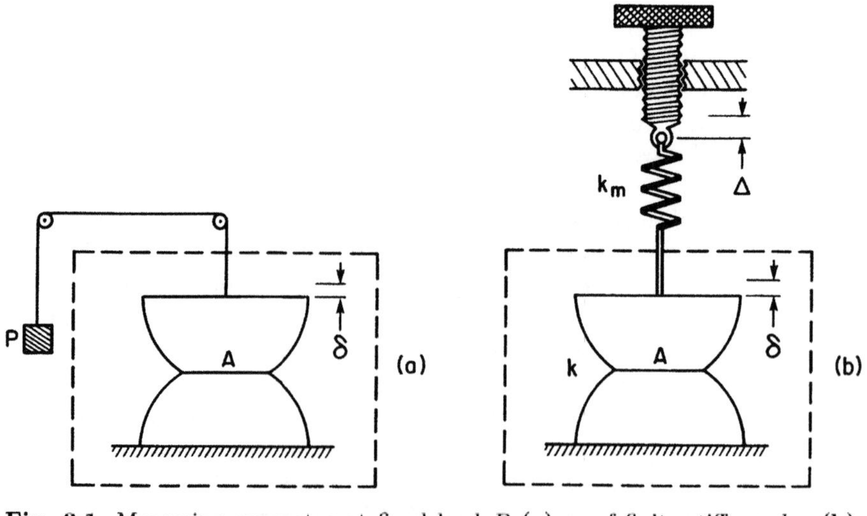

Fig. 3.1. Measuring apparatus at fixed load P (**a**) or of finite stiffness k_{m} (**b**). When the stiffness is infinite, the measuring apparatus is said to be at "fixed grips"

accordingly as the area of contact increases or decreases. Our aim is to study the equilibrium and the stability of the system under different internal constraints. We will follow the general method of the thermodynamics, as exposed for example in the excellent book by Callen [4]. This method is as set out below.

The thermodynamic state of a system is defined by a number of independent *extensive parameters*, such as entropy S, volume V, and other X_i (as the number of moles of each constituent). The equation

$$U = U(S, V, X_i, \dots)$$

defining the internal energy of the system, is called the *fundamental equation* and its differential

$$\mathrm{d}U = \left(\frac{\partial U}{\partial S}\right)_{V,X_i} \mathrm{d}S + \left(\frac{\partial U}{\partial V}\right)_{S,X_i} \mathrm{d}V + \left(\frac{\partial U}{\partial X_i}\right)_{S,V} \mathrm{d}X_i + \dots$$

can be written as

$$\mathrm{d}U = T\,\mathrm{d}S - p\,\mathrm{d}V + Y_i\,\mathrm{d}X_i + \dots ,$$

with the following *intensive parameters*:

$$T = \left(\frac{\partial U}{\partial S}\right)_{V,X_i} \qquad \text{(the temperature)} ,$$

$$-p = \left(\frac{\partial U}{\partial V}\right)_{S,X_i} \qquad \text{(the pressure)} ,$$

$$Y_i = \left(\frac{\partial U}{\partial X_i}\right)_{S,V} .$$

These intensive parameters are functions of the extensive parameters S, V, X_i, ... and knowledge of the *equations of state*

$$T = T(S, V, X_i, \dots) ,$$

$$p = p(S, V, X_i, \dots) ,$$

$$Y_i = Y_i(S, V, X_i, \dots)$$

is equivalent to knowledge of the fundamental equation.

An isolated system is in equilibrium if virtual perturbations of extensive parameters let its internal energy U remain constant. However, an equilibrium is often studied with various constraints, such as constant temperature, constant pressure, constant volume, etc. The elegant solution used in thermodynamics is to use Legendre transformations of the internal energy U which exchange any variable X_j with its partial derivative $Y_j \equiv (\partial U/\partial X_j)$. The equilibrium of the system corresponds thus to an extremum of the Legendre transform

$$\Psi = U - Y_j X_j .$$

These Legendre transforms are termed *thermodynamic potentials*. So, for the *Helmholtz free energy* F the temperature T replaces the entropy S as an independent variable:

$$\begin{aligned} F &= U - TS \\ \mathrm{d}F &= -S\ \mathrm{d}T - p\ \mathrm{d}V + Y_i\ \mathrm{d}X_i + \dots . \end{aligned}$$

In the *enthalpy* H, the pressure p replaces the volume V as an independent variable:

$$\begin{aligned} H &= U + pV \\ \mathrm{d}H &= T\ \mathrm{d}S + V\ \mathrm{d}p + Y_i\ \mathrm{d}X_i + \dots . \end{aligned}$$

In the *Gibbs free energy* $\mathcal{G}$ (or *free enthalpy*) the temperature T replaces entropy S, and pressure p replaces volume V:

$$\begin{aligned} \mathcal{G} &= U - TS + pV \\ \mathrm{d}\mathcal{G} &= -S\ \mathrm{d}T + V\ \mathrm{d}p + Y_i\ \mathrm{d}X_i . \end{aligned}$$

Thus, every partial derivative Y_i can be defined by

$$Y_i = \left(\frac{\partial U}{\partial X_i}\right)_{S,V,X_j} = \left(\frac{\partial F}{\partial X_i}\right)_{T,V,X_j} = \left(\frac{\partial H}{\partial X_i}\right)_{S,p,X_j} = \left(\frac{\partial \mathcal{G}}{\partial X_i}\right)_{T,p,X_j} . \tag{3.1}$$

Let us return to our particular system. Its energy $U = U(S, \delta, A)$ is a function of the extensive parameters S (entropy), δ and A. It can be split in elastic energy U_E and surface energy $U_\mathrm{S} = -wA$. (Note that we make no hypothesis of linearity between force and displacement, and that U_S is a potential energy recoverable when the surfaces come back into contact.) The first differential of the energy can be written as

$$\begin{aligned} \mathrm{d}U &= \left(\frac{\partial U}{\partial S}\right)_{\delta,A} \mathrm{d}S + \left(\frac{\partial U}{\partial \delta}\right)_{S,A} \mathrm{d}\delta + \left(\frac{\partial U}{\partial A}\right)_{S,\delta} \mathrm{d}A \\ &= T\ \mathrm{d}S + P\ \mathrm{d}\delta + (G - w)\ \mathrm{d}A \end{aligned} \tag{3.2}$$

with

$$\begin{aligned} &\left(\frac{\partial U}{\partial S}\right)_{\delta,A} = T , \\ &\left(\frac{\partial U}{\partial \delta}\right)_{S,A} = \left(\frac{\partial U_\mathrm{E}}{\partial \delta}\right)_{S,A} = P , \\ &\left(\frac{\partial U}{\partial A}\right)_{S,\delta} = \left(\frac{\partial U_\mathrm{E}}{\partial A}\right)_{S,\delta} + \left(\frac{\partial U_\mathrm{S}}{\partial A}\right)_{S,\delta} = G - w . \end{aligned}$$

The quantity G, which describes the variations of elastic energy with A at constant δ, is the (elastic) *energy release rate*. The three relationships

$$T = T(\delta, A, S) \tag{3.3a}$$
$$P = P(\delta, A, S) \tag{3.3b}$$
$$G = G(\delta, A, S) \tag{3.3c}$$

expressing the intensive parameters as functions of the independent extensive parameters are the three equations of state of the system, giving a complete description of it.

In the present case, where we want to study the equilibrium at constant temperature T by making perturbations of the area of contact, at fixed load P or at fixed displacement δ, we will use as thermodynamic potential the Helmholtz free energy

$$F = U - TS$$

or the Gibbs free energy:

$$\mathcal{G} = U - P\delta - TS$$

whose first differentials, using (3.2), are

$$\mathrm{d}F = P\,\mathrm{d}\delta + (G - w)\,\mathrm{d}A - S\,\mathrm{d}T\,, \tag{3.4}$$
$$\mathrm{d}\mathcal{G} = -\delta\,\mathrm{d}P + (G - w)\,\mathrm{d}A - S\,\mathrm{d}T\,. \tag{3.5}$$

The distinction between fixed load and fixed grips is similar to that between constant pressure and constant volume in the thermodynamics of gases.

Let us write the internal energy in the form $U = U_\mathrm{E} - wA$ and compare its differential to (3.2). Neglecting the thermal effects ($T\,\mathrm{d}S = 0$), we get

$$\boxed{\mathrm{d}U_\mathrm{E} = P\,\mathrm{d}\delta + G\,\mathrm{d}A}\,. \tag{3.4a}$$

Similarly, writing the Gibbs free energy as

$$\mathcal{G} = -U_\mathrm{C} - wA - TS\,,$$

where $U_\mathrm{C} \equiv P\delta - U_\mathrm{E}$ is the *complementary energy* , and comparing its differential to (3.5), we get, neglecting the thermal effects ($T\,\mathrm{d}S = 0$),

$$\mathrm{d}U_\mathrm{C} = \delta\,\mathrm{d}P - G\,\mathrm{d}A\,. \tag{3.5a}$$

From this we have

$$\delta = \left(\frac{\partial U_\mathrm{C}}{\partial P}\right)_A\,. \tag{3.6}$$

This last equation, which gives the displacement in the direction of the force P, is known as *Castigliano's generalized theorem* . It is only in the particular case of linear force-displacement relations, where $U_\mathrm{E} = 1/2P\delta$ and thus $U_\mathrm{C} = U_\mathrm{E}$, that one can use the classical *Castigliano's theorem* $\delta = (\partial U_\mathrm{E}/\partial P)_A$.

From (3.4a) and (3.5a) we can deduce

$$\boxed{G = \left(\frac{\partial U_\mathrm{E}}{\partial A}\right)_\delta = -\left(\frac{\partial U_\mathrm{C}}{\partial A}\right)_P = \left(\frac{\partial U_\mathrm{E}}{\partial A} + \frac{\partial U_\mathrm{P}}{\partial A}\right)_P}\,, \tag{3.7}$$

where $U_\mathrm{P} \equiv -P\delta$ represents the potential energy in the case where P is constant.

As noted by Burns *et al.* [5], the *Maxwell relations*, obtained by equating the crossed second derivatives of $\mathcal{G}$ or F, give

$$\left(\frac{\partial G}{\partial \delta}\right)_A = \left(\frac{\partial P}{\partial A}\right)_\delta\,, \tag{3.8}$$

$$\left(\frac{\partial G}{\partial P}\right)_A = -\left(\frac{\partial \delta}{\partial A}\right)_P\,, \tag{3.9}$$

with

$$\left(\frac{\partial P}{\partial A}\right)_\delta \left(\frac{\partial A}{\partial \delta}\right)_P \left(\frac{\partial \delta}{\partial P}\right)_A = -1\,. \tag{3.10}$$

The equilibrium at constant temperature and at fixed load ($\mathrm{d}T = 0$, $\mathrm{d}P = 0$) corresponds thus to an extremum of $\mathcal{G}$, whereas the equilibrium at constant temperature and at fixed grips ($\mathrm{d}T = 0$, $\mathrm{d}\delta = 0$) corresponds to an extremum of F. In both cases the equilibrium is given by

$$\boxed{G = w}\,, \tag{3.11}$$

which we name *Griffith's criterion of equilibrium.*

If $G \neq w$, the area of contact changes spontaneously to decrease the thermodynamic potential. If $G < w$, (3.4) and (3.5) show that A must increase for the potential to decrease: the crack recedes. If $G > w$, the contact area must decrease to have $\mathrm{d}\mathcal{G} < 0$ or $\mathrm{d}F < 0$: the crack advances. $G\,\mathrm{d}A$ is the mechanical energy recovered when the crack spreads by $\mathrm{d}A$. The rupture of bonds needing work $w\,\mathrm{d}A$, the excess $(G - w)\,\mathrm{d}A$ is changed into kinetic energy if there is no dissipation. $(G - w)$ is the *crack extension force* per unit length of crack. It is zero at equilibrium.

3.2.1 Stability of the Equilibrium and Adherence Force

The equilibrium given by $G = w$ can be stable, unstable or neutral. A thermodynamic system under a given constraint is stable if the corresponding potential is at a minimum, i.e. if its second derivative is positive. From (3.4) we can see that the stability at constant temperature ($\mathrm{d}T = 0$) and at fixed grips ($\mathrm{d}\delta = 0$) is given by

$$\left(\frac{\partial^2 F}{\partial A^2}\right)_{\delta,T} = \left(\frac{\partial G}{\partial A}\right)_{\delta} > 0 \, .$$

Similarly, at fixed load the stability is defined by

$$\left(\frac{\partial^2 \mathcal{G}}{\partial A^2}\right)_{P,T} = \left(\frac{\partial G}{\partial A}\right)_{P} > 0 \, .$$

It must be emphasized that the stability of equilibrium is generally not the same at fixed load and at fixed grips, and that at fixed Δ it depends on the stiffness k_m of the measuring apparatus. Figure 3.1b is a schematic display of such an apparatus. Its stiffness k_m is that of the spring. The cross-head displacement Δ can be imposed by turning the micrometric screw (which is thermodynamically equivalent to placing a load P onto the spring and then clamping it without external work) and can be divided into elastic displacement δ_m of the spring and elastic displacement δ of the two solids in contact. The spring exerts a force $P = k_\mathrm{m}\delta_\mathrm{m}$ on the two bodies, and from that

$$\Delta = \delta + \delta_\mathrm{m} = \delta + \frac{P}{k_\mathrm{m}} = \delta + C_\mathrm{m} P \, , \tag{3.12}$$

where C_m is the compliance of the mechanism. Let us study the stability of such a system. Its energy now includes the elastic energy U_m of the spring, given by

$$U = U_\mathrm{E}(A,\delta) + U_\mathrm{m}(\delta_\mathrm{m}) + U_S(A)$$

and its differential is

$$\begin{aligned}
\mathrm{d}U &= \left(\frac{\partial U_\mathrm{E}}{\partial A}\right)_{\delta} \mathrm{d}A + \left(\frac{\partial U_\mathrm{E}}{\partial \delta}\right)_{A} \mathrm{d}\delta + \frac{\mathrm{d}U_\mathrm{m}}{\mathrm{d}\delta_\mathrm{m}}\,\mathrm{d}\delta_\mathrm{m} + \frac{\mathrm{d}U_S}{\mathrm{d}A}\,\mathrm{d}A \\
&= G\,\mathrm{d}A + P\,\mathrm{d}\delta + P\,\mathrm{d}\delta_\mathrm{m} - w\,\mathrm{d}A \\
&= P\,\mathrm{d}\Delta + (G - w)\,\mathrm{d}A \, .
\end{aligned}$$

The equilibrium at fixed Δ is still given by $G = w$, and its stability is given by

$$\left(\frac{\partial G}{\partial A}\right)_{\Delta} > 0 \, , \tag{3.13}$$

but this stability depends on the stiffness k_m. Intituively, one can imagine the spring as a reservoir which can provide energy to the crack for a propagation at constant Δ as soon as the equilibrium is left. Let us compute $(\partial G/\partial A)_\Delta$

as a function of $(\partial G/\partial A)_\delta$ by writing $\delta = \delta(\Delta, A)$ and applying the rules of differentiation of composed functions. For the function $G(A, \delta(A, \Delta))$ we get

$$\left(\frac{\partial G}{\partial A}\right)_\Delta = \left(\frac{\partial G}{\partial A}\right)_\delta + \left(\frac{\partial G}{\partial \delta}\right)_A \left(\frac{\partial \delta}{\partial A}\right)_\Delta . \tag{3.14}$$

The relation

$$\left(\frac{\partial \delta}{\partial A}\right)_\Delta \left(\frac{\partial A}{\partial \Delta}\right)_\delta \left(\frac{\partial \Delta}{\partial \delta}\right)_A = -1$$

allows to write, with the help of (3.12)

$$\left(\frac{\partial \delta}{\partial A}\right)_\Delta = -\left(\frac{\partial P}{\partial A}\right)_\delta \frac{1}{k_\mathrm{m} + (\partial P/\partial \delta)_A} . \tag{3.15}$$

Using (3.15) and (3.8), (3.14) becomes

$$\left(\frac{\partial G}{\partial A}\right)_\Delta = \left(\frac{\partial G}{\partial A}\right)_\delta - \left(\frac{\partial P}{\partial A}\right)_\delta^2 \frac{1}{k_\mathrm{m} + (\partial P/\partial \delta)_A} , \tag{3.16}$$

or, equivalently,

$$\left(\frac{\partial G}{\partial A}\right)_\Delta = \left(\frac{\partial G}{\partial A}\right)_P + \left(\frac{\partial \delta}{\partial A}\right)_P^2 \frac{1}{C_\mathrm{m} + (\partial \delta/\partial P)_A} . \tag{3.17}$$

The quantity $(\partial P/\partial \delta)_A = k$ is the stiffness of the two elastic bodies in contact, and is positive. So, $(\partial G/\partial A)_\Delta$ and $(\partial G/\partial A)_\delta$ can be of opposite sign. Equation (3.16) shows that the stability range increases monotonically with the stiffness, from the fixed load case ($k_\mathrm{m} = 0$) up to the fixed grips case ($k_\mathrm{m} \to \infty$). Thus, the equilibrium can be unstable at fixed load and stable at fixed grips. When $k_\mathrm{m} \to 0$, we approach the fixed-load conditions and (3.16) becomes

$$\left(\frac{\partial G}{\partial A}\right)_P = \left(\frac{\partial G}{\partial A}\right)_\delta - \left(\frac{\partial P}{\partial A}\right)_\delta^2 \left(\frac{\partial \delta}{\partial P}\right)_A ,$$

a result which could be directly obtained by writing

$$\mathrm{d}G = \left(\frac{\partial G}{\partial A}\right)_P \mathrm{d}A + \left(\frac{\partial G}{\partial P}\right)_A \mathrm{d}P = \left(\frac{\partial G}{\partial A}\right)_\delta \mathrm{d}A + \left(\frac{\partial G}{\partial \delta}\right)_A \mathrm{d}\delta$$

and using (3.8–10).

Another important concept is that of *controlled rupture*. Let us consider a stable equilibrium, defined by $G = w$ and $(\partial G/\partial A) > 0$. If a fluctuation decreases the area of contact ($\mathrm{d}A < 0$), G decreases and we have $G < w$: the crack recedes to return to its equilibrium position. The same return to equilibrium occurs when the fluctuation increases the area of contact (the loading system being constant). The crack can move only if the load P or the displacement δ is modified, and it moves so as to return G to its value w. We are then dealing with a controlled rupture, but for the crack to be in equilibrium with the loading requires the variations of P or δ are very slow, and we are concerned with a quasistatic transformation. The fact that G remains constantly equal to w during this propagation is written as

$$\mathrm{d}G = \left(\frac{\partial G}{\partial A}\right)_{\Delta} \mathrm{d}A + \left(\frac{\partial G}{\partial \Delta}\right)_{A} \mathrm{d}\Delta = 0 \,, \tag{3.18}$$

with, as particular cases,

$$\mathrm{d}G = \left(\frac{\partial G}{\partial A}\right)_{\delta} \mathrm{d}A + \left(\frac{\partial G}{\partial \delta}\right)_{A} \mathrm{d}\delta = 0 \,, \qquad (k_\mathrm{m} \to \infty) \,,$$

$$\mathrm{d}G = \left(\frac{\partial G}{\partial A}\right)_{P} \mathrm{d}A + \left(\frac{\partial G}{\partial P}\right)_{A} \mathrm{d}P = 0 \,, \quad (k_\mathrm{m} = 0) \,.$$

Starting from a position of stable equilibrium with the two solids compressed ($P > 0, \delta > 0$), let us decrease progressively the cross-head displacement Δ. One generally encounters a progressive reduction of the area of contact A, i.e. a controlled rupture with $G = w$ and $(\partial G/\partial A)_{\Delta} > 0$ until a critical displacement Δ_c where $(\partial G/\partial A)_{\Delta} = 0$. Beyond that point, the equilibrium becomes unstable and the crack propagates spontaneously at Δ_c = const. until the separation is complete. It propagates with a crack extension force $G - w$, which increases as A decreases (since $(\partial G/\partial A)_{\Delta} < 0$). This instability point on the equilibrium curve $P(\delta)$ corresponds to the point where $\mathrm{d}\Delta/\,\mathrm{d}\delta = 0$, in other words, in accordance with (3.12) to

$$\boxed{k_\mathrm{m} + \frac{\mathrm{d}P}{\mathrm{d}\delta} = 0} \,. \tag{3.19}$$

On the equilibrium curve $P(\delta)$, stability ends with an horizontal tangent in experiments at fixed load P ($k_\mathrm{m} = 0$), and with a vertical tangent in experiments at fixed grips ($k_\mathrm{m} \to \infty$).

The critical load P_c corresponding to the limit of stability of the system given by

$$\boxed{\begin{array}{l} G = w \\ \left(\dfrac{\partial G}{\partial A}\right) \leq 0 \end{array}}$$

defines the *adherence force*[1] (or pull off force) of the two solids in contact, a force which generally depends on the stiffness k_m of the measuring apparatus. In some geometries and loading conditions, the equilibrium is always

[1] It must be noted that the *adherence force*, which depends on the geometry of the system and the stiffness of the measuring apparatus, should not be confused with *adhesion forces*, which represent the interaction forces between the surfaces.

unstable; as soon as it is reached, the crack propagates spontaneously (at constant loading conditions) and the criterion defining the adherence force is simply reduced to $G = w$. This case applies for the double cantilever at fixed load, or for a flat punch on an elastic half-space. It is also the case for the classical Griffith crack, as we will see below.

3.2.2 Berry's Representation

Berry [6] showed in 1960 that by plotting the locus of equilibrium points ($G = w$) and the loading path on a diagram (P, δ) it is possible to obtain a simple geometric interpretation of the various energies that come into play.

We have seen that the Griffith relation $G = w$ links two of three variables δ, P, A via (3.3), and allows the plotting of the equilibrium curves $P(\delta), \delta(A), A(P)$ which are functions of w. The diagram $P(\delta)$ is particularly interesting since it allows us to visualize easily the elastic energy of the system.

To compute the elastic energy stored at any point of the diagram $P(\delta)$, we will make use of (3.4a), choosing a *loading path* starting from the origin and having a loading branch $G = 0$ and a rectilinear unloading branch at $A =$ const., so that U_{E} is simply the area under this loading path. On the other hand, if we follow the *equilibrium curve* $G = w$, integration of (3.4a) gives

$$U_{\mathrm{E}} = \int_{eq} P \, \mathrm{d}\delta + \int_{eq} w \, \mathrm{d}A \, ,$$

and hence

$$\int_{eq} P \, \mathrm{d}\delta = U_{\mathrm{E}} + U_{\mathrm{S}} = U \, .$$

The internal energy of the system is thus the area under the equilibrium curve and its surface energy, $U_{\mathrm{S}} = -wA$, is, except for the sign, the area between the loading curve and the equilibrium curve.

Figure 3.2a is a representation of such a Berry's diagram in the simple case where the loading path is reduced to the branch at $A =$ const. (i.e. $U_{\mathrm{E}} = P\delta/2$). Starting from an equilibrium point M_0 (stable or unstable) corresponding to an area of contact A_0, the crack can evolve either following the equilibrium curve $G = w$ (quasistatic transformation), or at $P =$ const., or $\delta =$ const., or in a more complex manner (if the velocity of displacement $\dot{\delta}$ is imposed, for example). At a point M in the trajectory in the diagram for $P(\delta)$ (Fig. 3.2b), it is possible to evaluate U_{E} and U_{S} as previously, but

a)

b)

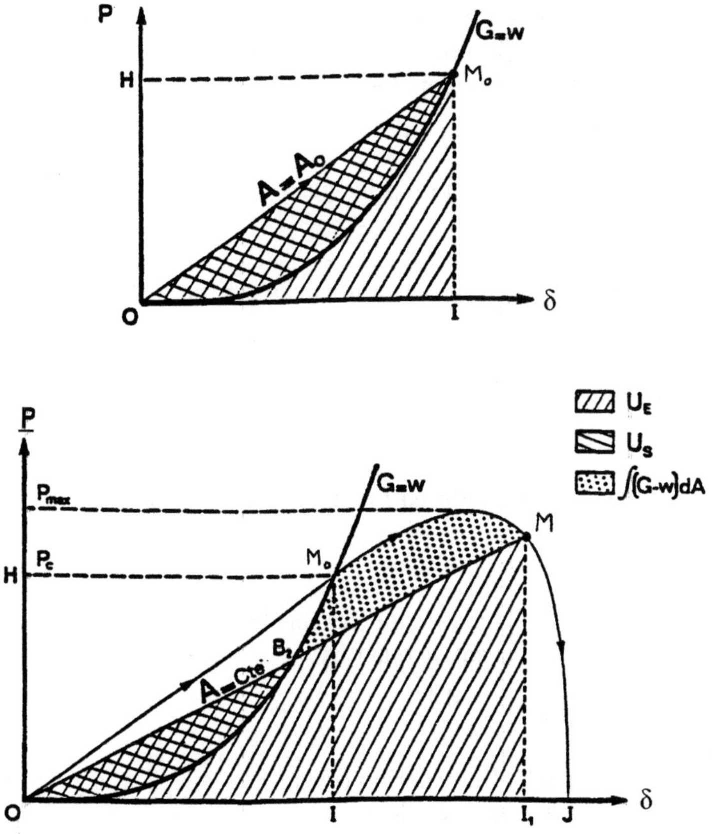

Fig. 3.2. Berry's diagrams

moreover the dotted area has a definite signification. The value of this area is

$$\Sigma = \int_{A_0}^{A} P \,\mathrm{d}\delta - \Delta U_{\mathrm{E}} + \int_{A_0}^{A} w \,\mathrm{d}A \,.$$

With

$$\Delta U_{\mathrm{E}} = \int_{A_0}^{A} P \,\mathrm{d}\delta + \int_{A_0}^{A} G \,\mathrm{d}A \,,$$

we get

$$|\Sigma| = \int_{A_0}^{A} (G - w) \,\mathrm{d}A \,.$$

The last expression represents the energy stored by the system in excess to the energy needed to break the bonds. This excess is transformed into kinetic energy U_K if the solid is purely elastic. If the solid is dissipative, and if one can neglect the kinetic energy, this area is representative of the entropy produced in the system. In effect, if one compares (3.2), namely

$$\mathrm{d}U = T\,\mathrm{d}S + P\,\mathrm{d}\delta + (G - w)\,\mathrm{d}A\,,$$

with that of the energy conservation principle,

$$\mathrm{d}U + \mathrm{d}U_K = P\,\mathrm{d}\delta + \mathrm{d}Q\,,$$

where $\mathrm{d}Q$ is the heat received by the system, we get (neglecting U_K)

$$T\,\mathrm{d}S = \mathrm{d}Q - (G - w)\,\mathrm{d}A\,.$$

Comparing this result with the classical equation

$$T\,\mathrm{d}S = \mathrm{d}Q + T\,\mathrm{d}_i S$$

it appears that the entropy produced by the system is given by

$$\mathrm{d}_i S = -\frac{1}{T}(G - w)\,\mathrm{d}A\,.$$

The propagation at $G = w$ is indeed a reversible transformation. The rate of entropy production (a non-negative quantity) is

$$P[S] = \frac{\mathrm{d}_i S}{\mathrm{d}t} = -\frac{1}{T}(G - w)\frac{\mathrm{d}A}{\mathrm{d}t}\,, \tag{3.20}$$

as shown by Rice [7] in 1978. One can thus consider $(G-w)/T$ as a generalized force and $-\mathrm{d}A/\mathrm{d}t$ as a generalized flow.

3.2.3 Multiple Loadings and Mutual Energy

Hitherto, we have reasoned on a single force P which moves its application point by δ in its own direction. When several forces act simultaneously, the elastic energy of the system is not necessarily equal to the sum of elastic energies corresponding to each force applied alone. In effect, after a force P_1 has provided the work $W_1 = P_1\delta_1/2$, the application of a second force P_2, which provides the work $W_2 = P_2\delta_2/2$, moves the application point of P_1 already applied, and therefore introduces a supplementary work $W_{12} = P_1\delta_{12}$, termed *mutual work* or *indirect work* for the two forces. The order of application being immaterial, we have $W_{12} = W_{21}$, which is *Betti's reciprocity theorem* (1872). Note that this mutual work is implicit in the Clapeyron formula

$$W = \sum_i \frac{1}{2} P_i \delta_i\,,$$

where the δ_i are $\delta_{ii} + \sum_j \delta_{ij}$.

This mutual work is zero only when the displacement δ_{12} is perpendicular to P_1 such that it does not work. One has to take care to this point when calculating G with several forces applied.

3.3 The Calculation of Griffith

It is from Inglis's results, (2.103–105), that Griffith [1] elaborated in 1920 his famous theory of fracture. He noted that the stress concentration at the tip of a defect depends only on the *shape* of that defect. Thus any theory of fracture based on a simple criterion of critical tensile stress leads to an independence of the *size* of the defect on the brittleness, which is contradictory to experimental results. He then expressed the idea that the equilibrium of a solid with a crack should correspond to a minimum of energy, taking into account the potential energy due to the creation of new surfaces in the solid.

Inglis had given, in a simple form, the stresses and displacements in the case of a plate traversed by an elliptical cavity (semi-axes a and b, with $c^2 = a^2 - b^2$ and defined by $\xi = \xi_0$ in elliptical coordinates) and subjected to a tensile stress σ uniform in all directions (an isotropic loading). Griffith computed the elastic energy per unit thickness of the plate, in an elliptical zone around the cavity, and found that for large ξ this energy tends towards

$$U_{\mathrm{E}} = \frac{\pi c^2 \sigma^2 (1+\nu)}{4E} \left[\frac{1}{2}(\kappa - 1)\mathrm{e}^{2\xi} + (3-\kappa)\cosh 2\xi_0 \right] ,$$

where κ has been defined in Sect. 2.5.1. The increase of the elastic energy due to the presence of the cavity is thus:

$$\Delta U_{\mathrm{E}} = \frac{\pi c^2 \sigma^2 (1+\nu)}{4E}(3-\kappa)\cosh 2\xi_0 ,$$

and for a very flat cavity ($b \ll a$):

$$\Delta U_{\mathrm{E}} = \frac{\pi a^2 \sigma^2 (1+\nu)}{4E}(3-\kappa) .$$

The potential energy due to the surface being $U_{\mathrm{S}} = 4\gamma a$, Griffith states that a crack is in equilibrium if $\partial(\Delta U_{\mathrm{E}} - U_{\mathrm{S}})/\partial a = 0$, i.e. for

$$\sigma = \sqrt{\frac{2E\gamma}{\nu(1+\nu)\pi a}} \qquad \text{(in plane strain)} ,$$

$$\sigma = \sqrt{\frac{2E\gamma}{\nu \pi a}} \qquad \text{(in plane stress)} .$$

These formulae are obviously wrong, since they make σ infinite when $\nu = 0$. In 1924 Griffith [8] modified his calculation in a rather obscure manner to get

$$\Delta U_{\mathrm{E}} = \frac{\pi a^2 \sigma^2}{4E}(1+\nu)(\kappa+1) ,$$

and hence

$$\sigma = \sqrt{\frac{2E\gamma}{(1-\nu^2)\pi a}} \quad \text{(in plane strain)}\,, \tag{3.21a}$$

$$\sigma = \sqrt{\frac{2E\gamma}{\pi a}} \quad \text{(in plane stress)}\,. \tag{3.21b}$$

The geometry studied by Griffith being an unstable one, the crack propagates spontaneously as soon as equilibrium is reached, so that (3.21a) and (3.21b) have been confused for a long time with a criterion of unstability. The first examples of stable cracks advancing as the load increases were given by Obreimov [9] in 1930 for cleavage by a wedge, and by Roesler [10] in 1956 for a Hertzian fracture under a rigid punch.

The *Griffith criterion* (3.21a) and (3.21b) introduced for the first time into the mechanics of solids the concept of surface energy hitherto reserved to physicochemists. Orowan [11] showed in 1945 how Griffith's results could be recovered by assuming that the cracks had a very small radius of curvature, of the order of interatomic distances Z_0. Making $\varrho = Z_0 \ll c$ and $\sigma_y = \sigma_{\text{th}}$ in (2.109), we obtain

$$\sigma_y = 2\sqrt{\frac{a}{\varrho}}\sigma\,,$$

and comparing this result with (1.51), which gave us

$$\sigma_{\text{th}} \cong \sqrt{\frac{E\gamma}{Z_0}}\,,$$

we get

$$\sigma = \sqrt{\frac{E\gamma}{4a}}\,.$$

This result is close to (3.21). Indeed, observations made with an electron microscope show that, in brittle materials, such as glass, the cracks are extremely acute. However, for cracks in this form, the two surfaces are in molecular interaction and Inglis's hypothesis of a cavity with $\sigma_\xi = 0$ everywhere is no longer valid. Griffith had correctly noted this point: "The molecular attractions across such a crack must be small except very near its ends; it may therefore be said that the application of the mathematical theory of elasticity, on the basis that the crack is assumed to be a traction-free surface, must give the stresses correctly at all points of the body, with the exception of those near the ends of the crack. In a sufficiently large crack the error in the strain energy so calculated must be negligible. Subject to the validity of other assumptions involved, the strength of smaller cracks calculated on this basis must evidently be too low." We will return to these important points in connection with the models of Barenblatt and Dugdale (Sect. 3.8).

3.4 Study of Cracks: Stress Intensity Factors

After Griffith's work, fracture mechanics suffered a long eclipse and revived only after the Second World War to understand the sudden and catastrophic fractures that occured, for example, on Liberty Ships.[2] It became necessary to know the stress field correctly in the neighbourhood of a crack tip in a material subjected to mechanical loadings.

We will study, as an example, the case of a linear crack aligned with z in a plate of infinite dimensions along x and y and subjected to a biaxial loading along x and y.

3.4.1 Exact Solution for a Straight Internal Crack in a Plate

There is no mathematical difficulty in making $b = 0$ (i.e. $\xi_0 = 0$) in (2.98–102) established for an elliptical cavity with semi-axes a and b in a plate subjected at infinity to a biaxial loading σ and $k\sigma$. We get, in elliptical coordinates (ξ, η) with $m = 1 + k$, and $n = 1 - k$, where β is the angle of the crack with the stress σ,

$$\frac{\sigma_\eta + \sigma_\xi}{\sigma} = n\cos 2\beta + \lambda\left[(m - n\cos 2\beta)\sinh 2\xi - n\sin 2\beta \sin 2\eta\right] , \tag{3.22}$$

$$\begin{aligned}\frac{\sigma_\eta - \sigma_\xi}{\sigma} &= \lambda^2(m - n\cos 2\beta)\sinh 2\xi(1 - \cos 2\eta) \\ &\quad - \lambda n\cos 2\beta(\cosh 2\xi \cos 2\eta - 1) \\ &\quad - \lambda n \sin 2\beta\left[\cosh 2\xi - \lambda(\cosh 2\xi - 1)\right]\sin 2\eta ,\end{aligned} \tag{3.23}$$

$$\begin{aligned}\frac{2\tau_{\xi\eta}}{\sigma} &= \lambda^2(m - n\cos 2\beta)(\cosh 2\xi - 1)\sin 2\eta \\ &\quad - \lambda n\cos 2\beta \sinh 2\xi \sin 2\eta \\ &\quad + \lambda n\sin 2\beta \sinh 2\xi\left[\cos 2\eta - \lambda(1 - \cos 2\eta)\right] ,\end{aligned} \tag{3.24}$$

$$\begin{aligned}u_x = \frac{\sigma a}{8\mu}\Bigg\{&(\kappa + 1)\left(\frac{x}{a}n\cos 2\beta - \frac{y}{a}n\sin 2\beta\right) \\ &+ (m - n\cos 2\beta)[\kappa + 1 - 2\lambda(\cosh 2\xi - 1)]\sinh\xi\cos\eta \\ &+ n\sin 2\beta[\kappa + 1 + 2\lambda(\cosh 2\xi - 1)]\cosh\xi\sin\eta\Bigg\} ,\end{aligned} \tag{3.25}$$

$$\begin{aligned}u_y = \frac{\sigma a}{8\mu}\Bigg\{&(\kappa - 3)\frac{y}{a}n\cos 2\beta + (\kappa + 1)\frac{x}{a}n\sin 2\beta \\ &+ (m - n\cos 2\beta)[\kappa + 1 - 2\lambda(\cosh 2\xi - 1)]\cosh\xi\sin\eta \\ &- n\sin 2\beta[\kappa - 3 + 2\lambda(\cosh 2\xi - 1)]\sinh\xi\cos\eta\Bigg\} .\end{aligned} \tag{3.26}$$

[2] The *Schenectady* in particular brokes suddenly in two, with a noise of detonation, from the superstructure to the keel, at its wharf, on January 16, 1942 at 10.30 pm, even though she just made satisfactory tests at sea.

Changing the coordinates according to (2.84) and (2.85), the stresses can be written in rectangular coordinates:

$$\begin{aligned}\sigma_y + \sigma_x &= \sigma_\eta + \sigma_\xi \, , \\ \sigma_y - \sigma_x &= u(\sigma_\eta - \sigma_\xi) + 2v\tau_{\xi\eta} \, , \\ 2\tau_{xy} &= -v(\sigma_\eta - \sigma_\xi) + 2u\tau_{\xi\eta} \, ,\end{aligned}$$

with

$$\begin{aligned}u &= \lambda(\cosh 2\xi \cos 2\eta - 1) \, , \\ v &= \lambda \sinh 2\xi \sin 2\eta \, ,\end{aligned}$$

and

$$\frac{1}{\lambda} = \cosh 2\xi - \cos 2\eta \, .$$

After some calculations, and with the help of

$$\Delta = \frac{1}{4\lambda^2} = (X^2 + Y^2 + 1)^2 - 4X^2 = (X^2 + Y^2 - 1)^2 + 4Y^2 \, , \tag{3.27}$$

$$\begin{aligned}U = \cosh\xi \sin\eta &= \operatorname{sign}(Y) \left[\frac{\sqrt{\Delta} - (X^2 - Y^2 - 1)}{2} \right]^{1/2} && (Y \neq 0) \, , \\ &= \pm\sqrt{1 - X^2} && (Y = 0 \text{ and } |X| < 1) \, , \\ &= 0 && (Y = 0 \text{ and } |X| \geq 1) \, ,\end{aligned} \tag{3.28}$$

$$V = \sinh\xi \cos\eta = \operatorname{sign}(X) \left[\frac{\sqrt{\Delta} + X^2 - Y^2 - 1}{2} \right]^{1/2} \, , \tag{3.29}$$

$$2\lambda(\cosh 2\xi - 1) = 1 + \frac{X^2 + Y^2 - 1}{\sqrt{\Delta}} \, , \tag{3.30}$$

(where $X = x/a$ and $Y = y/a$), we get [12]

$$\frac{\sigma_y + \sigma_x}{\sigma} = n \cos 2\beta + \frac{R}{2}(m - n\cos 2\beta) - \frac{T}{2} n \sin 2\beta \, , \tag{3.31}$$

$$\frac{\sigma_y - \sigma_x}{\sigma} = -n\cos 2\beta + P(m - n\cos 2\beta) + \left(\frac{T}{2} + Q\right) n \sin 2\beta \, , \tag{3.32}$$

$$\frac{2\tau_{xy}}{\sigma} = Q(m - n\cos 2\beta) + \left(\frac{R}{2} - P\right) n \sin 2\beta \, , \tag{3.33}$$

$$u_x = \frac{\sigma a}{8\mu}\left\{(\kappa+1)(Xn\cos 2\beta - Yn\sin 2\beta) + (m - n\cos 2\beta)\left[\kappa - \frac{X^2+Y^2-1}{\sqrt{\Delta}}\right]V + n\sin 2\beta\left[\kappa + 2 + \frac{X^2+Y^2-1}{\sqrt{\Delta}}\right]U\right\}, \quad (3.34)$$

$$u_y = \frac{\sigma a}{8\mu}\left\{(\kappa-3)Yn\cos 2\beta + (\kappa+1)Xn\sin 2\beta + (m - n\cos 2\beta)\left[\kappa - \frac{X^2+Y^2-1}{\sqrt{\Delta}}\right]U - n\sin 2\beta\left[\kappa - 2 + \frac{X^2+Y^2-1}{\sqrt{\Delta}}\right]V\right\}, \quad (3.35)$$

with

$$R = 2\lambda\sinh 2\xi = \left[\frac{(X^2+Y^2+\sqrt{\Delta})^2-1}{\Delta}\right]^{1/2}, \quad (3.36)$$

$$T = 2\lambda\sin 2\eta = \operatorname{sign}(XY)\left[\frac{1-(X^2+Y^2-\sqrt{\Delta})^2}{\Delta}\right]^{1/2}, \quad (3.37)$$

$$P = \lambda^2\sinh 2\xi(1-\cos 2\eta)[2\lambda(\cosh 2\xi\cos 2\eta - 1)+1] = \frac{YU}{\Delta^{3/2}}[2(X^2-Y^2-1)+\sqrt{\Delta}], \quad (3.38)$$

$$Q = \lambda^2\sin 2\eta(\cosh 2\xi - 1)[2\lambda(\cosh 2\xi\cos 2\eta - 1)-1] = \frac{YV}{\Delta^{3/2}}[2(X^2-Y^2-1)-\sqrt{\Delta}]. \quad (3.39)$$

Note that on the X-axis we have $P = Q = 0$ and

$$\left.\begin{array}{l} U = T = 0 \\ V = \operatorname{sign}(X)\sqrt{X^2-1} \\ R = \dfrac{2|X|}{\sqrt{X^2-1}} \end{array}\right\} \quad \text{if } |X| > 1,$$

$$\left.\begin{array}{l} V = R = 0 \\ U = \pm\sqrt{1-X^2} \\ T = \pm\dfrac{2X}{\sqrt{1-X^2}} \end{array}\right\} \quad \text{if } |X| < 1.$$

Note also that for $k = 0$ (uniaxial loading) and $\beta = 0°$, we have everywhere $\sigma_x = \sigma$ and $\sigma_y = \tau_{xy} = 0$—as if the crack did not exist.

These stresses and displacements are displayed in Figs. 3.3–7, which can be compared to corresponding Figs. 2.4 and 2.6–9 for $b/a = 0.5$.

Particular Cases. In the particular case where $b = 90°$ (*mode I*), the expressions are especially simple. In effect, we have:

$$\frac{\tau_m}{\sigma} = \left[n\left(\frac{n}{4} + P\right) + \frac{Y^2}{\Delta^{3/2}}\right]^{1/2} \quad \text{(isochromatic lines)}\,, \tag{3.40}$$

$$\frac{\sigma_1}{\sigma} = \frac{R-n}{2} + \frac{\tau_m}{\sigma}\,, \tag{3.41}$$

$$\frac{\sigma_2}{\sigma} = \frac{R-n}{2} - \frac{\tau_m}{\sigma}\,, \tag{3.42}$$

$$\frac{\sigma_z}{\sigma} = \nu(R-n) \quad \text{(plane strain)}\,, \tag{3.43}$$

$$\frac{2E}{\nu\sigma d}u_z = -(R-n) \quad \text{(isopachics (plane stress))}\,, \tag{3.44}$$

$$\frac{\sigma_{eq}^2}{\sigma^2} = 3\left(\frac{\tau_m}{\sigma}\right)^2 + \left(\frac{R-n}{2}\right)^2, \quad \text{(von Mises (plane stress))}\,, \tag{3.45}$$

$$\frac{\sigma_{eq}^2}{\sigma^2} = 3\left(\frac{\tau_m}{\sigma}\right)^2 + (1-2\nu)^2\left(\frac{R-n}{2}\right)^2, \tag{3.46}$$

(von Mises (plane strain)),

$$\frac{\mathrm{d}Y}{\mathrm{d}X} = \frac{n+2P}{2Q} \pm \operatorname{sign}(Q)\left[\left(\frac{n+2P}{2Q}\right)^2 + 1\right]^{1/2} \tag{3.47}$$

(stress trajectories).

Note that in uniaxial loading the faces of a crack are subjected to a compressive stress given by $\sigma_x = -\sigma$. For thin plates, this compression leads to buckling: the faces of the crack can move out of the plane of the plate either on the same side, or each in the opposite side. More precisely, as shown in Fig. 3.8, there is a large zone of such a plate, at 45° to the crack, which undergoes a compressive stress σ_2, as observed by Gilabert *et al.* [13]. This compression on thin plates provokes periodic buckling folds which closely follow the stress trajectories in a zone resembling a Maltese cross (note the contours of σ_2 in Fig. 3.4).

In a second case, where $\beta = 45°$ and $k = -1$ (*mode II*), calculations can be simplified by noting that

$$\frac{\tau_m^2}{\sigma} = \frac{Y^2}{\Delta^{3/2}}(3 - 6X^2 + 2Y^2) + \frac{X^2+Y^2}{\sqrt{\Delta}}\,,$$

which directly gives the isochromatic lines.

Stresses and Displacements on the Axes. Stresses and displacements take very simple forms on the x-axis and the y-axis.

(a) (b)

(c) (d)

(e) (f)

(g) (h)

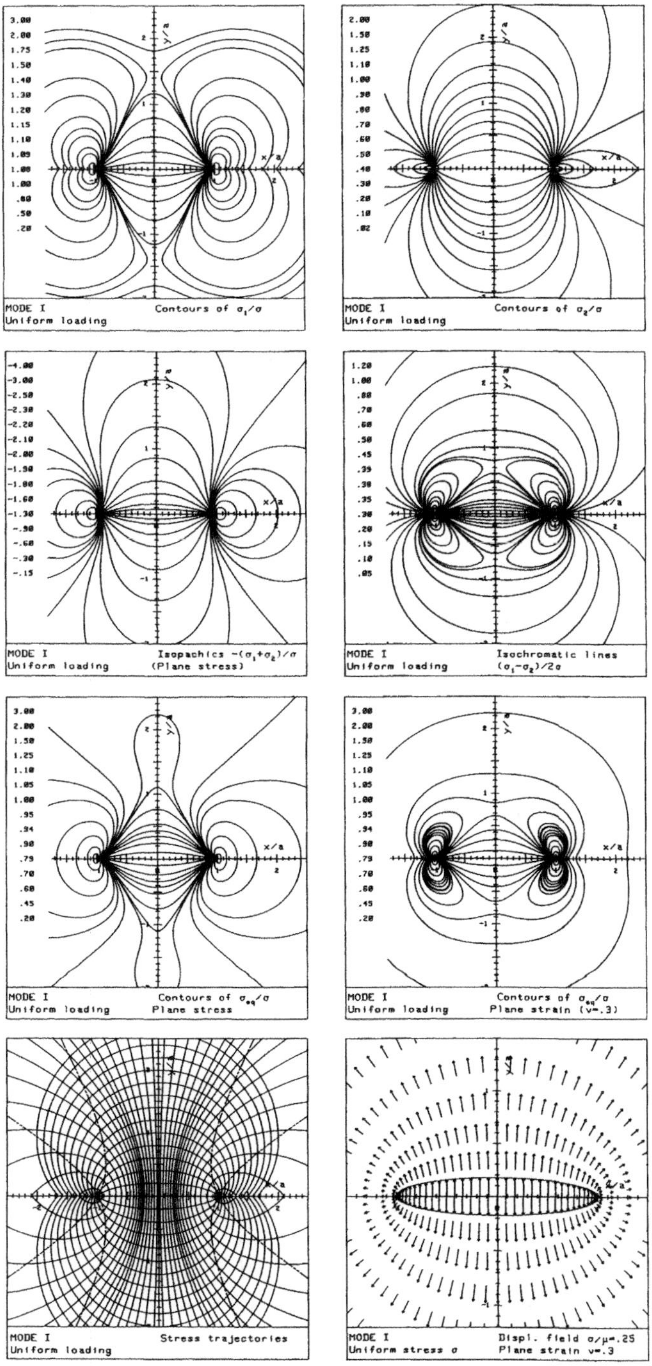

Fig. 3.3. Contours, under uniform loading, of: **(a)** and **(b)** the two principal stresses; **(c)** the isopachics; **(d)** the isochromatic lines; **(e)** von Mises criterion under plane stress; **(f)** von Mises criterion under plane strain; **(g)** stress trajectories; and **(h)** the displacements field

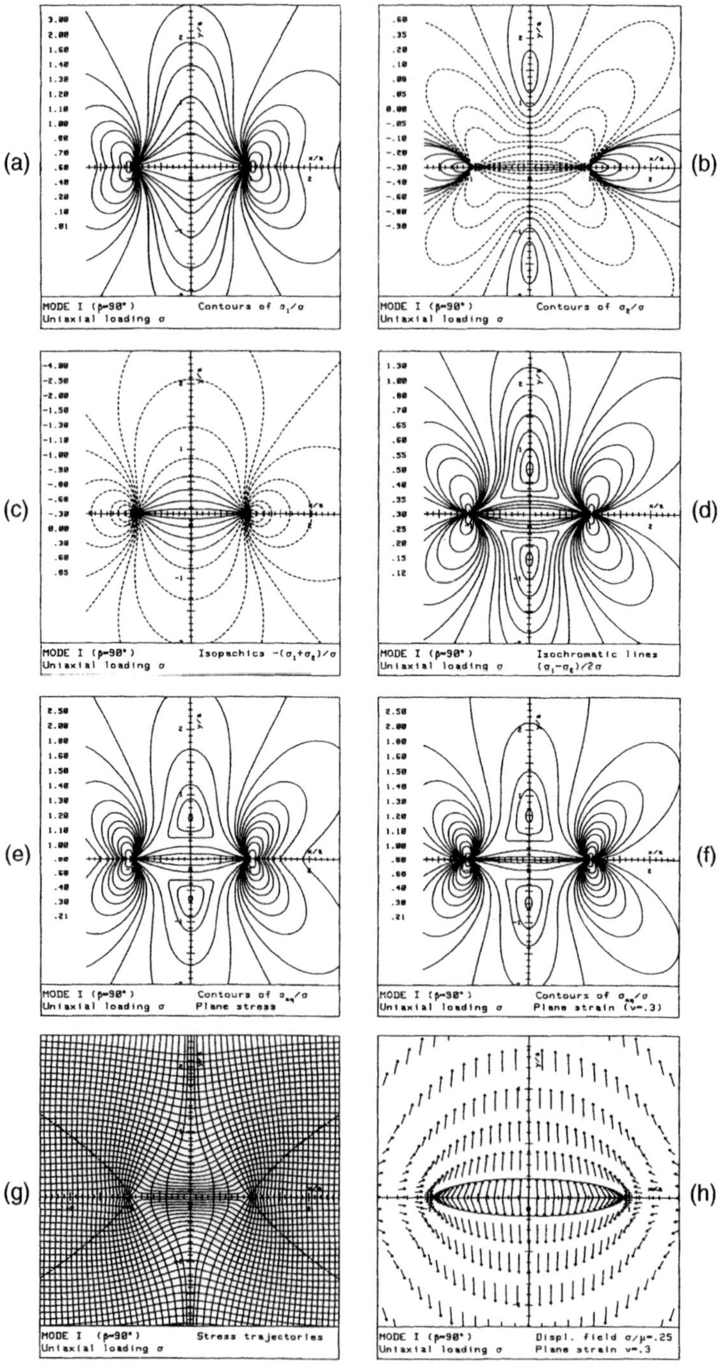

Fig. 3.4. Contours, under uniaxial loading, for: **(a)** and **(b)** the two principal stresses; **(c)** the isopachics; **(d)** the isochromatic lines; **(e)** von Mises criterion under plane stress; **(f)** von Mises criterion under plane strain; **(g)** stress trajectories; and **(h)** the displacements field

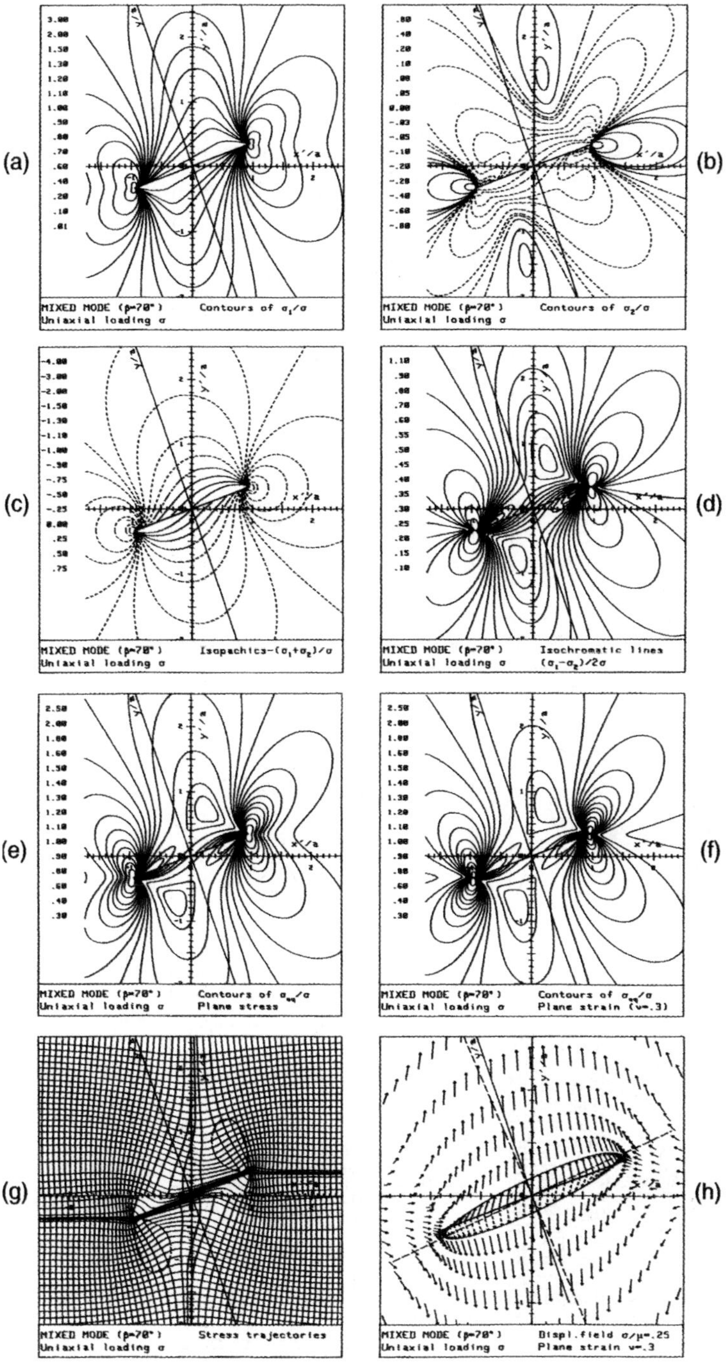

Fig. 3.5. Contours, under inclined uniaxial loading, for: **(a)** and **(b)** the two principal stresses; **(c)** the isopachics; **(d)** the isochromatic lines; **(e)** von Mises criterion under plane stress; **(f)** von Mises criterion under plane strain; **(g)** stress trajectories; and **(h)** the displacements field

(a) (b)

(c) (d)

(e) (f)

(g) (h)

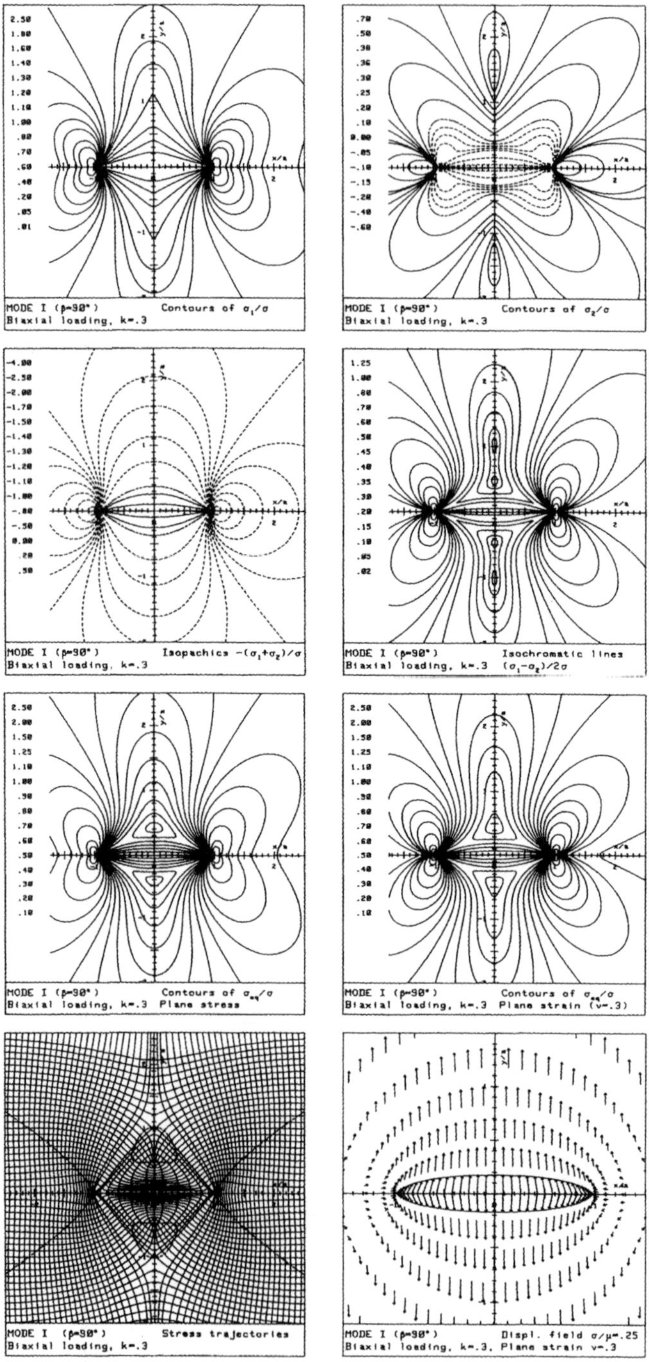

Fig. 3.6. Contours, under biaxial loading, for: **(a)** and **(b)** the two principal stresses; **(c)** the isopachics; **(d)** the isochromatic lines; **(e)** von Mises criterion under plane stress; **(f)** von Mises criterion under plane strain; **(g)** stress trajectories; and **(h)** the displacements field

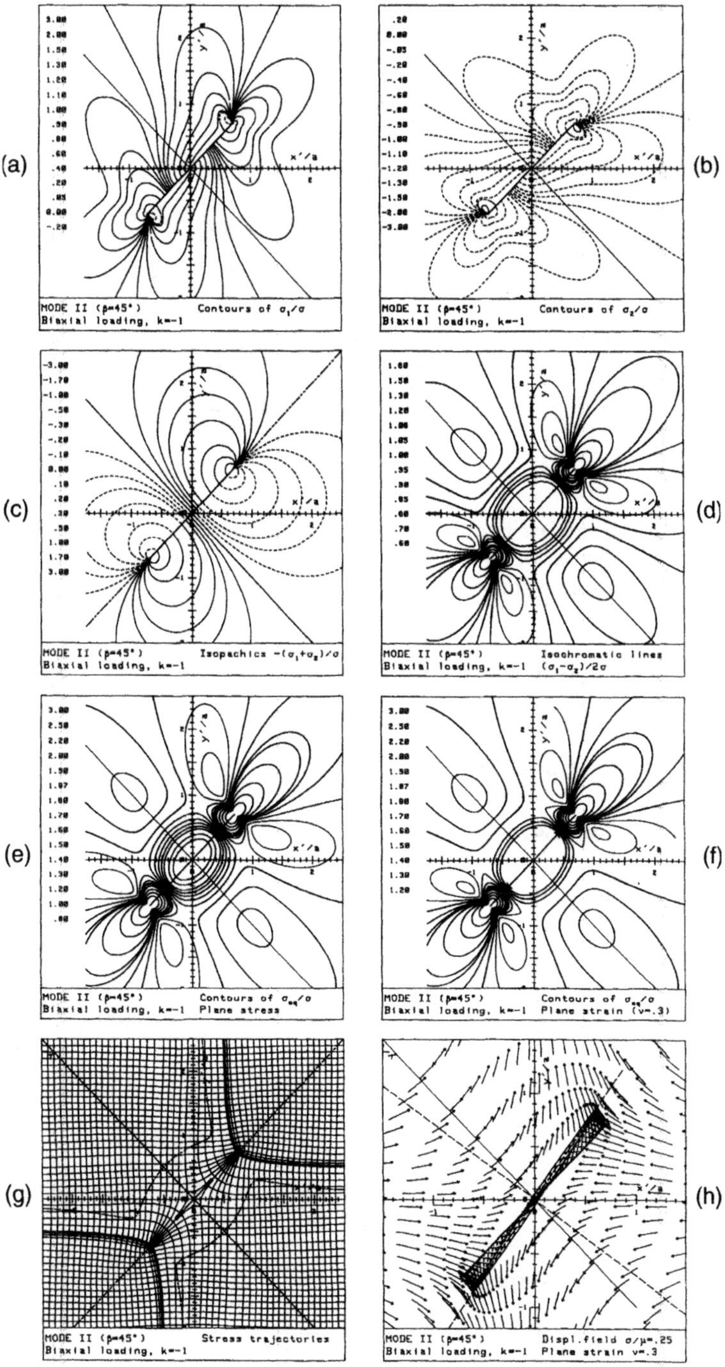

Fig. 3.7. Contours, under pure shear, for: **(a)** and **(b)** the two principal stresses; **(c)** the isopachics; **(d)** the isochromatic lines; **(e)** von Mises criterion under plane stress; **(f)** von Mises criterion under plane strain; **(g)** stress trajectories; and **(h)** the displacements field

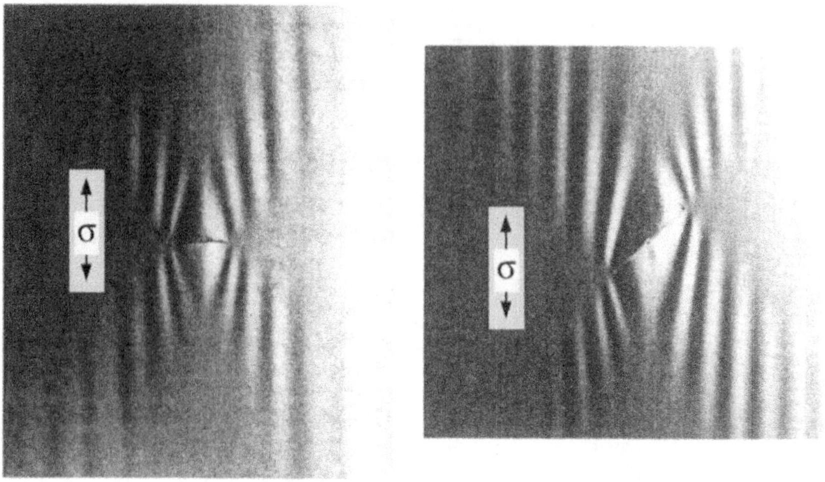

Fig. 3.8. Buckling of a thin plate in uniaxial tension and containing a crack oriented at 90° or at 45°. The orientation of the folds follow closely the stress trajectories, in zones where σ_2 is a compression

On the x-axis (the axis of the crack), and for $|x| > a$, we have

$$(\sigma_y)_{y=0} = \frac{\sigma}{2}(m - n\cos 2\beta)\frac{|x|}{(x^2 - a^2)^{1/2}}\,, \tag{3.48}$$

$$(\sigma_x)_{y=0} = \sigma_y + \sigma n \cos 2\beta\,, \tag{3.49}$$

$$(\tau_{xy})_{y=0} = \frac{\sigma}{2} n \sin 2\beta \frac{|x|}{(x^2 - a^2)^{1/2}}\,, \tag{3.50}$$

$$(u_x)_{y=0} = \mathrm{sign}(x)\frac{\sigma}{8\mu}\Big[(n\cos 2\beta)(\kappa+1)\,|x| + (m - n\cos 2\beta)(\kappa-1)\sqrt{x^2-a^2}\Big]\,, \tag{3.51}$$

$$(u_y)_{y=0} = \mathrm{sign}(x)\frac{\sigma}{8\mu} n \sin 2\beta \Big[(\kappa+1)\,|x| - (\kappa-1)\sqrt{x^2-a^2}\Big]\,. \tag{3.52}$$

For $|x| < a$ (the faces of the crack) we have

$$(\sigma_y)_{y=0} = (\tau_{xy})_{y=0} = 0\,, \tag{3.53}$$

$$(\sigma_x)_{y=0} = \sigma n \left(\cos 2\beta - \frac{|x|}{\sqrt{a^2 - x^2}} \sin 2\beta\right)\,, \tag{3.54}$$

$$(u_x)_{y=0} = \frac{\sigma}{8\mu}(\kappa+1)\left[xn\cos 2\beta \pm n\sin 2\beta\sqrt{a^2-x^2}\right], \tag{3.55}$$

$$(u_y)_{y=0} = \frac{\sigma}{8\mu}(\kappa+1)\left[xn\sin 2\beta \pm (m - n\cos 2\beta)\sqrt{a^2-x^2}\right], \tag{3.56}$$

with positive sign in the last two equations for the upper face, and negative sign for the lower face. Equations (3.55) and (3.56) were directly obtained by Theocaris *et al.* [14] in 1986.

On the y-axis perpendicular to a crack, where $Q = T = V = 0$, we have:

$$(\sigma_y)_{x=0} = \frac{\sigma}{2}(m - n\cos 2\beta)\left(\frac{y^2}{y^2+a^2}\right)^{3/2}, \tag{3.57}$$

$$(\sigma_x)_{x=0} = \frac{\sigma}{2}(m - n\cos 2\beta)\frac{|y|}{(y^2+a^2)^{3/2}}(y^2+2a^2) + \sigma n\cos 2\beta\,, \tag{3.58}$$

$$(\tau_{xy})_{x=0} = \frac{\sigma}{2}n\sin 2\beta\frac{|y|}{(y^2-a^2)^{3/2}}(y^2+2a^2)\,, \tag{3.59}$$

$$(u_x)_{x=0} = \operatorname{sign}(y)\frac{\sigma}{8\mu}n\sin 2\beta \times\left[-(\kappa+1)|y| + \left(\kappa+1+\frac{2y^2}{y^2+a^2}\right)\sqrt{y^2+a^2}\right], \tag{3.60}$$

$$(u_y)_{x=0} = \operatorname{sign}(y)\frac{\sigma}{8\mu}\left[(n\cos 2\beta)(\kappa-3)|y| + (m-n\cos 2\beta)\left(\kappa+1-\frac{2y^2}{y^2+a^2}\right)\sqrt{y^2+a^2}\right]. \tag{3.61}$$

The stresses on the x-axis and the y-axis have been displayed for $\beta = 90°$ in Figs. 2.2 and 2.4, with respectively $k = 1$ and $k = 0$. The stresses are infinite at the crack tip and return to the boundary conditions at large distances.

Deformation of a Crack. The discontinuity of displacements u_x and u_y of a point on the faces of the crack (between $y = 0_+$ and $y = 0_-$) is given by (3.55) and (3.56) as

$$[u_x] = \frac{\kappa+1}{4\mu}\sigma n\sin 2\beta\sqrt{a^2-x^2}\,, \tag{3.62}$$

$$[u_y] = \frac{\kappa+1}{4\mu}\sigma(m-n\cos 2\beta)\sqrt{a^2-x^2}\,. \tag{3.63}$$

It is possible to show (within the framework of linearized elasticity) that the crack, initially flat, takes the shape of an ellipse with semi-axes given by:

$$a' = a(1 + Sn\cos 2\beta)\,, \tag{3.64}$$

$$b' = aS(m - n\cos 2\beta)\,, \tag{3.65}$$

turned through an angle α given by

$$\sin 2\alpha = 2Sn \sin 2\beta \,, \tag{3.66}$$

with

$$S = \frac{\kappa + 1}{8\mu}\sigma \,.$$

There is thus no rotation of the crack if $n = 0$ (isotropic loading) or if $\beta = 0°$ or $90°$.

3.4.2 Power Expansion

Let us study the stresses and displacements near the crack tip, by letting, in (3.31–35),

$$x = a + r\cos\theta$$
$$y = r\sin\theta$$

with $0 < r/a \ll 1$. We get from these assumptions the following:

$$\sqrt{\Delta} \approx \frac{2r}{a} + \frac{r^2}{a^2}\cos\theta \,,$$

$$U \approx \sqrt{\frac{2r}{a}}\sin\frac{\theta}{2}\left[1 + \frac{r}{4a}(2\cos\theta + 1)\right] \,,$$

$$V \approx \sqrt{\frac{2r}{a}}\cos\frac{\theta}{2}\left[1 + \frac{r}{4a}(2\cos\theta - 1)\right] \,,$$

$$2\lambda(\cosh 2\xi - 1) \approx 2\cos^2\frac{\theta}{2} + \frac{r}{2a}\sin^2\frac{\theta}{2} \,,$$

$$R \approx 2\sqrt{\frac{a}{2r}}\cos\frac{\theta}{2} + \frac{3}{2}\sqrt{\frac{r}{2a}}\cos\frac{\theta}{2} \,,$$

$$T \approx 2\sqrt{\frac{a}{2r}}\sin\frac{\theta}{2} - \frac{3}{2}\sqrt{\frac{r}{2a}}\sin\frac{\theta}{2} \,,$$

$$P \approx \sqrt{\frac{a}{2r}}\sin\frac{\theta}{2}\cos\frac{\theta}{2}\sin\frac{3\theta}{2} - \frac{3}{4}\sqrt{\frac{r}{2a}}\cos\frac{\theta}{2}\sin^2\frac{\theta}{2} \,,$$

$$Q \approx \sqrt{\frac{a}{2r}}\sin\frac{\theta}{2}\cos\frac{\theta}{2}\cos\frac{3\theta}{2} - \frac{3}{4}\sqrt{\frac{r}{2a}}\sin\frac{\theta}{2}\cos^2\frac{\theta}{2} \,.$$

Letting:

$$K_{\mathrm{I}} = \frac{1}{2}(m - n\cos 2\beta)\sigma\sqrt{\pi a} \,, \tag{3.67}$$

$$K_{\mathrm{II}} = \frac{1}{2}n\sin 2\beta\sigma\sqrt{\pi a} \,, \tag{3.68}$$

we easily obtain

$$
\begin{aligned}
\sigma_x = & \frac{1}{\sqrt{2\pi r}}\left[K_{\mathrm{I}}\cos\frac{\theta}{2}\left(1-\sin\frac{\theta}{2}\sin\frac{3\theta}{2}\right)\right. \\
& \left.-K_{\mathrm{II}}\sin\frac{\theta}{2}\left(2+\cos\frac{\theta}{2}\cos\frac{3\theta}{2}\right)\right] + \sigma n\cos 2\beta \\
& +\frac{3}{8a}\sqrt{\frac{r}{2\pi}}\left[K_{\mathrm{I}}\cos\frac{\theta}{2}(3-\cos\theta)+K_{\mathrm{II}}\sin\frac{\theta}{2}(5+\cos\theta)\right] ,
\end{aligned} \tag{3.69}
$$

$$
\begin{aligned}
\sigma_y = & \frac{1}{\sqrt{2\pi r}}\cos\frac{\theta}{2}\left[K_{\mathrm{I}}\left(1+\sin\frac{\theta}{2}\sin\frac{3\theta}{2}\right)+K_{\mathrm{II}}\sin\frac{\theta}{2}\cos\frac{3\theta}{2}\right] \\
& +\frac{3}{8a}\sqrt{\frac{r}{2\pi}}\cos\frac{\theta}{2}\left[K_{\mathrm{I}}(1+\cos\theta)-K_{\mathrm{II}}\sin\theta\right] ,
\end{aligned} \tag{3.70}
$$

$$
\begin{aligned}
\tau_{xy} = & \frac{1}{\sqrt{2\pi r}}\cos\frac{\theta}{2}\left[K_{\mathrm{I}}\sin\frac{\theta}{2}\cos\frac{3\theta}{2}+K_{\mathrm{II}}\left(1-\sin\frac{\theta}{2}\sin\frac{3\theta}{2}\right)\right] \\
& -\frac{3}{8a}\sqrt{\frac{r}{2\pi}}\cos\frac{\theta}{2}\left[K_{\mathrm{I}}\sin\theta-K_{\mathrm{II}}(3-\cos\theta)\right] ,
\end{aligned} \tag{3.71}
$$

$$
\begin{aligned}
2\mu u_x = & \frac{\sigma}{4}(\kappa+1)(xn\cos 2\beta - yn\sin 2\beta) \\
& +\sqrt{\frac{r}{2\pi}}\left[K_{\mathrm{I}}(\kappa-\cos\theta)\cos\frac{\theta}{2}+K_{\mathrm{II}}(\kappa+\cos\theta+2)\sin\frac{\theta}{2}\right] \\
& +\frac{\pi}{2a}\left(\frac{r}{2\pi}\right)^{3/2}\left\{K_{\mathrm{I}}\left[(\kappa+1)\cos\frac{3\theta}{2}-2\cos^3\frac{\theta}{2}\right]\right. \\
& \left.+K_{\mathrm{II}}\left[(\kappa+1)\sin\frac{3\theta}{2}+3\sin\theta\cos\frac{\theta}{2}\right]\right\} ,
\end{aligned} \tag{3.72}
$$

$$
\begin{aligned}
2\mu u_y = & \frac{\sigma}{4}[(\kappa+1)xn\sin 2\beta + (\kappa-3)yn\cos 2\beta] \\
& +\sqrt{\frac{r}{2\pi}}\left[K_{\mathrm{I}}(\kappa-\cos\theta)\sin\frac{\theta}{2}-K_{\mathrm{II}}(\kappa+\cos\theta-2)\cos\frac{\theta}{2}\right] \\
& +\frac{\pi}{2a}\left(\frac{r}{2\pi}\right)^{3/2}\left\{K_{\mathrm{I}}\left[(\kappa-1)\sin\frac{3\theta}{2}-2\sin^3\frac{\theta}{2}\right]\right. \\
& \left.-K_{\mathrm{II}}\left[(\kappa-1)\cos\frac{3\theta}{2}+3\sin\theta\sin\frac{\theta}{2}\right]\right\} .
\end{aligned} \tag{3.73}
$$

The discontinuities of displacements u_x and u_y between the upper face ($\theta = +\pi$) and the lower face ($\theta = -\pi$) of a point on a crack are given by

$$[u_y] = \frac{\kappa+1}{\mu} K_{\mathrm{I}} \left[\sqrt{\frac{r}{2\pi}} - \frac{1}{2}\sqrt{\frac{a}{\pi}} \left(\frac{r}{2a}\right)^{3/2} \right] , \tag{3.74}$$

$$[u_x] = \frac{\kappa+1}{\mu} K_{\mathrm{II}} \left[\sqrt{\frac{r}{2\pi}} - \frac{1}{2}\sqrt{\frac{a}{\pi}} \left(\frac{r}{2a}\right)^{3/2} \right] . \tag{3.75}$$

The stresses can also be expressed in polar coordinates centred on the crack tip. The following can be obtained:

$$\begin{aligned}
\sigma_r &= \sigma n \cos 2\beta \cos^2\theta \\
&+ \frac{1}{2\sqrt{2\pi r}} \left[K_{\mathrm{I}} \cos\frac{\theta}{2}(3-\cos\theta) + K_{\mathrm{II}} \sin\frac{\theta}{2}(3\cos\theta - 1) \right] \\
&+ \frac{3}{8a}\sqrt{\frac{r}{2\pi}} \left[K_{\mathrm{I}} \cos\frac{\theta}{2}(1-\cos\theta + 2\cos^2\theta) \right. \\
&\qquad \left. + K_{\mathrm{II}} \sin\frac{\theta}{2}(10\cos^2\theta + 5\cos\theta - 1) \right] ,
\end{aligned} \tag{3.76}$$

$$\begin{aligned}
\sigma_\theta &= \sigma n \cos 2\beta \sin^2\theta \\
&+ \frac{1}{2\sqrt{2\pi r}} \cos\frac{\theta}{2} \left[K_{\mathrm{I}}(1+\cos\theta) - 3K_{\mathrm{II}} \sin\theta \right] \\
&+ \frac{3}{8a}\sqrt{\frac{r}{2\pi}} \cos\frac{\theta}{2} \left[K_{\mathrm{I}}(1+\cos\theta)(3-2\cos\theta) \right. \\
&\qquad \left. + 5K_{\mathrm{II}}(1-2\cos\theta)\sin\theta \right] ,
\end{aligned} \tag{3.77}$$

$$\begin{aligned}
\tau_{r\theta} &= -\sigma n \cos 2\beta \sin\theta \cos\theta \\
&+ \frac{1}{2\sqrt{2\pi r}} \cos\frac{\theta}{2} \left[K_{\mathrm{I}} \sin\theta + K_{\mathrm{II}}(3\cos\theta - 1) \right] \\
&+ \frac{3}{8a}\sqrt{\frac{r}{2\pi}} \cos\frac{\theta}{2} \left[K_{\mathrm{I}} \sin\theta (1-2\cos\theta) \right. \\
&\qquad \left. + K_{\mathrm{II}}(10\cos^2\theta - 5\cos\theta - 3) \right] .
\end{aligned} \tag{3.78}$$

The symbols K_{I}, K_{II} (and in fact K_{III}), characterizing the force of the singularity (in $1/\sqrt{r}$) of the stresses, were introduced by Irwin [15] in 1958 (with K for Kies, one of his collaborators) and termed by him *stress intensity factors* [16] for the three components of displacement at a crack tip (Fig. 3.9): *mode I*, or opening mode; *mode II*, or sliding mode; *mode III*, or tearing mode. We must take care that these modes are defined by the local state at the crack tip and cannot be deduced directly from any external loading. For example, when one presses vertically an axisymmetric punch on the surface of a brittle solid, a vertical crack appears at the edge of the contact (a Hertzian

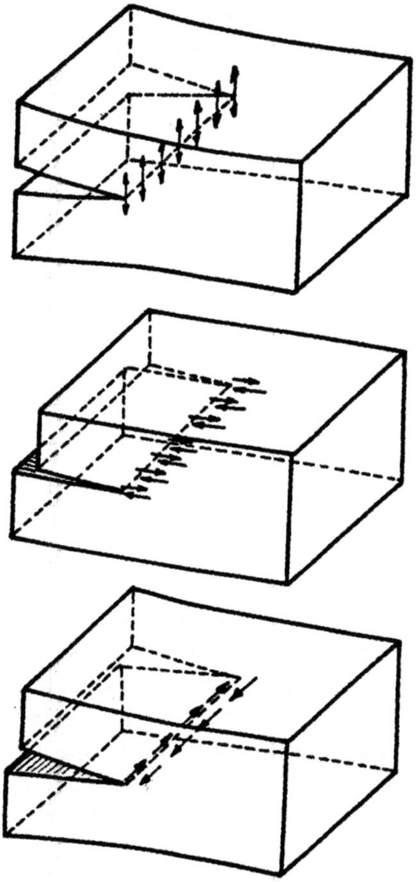

Fig. 3.9. The three modes of displacement at a crack tip (after Irwin [16])

fracture), which opens in mode I and not in mode II as could be suggested by the geometry.

Equations (3.67) and (3.68) show that stress intensity factors should be expressed in units of Pa $\sqrt{\text{m}}$. They should not be confused with stress concentration factors, which are nondimensional.

As shown by (3.66), there is no rotation of the crack in pure mode I ($\beta = 0°$, $\beta = 90°$, or isotropic loading), that is, when the crack is parallel to the principal axis of the loading. It is the component of mode II which produces this rotation. For $\beta = 90°$ (pure mode I), the value of K_{I} is independent of any biaxiality (i.e. independent of the stress applied parallel to the crack), because $m + n = 2$ and hence

$$K_{\text{I}} = \sigma\sqrt{\pi a}\ .$$

However, the distribution of stress depends on the biaxiality: cf Figs. 3.3 and 3.4. More generally, K_{I} and K_{II} depend respectively only on the components of the imposed loading which are normal and tangential to the crack. In effect, as shown by Mohr circle construction, the normal stress for a face parallel to the crack plane is

$$\sigma_y = \frac{\sigma + k\sigma}{2} - \frac{\sigma - k\sigma}{2}\cos 2\beta = \frac{1}{2}(m - n\cos 2\beta)\sigma \; ,$$

and hence

$$K_{\mathrm{I}} = \sigma_y\sqrt{\pi a} \; ,$$

whereas the shear stress is

$$\tau = \frac{\sigma - \kappa\sigma}{2}\sin 2\beta = \frac{1}{2}\sigma n \sin 2\beta \; ,$$

and hence

$$K_{\mathrm{II}} = \tau\sqrt{\pi a} \; .$$

This problem of biaxiality, and of the constant term $\sigma n\cos 2\beta$ in the power expansion of σ_x (the T-stress), (3.69), has given rise to many debates.

Sneddon [17], in 1946, gave the first power expansion of stresses for an isotropic loading ($n = 0$, hence mode I) by developing the solution given by Westergaard [18] for a crack subjected to an inner pressure p_0: he was thus not concerned with the problem of the constant term. In 1957, Williams [19], independently of Sneddon, gave the power expansion for mode I, but without the constant term. It rapidly appeared that the singular terms of the power expansion alone were insufficient to evaluate the stress intensity factors from the isochromatic fringes [20], and Irwin [21] proposed to use the second term (constant) of the power expansion to improve the precision: for a uniaxial loading in mode I [2], he substracted from σ_x, in a rather obscure manner, the quantity σ (the term $\sigma n\cos 2\beta$ of (3.69)). It was only in 1966 that Sih [22] noticed that a constant term A was missing in the Westergaard equations used hitherto. Eftis and Liebowitz [23] showed in 1972 that this term was $\sigma n/2$ and that, for a biaxial loading with $\beta = 90°$, one must substract the quantity σn from the power expansion of σ_x, (and add terms to the power expansion of u_x and u_y), which modifies considerably the aspect of isochromatic lines so calculated [24].

The inclined crack under biaxial loading was studied by Eftis and Subramonian [25] in 1978. They showed that stress intensity factors generally depend on the biaxiality factor k (see (3.67) and (3.68)), and, limited to two terms (singular term plus constant term for the stresses, term in $\sqrt{r}$ plus term in r for the displacements), they gave the power expansions (3.69–73) (with, however, an error on u_x later corrected by Smith in 1987 [26]). Finally the power expansions of the stresses with three terms, as per (3.69–71), were obtained by Theocaris and Spyropoulos [27] in 1983.

Modes II and III of Irwin's analysis are somewhat questionable. A crack can be loaded in mode I+II for example, but experiments seem to show that cracks propagate only in mode I (at least in homogeneous and isotropic materials), so that in a complex stress field a crack turns to follow stress trajectories. This phenomenon has been used in the method of cracking varnish. A thin film of varnish, brittle after drying, is deposited onto a test piece. When it is under tension, the network of cracks in the varnish reveals the stress trajectories of the tensile principal stress at the surface of the test piece. It is possible to obtain the perpendicular network, when the two principal stresses are of opposite sign, by first placing the test piece under tension and then depositing the varnish before relaxing the tension.

This phenomenon is also used to measure the crack speed in glass [28]. By sending ultrasonic waves of frequency ν perpendicularly to the crack plane, one can slightly modulate the slope of the plane of principal stress in front of the crack. As such a crack constantly orients itself to remain perpendicular to tensile maximum stresses, the surface of fracture displays undulations of wavelength λ, from which one can deduce the velocity $v = \lambda\nu$ of the crack.

The fact that fractures open along a plane perpendicular to the direction of maximum tensile stresses at the crack tip leads to the phenomenon of *fracture lances* [28, 29] when loading in traction and torsion are superimposed. As the fracture plane cannot ajust on the plane of principal stress all along the crack line, it decomposes into a series of correctly oriented planes separated by steps (or lances) elongated in the direction of the crack.

3.4.3 Comparison with the Exact Solution

To visualize the approximation given by the power expansions, let us compare, for a biaxial loading in mode I ($\beta = 90°$) and in plane strain, the exact solutions on the x-axis, as shown by (3.48), (3.49) and (3.56), and the first terms of the power expansion at $r = a + x$ or $r = a - x$

$$\frac{\sigma_y(x,0)}{\sigma} = \left(\frac{2r}{a}\right)^{-1/2} + \frac{3}{8}\left(\frac{2r}{a}\right)^{1/2} - \frac{5}{128}\left(\frac{2r}{a}\right)^{3/2} + \dots ,$$

$$\frac{\sigma_x(x,0)}{\sigma} = \frac{\sigma_y(x,0)}{\sigma} - n ,$$

$$\frac{u_y(x,0)}{a} = \sigma\frac{\kappa+1}{2\mu}\left[\left(\frac{r}{2a}\right)^{1/2} - \frac{1}{2}\left(\frac{r}{2a}\right)^{3/2} + \dots\right] .$$

The exact solution and the different terms of the power expansion are given in Fig. 3.10 for $k = -1/2$. The sum of the first terms is a dashed-dotted line in the figure.

Figure 3.11 displays the stress tensor in mode I ($K_{\mathrm{II}} = 0$) reduced to the singular terms only. The principal stresses are thus written as

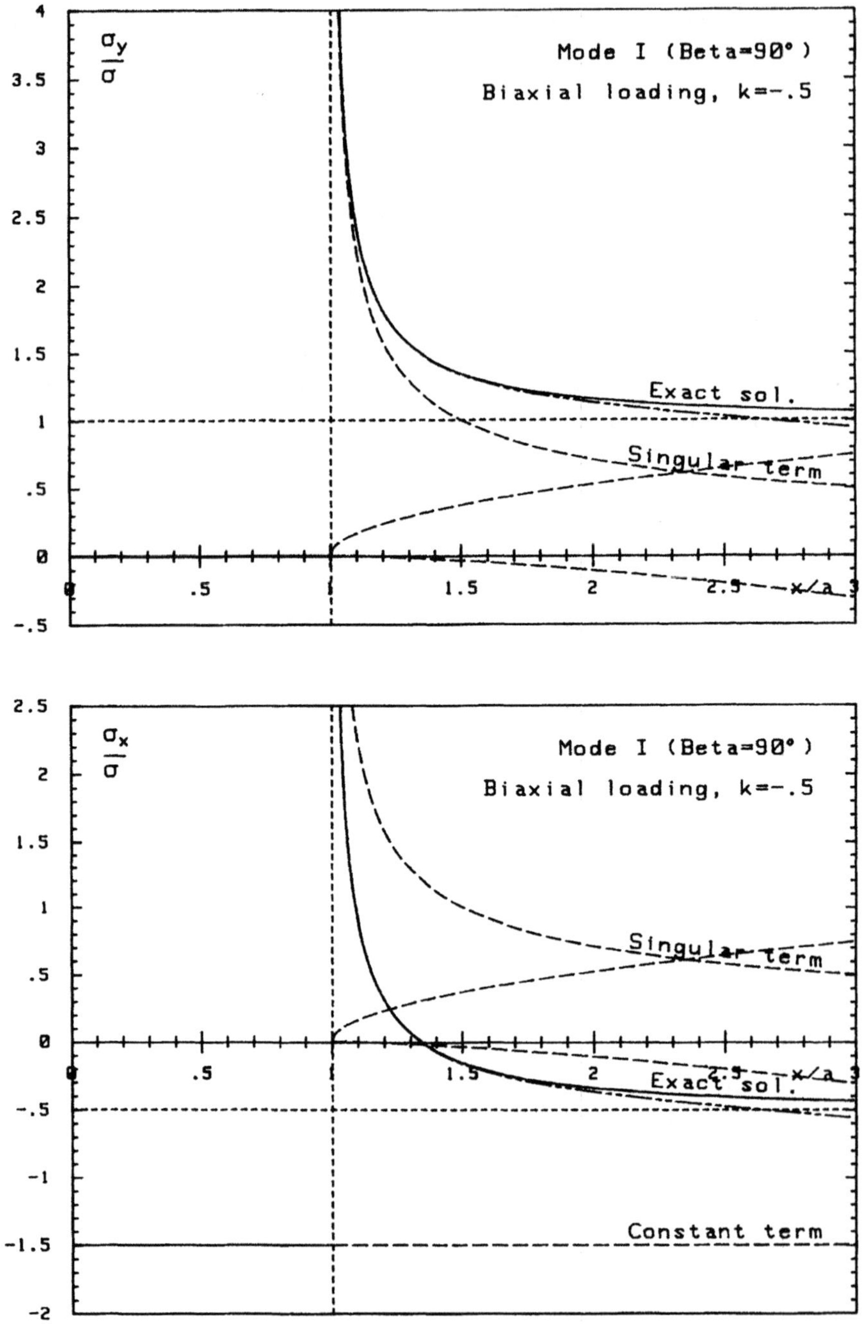

Fig. 3.10. Stresses on the x-axis in mode I: comparison of the power expansion with the exact solution

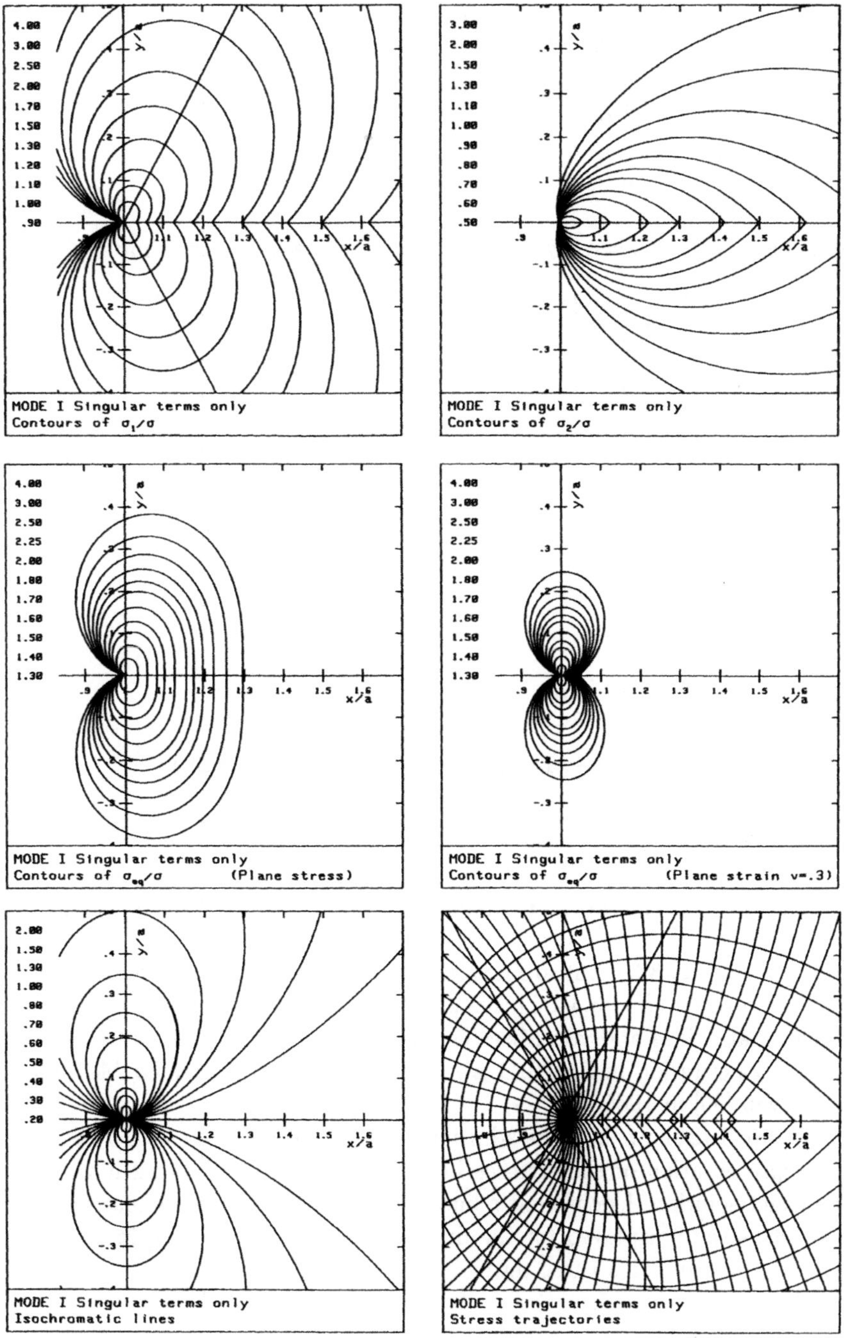

Fig. 3.11. Mode I: stress tensor reduced to singular terms only

$$\sigma_1 = \frac{K_{\rm I}}{\sqrt{2\pi r}} \left(\cos\frac{\theta}{2} + \frac{1}{2}|\sin\theta| \right) , \qquad (3.79)$$

$$\sigma_2 = \frac{K_{\rm I}}{\sqrt{2\pi r}} \left(\cos\frac{\theta}{2} - \frac{1}{2}|\sin\theta| \right) , \qquad (3.80)$$

$$\sigma_z = \frac{K_{\rm I}}{\sqrt{2\pi r}} 2\nu \cos\frac{\theta}{2} \quad \text{(in plane strain)} , \qquad (3.81)$$

and the contours $m = \sigma_j/\sigma$ of the principal stresses are very simply given by

$$\frac{r}{a} = \frac{1}{2m^2} \left(\cos\frac{\theta}{2} + \frac{1}{2}|\sin\theta| \right)^2 \quad \text{for } \sigma_1 ,$$

$$\frac{r}{a} = \frac{1}{2m^2} \left(\cos\frac{\theta}{2} - \frac{1}{2}|\sin\theta| \right)^2 \quad \text{for } \sigma_2 ,$$

$$\frac{r}{a} = \frac{1}{2m^2} \left(\nu \cos\frac{\theta}{2} \right)^2 \quad \text{for } \sigma_z .$$

For isochromatic lines ($m = (\sigma_1 - \sigma_2)/2\sigma$), we have

$$\frac{r}{a} = \frac{\sin^2\theta}{8m^2} ,$$

and for the von Mises criterion ($m = \sigma_{\rm eq}/\sigma$) we have

$$\frac{r}{a} = \frac{1}{2m^2} \left[\frac{3}{4}\sin^2\theta + (1-2\nu)^2 \cos\frac{\theta}{2} \right] \quad \text{(in plane strain)} ,$$

$$\frac{r}{a} = \frac{1}{2m^2} \left[\frac{3}{4}\sin^2\theta + \cos^2\frac{\theta}{2} \right] \quad \text{(in plane stress)} .$$

We can see that these power expansions, limited as they are to the singular term, give a rather good description of the behaviour under isotropic loading ($n = 0$)—cf Fig. 3.3—but are not in agreement with the behaviour found in uniaxial loading (Fig. 3.4). This is particularly evident on stress trajectories, isochromatic lines and von Mises criterion. (The well-known contours of the von Mises criterion in Fig. 3.11 were given for the first time by McClintock and Irwin [30] in 1965.)

Figure 3.12 shows how the addition of the constant term $-\sigma$ to σ_x modifies the representation of the stresses in uniaxial mode I and bring it nearer of the exact solution. Isochromatic lines deduced from a power expansion with two terms were displayed for the first time by Eftis *et al.* [24] in 1977. The power expansion with three terms brings little improvement.

3.5 Direct Calculation of the Stress Intensity Factors

It can be shown that the terms in $1/\sqrt{r}$ for the stresses and in $\sqrt{r}$ for the displacements are independent of the studied geometry (only $K_{\rm I}$ and $K_{\rm II}$

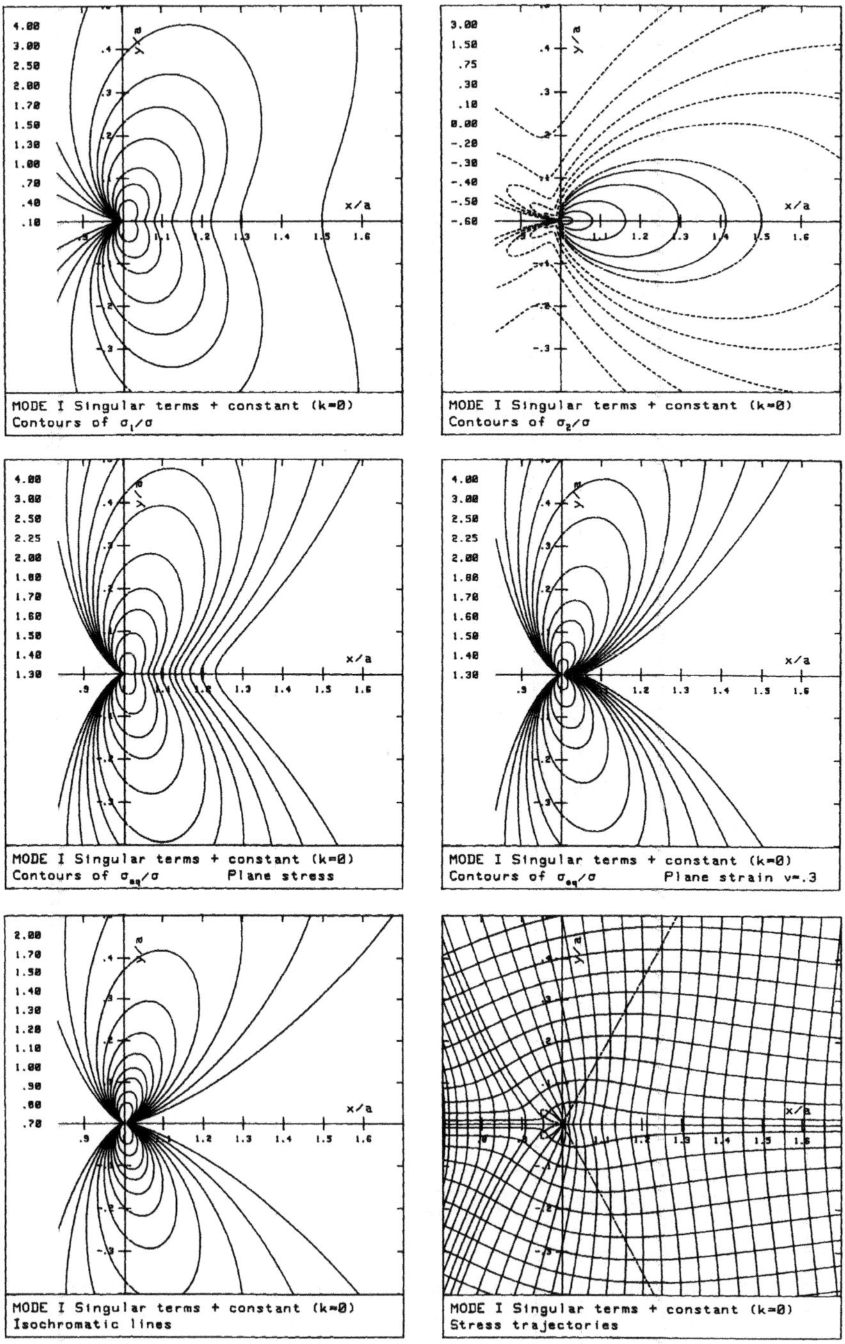

Fig. 3.12. Uniaxial mode I: singular terms + constant term $-\sigma$ on σ_x

depend on the geometry). The other terms are specific of the problem studied. These stress intensity factors play a great role in fracture mechanics because of an interesting property arising from the principle of superposition: for a given mode, the stress intensity factors are additive.

Stress intensity factors can be obtained directly from the Westergaard Z functions or from the Muskhelishvili complex potentials. We will develop only the second method (cf Paris and Sih [31]).

Limited to the singular terms, (3.69) and (3.70) give

$$\sigma_x + \sigma_y = 2\frac{K_{\mathrm{I}}}{\sqrt{2\pi r}}\cos\frac{\theta}{2} - 2\frac{K_{\mathrm{II}}}{\sqrt{2\pi r}}\sin\frac{\theta}{2}\,, \tag{3.82}$$

an expression whose form is independent of the geometry. Let $z_0 = x_0 + \mathrm{i}y_0$ represent the position of the crack tip and let

$$z - z_0 = r\mathrm{e}^{\mathrm{i}\theta}\,.$$

We deduce from this that

$$\frac{1}{\sqrt{z-z_0}} = \frac{1}{\sqrt{r}}\left(\cos\frac{\theta}{2} - \mathrm{i}\sin\frac{\theta}{2}\right)\,,$$

i.e. that

$$\frac{1}{\sqrt{r}}\cos\frac{\theta}{2} = \mathrm{Re}\,\frac{1}{\sqrt{z-z_0}}\,,$$
$$\frac{1}{\sqrt{r}}\sin\frac{\theta}{2} = \mathrm{Re}\,\frac{\mathrm{i}}{\sqrt{z-z_0}}\,.$$

Equation (3.82) can be thus written as

$$\sigma_x + \sigma_y = \mathrm{Re}\left[\frac{2}{\sqrt{2\pi(z-z_0)}}(K_{\mathrm{I}} - \mathrm{i}K_{\mathrm{II}})\right]$$

for z near z_0. Comparing this result with the first equation of Kolossov–Muskhelishvili, shown in (2.54) as

$$\sigma_x + \sigma_y = 4\,\mathrm{Re}\,\varphi'(z)\,,$$

we can derive

$$\boxed{K_{\mathrm{I}} - \mathrm{i}K_{\mathrm{II}} = 2\sqrt{2\pi}\lim_{z\to z_0}\sqrt{z-z_0}\,\varphi'(z)}\,. \tag{3.83}$$

Knowing the function $\varphi'(z)$ for a given problem, it is possible to calculate directly K_{I} and K_{II} without calculation of the stress tensor. Let us return to the example of the elliptic cavity in an infinite plate subjected to biaxial loading. We have seen in Sect. 2.7.2 that $\varphi'(z)$ was given by:

$$\varphi'(z) = \frac{\sigma}{4}\left[n\mathrm{e}^{2\xi_0}\mathrm{e}^{2\mathrm{i}\beta} + \left(m - n\mathrm{e}^{2\xi_0}\mathrm{e}^{2\mathrm{i}\beta}\right)\frac{\cosh\zeta}{\sinh\zeta}\right]\,.$$

We thus have for a crack with $\xi_0 = 0$ and $z = a\cosh\zeta$ (cf (2.87)),

$$\varphi'(z) = \frac{\sigma}{4}\left[ne^{2i\beta} + \left(m - ne^{2i\beta}\right)\frac{z}{\sqrt{z^2 - a^2}}\right] .$$

With $z_0 = a$, (3.83) immediately gives us

$$K_{\mathrm{I}} - \mathrm{i}K_{\mathrm{II}} = \frac{1}{2}\sigma\sqrt{\pi a}\left(m - ne^{2i\beta}\right) ,$$

i.e. (3.67) and (3.68), without developping the exact solution. It is in this manner that Eftis and Subramonian [25] obtained the above stress intensity factors, generalizing the results of Sih *et al.* [32] for uniaxial loading. One can find in the paper by Paris and Sih [31] expressions of stress intensity factors for a number of geometries.

Let us return also to the case of the elliptical cavity subjected to an inner pressure p_0 which is isotropic and uniform. For $b = 0$, the complex potential, (2.143), is written as

$$\varphi(z) = \frac{p_0}{2}(a\sinh\zeta - z) ,$$

and hence

$$\varphi'(z) = \frac{p_0}{2}\left(\frac{z}{\sqrt{z^2 - a^2}} - 1\right) ,$$

so that, using (3.83)

$$K_{\mathrm{I}} = p_0\sqrt{\pi a} \qquad (3.84)$$

without any calculation of σ_y and u_y, which, from (2.153) and (2.151), are known to be

$$\sigma_y(x,0) = p_0\left(\frac{x}{\sqrt{x^2 - a^2}} - 1\right) , \quad (x > a) , \qquad (3.85)$$

$$u_y(x,0) = \frac{\kappa + 1}{4\mu}p_0\sqrt{a^2 - x^2} , \quad (x < a) . \qquad (3.86)$$

Let us give some other examples, this time in axisymmetric geometry, which we will use later. The case of a penny-shaped crack of radius a in an infinite medium submitted to an uniaxial loading perpendicular to the crack plane was studied by Sneddon [17] in 1946. He obtained

$$K_{\mathrm{I}} = \frac{2}{\pi}\sigma\sqrt{\pi a} . \qquad (3.87)$$

Irwin [15] has shown that for this penny-shaped crack there is locally a state of *plane strain*, a result that was generalized later by Kassir and Sih [33] for any tridimensional crack.

The case of an *external circular crack* of radius a (for instance, a deeply notched bar subjected to a load P) has been studied by Kassir and Sih [34] who obtained

$$K_{\mathrm{I}} = \frac{P}{2a\sqrt{\pi a}} = \frac{1}{2}\sigma\sqrt{\pi a}\,. \tag{3.88}$$

We will use this formula in the adherence of a flat punch on an elastic half-plane.

There exists, of course, a relation between the stress intensity factor K_{I} and the stress concentration factor K_t, as shown by Irwin [15] in 1958. We have seen in Sect. 2.7.5, via (2.133), that at the end of the major axis of an ellipse the stress σ_y is given by:

$$\sigma_y = -\sigma n \cos 2\beta + (m - n\cos 2\beta)\sigma\sqrt{\frac{a}{\varrho}}\,,$$

where $\varrho \equiv b^2/a$ is the radius of curvature. We see that when $\varrho \to 0$,

$$\frac{1}{2}\sigma_y\sqrt{\pi\varrho} \to \frac{1}{2}(m - n\cos 2\beta)\sigma\sqrt{\pi a}\,,$$

which is the expression for K_{I}. Hence

$$\boxed{K_{\mathrm{I}} = \lim_{\varrho\to 0}\frac{1}{2}\sigma_y\sqrt{\pi\varrho}}\,. \tag{3.89}$$

There is, however, a problem of physical order when ϱ tends towards zero. All the foregoing expressions have been established on the hypothesis that there is no attraction force between the walls of the cavity. When ϱ decreases, the walls approach closer and σ_y increases. But σ_y cannot exceed the theoretical stress: when this value is reached, bonds begin to rupture and a crack to form. In mode I this arises for an applied stress:

$$\sigma \cong \frac{1}{2}\sigma_{\mathrm{th}}\sqrt{\frac{\varrho}{a}}\,,$$

that is

$$K(\varrho) = \sigma\sqrt{\pi a} = \frac{1}{2}\sigma_{\mathrm{th}}\sqrt{\pi\varrho}\,,$$

where the smaller is ϱ the weaker is $K(\varrho)$. But experiments show that when ϱ decreases, $K(\varrho)$ reaches a limit value K_{I0} and no longer varies. This result is often interpreted by saying that cracks have a "natural" radius of curvature ϱ_{c} (of the order of interatomic distances) and hence

$$\frac{K(\varrho)}{K_{\mathrm{I0}}} = \sqrt{\frac{\varrho}{\varrho_{\mathrm{c}}}}\,.$$

In fact, when ϱ reaches ϱ_c a purely elastic description is no longer valid, and one cannot neglect the strong interaction forces—inner forces of the system—in the description of stresses in the neighbourhood of the crack tip. This is what Griffith had well understood when introducing the concept of surface energy. As we will see below, it is Barenblatt [35] which has shown in 1962 how these forces deform the crack tip, and how the stress singularities disappear.

It follows that it is necessary to study more closely cavities and cracks no longer with a free surface, but with surfaces submitted to a system of stresses.

3.6 Cracks Carrying Interior Tractions

3.6.1 Straight Crack in Plate

Arbitrary Pressure Distribution. Westergaard [18] gave in 1939 the stress function for a crack lying from $x = -a$ to $x = a$, open under the action of two forces of intensity P per unit thickness and situated at $x = 0$. Irwin [15] later generalized the result for two forces applied at a point x $(0 < x < a)$: the stress intensity factor at point a is

$$K_{\mathrm{I}} = \frac{P}{\sqrt{\pi a}}\sqrt{\frac{a+x}{a-x}}\,, \tag{3.90}$$

and at point $-a$:

$$K_{\mathrm{I}} = \frac{P}{\sqrt{\pi a}}\sqrt{\frac{a-x}{a+x}}\,.$$

For a pressure distribution $p(x) > 0$, acting inside the crack along Oy, we have thus at point $x = a$:

$$\begin{aligned} K_{\mathrm{I}} &= \frac{1}{\sqrt{\pi a}}\int_{-a}^{+a} p(x)\left(\frac{a+x}{a-x}\right)^{1/2} \mathrm{d}x \\ &= \sqrt{\frac{a}{\pi}}\int_{-a}^{+a}\frac{p(x)\,\mathrm{d}x}{\sqrt{a^2-x^2}} + \frac{1}{\sqrt{\pi a}}\int_{-a}^{+a}\frac{p(x)x\,\mathrm{d}x}{\sqrt{a^2-x^2}}\,. \end{aligned}$$

If the pressure distribution is symmetrical ($p(x)$ even), we have:

$$\boxed{K_{\mathrm{I}} = 2\sqrt{\frac{a}{\pi}}\int_{0}^{a}\frac{p(x)\,\mathrm{d}x}{\sqrt{a^2-x^2}}}\,. \tag{3.91}$$

For a uniform pressure $p(x) = p$, we obtain $K_{\mathrm{I}} = p\sqrt{\pi a}$, as for hydrostatic pressure as shown in (3.84). (The difference between an isotropic uniform pressure and an uniaxial uniform pressure acting along Oy will be made clear through the constant term in the power expansion of σ_x.)

The exact solution for the opening $u_y(x,0)$ and the stress distribution $\sigma_y(x,0)$ ahead of a crack subjected to a pressure distribution $p(x) = p(-x)$ has been studied by numerous authors. It is possible to obtain

$$u_y(x,0) = \frac{\kappa+1}{2\pi\mu} \int_x^a \frac{t g(t)\,\mathrm{d}t}{\sqrt{t^2-x^2}}, \qquad x < a \tag{3.92a}$$

$$= \frac{\kappa+1}{2\pi\mu} \left[g(a)\sqrt{a^2-x^2} - \int_x^a g'(t)\sqrt{t^2-x^2}\,\mathrm{d}t \right], \tag{3.92b}$$

$$\frac{\mathrm{d}u_y}{\mathrm{d}x} = \frac{\kappa+1}{2\pi\mu} \left[-g(a)\frac{x}{\sqrt{a^2-x^2}} + x\int_x^a \frac{g'(t)\,\mathrm{d}t}{\sqrt{t^2-x^2}} \right], \tag{3.93}$$

$$\sigma_y(x,0) = -\frac{2}{\pi}\frac{\mathrm{d}}{\mathrm{d}x} \int_0^a \frac{t g(t)\,\mathrm{d}t}{\sqrt{x^2-t^2}}, \qquad x > a \tag{3.94a}$$

$$= \frac{2x}{\pi} \int_0^a \frac{t g(t)\,\mathrm{d}t}{(x^2-t^2)^{3/2}} \tag{3.94b}$$

$$= \frac{2x}{\pi} \left[\frac{g(a)}{\sqrt{x^2-a^2}} - \int_0^a \frac{g'(t)\,\mathrm{d}t}{\sqrt{x^2-t^2}} \right] \tag{3.94c}$$

$$= \frac{2x}{\pi\sqrt{x^2-a^2}} \int_0^a \frac{p(s)\sqrt{a^2-s^2}}{x^2-s^2}\,\mathrm{d}s, \tag{3.94d}$$

with

$$g(t) = \int_0^t \frac{p(s)\,\mathrm{d}s}{\sqrt{t^2-s^2}} \tag{3.95a}$$

$$= \frac{\pi}{2}p(t) - \int_0^t p'(s)\sin^{-1}\left(\frac{s}{t}\right)\mathrm{d}s\,. \tag{3.95b}$$

After differentiation under the integral, we get

$$g'(t) = \frac{1}{t}\int_0^t \frac{s p'(s)\,\mathrm{d}s}{\sqrt{t^2-s^2}}\,. \tag{3.96}$$

Equation (3.92a) was obtained by England and Green [36] in 1963 and by Lowengrub [37] in 1966. Equation (3.94c) was also given by Lowengrub (albeit with an error of sign), but (3.94d) was obtained previously by Keer [38] in 1964. This last equation (3.94d) is obtained here by changing the order of integration in (3.94b)

In (3.94c), the first term is a singular term, and its power expansion leads to

$$\sigma_y(x,0) \cong 2\sqrt{\frac{a}{\pi}}\frac{g(a)}{\sqrt{2\pi r}} = \frac{K_\mathrm{I}}{\sqrt{2\pi r}}, \tag{3.97}$$

with

$$K_{\mathrm{I}} = 2\sqrt{\frac{a}{\pi}}g(a) = 2\sqrt{\frac{a}{\pi}}\int_0^a \frac{p(s)\,\mathrm{d}s}{\sqrt{a^2 - s^2}}\,, \tag{3.98}$$

which is (3.91). Similarly the power expansion of (3.92b) gives

$$u_y(x,0) = \frac{\kappa + 1}{2\mu}K_{\mathrm{I}}\sqrt{\frac{r}{2\pi}}\,. \tag{3.99}$$

In the Dugdale and Barenblatt models, as we shall see below, the singular term due to the stress distribution inside the crack (intermolecular forces) is cancelled by the singular term due to external loading. It is thus necessary to proceed further in the power expansions of σ_y and u_y.

Goodier [39] has shown by a rather tedious calculation that the power expansion of (3.94d) can be written as

$$\sigma_y(x,0) = \frac{K_{\mathrm{I}}}{\sqrt{2\pi r}} - p(a) + \cdots\,.$$

Let us prove this result. Multiply above and below by $\sqrt{a^2 - s^2}$ in (3.94d) and write $(a^2 - s^2) = (a^2 - x^2 + x^2 - s^2)$. We then get

$$\begin{aligned}\sigma_y(x,0) &= \frac{K_{\mathrm{I}}}{\sqrt{\pi a}}\frac{x}{\sqrt{x^2 - a^2}} \\ &\quad - \frac{2x}{\pi}\sqrt{x^2 - a^2}\int_0^a \frac{p(s)\,\mathrm{d}s}{(x^2 - s^2)\sqrt{a^2 - s^2}}\,.\end{aligned} \tag{3.100a}$$

An integration by parts leads to:

$$\begin{aligned}\sigma_y(x,0) &= \frac{K_{\mathrm{I}}}{\sqrt{\pi a}}\frac{x}{\sqrt{x^2 - a^2}} - p(a) \\ &\quad + \frac{2}{\pi}\int_0^a p'(s)\tan^{-1}\left(\frac{s\sqrt{x^2 - a^2}}{x\sqrt{a^2 - s^2}}\right)\mathrm{d}s\,.\end{aligned} \tag{3.100b}$$

As the last term tends towards zero when x tends towards a, we have the searched result.

In the expression of u_y in (3.92b), t is very close to a when $x \to a$, and $g'(t)$ is close to $g'(a)$; hence:

$$\int_x^a g'(t)\sqrt{t^2 - x^2}\;\mathrm{d}t \approx g'(a)\sqrt{2a}\int_x^a \sqrt{t - x}\;\mathrm{d}t\,,$$

i.e.

$$u_y(x,0) \approx \frac{\kappa + 1}{2\pi\mu}\left[g(a)\sqrt{a^2 - x^2} + \frac{2}{3}g'(a)\sqrt{2a}(a - x)^{3/2}\right]\,. \tag{3.101}$$

In the Barenblatt model, $p(x)$ represents the distribution of intermolecular forces and $-p(a)$ the theoretical stress σ_{th}.

Constant Pressure at the Crack Tip. Consider first the simple case of a constant pressure p, acting along the y-axis, in the entire crack. We get $g(t) = p\pi/2$ and, using (3.94b) and (3.92a) we have

$$\sigma_y(x,0) = p\left(\frac{x}{\sqrt{x^2-a^2}} - 1\right) , \qquad x > a , \tag{3.102}$$

$$u_y(x,0) = \frac{\kappa+1}{4\mu} p\sqrt{a^2-x^2} , \qquad x < a , \tag{3.103}$$

similar to the equations for hydrostatic pressure, (see (3.85) and (3.86)).

Consider now the case of a linear crack subjected to a constant pressure p (positive) between c and a, and $-c$ and $-a$, but zero between $-c$ and c. Equations (3.95a) and (3.98) become:

$$\begin{aligned} g(t) &= p\int_c^t \frac{ds}{\sqrt{t^2-s^2}} = p\cos^{-1}\frac{c}{t} \quad \text{if } t > c , \\ &= 0 \qquad\qquad\qquad\qquad\qquad \text{if } t < c , \\ K_\mathrm{I} &= 2\sqrt{\frac{a}{\pi}} p\cos^{-1}\frac{c}{a} = 2\sqrt{\frac{a}{\pi}} g(a) , \end{aligned} \tag{3.104}$$

and we then have

$$g'(t) = \frac{pc}{t\sqrt{t^2-c^2}} \qquad \text{if } t > c . \tag{3.105}$$

If we note that the pressure is a step function, and that its derivative is $p' = p\delta(c)$, where $\delta(c)$ is the Dirac function, (3.96) immediately gives (3.105), and (3.100b) gives:

$$\boxed{\begin{aligned} \sigma_y(x,0) = {} & \frac{K_\mathrm{I}}{\sqrt{\pi a}} \frac{x}{\sqrt{x^2-a^2}} \\ & - p + \frac{2p}{\pi}\tan^{-1}\frac{c\sqrt{x^2-a^2}}{x\sqrt{a^2-c^2}} , \qquad x > a \end{aligned}} , \tag{3.106}$$

whose power expansion at $r = x - a$ is

$$\sigma_y(x,0) \approx \frac{K_\mathrm{I}}{\sqrt{2\pi r}} - p + \left(\frac{2p}{\pi\sqrt{\pi a}} \frac{c}{\sqrt{a^2-c^2}} + \frac{3K_\mathrm{I}}{8\pi a}\right)\sqrt{2\pi r} . \tag{3.107}$$

Using (3.105), (3.92b) can be written as

$$u_y(x,0) = \frac{\kappa+1}{2\pi\mu}\left[g(a)\sqrt{a^2-x^2} - pc\int_{\max(x,c)}^a \frac{\sqrt{t^2-x^2}}{t\sqrt{t^2-c^2}}\,\mathrm{d}t\right] . \tag{3.108}$$

For $x = c$ we immediately have

$$u_y(c,0) = \frac{\kappa+1}{2\pi\mu}\left[g(a)\sqrt{a^2-c^2} - pc\ln\frac{a}{c}\right] . \tag{3.109}$$

By multiplying above and below by $\sqrt{t^2-x^2}$ in (3.108) we obtain two simple integrals which give

$$\boxed{u_y(x,0) = \frac{\kappa+1}{4\pi\mu}\left[K_{\rm I}\sqrt{\frac{\pi}{a}}\sqrt{a^2-x^2} + pc\ln\frac{|\xi-1|}{\xi+1} - px\ln\frac{|c\xi-x|}{c\xi+x}\right]}\ , \tag{3.110}$$

with

$$\xi = \sqrt{\frac{a^2-x^2}{a^2-c^2}}\ .$$

The slope of the profile given by (3.93) is thus

$$\frac{\mathrm{d}u_y}{\mathrm{d}x} = \frac{\kappa+1}{4\pi\mu}\left[-K_{\rm I}\sqrt{\frac{\pi}{a}}\frac{x}{\sqrt{a^2-x^2}} + p\ln\frac{\xi+x/c}{|\xi-x/c|}\right]\ . \tag{3.111}$$

There is thus a vertical tangent at $x=a$ and at $x=c$. These displacements are displayed in Fig. 3.13, where dotted lines correspond to the crack opening at $x=c$, (3.109).

3.6.2 Penny-Shaped Crack

Arbitrary Pressure Distribution. The case of a penny-shaped crack of radius a in an infinite medium submitted to an axisymmetric pressure distribution $p(r)$ on $0<r<a$ was studied by Sneddon [17,40]. The opening of the crack and the stress σ_y ahead of the crack are given by

$$u_y(r,0) = \frac{4(1-\nu^2)}{\pi E}\int_r^a \frac{g(t)\,\mathrm{d}t}{\sqrt{t^2-r^2}}\ , \qquad r<a\ , \tag{3.112}$$

$$\sigma_y(r,0) = \frac{2}{\pi}\left[\frac{g(a)}{\sqrt{r^2-a^2}} - \int_0^a \frac{g'(t)\,\mathrm{d}t}{\sqrt{r^2-t^2}}\right]\ , \qquad r>a\ , \tag{3.113}$$

with

$$g(t) = \int_0^t \frac{sp(s)\,\mathrm{d}s}{\sqrt{t^2-s^2}} \tag{3.114a}$$

$$= p(0)t + \int_0^t p'(s)\sqrt{t^2-s^2}\ \mathrm{d}s\ , \tag{3.114b}$$

and after differentiation under the integral

$$g'(t) = p(0) + t\int_0^t \frac{p'(s)\,\mathrm{d}s}{\sqrt{t^2-s^2}}\ . \tag{3.115}$$

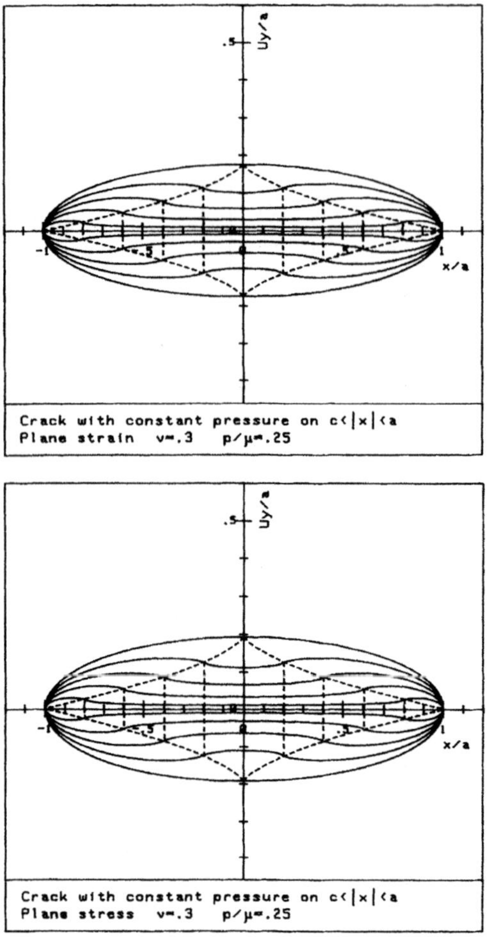

Fig. 3.13. Linear crack subjected to a constant pressure near its ends. Plane strain and plane stress cases

After integration by parts, and then changing the order of integration, (3.113) can be written:

$$\sigma_y(r,0) = \frac{2}{\pi}\frac{1}{\sqrt{r^2-a^2}}\int_0^a \frac{sp(s)\sqrt{a^2-s^2}}{r^2-s^2}\,\mathrm{d}s\,, \tag{3.116}$$

a result given by Keer [38] and Chan *et al.* [41].

In (3.113), the first term is a singular one, and its power expansion leads immediately to

$$\boxed{K_{\mathrm{I}} = \frac{2}{\sqrt{\pi a}}g(a) = \frac{2}{\sqrt{\pi a}}\int_0^a \frac{sp(s)\,\mathrm{d}s}{\sqrt{a^2-s^2}}}\,, \tag{3.117}$$

a result first given by Barenblatt [35].

Constant Pressure Inside the Crack. For a constant pressure $p > 0$ acting along the y-axis in the entire crack, we have $g(t) = pt$ and (3.117), (3.112) and (3.113) and boundary conditions reduce to

$$K_{\mathrm{I}} = \frac{2}{\pi} p\sqrt{\pi a} \,,$$

$$u_y(r,0) = \frac{4(1-\nu^2)}{\pi E} p\sqrt{a^2 - r^2} \,, \qquad r < a \,, \tag{3.118}$$

$$\begin{aligned} \sigma_y(r,0) &= \frac{2p}{\pi}\left(\frac{a}{\sqrt{r^2-a^2}} - \sin^{-1}\frac{a}{r}\right), && r > a \,, \\ \sigma_y(r,0) &= -p \,, && r < a \,, \end{aligned} \tag{3.119}$$

whose power expansion, respectively at $\varrho = a - r$ and $\varrho = r - a$ are

$$u_y(\varrho,0) \approx 4\frac{1-\nu^2}{E} K_{\mathrm{I}} \sqrt{\frac{\varrho}{2\pi}} \,, \tag{3.120}$$

$$\sigma_y(\varrho,0) \approx \frac{K_{\mathrm{I}}}{\sqrt{2\pi\varrho}} - p + \frac{7K_{\mathrm{I}}}{4a}\sqrt{\frac{\varrho}{2\pi}} \,. \tag{3.121}$$

Letting $u_0 = u_y(0,0)$, we can see that (3.118) can be written

$$\frac{u_y^2}{u_0^2} + \frac{r^2}{a^2} = 1 \,,$$

which shows that the penny-shaped crack under a constant pressure is deformed into a flat ellipsoid of revolution.

Let us now superimpose on the previous system a uniaxial stress $\sigma = p$ acting at infinity and on the crack faces. We get:

$$u_y(r,0) = \frac{4(1-\nu^2)}{\pi E} \sigma\sqrt{a^2 - r^2} \,, \qquad r < a \,, \tag{3.122}$$

$$\begin{aligned} \sigma_y(r,0) &= 0 \,, && r < a \,, \\ \sigma_y(r,0) &= \frac{2\sigma}{\pi}\left(\frac{a}{\sqrt{r^2-a^2}} + \cos^{-1}\frac{a}{r}\right), && r > a \,, \end{aligned} \tag{3.123}$$

$$K_{\mathrm{I}} = \frac{2}{\pi} \sigma\sqrt{\pi a} \,, \tag{3.124}$$

which is the solution for a penny-shaped crack in an infinite medium subjected to an uniaxial traction perpendicular to the crack plane.

Constant Pressure Near the Crack Tip. For a pressure p constant on $c < r < a$, zero on $r < c$, (3.114a) and (3.117) become

$$g(t) = \begin{cases} p \int_0^t \frac{s\,\mathrm{d}s}{\sqrt{t^2 - s^2}} = p\sqrt{t^2 - c^2}\,, & t > c\,, \\ 0\,, & t < c\,, \end{cases} \tag{3.125}$$

$$K_{\mathrm{I}} = \frac{2}{\sqrt{\pi a}} g(a) = \frac{2p}{\sqrt{\pi a}} \sqrt{a^2 - c^2}\,, \tag{3.126}$$

and we have

$$g'(t) = \begin{cases} \frac{pt}{\sqrt{t^2 - c^2}}\,, & t > c\,, \\ 0\,, & t < c\,. \end{cases} \tag{3.127}$$

Equation (3.113) can then be written as

$$\sigma_y(r, 0) = \frac{K_{\mathrm{I}}}{\sqrt{\pi a}} \frac{a}{\sqrt{r^2 - a^2}} - \frac{p}{\pi} \int_c^a \frac{2t\,\mathrm{d}t}{\sqrt{t^2 - c^2}\sqrt{r^2 - t^2}}\,,$$

i.e.

$$\boxed{\begin{aligned} \sigma_y(r, 0) = {} & \frac{K_{\mathrm{I}}}{\sqrt{\pi a}} \frac{a}{\sqrt{r^2 - a^2}} \\ & - p + \frac{2p}{\pi} \tan^{-1} \sqrt{\frac{r^2 - a^2}{a^2 - c^2}}\,, \qquad r > a\,, \end{aligned}} \tag{3.128}$$

whereas u_y, from (3.112), becomes

$$\begin{aligned} u_y(r, 0) &= 4\frac{1 - \nu^2}{\pi E} p \int_{\max(r,c)}^a \sqrt{\frac{t^2 - c^2}{t^2 - r^2}}\,\mathrm{d}t\,, & r < a\,, \\ &= 4\frac{1 - \nu^2}{\pi E} p \int_{\max(r,c)}^a \frac{\sqrt{t^2 - c^2}}{t}\,\mathrm{d}\left(\sqrt{t^2 - r^2}\right)\,, & r < a\,. \end{aligned}$$

Let us integrate by parts to reveal the term in $\sqrt{a^2 - r^2}$, thus:

$$\boxed{\begin{aligned} u_y(r, 0) = {} & 2\frac{1 - \nu^2}{E} \\ & \times \left[\frac{K_{\mathrm{I}}}{\sqrt{\pi a}} \sqrt{a^2 - r^2} - \frac{2p}{\pi} c^2 \int_{\max(r,c)}^a \frac{\sqrt{t^2 - r^2}\,\mathrm{d}t}{t^2 \sqrt{t^2 - c^2}} \right]\,. \end{aligned}} \tag{3.129}$$

This integral can be expressed as functions of elliptic integrals[3] $F(\varphi, k)$, $E(\varphi, k)$, $K(k) = F(\pi/2, k)$, and $E(k) = E(\pi/2, k)$.

For $r < c$ we obtain

$$u_y(r,0) = 2\frac{1-\nu^2}{E} \times \left\{ \frac{K_{\mathrm{I}}}{\sqrt{\pi a}}\sqrt{a^2 - r^2} + \frac{2p}{\pi} c \left[E\left(\varphi_2, \frac{r}{c}\right) - E\left(\frac{r}{c}\right) \right] \right\} , \tag{3.130}$$

with $\varphi_2 = \sin^{-1}(c/a)$; and for $c < r < a$ we obtain

$$u_y(r,0) = 2\frac{1-\nu^2}{E} \left\{ \frac{K_{\mathrm{I}}}{\sqrt{\pi a}}\sqrt{a^2 - r^2} + \frac{2p}{\pi} r \left[E\left(\varphi_1, \frac{c}{r}\right) - E\left(\frac{c}{r}\right) \right] - \frac{2p}{\pi}\frac{r^2 - c^2}{r} \left[F\left(\varphi_1, \frac{c}{r}\right) - K\left(\frac{c}{r}\right) \right] \right\} , \tag{3.131}$$

with $\varphi_1 = \sin^{-1}(r/a)$.

For $r = c$, these equations simplify, since $E(\varphi, 1) = \sin\varphi$:

$$\begin{aligned} u_y(c,0) &= 2\frac{1-\nu^2}{E} \left[\frac{K_{\mathrm{I}}}{\sqrt{\pi a}}\sqrt{a^2 - c^2} - \frac{2p}{\pi}\frac{c}{a}(a-c) \right] \\ &= 4\frac{1-\nu^2}{\pi E} p(a-c) . \end{aligned} \tag{3.132}$$

3.6.3 External Circular Crack

Arbitrary Pressure Distribution in the Crack. In the case of an external circular crack of radius a in an infinite solid (whose approximation can be a deeply notched bar) subjected to an axisymmetric pressure distribution, Lowengrub and Sneddon [42] obtained

$$u_y(r,0) = 4\frac{1-\nu^2}{\pi E} \int_a^r \frac{g(t)\,\mathrm{d}t}{\sqrt{r^2 - t^2}} , \qquad r > a , \tag{3.133}$$

$$\sigma_y(r,0) = \frac{2}{\pi} \left[\frac{g(a)}{\sqrt{a^2 - r^2}} + \int_a^\infty \frac{g'(t)\,\mathrm{d}t}{\sqrt{t^2 - r^2}} \right] , \qquad r < a , \tag{3.134}$$

[3] These elliptic integrals are defined by

$$E(\varphi, k) = \int_0^\varphi \sqrt{1 - k^2 \sin^2\varphi}\;\mathrm{d}\varphi$$

$$F(\varphi, k) = \int_0^\varphi \frac{\mathrm{d}\varphi}{\sqrt{1 - k^2 \sin^2\varphi}} .$$

with

$$g(t) = \int_t^\infty \frac{sp(s)\,ds}{\sqrt{s^2 - t^2}} . \tag{3.135}$$

Here also, the power expansion of the singular term leads immediately to

$$\boxed{K_{\rm I} = \frac{2}{\sqrt{\pi a}} \int_a^\infty \frac{rp(r)\,dr}{\sqrt{r^2 - a^2}}} , \tag{3.136}$$

a result also given by Fabrikant [43].

We can remark that the stresses in the plane $y = 0$ are not self-equilibrated, for the hypothesis implicitly done above is that of zero displacement when y is infinite: there thus exist forces at infinity for doing so. This problem has, of course, an infinity of solutions, depending on the boundary conditions chosen at infinity [44, 45].

Constant Pressure at the Crack Tip. For a constant pressure p, positive on $a < r < c$ and zero on $r > c$, we have

$$g(t) = \begin{cases} p\sqrt{c^2 - t^2} , & a < t < c , \\ 0 , & t \geq c , \end{cases} \tag{3.137}$$

$$g'(t) = \begin{cases} -\dfrac{pt}{\sqrt{c^2 - t^2}} , & a < t < c , \\ 0 , & t \geq c , \end{cases} \tag{3.138}$$

and (3.134) and (3.136) then become

$$\sigma_y(r, 0) = \frac{2p}{\pi} \left[\sqrt{\frac{c^2 - a^2}{a^2 - r^2}} - \tan^{-1} \sqrt{\frac{c^2 - a^2}{a^2 - r^2}} \right] , \tag{3.139}$$

$$K_{\rm I} = \frac{2p}{\sqrt{\pi a}} \sqrt{c^2 - a^2} . \tag{3.140}$$

However, as stated above, the stresses in the plane $y = 0$ are not self-equilibrated since the force

$$\int_0^a \sigma_y 2\pi r\,dr = p\pi(c^2 - a^2) - 2pa^2 \left[\frac{c^2}{a^2} \cos^{-1} \frac{a}{c} - \sqrt{\frac{c^2}{a^2} - 1} \right]$$

does not equilibrate with the force $p\pi(c^2 - a^2)$ applied inside the crack. A force

$$P' = 2pa^2 \left[\frac{c^2}{a^2} \cos^{-1} \frac{a}{c} - \sqrt{\frac{c^2}{a^2} - 1} \right] > 0 \tag{3.141}$$

is thus exerted at infinity which permits $u_y(r, \infty)$ to be zero.

Keeping the radius a constant, let us apply at infinity a force $-P'$ (tensile force). As we will see at Chapt. 4, this leads to a displacement like the Boussinesq flat punch and introduces in the ligament a stress distribution given by

$$\sigma_{\mathrm{B}} = \frac{P'}{2\pi a} \frac{1}{\sqrt{a^2 - r^2}} > 0 \,. \tag{3.142}$$

Let us add the two distributions (3.139) and (3.142). We obtain

$$\sigma_{\mathrm{T}}(r,0) = \frac{p}{\pi} \left[\frac{\sqrt{c^2 - a^2} + (c^2/a)\cos^{-1} a/c}{\sqrt{a^2 - r^2}} - 2\tan^{-1} \sqrt{\frac{c^2 - a^2}{a^2 - r^2}} \right] \tag{3.143}$$

and we can verify that this distribution equilibrates with the force $p\pi(c^2 - a^2)$, the force applied at infinity being zero.

Let us introduce the stress intensity factor

$$\boxed{K_{\mathrm{I}} = \frac{p}{\sqrt{\pi a}} \left[\sqrt{c^2 - a^2} + \frac{c^2}{a} \cos^{-1} \frac{a}{c} \right]} \,. \tag{3.144}$$

Equation (3.143) then becomes

$$\boxed{\sigma_{\mathrm{T}}(r,0) = \frac{K_{\mathrm{I}}}{\sqrt{\pi a}} \frac{a}{\sqrt{a^2 - r^2}} - \frac{2p}{\pi} \tan^{-1} \sqrt{\frac{c^2 - a^2}{a^2 - r^2}}} \,. \tag{3.145}$$

Integration of (3.133) leads to:

$$u_y(r,0) = 4\frac{1 - \nu^2}{\pi E} \left[\frac{p}{a} \sqrt{c^2 - a^2} \sqrt{r^2 - a^2} \right.$$
$$\left. - pc^2 \int_a^{\min(r,c)} \frac{\sqrt{r^2 - t^2}}{t^2 \sqrt{c^2 - t^2}} \, \mathrm{d}t \right] , \qquad r > a \,, \tag{3.146}$$

the remaining integral being an elliptic integral.

However, we must take into account the elastic displacement due to the flat punch motion under the load $-P'$. We will see in Boussinesq's theory that the profile of the surface (cf Fig. 3.14) is given by

$$u_{\mathrm{B}} = -\frac{1 - \nu^2}{\pi E} \frac{P'}{a} \sin^{-1} \frac{a}{r} \,, \qquad r \geq a \,, \tag{3.147}$$

and the "penetration" of the punch is defined by

$$\delta_{\mathrm{B}} = -\frac{1 - \nu^2}{\pi E} \frac{P'}{a} \frac{\pi}{2} < 0 \,. \tag{3.148}$$

Taking the origin of displacement at the tip of the crack (Fig. 3.14), we have

$$\begin{aligned} u_y' &= \delta_{\mathrm{B}} - u_{\mathrm{B}} \\ &= -\frac{1-\nu^2}{\pi E}\frac{P'}{a}\cos^{-1}\frac{a}{r}\,, \end{aligned}$$

i.e.

$$u_y'(r,0) = 2\frac{1-\nu^2}{\pi E}pa\left[\sqrt{\frac{c^2}{a^2}-1}-\frac{c^2}{a^2}\cos^{-1}\frac{a}{c}\right]\cos^{-1}\frac{a}{r} < 0\,. \tag{3.149}$$

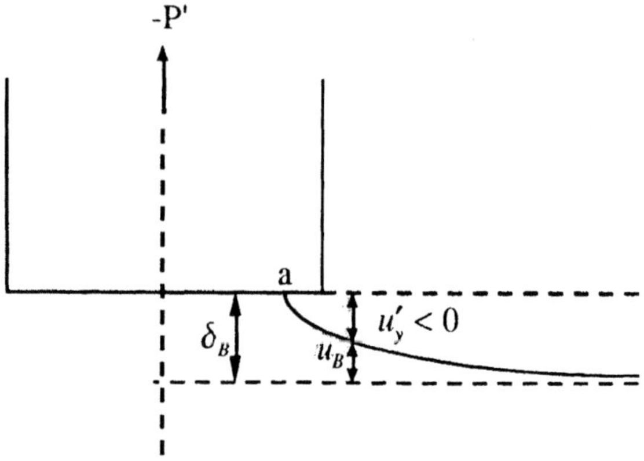

Fig. 3.14. Rigid body displacement

We must now add this displacement to the displacement given by (3.146) taking care with the sign. The force $-P'$ at infinity increases the opening u_y so that

$$u_{\mathrm{T}} = u_y - u_y'\,,$$

i.e.

$$\boxed{\begin{aligned} u_{\mathrm{T}}(r,0) = {}& 2\frac{1-\nu^2}{\pi E}K_{\mathrm{I}}\sqrt{\pi a}\cos^{-1}\frac{a}{r} \\ & + 4\frac{1-\nu^2}{\pi E}p\left[\sqrt{c^2-a^2}\left(\sqrt{\frac{r^2}{a^2}-1}-\cos^{-1}\frac{a}{r}\right)\right. \\ & \left. - c^2\int_a^{\min(r,c)}\frac{\sqrt{r^2-t^2}}{t^2\sqrt{c^2-t^2}}\,\mathrm{d}t\right], \end{aligned}} \tag{3.150}$$

which describes the opening of the crack due to the pressure p applied in the vicinity of the crack tip. We will make use of these equations in the study of the contact of a sphere on plane.

Note carefully that these different stress intensity factors only describe the singular stresses in the material in the vicinity of the crack tip. These singular stresses cancel on the walls of the crack, with $\theta = \pm\pi$. However, these walls remain subjected to the stresses $p(x)$ or $p(r)$ which have been imposed.

3.7 Calculation of G, Irwin's Formula, and the J Integral

3.7.1 Irwin's Formula

Irwin [3] showed in 1957 the equivalence between the thermodynamic aspect and the mechanical aspect, or more precisely between G and K since the concept of surface energy does not appear in the mechanical aspect.

Let us consider (Fig. 3.15) a crack in mode I whose tip is at $x = \mathrm{d}a$ with $\mathrm{d}a$ small. The final state (stress and displacement) is drawn in full line. Let us evaluate the work done, which is by definition $G\,\mathrm{d}a$ (per unit thickness) with limitation to singular terms only.

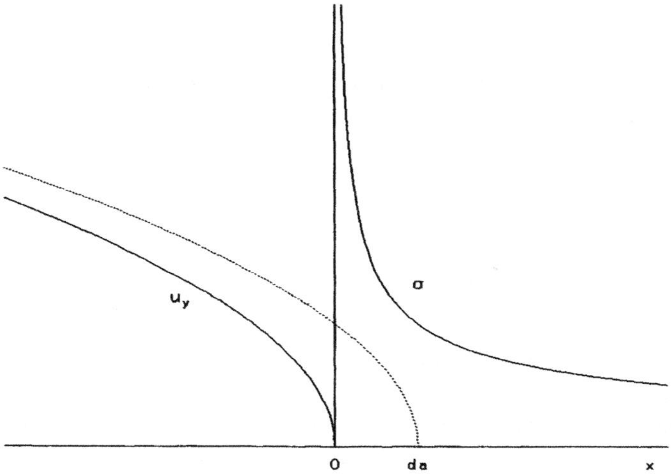

Fig. 3.15. Representation of a crack in mode I with the crack tip at $\mathrm{d}a$

This work is the integral on $\mathrm{d}a$ of the elementary work of the stresses $\sigma_y = K_\mathrm{I}/\sqrt{2\pi x}$ to displace their application points by the quantity:

$$[u_y] = \frac{\kappa + 1}{\mu} K_\mathrm{I} \sqrt{\frac{\mathrm{d}a - x}{2\pi}} ,$$

which represents the opening of the crack in its initial position, given by (3.74). Hence, applying the formula of *linear elasticity*, we get

$$G_{\mathrm{I}}\,\mathrm{d}a = \int_0^{\mathrm{d}a} \frac{1}{2}\sigma_y\,[u_y]\,\mathrm{d}x = \frac{\kappa+1}{4\pi\mu}{K_{\mathrm{I}}}^2 \int_0^{\mathrm{d}a} \left(\frac{\mathrm{d}a - x}{x}\right)^{1/2} \mathrm{d}x\,.$$

Letting $x = \mathrm{d}a \sin^2\alpha$, we obtain easily:

$$G_{\mathrm{I}} = \frac{\kappa+1}{8\mu}{K_{\mathrm{I}}}^2\,, \tag{3.151}$$

i.e.

$$G_{\mathrm{I}} = \frac{1-\nu^2}{E}{K_{\mathrm{I}}}^2 \qquad \text{in plane strain, and}$$

$$G_{\mathrm{I}} = \frac{{K_{\mathrm{I}}}^2}{E} \qquad \text{in plane stress}\,.$$

$G\,\mathrm{d}a$ thus appears as the work of singular stresses at the crack tip when it moves by $\mathrm{d}a$. The following terms in the power expansion lead to terms in $\mathrm{d}a$ of an order of magnitude larger, which disappear[4] when $\mathrm{d}a \to 0$. (Note that it is to obtain a simple formula for G that the $\sqrt{\pi}$ was introduced in K_{I}.)

The same reasoning applies for modes II and III, with the results

$$G_{\mathrm{II}} = \frac{\kappa+1}{8\mu}{K_{\mathrm{II}}}^2\,, \tag{3.152}$$

$$G_{\mathrm{III}} = \frac{1}{2\mu}{K_{\mathrm{III}}}^2\,. \tag{3.153}$$

Modes I, II and III being uncoupled at the crack tip, there is no mutual work, and for a mixed loading we will have:

$$\boxed{G = \frac{\kappa+1}{8\mu}\left({K_{\mathrm{I}}}^2 + {K_{\mathrm{II}}}^2\right) + \frac{1}{2\mu}{K_{\mathrm{III}}}^2}\,. \tag{3.154}$$

These equations, due to Irwin, have been established with the following hypotheses:

1. the propagation of the crack is in its plane (the crack does not turn, even in mixed mode);
2. there is a linear relation between stress and strain;
3. no interaction occurs between crack faces.

---

[4] It is possible to verify in the particular case of a crack traversing a plate that (3.151) is obtained using the exact solutions of (3.48) and (3.63) which in mode I are written:

$$\sigma_y = \frac{\sigma x}{\sqrt{x^2 - a^2}} \qquad x > a\,,$$

$$[u_y] = \frac{\kappa+1}{2\mu}\sigma\sqrt{a^2 - x^2} \qquad x < a\,,$$

where x is the abscissa of a point from the centre of the crack

3.7.2 Remarks

We must clearly see that (3.154) is not a definition of G, but a calculation of G, taking into account the above hypotheses. In particular, we will see with regard to the Barenblatt and Dugdale models that this equation is valid provided that the length over which the interaction forces act (cohesion zone) is negligible compared with the length of the crack or to a similar characteristic length of the system, as remarked by Griffith as early as 1920. In this cohesion zone, the stress–strain relations are strongly nonlinear, for example in the Lennard–Jones model. The range of validity of (3.154) is that of *Linear Elastic Fracture Mechanics* (LEFM).

In a number of cases the stress intensity factors can be written in the form $K_{\mathrm{I}} \approx \sigma\sqrt{\pi a}$. If, in absence of crack, the specimen is submitted to a constant stress σ, the density of elastic energy is in every point given by $W = \sigma^2/2E$, and we can write G in the form

$$G = kaW \ .$$

The Rice J integral will generalize this use of the density of elastic energy.

3.7.3 The J Integral

With the Irwin formula, it is the first time in this chapter that we meet the hypothesis of linear elasticity, which the definition of G as

$$G = \left(\frac{\partial U_{\mathrm{E}}}{\partial A} + \frac{\partial U_{\mathrm{P}}}{\partial A}\right)_{\mathrm{P}} = \left(\frac{\partial U_{\mathrm{E}}}{\partial A}\right)_{\delta}$$

does not imply. Rice [46, 47] showed in 1968 how to compute G as a function of stresses and deformations, free of any hypothesis of linearity. For this delicate demonstration we will follow Kanninen and Popelar [48].

Let us consider (Fig. 3.16) a solid of unit thickness and of area S having a linear crack of length a along $\mathrm{O}x_1$ and whose faces are assumed to be traction-free. We have then $\mathrm{d}A = -\mathrm{d}a$. We will define a density of elastic energy at any point of the solid by

$$W = W(x_1, x_2) = W(\varepsilon_{ij}) = \int_0^{\varepsilon_{ij}} \sigma_{ij}\,\mathrm{d}\varepsilon_{ij} \ . \tag{3.155}$$

The differential

$$\mathrm{d}W = \sigma_{ij}\,\mathrm{d}\varepsilon_{ij}$$

being a total differential since W depends only on the final state. (Sums on i and j are understood.) Let us recall that for an isotropic and linear elastic solid, (3.155) reduces to

$$W = \frac{1}{2}\left(\sigma_1\varepsilon_1 + \sigma_2\varepsilon_2 + 2\tau_{12}\varepsilon_{12}\right) \ .$$

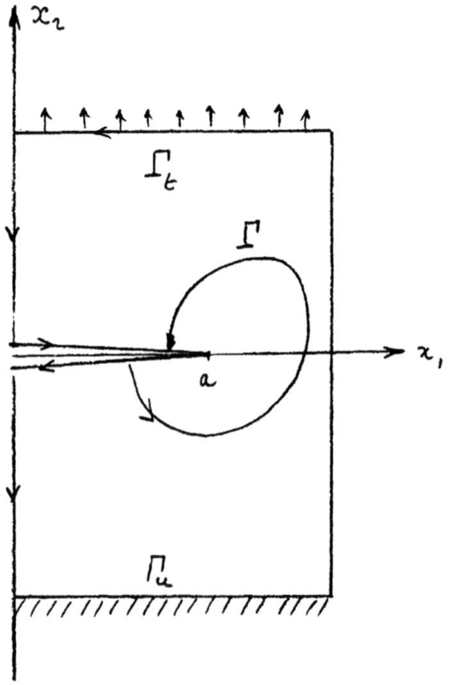

Fig. 3.16. A solid with a linear crack

The elastic energy of the solid will thus be

$$U_{\mathrm{E}} = \iint_S W \,\mathrm{d}S \,.$$

Call Γ_t the part of the contour of the solid on which are applied the imposed stress vectors $\boldsymbol{T}$, causing displacements $\boldsymbol{u}$, and call Γ_u the part of the contour on which the displacements $\boldsymbol{u}$ are imposed, the total contour (closed) of the solid being $\Gamma_0 = \Gamma_t + \Gamma_u$. (The faces of the cracks on which $\boldsymbol{T} = 0$ are included in Γ_t.)

The potential energy of the body is

$$U_{\mathrm{P}} = -\int_{\Gamma_t} T_i u_i \,\mathrm{d}s \,,$$

where $\mathrm{d}s$ is a length element on Γ_t. We have thus

$$U_{\mathrm{E}} + U_{\mathrm{P}} = \iint_S W \,\mathrm{d}S - \int_{\Gamma_t} T_i u_i \,\mathrm{d}s$$

and, assuming that the imposed stresses T_i are independent of the crack length, we have

$$G = -\left(\frac{\partial U_{\mathrm{E}}}{\partial a} + \frac{\partial U_{\mathrm{P}}}{\partial a}\right)_{T_i} = -\iint_S \frac{\mathrm{d}W}{\mathrm{d}a} \,\mathrm{d}S + \int_{\Gamma_t} T_i \frac{\mathrm{d}u_i}{\mathrm{d}a} \,\mathrm{d}s \,.$$

The displacements imposed on Γ_u being independent of a, we can extend the path of integrations to Γ_0, thus:

$$G = -\iint_S \frac{\mathrm{d}W}{\mathrm{d}a}\,\mathrm{d}S + \int_{\Gamma_0} T_i \frac{\mathrm{d}u_i}{\mathrm{d}a}\,\mathrm{d}s\,.$$

Placing the origin at the crack tip ($X_1 = x_1 - a$) we have

$$\begin{aligned}\frac{\mathrm{d}}{\mathrm{d}a} \equiv \left(\frac{\partial}{\partial a}\right)_{x_1} &= \left(\frac{\partial}{\partial a}\right)_{X_1} + \left(\frac{\partial}{\partial X_1}\right)_a \left(\frac{\partial X_1}{\partial a}\right)_{x_1}\\ &= \left(\frac{\partial}{\partial a}\right)_{X_1} - \left(\frac{\partial}{\partial x_1}\right)_a\,,\end{aligned}$$

i.e.

$$G = -\iint_S \left(\frac{\partial W}{\partial a} - \frac{\partial W}{\partial x_1}\right)\mathrm{d}S + \int_{\Gamma_0} T_i \left(\frac{\partial u_i}{\partial a} - \frac{\partial u_i}{\partial x_1}\right)\mathrm{d}s\,. \qquad (3.156)$$

However,

$$\begin{aligned}\frac{\partial W}{\partial a} &= \frac{\partial W}{\partial \varepsilon_{ij}}\frac{\partial \varepsilon_{ij}}{\partial a}\\ &= \sigma_{ij}\frac{\partial \varepsilon_{ij}}{\partial a}\\ &= \sigma_{ij}\left[\frac{1}{2}\frac{\partial}{\partial a}\left(\frac{\partial u_i}{\partial x_j} + \frac{\partial u_j}{\partial x_i}\right)\right]\\ &= \sigma_{ij}\frac{\partial}{\partial x_j}\left(\frac{\partial u_i}{\partial a}\right)\,, \qquad \text{since } \sigma_{ij} = \sigma_{ji}\,.\end{aligned}$$

This relation can be transformed, using the relation

$$\frac{\partial}{\partial x_j}\left(\sigma_{ij}\frac{\partial u_i}{\partial a}\right) = \frac{\partial \sigma_{ij}}{\partial x_j}\frac{\partial u_i}{\partial a} + \sigma_{ij}\frac{\partial}{\partial x_j}\left(\frac{\partial u_i}{\partial a}\right)$$

and noting that in the absence of body forces

$$\frac{\partial \sigma_{ij}}{\partial x_j} = \operatorname{div}\sigma_i = 0\,.$$

In the integral (3.156) we can write

$$\begin{aligned}\iint_S \frac{\partial W}{\partial a}\,\mathrm{d}S &= \iint_S \frac{\partial}{\partial x_j}\left(\sigma_{ij}\frac{\partial u_i}{\partial a}\right)\mathrm{d}S\\ &= \int_{\Gamma_0} \sigma_{ij}\frac{\partial u_i}{\partial a}n_j\,\mathrm{d}s\\ &= \int_{\Gamma_0} T_i\frac{\partial u_i}{\partial a}\,\mathrm{d}s\end{aligned}$$

(with $T_i = \sigma_{ij} n_j$), and (3.156) then reduces to

$$G = \iint_S \frac{\partial W}{\partial x}\, \mathrm{d}S - \int_{\Gamma_0} T_i \frac{\partial u_i}{\partial a}\, \mathrm{d}s\,. \tag{3.157}$$

The path Γ_0 being a closed path, let us transform the surface integral into a line integral by applying the Riemann formula:

$$\iint_S \left(\frac{\partial P}{\partial x_1} + \frac{\partial Q}{\partial x_2} \right) \mathrm{d}x_1\, \mathrm{d}x_2 = \int_{\Gamma^*} (P\, \mathrm{d}x_2 - Q\, \mathrm{d}x_1) \tag{3.158}$$

(Γ^* being the closed path enclosing the area S, followed counterclockwise). We get

$$\boxed{G = \int_{\Gamma_0} \left(W\, \mathrm{d}x_2 - T_i \frac{\partial u_i}{\partial x_1}\, \mathrm{d}s \right)}\,. \tag{3.159}$$

Rice has introduced the following integral:

$$\boxed{J = \int_{\Gamma} \left(W\, \mathrm{d}x_2 - T_i \frac{\partial u_i}{\partial x_1}\, \mathrm{d}s \right)}\,, \tag{3.160}$$

where Γ is an open contour starting from the lower crack face extending counterclockwise around the crack tip to a point on the upper face. $\boldsymbol{T}$ and $\boldsymbol{u}$ are stress traction vectors and displacements on the contour Γ, oriented along the outward normal.

We can show that the J integral is independent of the contour Γ. In effect, let us consider a closed contour Γ^* enclosing an area S^* in which there is no body force F_i. The Riemann formula (3.158) yields

$$\int_{\Gamma^*} \left(W\, \mathrm{d}x_2 - T_i \frac{\partial u_i}{\partial x_1}\, \mathrm{d}s \right) = \iint_{S^*} \left[\frac{\partial W}{\partial x_1} - \frac{\partial}{\partial x_j} \left(\sigma_{ij} \frac{\partial u_i}{\partial x_1} \right) \right] \mathrm{d}x_1\, \mathrm{d}x_2 \tag{3.161}$$

with, as above

$$\frac{\partial}{\partial x_j} \left(\sigma_{ij} \frac{\partial u_i}{\partial x_1} \right) = \sigma_{ij} \frac{\partial}{\partial x_j} \left(\frac{\partial u_i}{\partial x_1} \right).$$

Let us differentiate W with respect to x_1, as we have done with respect to a:

$$\begin{aligned} \frac{\partial W}{\partial x_1} &= \frac{\partial W}{\partial \varepsilon_{ij}} \frac{\partial \varepsilon_{ij}}{\partial x_1} \\ &= \sigma_{ij} \frac{\partial \varepsilon_{ij}}{\partial x_1} \\ &= \sigma_{ij} \frac{\partial}{\partial x_1} \left[\frac{1}{2} \left(\frac{\partial u_i}{\partial x_j} + \frac{\partial u_j}{\partial x_i} \right) \right] \\ &= \sigma_{ij} \frac{\partial}{\partial x_j} \left(\frac{\partial u_i}{\partial x_1} \right), \qquad \text{since} \sigma_{ij} = \sigma_{ji}\,. \end{aligned}$$

The integral (3.161) is thus identically null on a closed path.

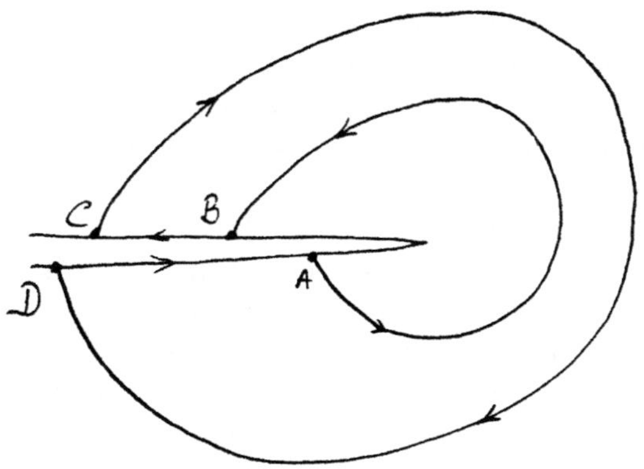

Fig. 3.17. Closed contours around a crack face

Consider now (Fig. 3.17) a closed contour $\Gamma^* = \text{ABCD}$ composed of two open paths Γ_1 and Γ_2 followed in opposite direction and linked by segments BC and DA of the crack faces, horizontal ($\mathrm{d}x_2 = 0$) and traction-free ($T_i = 0$). The integral

$$\int \left(W \,\mathrm{d}x_2 - T_i \frac{\partial u_i}{\partial x_1} \,\mathrm{d}s \right)$$

being zero on BC and AD, we can deduce that $J_1 - J_2 = 0$ whatever the contours. The J integral defined above is thus independent of the contour Γ.

It remains to identify J and G, (at least for a linear crack) as taking for the path the external contour of the solid. The path Γ of the J integral starts from a traction-free point of the crack and ends at another traction-free point of the opposite face. The path Γ_0 is the *external contour* (closed) of the solid on which are imposed either stresses or displacements (Fig. 3.18). There is no problem to take for path Γ the path Γ_0 cut off from crack faces. But, as with the latter, $\mathrm{d}x_2 = 0$ and $T_i = 0$, and we thus have $J = G$. However, if there exists at the crack tip a *cohesive zone* where cohesive or adhesion forces act, what is then the external contour of the solid? Does this cohesive zone belong to the interior of the solid? It seems that the answer

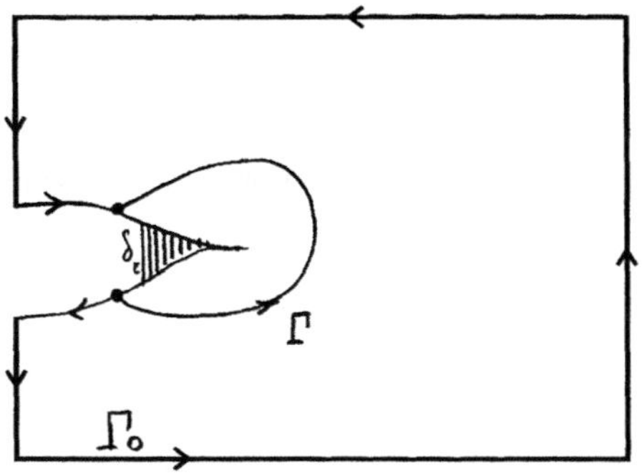

Fig. 3.18. External closed contour around a crack face

must be "yes": it does not belong to the contour Γ_t on which the stresses are imposed, nor to the contour Γ_u on which displacements are imposed. It is instead a zone of "elastic response" developed by the solid, and we have seen in Sect. 1.4.1 that the distinction between elasticity and attractive force (adhesion or cohesion) is a convention: we are concerned with a single curve of stress versus distance, displaying a maximum, the theoretical stress. It is because we have neglected the interactive stresses between the crack faces, while keeping the elastic stresses—which is physically a nonsense—that one obtains the unrealistic result of infinite stresses at the crack tip.

Letting, by convention, that cohesive zone to be in the interior of the solid, there is no longer problem to deform Γ to make it identical to Γ_0, and one has always $J = G$. The cohesion forces deform the solid elastically, give rise to a tangential junction (there is no longer infinite strain) and eliminate the singularities (in line with the models of Barenblatt and Dugdale). The elastic energy involved in the cohesive zone is thus taken into account in the density of elastic energy W of the solid.

Let us return, for a while, to the hypothesis in which the cohesive forces are ignored. In mixed mode I+II, the stress and displacement fields near the crack tip are given by (3.69–73). Take as path Γ a circle of radius r followed from $\theta = -\pi$ to $\theta = +\pi$. With $\mathrm{d}s = r\,\mathrm{d}\theta$ and $\mathrm{d}x_2 = r\cos\theta\,\mathrm{d}\theta$, the J integral is written

$$J = G = \int_{-\pi}^{+\pi} \left[W(r,\theta)\cos\theta - T_i(r,\theta)\frac{\partial u_i}{\partial x_1} \right] r\,\mathrm{d}\theta \,.$$

When $r \to 0$, only the singular terms remain and one obtains again Irwin's result, namely (3.154).

If we now introduce a cohesive zone of length d (as per the Barenblatt and Dugdale models), in which the stresses $\sigma(u)$ are tensile stresses along $\mathrm{O}x_2$ and depend upon the opening u, it suffices to take as path Γ a line enclosing the nonlinear cohesive zone since the nonsingular terms make no contribution. With $\mathrm{d}x_2 = 0$, and $T_2 = \sigma(u)$ on Γ, the J integral reduces to

$$\begin{aligned} J = G &= -\int_\Gamma T_i \frac{\partial u_i}{\partial x_1}\,\mathrm{d}s \\ &= -\int_0^d \left[\sigma(u)\frac{\partial u^+}{\partial x_1} - \sigma(u)\frac{\partial u^-}{\partial x_1} \right] \mathrm{d}x_1 \\ &= -\int_0^d \sigma(u)\frac{\partial \delta}{\partial x_1}\,\mathrm{d}x_1 \,, \end{aligned}$$

with $\delta = u^+ - u^-$ representing the opening of the crack. Denote by δ_t the value of the crack opening beyond which the cohesion forces become negligible. We then have

$$\boxed{J = G = \int_0^{\delta_t} \sigma(\delta)\,\mathrm{d}\delta} \,. \tag{3.162}$$

If we assume, for simplification, that $\sigma(u) = \sigma_0$ is constant in the cohesive zone, we get:

$$\boxed{J = G = \sigma_0 \delta_t} \,, \tag{3.163}$$

where δ_t is the *crack opening displacement* (COD).

3.8 The Dugdale Model

We have seen that when the minor axis of an elliptical cavity is made to tend towards zero to obtain a crack, the stresses in $x = a$ tend towards infinity. This unrealistic behaviour arises from the fact that we have assumed that the walls of the cavity are free surfaces. This hypothesis becomes untenable when the walls come sufficiently close to one another for molecular attractive forces to deform the walls elastically, which forces can no longer be neglected. It is thus necessary to treat the problem of a cavity or a crack subjected both to an external loading and to an internal distribution of tensile stresses (generally named "cohesion forces") such that the stresses at $x = a$ cannot exceed the theoretical stress σ_{th}. For a crack, this is equivalent to saying that singular terms must disappear, i.e. that the stress intensity factor K_i due to internal loading cancels with the K_{I} due to external loading. The terms in $\sqrt{r}$ of the power expansion of u_y also disappear, and the crack ends by an acute tip due to the next term, in $r^{3/2}$.

The first to have described this model is Barenblatt [35] in 1962. However, an exact theory is difficult for we have to deal with a self-consistent problem: the distribution of cohesion forces σ_i depends on the exact shape of the deformed crack, which itself depends on the distribution of these stresses. In a first approximation, it is convenient to assume that cohesion forces have a constant value σ_0 over a length $d = a - c$ (the cohesive zone) in the vicinity of the tip of a crack of length $2a$. This is the Dugdale model.

In fact Dugdale [49], in 1960, was interested by the plastic deformation at the tip of a crack of length $2c$ in an ideal elastoplastic material. Noting that the stress σ_y in the material cannot exceed the flow stress σ_0, he proposed to compute the length d of the plastic zone, by comparing this problem to that of a hypothetical crack of length $2a$ subjected to tensile stress σ_0 over a length $d = a - c$ at its two ends and by writing that the stresses are finite at $x = a$. To do that, he used the complex potentials given by Muskhelishvili [50] for an elliptical cavity subjected to constant pressure on a part of its surface. The shape of the crack, with its acute tip, was then computed by Goodier and Field [51] in 1963. In fact, it is easier to use the solutions of England and Green [36] and Lowengrub [37], given in Sect. 3.6.1, than the Muskhelishvili solution.

3.8.1 Straight Crack in a Plate

Let us return to the linear crack subjected to a biaxial loading σ and $k\sigma$ in mode I ($\beta = 90°$). In accordance with (3.67), the stress intensity factor is given by

$$K_{\mathrm{I}} = \sigma\sqrt{\pi a}\,. \tag{3.164}$$

The stress $\sigma_y(x,0)$ on the x-axis, and the opening $u_y(x,0)$ of the crack are given by (3.48) and (3.56):

$$\sigma_y(x,0) = \frac{\sigma x}{\sqrt{x^2-a^2}} = \frac{K_{\mathrm{I}}}{\sqrt{\pi a}}\frac{x}{\sqrt{x^2-a^2}}\,, \qquad x > a\,, \tag{3.165}$$

$$u_y(x,0) = \frac{\kappa+1}{4\mu}\sigma\sqrt{a^2-x^2} = \frac{\kappa+1}{4\mu}\frac{K_{\mathrm{I}}}{\sqrt{\pi a}}\sqrt{a^2-x^2}\,, \quad x < a\,. \tag{3.166}$$

Assume that the cohesion forces exert a constant traction σ_0 on a length $d = a - c$. We can thus use the results of Sect. 3.6.1 on linear cracks subjected to internal constant pressure near the crack tip with $p = -\sigma_0$. Using (3.106), these internal forces exert a stress[5] given by

$$\sigma_y(x,0) = \frac{K_i}{\sqrt{\pi a}}\frac{x}{\sqrt{x^2-a^2}} + \sigma_0 - \frac{2\sigma_0}{\pi}\tan^{-1}\frac{c\sqrt{x^2-a^2}}{x\sqrt{a^2-c^2}} \tag{3.167}$$

with

$$K_i = -2\sqrt{\frac{a}{\pi}}\sigma_0\cos^{-1}\frac{c}{a}\,. \tag{3.168}$$

Finally, the external loading and the internal forces lead to a stress

$$\sigma_y(x,0) = \frac{K_{\mathrm{I}}+K_i}{\sqrt{\pi a}}\;\frac{x}{\sqrt{x^2-a^2}} + \sigma_0 - \frac{2\sigma_0}{\pi}\tan^{-1}\frac{c\sqrt{x^2-a^2}}{x\sqrt{a^2-c^2}}\,. \tag{3.169}$$

The stress is finite at $x = a$ only if

$$K_{\mathrm{I}} + K_i = 0\,,$$

i.e.

$$\boxed{\frac{c}{a} = \cos\left(\frac{\pi}{2}\frac{\sigma}{\sigma_0}\right)}\,, \tag{3.170a}$$

[5] It is evident that stress intensity factors in mode I can only be positive; otherwise we would have interpenetration of matter. But the negative value we have will algebraically added to K_{I}.

or, equivalently,

$$\frac{d}{a} = 2\sin^2\left(\frac{\pi}{4}\frac{\sigma}{\sigma_0}\right) , \tag{3.170b}$$

the result given by Dugdale [49]. The stress σ_y is then

$$\sigma_y(x,0) = \sigma_0 - \frac{2\sigma_0}{\pi}\tan^{-1}\frac{c\sqrt{x^2-a^2}}{x\sqrt{a^2-c^2}} , \qquad x > a , \tag{3.171a}$$

or, using (3.170a),

$$\boxed{\sigma_y(x,0) = \sigma_0 - \frac{2\sigma_0}{\pi}\tan^{-1}\left[\frac{\sqrt{1-a^2/x^2}}{\tan(\pi\sigma/2\sigma_0)}\right] , \qquad x > a} . \tag{3.171b}$$

In the same manner, the addition of displacements, (3.166) and (3.110) gives

$$\begin{aligned} u_y(x,0) &= \frac{\kappa+1}{4\mu}\frac{K_{\mathrm{I}}+K_i}{\sqrt{\pi a}}\sqrt{a^2-x^2} \\ &\quad - \frac{\kappa+1}{4\pi\mu}\sigma_0\left[c\ln\frac{|\xi-1|}{\xi+1} - x\ln\frac{|c\xi-x|}{c\xi+x}\right] . \end{aligned}$$

The terms in $\sqrt{a^2-x^2}$ also disappear when $K_{\mathrm{I}}+K_i=0$, and the profile of the crack is given by

$$\boxed{u_y(x,0) = \frac{\kappa+1}{4\pi\mu}\sigma_0\left[c\ln\frac{\xi+1}{|\xi-1|} - x\ln\frac{\xi+x/c}{|\xi-x/c|}\right] , \qquad x < a} , \tag{3.172}$$

with

$$\xi = \sqrt{\frac{a^2-x^2}{a^2-c^2}} = \frac{\sqrt{1-x^2/a^2}}{\sin(\pi\sigma/2\sigma_0)}$$

($\xi < 1$ if $x > c$ and $\xi > 1$ if $x < c$). The crack opening displacement $\delta_t = 2u_y$ at $x = c$, at the limit of the cohesion zone, is given directly by the addition of (3.166) and (3.109) with $K_{\mathrm{I}}+K_i=0$:

$$\boxed{\delta_t = \frac{\kappa+1}{\pi\mu}\sigma_0 c\ln\frac{a}{c}} . \tag{3.173}$$

Equations (3.172) and (3.173) were obtained for the first time by Goodier and Field [51] in 1963, in a slightly different form ($x = a\cos\theta, c = a\cos\theta_2$).

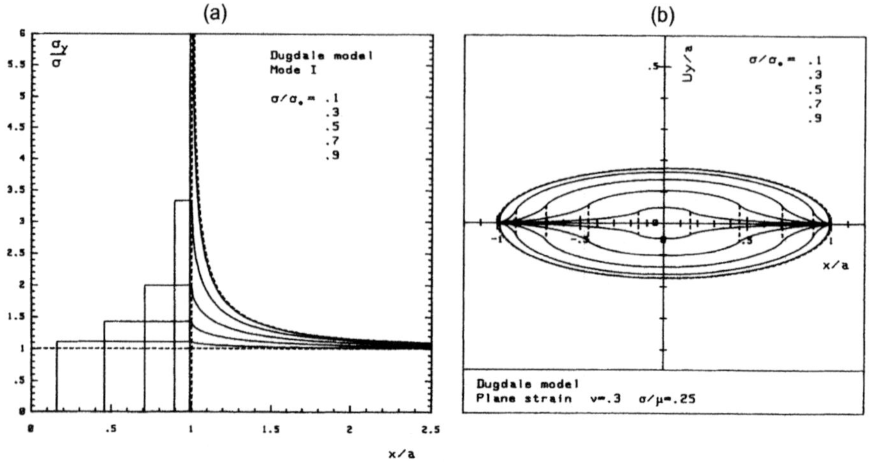

Fig. 3.19. Dugdale model for a linear crack. Stress distribution and shape of the crack for various values of σ/σ_0

Equation (3.173) was obtained independently by Burdekin and Stone [52] in 1966. The slope of the profile is, from (3.172) or (3.111), given by

$$\boxed{\frac{\mathrm{d}u_y}{\mathrm{d}x} = -\frac{\kappa+1}{4\pi\mu}\sigma_0 \ln\frac{\xi + x/c}{|\xi - x/c|}} , \tag{3.174}$$

i.e. zero at $x = 0$ and $x = a$ and infinite at $x = c$.

Figure 3.19 represents the stress distribution and the shape of the crack for different values of σ/σ_0. Note that for $\sigma/\sigma_0 = 0.1$ the results are almost identical to those of the Griffith model shown by (3.165) and (3.166), as displayed by the dotted line in the figure.

Figure 3.20 displays the variations of δ_t/a and δ_t/c with σ/σ_0. Note the presence of a maximum of δ_t/a for $c/a = \mathrm{e}^{-1}$, i.e. $\sigma/\sigma_0 = 0.76$. The variations of d/a and d/c with σ/σ_0 are displayed in Fig. 3.21.

We have seen, concerning the J integral, that for the Dugdale model we have

$$G = \sigma_0\delta_t ,$$

i.e.

$$G = -\frac{\kappa+1}{\pi\mu}\sigma_0^2 c \ln\left[\cos\left(\frac{\pi}{2}\frac{\sigma}{\sigma_0}\right)\right] \tag{3.175a}$$

$$= -\frac{\kappa+1}{\pi\mu}\sigma_0^2 a \cos\left(\frac{\pi}{2}\frac{\sigma}{\sigma_0}\right) \ln\left[\cos\left(\frac{\pi}{2}\frac{\sigma}{\sigma_0}\right)\right] . \tag{3.175b}$$

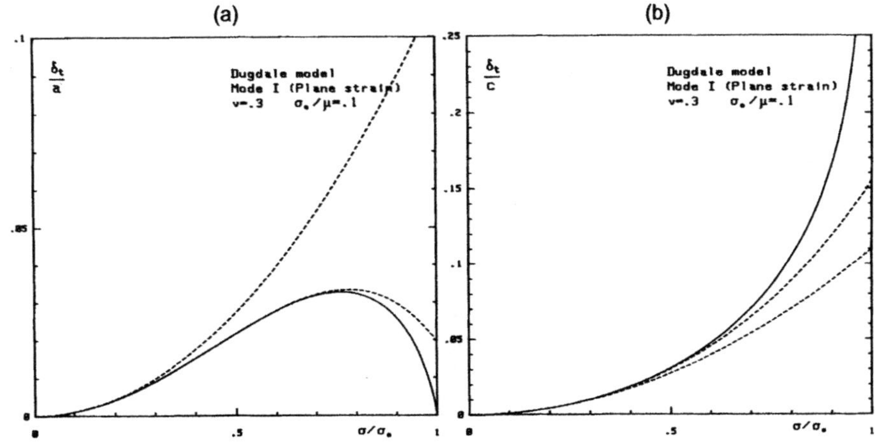

Fig. 3.20. Dugdale model for a linear crack. Crack opening displacement δ_t as a function of σ/σ_0 (*dashed lines* displaying the approximations with one or two terms, as per (3.179) and (3.180))

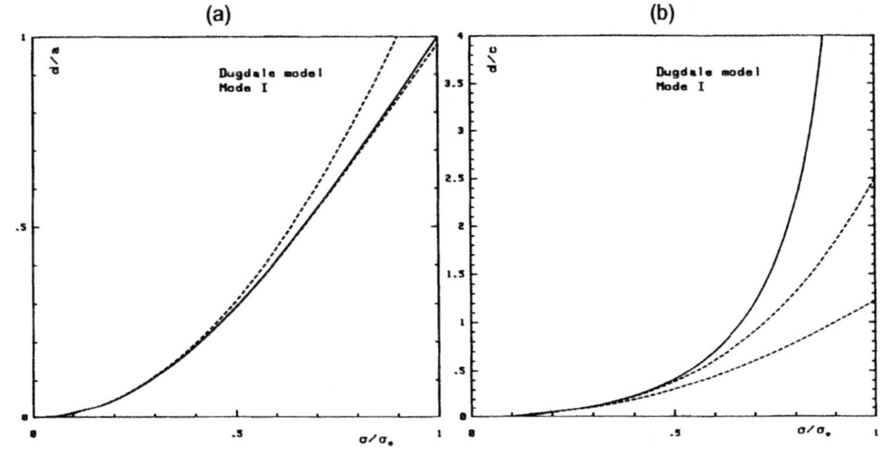

Fig. 3.21. Dugdale model for a linear crack. Relative length of the cohesion zone as a function of σ/σ_0 (*dashed lines* displaying the approximations with one or two terms)

When $\sigma/\sigma_0 \to 1, G \to 0$ at $a = \text{const.}$, whereas $G \to \infty$ at $c = \text{const.}$ Let

$$G_{\mathrm{L}}(a) = \frac{\kappa + 1}{8\mu}\sigma^2 \pi a = \frac{\kappa + 1}{8\mu} {K_{\mathrm{I}}}^2(a) \,,$$
$$G_{\mathrm{L}}(c) = \frac{\kappa + 1}{8\mu}\sigma^2 \pi c = \frac{\kappa + 1}{8\mu} {K_{\mathrm{I}}}^2(c)$$

be the expressions of G in linear elasticity, after the calculation of Irwin, for a crack of length $2a$ or $2c$. The ratios $G/G_{\mathrm{L}}(a)$ and $G/G_{\mathrm{L}}(c)$ are displayed in Fig. 3.22 as a function of σ/σ_0.

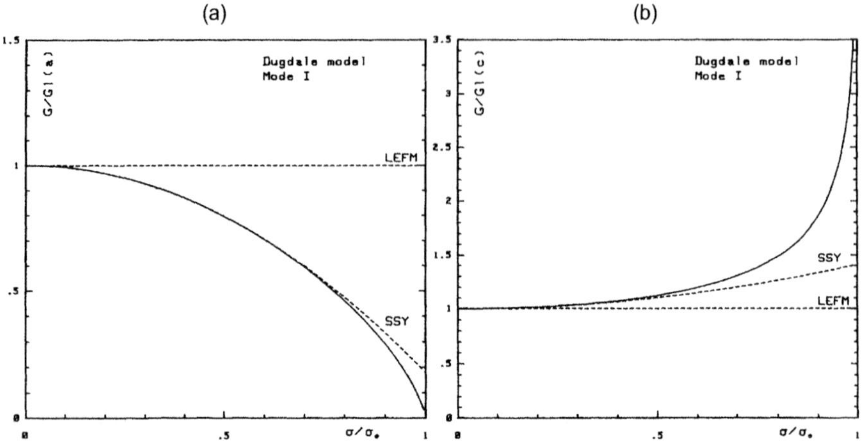

Fig. 3.22. Dugdale model for a linear crack. Relative variation of the energy release rate with σ/σ_0 (*dashed lines* displaying the approximations with one or two terms

Usual Approximations. The above solutions are exact. They can be simplified in the case where $\sigma/\sigma_0 \ll 1$, and we obtain the following approximations (known as small scale yielding (SSY) approximations), taking care to develop the logarithms to the second order:

$$\frac{c}{a} \cong 1 - \frac{\pi^2\sigma^2}{8\sigma_0^2}\left(1 - \frac{\pi^2\sigma^2}{48\sigma_0^2}\right), \tag{3.176}$$

$$\begin{aligned} d &\cong a\frac{\pi^2\sigma^2}{8\sigma_0^2}\left(1 - \frac{\pi^2\sigma^2}{48\sigma_0^2}\right) \\ &= \frac{\pi {K_\mathrm{I}}^2(a)}{8\sigma_0^2}\left(1 - \frac{\pi^2\sigma^2}{48\sigma_0^2}\right) \\ &= \frac{\pi {K_\mathrm{I}}^2(c)}{8\sigma_0^2}\left(1 + \frac{5\pi^2\sigma^2}{48\sigma_0^2}\right). \end{aligned} \tag{3.177}$$

For d, the SSY approximation generally given is

$$\boxed{d \cong \frac{\pi {K_\mathrm{I}}^2}{8\sigma_0^2}};$$

hence

$$\frac{d}{c} \cong \frac{\pi^2\sigma^2}{8\sigma_0^2}\left(1 + \frac{5\pi^2\sigma^2}{48\sigma_0^2}\right), \tag{3.178}$$

$$\frac{\delta_t}{c} \cong \frac{\kappa+1}{\pi\mu}\sigma_0\frac{\pi^2\sigma^2}{8\sigma_0^2}\left(1 + \frac{\pi^2\sigma^2}{24\sigma_0^2}\right), \tag{3.179}$$

$$\frac{\delta_t}{a} \cong \frac{\kappa+1}{\pi\mu}\sigma_0\frac{\pi^2\sigma^2}{8\sigma_0^2}\left(1 - \frac{\pi^2\sigma^2}{12\sigma_0^2}\right), \tag{3.180}$$

$$\delta_t \cong \frac{\kappa+1}{8\mu}\frac{K_{\mathrm{I}}^2(a)}{\sigma_0}\left(1-\frac{\pi^2\sigma^2}{12\sigma_0^2}\right)$$

$$\cong \frac{\kappa+1}{8\mu}\frac{K_{\mathrm{I}}^2(c)}{\sigma_0}\left(1+\frac{\pi^2\sigma^2}{24\sigma_0^2}\right) , \tag{3.181}$$

$$\frac{\delta_t}{d} \cong \frac{\kappa+1}{\pi\mu}\sigma_0\left(1-\frac{\pi^2\sigma^2}{16\sigma_0^2}\right) , \tag{3.182}$$

$$G \cong \frac{\kappa+1}{8\mu}K_{\mathrm{I}}^2(a)\left(1-\frac{\pi^2\sigma^2}{12\sigma_0^2}\right) = \frac{\kappa+1}{8\mu}K_{\mathrm{I}}^2(a)\left(1-\frac{2d}{3a}\right)$$

$$\cong \frac{\kappa+1}{8\mu}K_{\mathrm{I}}^2(c)\left(1+\frac{\pi^2\sigma^2}{24\sigma_0^2}\right) = \frac{\kappa+1}{8\mu}K_{\mathrm{I}}^2(c)\left(1+\frac{d}{3a}\right) . \tag{3.183}$$

These approximations with one or two terms are displayed in the dashed lines in Figs. 3.20–3.22.

To develop σ_y in the neighbourhood of $x = a$, let us start from (3.171a) with $\sqrt{x^2 - a^2} \cong \sqrt{2ar}$. We then get

$$\sigma_y(x,0) \cong \sigma_0 - \frac{2\sigma_0}{\pi}\tan^{-1}\sqrt{\frac{r}{d}} ,$$

and, for $r \ll d$,

$$\boxed{\sigma_y(x,0) \cong \sigma_0 - \frac{2\sigma_0}{\pi}\sqrt{\frac{r}{d}}} . \tag{3.184}$$

Equation (3.172), giving the profile, can be written as

$$u_y(x,0) = \frac{\kappa+1}{4\pi\mu}\sigma_0\left[x\ln\frac{|c\xi - x|\,(\xi+1)}{(c\xi+x)\,|\xi-1|} - (x-c)\ln\frac{\xi+1}{|\xi-1|}\right]$$

$$= \frac{\kappa+1}{4\pi\mu}\sigma_0\left[2x\ln\frac{|c\xi - x|\,(\xi+1)(a^2-c^2)}{a\,|c^2-x^2|} - (x-c)\ln\frac{\xi+1}{|\xi-1|}\right]$$

$$= \frac{\kappa+1}{2\pi\mu}\sigma_0\left[x\ln\frac{a^2+xc+\xi(a^2-c^2)}{a(c+x)} - \frac{(x-c)}{2}\ln\frac{\xi+1}{|\xi-1|}\right]$$

$$= \frac{\kappa+1}{2\pi\mu}\sigma_0\left[x\ln\left(1+d\frac{1-x/a+\xi\,(1+c/a)}{c+x}\right) - \frac{(x-c)}{2}\ln\frac{\xi+1}{|\xi-1|}\right] .$$

If $c/a \cong 1$ and for x near c, we get, letting $x_1 = x - c$:

$$u_y(x,0) = \frac{\kappa+1}{2\pi\mu}\sigma_0 d\left[\xi - \frac{x_1}{2d}\ln\frac{\xi+1}{|\xi-1|}\right] , \tag{3.185}$$

an approximation given by Rice [47] in 1968 (but for $\xi < 1$ only) where

$$\xi \cong \sqrt{1 - \frac{x_1}{d}} \,. \tag{3.186}$$

If we let $\delta = 2u_y$ and if we take the one-term approximation of δ_t/d given in (3.182), we get

$$\boxed{\frac{\delta}{\delta_t} \cong \xi - \frac{x_1}{2d} \ln \frac{1+\xi}{|1-\xi|}} \,, \tag{3.187}$$

an expression used (but for $\xi < 1$ only) by Morgan and Ward [53] in 1977 to compare their experimental results on the crazing of polymers with the Dugdale model.

In Fig. 3.23 we have the ratio δ/δ_t as a function of the parameter x_1/d for various values of σ/σ_0 using (3.172). The approximation given by (3.186) and (3.187) is in fact indistinguishable (at least for $x_1/d > -0.6$) from the exact solution for $\sigma/\sigma_0 = 0.1$, and still remains excellent for $\sigma/\sigma_0 < 0.7$.

Let us show that the profile near the crack tip is indeed in $r^{3/2}$ when the singularity has been eliminated. With $r = (a - x) \to 0$ ($\xi \to 0$), and by developing up to ξ^3 (3.172) is written

$$u_y(r, 0) \cong \frac{\kappa + 1}{6\pi\mu} \sigma_0 c \xi^3 \left(1 - \frac{c^2}{a^2}\right) ,$$

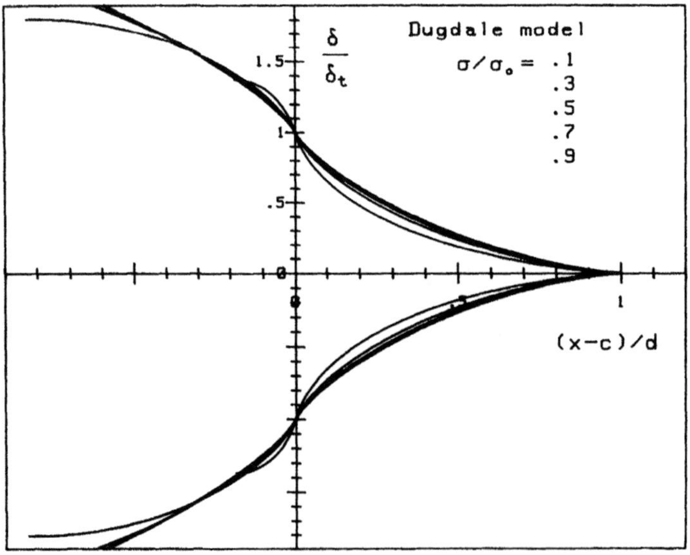

Fig. 3.23. Dugdale model for a linear crack. Opening $\delta = 2u_y$ as a function of the parameter x_1/d for various values of σ/σ_0. The symmetry with respect to Ox allows better visualization of the profile

with

$$\xi = \sqrt{\frac{2ar}{a^2 - c^2}}$$

and, for $d \ll a$,

$$u_y(r, 0) \cong \frac{\kappa + 1}{3\pi\mu} \sigma_0 d \left(\frac{r}{d}\right)^{3/2} . \tag{3.188}$$

3.8.2 Penny-Shaped Crack

Let us now consider a penny-shaped crack of radius a in an infinite medium submitted to an uniaxial stress σ perpendicular to the crack plane. We have seen in Sect. 3.6.2 via (3.123) that stresses and displacements in $y = 0$ were given by

$$\sigma_y(r, 0) = \frac{K_\mathrm{I}}{\sqrt{\pi a}} \left(\frac{a}{\sqrt{r^2 - a^2}} + \cos^{-1} \frac{a}{r} \right) , \qquad r > a , \tag{3.189}$$

$$u_y(r, 0) = \frac{2(1 - \nu^2)}{E} \frac{K_\mathrm{I}}{\sqrt{\pi a}} \sqrt{a^2 - r^2} , \qquad r < a , \tag{3.190}$$

with

$$K_\mathrm{I} = \frac{2\sigma}{\pi} \sqrt{\pi a} . \tag{3.191}$$

Let us assume that cohesion forces exert a constant traction σ_0 on a length $d = a - c$, and use the results of Sect. 3.6.2 on penny-shaped cracks subjected to a constant internal pressure near their ends, with $p = -\sigma_0$. These cohesion forces thus cause stresses and displacements as follows:

$$\sigma_y(r, 0) = \frac{K_i}{\sqrt{\pi a}} \frac{a}{\sqrt{r^2 - a^2}} + \frac{2\sigma_0}{\pi} \tan^{-1} \sqrt{\frac{a^2 - c^2}{r^2 - a^2}} , \qquad r > a , \tag{3.192}$$

$$u_y(r, 0) = 2 \frac{1 - \nu^2}{E} \tag{3.193}$$

$$\times \left[\frac{K_i}{\sqrt{\pi a}} \sqrt{a^2 - r^2} + \frac{2\sigma_0}{\pi} c^2 \int\limits_{\max(r,c)}^{a} \frac{\sqrt{t^2 - r^2}\,\mathrm{d}t}{t^2 \sqrt{t^2 - c^2}} \right] , \qquad r < a ,$$

with

$$K_i = -\frac{2\sigma_0}{\sqrt{\pi a}} \sqrt{a^2 - c^2} . \tag{3.194}$$

Finally, the stresses due to both external loading and internal forces, given by

$$\sigma_y(r, 0) = \frac{K_\mathrm{I} + K_i}{\sqrt{\pi a}} \frac{a}{\sqrt{r^2 - a^2}} + \frac{K_\mathrm{I}}{\sqrt{\pi a}} \cos^{-1} \frac{a}{r} + \frac{2\sigma_0}{\pi} \tan^{-1} \sqrt{\frac{a^2 - c^2}{r^2 - a^2}} ,$$

will be finite at $r = a$ only if

$$K_{\mathrm{I}} + K_i = 0 \,,$$

i.e. if

$$\boxed{\frac{c}{a} = \sqrt{1 - \frac{\sigma^2}{\sigma_0^2}}} \,. \tag{3.195}$$

The total stress then reduces to:

$$\boxed{\sigma_{y\mathrm{T}}(r,0) = \frac{2\sigma}{\pi}\cos^{-1}\frac{a}{r} + \frac{2\sigma_0}{\pi}\tan^{-1}\sqrt{\frac{a^2-c^2}{r^2-a^2}} \,, \qquad r > a} \,, \tag{3.196}$$

which correctly varies from σ_0 to σ when r varies from a to infinity. Similarly the displacement reduces to:

$$\boxed{u_{y\mathrm{T}}(r,0) = 4\frac{1-\nu^2}{\pi E}\sigma_0 c^2 \int\limits_{\max(r,c)}^{a} \frac{\sqrt{t^2-r^2}}{t^2\sqrt{t^2-c^2}}\,\mathrm{d}t \,, \qquad r < a} \,, \tag{3.197}$$

and the crack opening displacement $\delta_t = 2u_{y\mathrm{T}}(c,0)$ is given by

$$\delta_t = 8\frac{1-\nu^2}{\pi E}\sigma_0 c \left(1 - \frac{c}{a}\right) . \tag{3.198}$$

This formula was given by Rice [47] in 1968.

The energy rate is written thus

$$G = 8\frac{1-\nu^2}{\pi E}\sigma_0^2 c \left(1 - \sqrt{1 - \frac{\sigma^2}{\sigma_0^2}}\right) . \tag{3.199}$$

When $\sigma/\sigma_0 \to 0$, we have

$$G \cong 4\frac{1-\nu^2}{\pi E}c\sigma^2 \,,$$

and when $\sigma/\sigma_0 \to 1$, we have

$$G \cong 8\frac{1-\nu^2}{\pi E}c\sigma^2 \,.$$

In contrast to the two-dimensional case, G remains finite.

References

1. A.A. Griffith: Phil. Trans. Roy. Soc. A **221**, 163 (1920)
2. G.R. Irwin, J.A. Kies: Welding Journal **31**, Res. Suppl. 95 (1952)
3. G.R. Irwin: J. Appl. Mech. **24**, 361 (1957)
4. H.B. Callen: *Thermodynamics* (Wiley, 1960)
5. S.J. Burns, J.C. Pollet, C. Lun Chow: Int. J. Fracture **14**, 311 (1978)
6. J.P. Berry: J. Mech. Phys. Solids **8**, 194 (1960)
7. J.R. Rice: J. Mech. Phys. Solids **26**, 61 (1978)
8. A.A. Griffith: Proc. First Int. Congress of Appl. Mechanics (Delft 1924) pp. 55
9. I.B. Obreimov: Proc. Roy. Soc. A **127**, 290 (1930)
10. F.C. Roesler: Proc. Phys. Soc. B **69**, 981 (1956)
11. E. Orowan: Trans. Inst. Eng. Shipbuilders Scotland **89**, 165 (1945)
12. D. Maugis: Eng. Fracture Mech. **43**, 217 (1992)
13. A. Gilabert, P. Sibillot, D. Sornette, C. Vanneste, D. Maugis, S. Muttin: Eur. J. Mech. A/Solids **11**, 65 (1992)
14. P.S. Theocaris, D. Pazis, B.D. Constantellos: Int. J. Fracture **30**, 135 (1986)
15. G.R. Irwin: *Fracture. Encyclopedia of Physics, vol. IV: Elasticity and Plasticity*, ed. by Flugge (Springer, Berlin 1958) pp. 551–590
16. G.R. Irwin, J.A. Kies, H.L. Smith: ASTM Proc. **58**, 640 (1958)
17. I.N. Sneddon: Proc. Roy. Soc. A **187**, 229 (1946)
18. H.W. Westergaard: Trans. ASME **61**, A49 (1939)
19. M.L. Williams: J. Appl. Mech. **24**, 109 (1957)
20. A.A. Wells, D. Post: Proc. SESA **16**, 69 (1958)
21. G.R. Irwin: Proc. SESA **16**, 92 (1958)
22. G.C. Sih: Int. J. Fracture **2**, 628 (1966)
23. J. Eftis, H. Liebowitz: Int. J. Fracture **2-8**, 383 (1972)
24. J. Eftis, N. Subramonian, H. Liebowitz: Eng. Fracture Mech. **9**, 189 (1977)
25. J. Eftis, N. Subramonian: Eng. Fracture Mech. **10**, 43 (1978)
26. R.N.L. Smith: Eng. Fracture Mech. **26**, 463 (1987)
27. P.S. Theocaris, C.P. Spyropoulos: Acta Mech. **48**, 57 (1983)
28. F. Kherkhof: *Fracture des verres* (1970, translation by Saint Gobain Industries 1974)
29. E. Sommer: Eng. Fracture Mech. **1**, 539 (1969)
30. F.A. McClintock, G.R. Irwin: 'Plasticity effects in fracture mechanics'. In: *Fracture toughness testing and its applications*, ASTM STP 381. (Philadelphia 1965) pp. 84–113
31. P.C. Paris, G.C. Sih: 'Stress analysis of cracks'. In: *Fracture toughness testing and its applications*, ASTM STP 381. (Philadelphia 1965) pp. 30–83
32. G.C. Sih, P.C. Paris, F. Erdogan: J. Appl. Mech. **29**, 306 (1962)
33. M.K. Kassir, G.C. Sih: J. Appl. Mech. **33**, 601 (1966)
34. M.K. Kassir, G.C. Sih: Int. J. Fracture **4**, 347 (1968)
35. G.I. Barenblatt: Adv. Appl. Mech. **7**, 55 (1962)
36. A.H. England, A.E. Green: Proc. Camb. Phil. Soc. **59**, 489 (1963)
37. M. Lowengrub: Proc. Edinburgh Math. Soc. **15**, 131 (1966)
38. L.M. Keer: J. Mech. Phys. Solids **12**, 149 (1964)
39. J.N. Goodier: 'Mathematical theory of equilibrium cracks'. In: *Fracture, vol. II*, ed. by H. Liebowitz (Academic Press, New York 1968) pp. 1–66
40. I.N. Sneddon: Proc. Camb. Phil. Soc. **61**, 609 (1965)
41. D.Y.C. Chan, B.D. Hughes, L.R. White: J. Colloid Interface Sci. **115**, 240 (1987)
42. M. Lowengrub, I.N. Sneddon: Int. J. Eng. Sci. **3**, 451 (1965)
43. V.I. Fabrikant: Eng. Fracture Mech. **22**, 855 (1985)

44. H.F. Bueckner: 'Field singularities and related integral representation'. In: *Methods of analysis and solution of crack problems*, ed. by G.C. Sih (Noordhoff, Leyden 1973) pp. 239–314
45. V.I. Fabrikant, T.S. Sankar, G.D. Xistris: Eng. Fracture Mech. **23**, 921 (1986)
46. J.R. Rice: J. Appl. Mech. Trans. ASME **90**, 379 (1968)
47. J.R. Rice: 'Mathematical analysis in the mechanics of fracture'. In: *Fracture, vol. II*, ed. by H. Liebowitz (Academic Press, New York 1968) pp. 191–311
48. M.F. Kanninen, C.H. Popelar: *Advanced fracture mechanics* (Oxford University Press, New York 1985)
49. D.S. Dugdale: J. Mech. Phys. Solids **8**, 100 (1960)
50. N.I. Muskhelishvili: *Some basic problems of the mathematical theory of elasticity* (Noordhoff, Leyden 1975)
51. J.N. Goodier, F.A. Field: 'Plastic energy dissipation in crack propagation'. In: *Fracture of solids*, ed. by D.C. Drucker, J.J. Gilman (Interscience Publ., New York 1963) pp. 103–118
52. F.M. Burdekin, D.E.W. Stone: J. Strain Anal. **1**, 145 (1966)
53. G.P. Morgan, I.M. Ward: Polymer **18**, 87 (1977)

4. Frictionless Elastic Contact

4.1 Introduction

In problems of indentation of an elastic half-space by a rigid axisymmetric frictionless punch (see Fig. 4.1), we have mixed boundary conditions which include both stresses and displacements, as follows:

$$u_z(\varrho, 0) = \delta - f(\varrho)\,, \qquad 0 \le \varrho \le 1\,, \tag{4.1}$$
$$\sigma_z(\varrho, 0) = 0\,, \qquad \varrho > 1\,, \tag{4.2}$$
$$\tau_{rz}(\varrho, 0) = 0\,, \tag{4.3}$$

where δ is the penetration depth of the punch, a is the radius of contact, $\varrho = r/a$, and $z = f(\varrho)$ is its shape function (with $f(0) = 0$), which must conform with the surface of the half-space inside the contact area. Furthermore, all stresses must approach zero at infinity. Note that the function f depends on a for a given shape of punch.

The first solution of such a problem is due to Hertz [1] in 1882, for the case of contact of a sphere on a plane. Hertz assumed a hemispherical pressure distribution in the contact area, and showed that it provides a complete solution to the problem. In 1885, Boussinesq [2] pointed out that, except for the case of a flat punch, the radius of contact a is generally unknown beforehand, and that its calculation needs the introduction of a supplementary hypothesis: that the normal stress falls to zero at the edge of the contact. This condition leads to tangential contact between the punch and the elastic half-space, and determines the penetration δ_0 of the punch. Boussinesq pointed out further that a difference in the penetration depth corresponds to the superposition of a rigid-body displacement $\delta_0 - \delta$ of the punch, which gives rise to the singular stresses and to the discontinuities of displacements characteristic of a flat punch. Compressive singular stresses must be ruled out, for they lead to negative discontinuities of displacements and thus to an interpenetration of the solids. On the other hand, the tensile singular stresses seemed also to be impossible, unless the solids are capable of adhesion. So, the necessity of cancellation of stresses on the edge of a rounded punch seemed well established for about a century.

This problem was reconsidered by K.L. Johnson in 1958. At that time it was commonly thought that adhesion plays a central role in friction, giving rise to cold-welded junctions which must then be sheared. Unfortunately, experiments failed to show any adhesion when the load was removed (except for soft metals like indium or lead, which are near their melting temperature at room temperature). It was known from Bradley [3] and Deryagin [4] that the adherence force between a rigid sphere and a rigid plane should be equal to $2\pi wR$. It was then suggested that junctions were broken by the elastic stress relaxation as the applied load was reduced. To analyse this point, Johnson [5] considered two spheres pressed under a load P, and adhering on a contact area given by the Hertz theory. The sphere's relative approach is δ_0. When the load is reduced at constant area of contact, this approach becomes δ, and this rigid-body displacement $\delta_0 - \delta$ gives rise immediately, according to the Boussinesq theory of the flat punch, to infinite tensile stresses on the edge of the contact. As, evidently, these infinite stresses cannot be sustained, the surfaces separate from the periphery towards the interior. So, Johnson concluded, two elastic spheres pressed under a force P cannot physically maintain their contact over an area larger than that given by the Hertz theory, which is exactly the remark made 73 years before by Boussinesq!

About a dozen years later, however, it became evident that, for soft elastic solids such as rubber, the area of contact could be larger than in the Hertz theory and that an adherence force could be measured. By a balance of elastic, potential and surface energies, Johnson, Kendall and Roberts [6], in their *JKR theory*, resolved the paradox. In the presence of surface energy, the area of contact is larger than in the Hertz theory, sustains infinite stresses on its periphery, and remains finite under negative loads until a critical traction force $F = \frac{3}{2}\pi wR$ is reached at which the surfaces separate abruptly; furthermore the connection to the surface is normal rather than tangential. Kendall [7] applied the same method to the adherence of a flat punch, and showed [8] that the geometry was similar to that of a solid with a deep external circular crack. The reason why rough surfaces may have negligible adherence, although individual spheres adhere, was explained by Johnson [9] and Fuller and Tabor [10]: some asperities (the highest) are compressed, while others are stretched. Assuming the same radius of curvature R for all asperities and a Gaussian distribution of heights with a standard deviation σ, Fuller and Tabor introduced an adhesion index θ which is the ratio of the adherence of a sphere to the force needed to push this sphere at a depth σ into the solid. For rough solids of high modulus, adherence can disappear.

Maugis *et al.* [11], regarding the edge of the contact area as a crack in mode I that recedes or advances as the area of contact A increases or decreases, showed that the JKR theory was consistent with linear elastic fracture mechanics, with the equilibrium point defined by $G = w$ and its stability by $(\partial G/\partial A) > 0$. The adherence force is the limit of stability of the system and depends on the stiffness of the measuring apparatus. Its value is $F = \frac{3}{2}\pi wR$ at fixed load, but $F = \frac{5}{6}\pi wR$ at fixed grips with a smaller area of contact. (Later they showed that the kinetics of crack propagation was in complete agreement with the expression of G deduced from the JKR theory.) The equilibrium condition $G = w$ corresponds to the supplementary hypothesis discussed by Boussinesq, and enables the penetration δ of the punch to be determined. For $w = 0$, the JKR theory reduces to that of Hertz.

However, this adherence force, being independent of Young's modulus, was irreconciliable with the adherence force $2\pi wR$ computed by Bradley and Derjaguin for a rigid sphere on a rigid plane. In 1975, Derjaguin, Muller and Toporov [12] presented a completely different theory, the DMT theory, in which molecular forces act in a ring-shaped zone around the contact but are assumed unable to change their profile, which remains Hertzian (Young's modulus not too small). By a "thermodynamic approach" they found that the attractive force is $2\pi wR$ at the point of contact but decreases rapidly to πwR when the approach increases. In this theory there is no stress singularity at the edge, the air gap closes smoothly, and separation occurs when the contact is reduced to a point (thus without instability) under an adherence force $F = 2\pi wR$.

A long dispute began when Tabor [13], comparing both theories, pointed out that the main error in the DMT theory is the neglect of the deformation due to attractive forces close to the edge of the contact, whereas that in the JKR theory is to neglect attractive forces outside the contact. By considering the expression of the height h of the discontinuity of displacement in the JKR theory, Tabor proposed that the behaviour should vary continuously from one theory to the other as a function of a parameter $w^2R/E^2Z_0^3$, where Z_0 is the equilibrium distance between atoms. After an exchange of acid notes with Tabor, Muller *et al.* [14] presented in 1980 a self-consistent numerical calculation using a Lennard–Jones potential and abandoned the hypothesis that adhesion forces do not change the Hertzian profile. The result was a continuous transition from the DMT to the JKR theory when a single parameter μ, proportional to the parameter introduced earlier by Tabor, increases. The DMT theory applies for $\mu \ll 1$ (hard solids, small radius of curvature and low energy of adhesion), and that of JKR for $\mu \gg 1$ (soft solids, large radius, large energy of adhesion). As shown later [15], these two extreme theories do not depend on the exact form of the interaction potential between surfaces.

In 1982, Barquins and Maugis [16], using the Sneddon [17] theory of the contact of axisymmetric punches in which an arbitrary rigid-body displacement $\chi(1)$ was cancelled such as to have zero stress at the edge of the contact, showed that if $\chi(1) \neq 0$ the stress singularities and discontinuity of displacement that appear are those of fracture mechanics, with a stress intensity factor K_{I} proportional to $\chi(1)$. Using the LEFM relation between G and $K_{\mathrm{I}}{}^2$, the JKR results are immediately obtained, but not those of DMT. In fact, this relation between G and $K_{\mathrm{I}}{}^2$ is valid only in linear fracture mechanics, when the size of the interaction zone is negligible compared with characteristic dimensions of the system (here the radius of contact). By applying a Dugdale model, where interaction forces are assumed constant on an annulus of width d around the contact [18], the JKR–DMT transition can be achieved, but this time with analytic formulae.

This brief historical background shows the striking analogy between the evolution of ideas in fracture mechanics and in the mechanics of contacts.

4.1.1 Method of Hankel Transforms

As proposed by Harding and Sneddon [19] in 1945, we will use the Hankel transform of zeroth order, $\Phi_0^{\mathrm{H}}(\xi, z)$, of the biharmonic function $\Phi(r, z)$. Expressions of stresses and displacements, and of their Hankel transforms, have been given as a function of $\Phi_0^{\mathrm{H}}(\xi, z)$ in (2.73–79). As the stresses approach zero at infinity, (2.71) can be written as

$$\Phi_0^{\mathrm{H}} = (A + B\xi z)\,\mathrm{e}^{-\xi z} .$$

The boundary condition $\tau_{rz}(r, 0) = 0$ is written, after (2.76), as

$$(1-\nu)\xi^3 \Phi_0^{\mathrm{H}}(\xi, 0) + \nu\xi \left(\frac{\mathrm{d}^2\Phi_0^{\mathrm{H}}}{\mathrm{d}z^2}\right)_{z=0} = 0$$

and leads to $A = 2\nu B$. Hence

$$\boxed{\Phi_0^{\mathrm{H}} = B(2\nu + \xi z)\mathrm{e}^{-\xi z}} . \tag{4.4}$$

The boundary condition (4.1) is written, from (2.73), as

$$\int_0^\infty \xi \left[(1-2\nu)\frac{\mathrm{d}^2\Phi_0^{\mathrm{H}}}{\mathrm{d}z^2} - 2(1-\nu)\xi^2\Phi_0^{\mathrm{H}}\right]_{z=0} J_0(\xi r)\,\mathrm{d}\xi = 2\mu\left[\delta - f\left(\frac{r}{a}\right)\right] .$$

Similarly, condition (4.2) is written as

$$\int_0^\infty \xi \left[(1-\nu)\frac{\mathrm{d}^3\Phi_0^{\mathrm{H}}}{\mathrm{d}z^3} - (2-\nu)\xi^2\frac{\mathrm{d}\Phi_0^{\mathrm{H}}}{\mathrm{d}z}\right]_{z=0} J_0(\xi r)\,\mathrm{d}\xi = 0 .$$

Replacing Φ_0^{H} by its expression given in (4.4), we can obtain the two integral equations

$$\int_0^\infty B\xi^3 J_0(\xi r)\,\mathrm{d}\xi = -\frac{\mu}{1-\nu}\left[\delta - f\left(\frac{r}{a}\right)\right] , \qquad 0 \le r \le a ,$$

$$\int_0^\infty B\xi^4 J_0(\xi r)\,\mathrm{d}\xi = 0 , \qquad r > a ,$$

which can be written as

$$\boxed{\begin{aligned} &\int_0^\infty \varphi(p) J_0(p\varrho)\,\mathrm{d}p = -\frac{E}{2(1-\nu^2)} a^4 \left[\delta - f(\varrho)\right], \quad 0 \le \varrho \le 1, && (4.5)\\ &\int_0^\infty p\varphi(p) J_0(p\varrho)\,\mathrm{d}p = 0, \qquad \varrho > 1, && (4.6) \end{aligned}}$$

letting $\varrho = r/a$, $p = a\xi$ and $\varphi(p) = p^3 B$.

Sneddon [20] showed in 1960 that if one let

$$\boxed{\varphi(p) = -\frac{Ea^4}{2(1-\nu^2)} \int_0^1 \chi(t) \cos(pt)\,\mathrm{d}t}, \tag{4.7}$$

then the second integral equation, (4.6), which is the Hankel transform of $\varphi(p)$, is identically zero. Sneddon also showed that the first becomes

$$\int_0^\varrho \frac{\chi(t)}{\sqrt{\varrho^2 - t^2}}\,\mathrm{d}t = \left[\delta - f(\varrho)\right] \tag{4.7a}$$

which is an Abel-type integral equation, whose solution (see Appendix B) is

$$\boxed{\begin{aligned} \chi(t) &= \frac{2}{\pi}\left(\delta - \frac{\mathrm{d}}{\mathrm{d}t}\int_0^t \frac{xf(x)}{\sqrt{t^2 - x^2}}\,\mathrm{d}x\right)\\ &= \frac{2}{\pi}\left(\delta - t\int_0^t \frac{f'(x)}{\sqrt{t^2 - x^2}}\,\mathrm{d}x\right) \end{aligned}}. \tag{4.8}$$

Let us demonstrate first that (4.6) is identically zero if we take for $\varphi(p)$ the expression given by (4.7). Let

$$I = \int_{p=0}^\infty J_0(p\varrho) \int_{t=0}^1 \chi(t) \cos(pt) p\,\mathrm{d}t\,\mathrm{d}p$$

and let us integrate the second integral by parts, giving

$$I = \chi(1) \int_{p=0}^\infty J_0(p\varrho) \sin p\,\mathrm{d}p - \int_{p=0}^\infty J_0(p\varrho) \int_{t=0}^1 \sin(pt) \chi'(t)\,\mathrm{d}t\,\mathrm{d}p\,.$$

After interchange of the order of integration in the second term, we have

$$I = \chi(1) \int_{p=0}^\infty J_0(p\varrho) \sin p\,\mathrm{d}p - \int_{t=0}^1 \chi'(t) \int_{p=0}^\infty J_0(p\varrho) \sin(pt)\,\mathrm{d}p\,\mathrm{d}t\,.$$

Using relation (A6) of Appendix A, we can verify that $I = 0$ if $\varrho > 1$, since $t \le 1$.

Let us show now that (4.5) can be reduced to an Abel-type integral. Using (4.7), (4.5) becomes

$$\int_{p=0}^{\infty} J_0(p\varrho) \int_{t=0}^{1} \chi(t)\cos(pt)\,\mathrm{d}t\,\mathrm{d}p = [\delta - f(\varrho)] \ .$$

With another interchange to the order of integration, we obtain

$$\int_{t=0}^{1} \chi(t) \int_{p=0}^{\infty} J_0(p\varrho)\cos(pt)\,\mathrm{d}p\,\mathrm{d}t = [\delta - f(\varrho)] \ ,$$

and we can use the relation (A5) with $\varrho < 1$. This integral equation then becomes

$$\int_0^{\varrho} \frac{\chi(t)}{\sqrt{\varrho^2 - t^2}}\,\mathrm{d}t = [\delta - f(\varrho)] \ ,$$

which is indeed (4.7a).

Knowing the function $f(r/a)$, we can then deduce $\chi(t)$, then the function $\varphi(p)$, hence B and then Φ_0^{H}. We can thus return to stresses and displacements, (2.73–79). With

$$\begin{aligned}
\Phi_0^{\mathrm{H}} &= B(2\nu + \xi z)\mathrm{e}^{-\xi z} \ , \\
\frac{\mathrm{d}\Phi_0^{\mathrm{H}}}{\mathrm{d}z} &= -B\xi(2\nu - 1 + \xi z)\mathrm{e}^{-\xi z} \ , \\
\frac{\mathrm{d}^2\Phi_0^{\mathrm{H}}}{\mathrm{d}z^2} &= B\xi^2(2\nu - 2 + \xi z)\mathrm{e}^{-\xi z} \ , \\
\frac{\mathrm{d}^3\Phi_0^{\mathrm{H}}}{\mathrm{d}z^3} &= -B\xi^3(2\nu - 3 + \xi z)\mathrm{e}^{-\xi z} \ ,
\end{aligned}$$

and introducing the integrals

$$\begin{aligned}
I_n^m &= \int_0^{\infty} p^n \varphi(p)\mathrm{e}^{-p\zeta} J_m(p\varrho)\,\mathrm{d}p \\
&= a^{n+4} \int_0^{\infty} \xi^{n+3} B\mathrm{e}^{-\xi z} J_m(\xi r)\,\mathrm{d}\xi \ ,
\end{aligned} \tag{4.9}$$

where $\zeta = z/a$, we obtain

$$2\mu u_r = \frac{1}{a^4}\left[(1-2\nu)I_0^1 - \zeta I_1^1\right] \ , \tag{4.10}$$

$$2\mu u_z = -\frac{1}{a^4}\left[2(1-\nu)I_0^0 + \zeta I_1^0\right] \ , \tag{4.11}$$

$$\sigma_r = \frac{1}{a^5}\left[I_1^0 - \zeta I_2^0 - (1-2\nu)\frac{I_0^1}{\varrho} + \frac{\zeta I_1^1}{\varrho}\right], \tag{4.12a}$$

$$\sigma_\theta = \frac{1}{a^5}\left[2\nu I_1^0 + (1-2\nu)\frac{I_0^1}{\varrho} - \frac{\zeta I_1^1}{\varrho}\right], \tag{4.12b}$$

$$\sigma_z = \frac{1}{a^5}\left[I_1^0 + \zeta I_2^0\right], \tag{4.12c}$$

$$\tau_{rz} = \frac{1}{a^5}\zeta I_2^1 . \tag{4.12d}$$

The hydrostatic stress is then simply

$$\sigma_0 = \frac{1}{a^5}\frac{2(1+\nu)}{3}I_1^0 . \tag{4.12e}$$

4.1.2 Some General Theorems

Sneddon [17] showed in 1965 that the penetration δ of the punch, the load P, the stress $\sigma_z(r,0)$ in the contact area ($\varrho < 1$), and the displacement $u_z(r,0)$ of the points of the surface for $\varrho > 1$, could be expressed as a function of $\chi(t)$ or of the shape $f(\varrho)$ of the punch:

$$\delta = \int_0^1 \frac{f'(x)\,\mathrm{d}x}{\sqrt{1-x^2}} + \frac{\pi}{2}\chi(1), \tag{4.13}$$

$$P = \frac{\pi a E}{1-\nu^2}\int_0^1 \chi(t)\,\mathrm{d}t = \frac{2aE}{1-\nu^2}\left[\delta - \int_0^1 \frac{x f(x)}{\sqrt{1-x^2}}\,\mathrm{d}x\right], \tag{4.14}$$

$$\sigma_z(\varrho,0) = -\frac{E}{2a(1-\nu^2)}\left[\frac{\chi(1)}{\sqrt{1-\varrho^2}} - \int_\varrho^1 \frac{\chi'(t)\,\mathrm{d}t}{\sqrt{t^2-\varrho^2}}\right], \quad \varrho < 1, \tag{4.15}$$

$$u_z(\varrho,0) = \int_0^1 \frac{\chi(t)\,\mathrm{d}t}{\sqrt{\varrho^2-t^2}}, \qquad \varrho > 1, \tag{4.16a}$$

$$= \chi(1)\sin^{-1}\frac{1}{\varrho} - \int_0^1 \chi'(t)\sin^{-1}\frac{t}{\varrho}\,\mathrm{d}t . \tag{4.16b}$$

Recall that the function f is known for a given punch only when the radius of contact is known. As we will see later, it is the chosen value for $\chi(1)$ which determines this radius and permits the complete solution of the problem.

Derivation of Equation for Load. Equation (4.13) gives the penetration δ and is directly deduced from (4.8). The force P needed to obtain this penetration is by definition

$$P = -2\pi\int_0^a r\sigma_z(r,0)\,\mathrm{d}r$$

with, in accordance with (4.12),

$$\sigma_z(r,0) = \frac{1}{a^5} I_1^0 = \frac{1}{a^5} \int_0^\infty p\varphi(p) J_0(p\varrho)\,\mathrm{d}p \,.$$

Hence

$$P = -\frac{2\pi}{a^3} \int_{\varrho=0}^1 \varrho \int_{p=0}^\infty p\varphi(p) J_0(p\varrho)\,\mathrm{d}p\,\mathrm{d}\varrho \,.$$

Taking $p\varrho$ as the variable and changing the order of integration, we obtain

$$P = -\frac{2\pi}{a^3} \int_{p=0}^\infty \frac{\varphi(p)}{p} \left[\int_{p\varrho=0}^p p\varrho J_0(p\varrho)\,\mathrm{d}(p\varrho) \right] \mathrm{d}p \,.$$

Using the relation (A2) from Appendix A in the form

$$\int_{p\varrho=0}^p p\varrho J_0(p\varrho)\,\mathrm{d}(p\varrho) = [(p\varrho)J_1(p\varrho)]_0^p = pJ_1(p) \,,$$

we get

$$P = -\frac{2\pi}{a^3} \int_0^\infty \varphi(p) J_1(p)\,\mathrm{d}p \,.$$

Replacing $\varphi(p)$ by its expression in (4.7), this yields

$$P = \frac{\pi a E}{1-\nu^2} \int_{p=0}^\infty \int_{t=0}^1 \chi(t) \cos(pt) J_1(p)\,\mathrm{d}t\,\mathrm{d}p \,,$$

which with a change in the order of integration gives

$$P = \frac{\pi a E}{1-\nu^2} \int_{t=0}^1 \chi(t) \int_{p=0}^\infty J_1(p) \cos(pt)\,\mathrm{d}p\,\mathrm{d}t \,.$$

Noting that according to relation (A7) we have

$$\int_0^\infty J_1(p)\cos(pt)\,\mathrm{d}p = 1 \,, \qquad 0 \le t \le 1 \,,$$

we finally obtain (4.14) for P.

Derivation of Equation for Stress. The relation given in (4.15) is less easy to derive. In $\sigma_z(r,0)$ only appears

$$I_1^0 = \int_0^\infty p\varphi(p) J_0(p\varrho)\,\mathrm{d}p$$

appears, which is the Hankel transform of zeroth order of $\varphi(p)$. We use the relation

$$\mathcal{H}_0\left[\varphi(p);\varrho\right] = \frac{1}{\varrho}\frac{\mathrm{d}}{\mathrm{d}\varrho}\varrho\mathcal{H}_1\left[\frac{1}{p}\varphi(p);\varrho\right]$$

between Hankel transforms. Replacing $\varphi(p)$ by its expression given in (4.7), we thus have

$$I_1^0 = -\frac{Ea^4}{2(1-\nu^2)}\frac{1}{\varrho}\frac{\mathrm{d}}{\mathrm{d}\varrho}\varrho\int_0^\infty p\left[\frac{1}{p}\int_0^1 \chi(t)\cos(pt)\,\mathrm{d}t\right]J_1(p\varrho)\,\mathrm{d}p\,.$$

With a change in the order of integration, we obtain

$$I_1^0 = -\frac{Ea^4}{2(1-\nu^2)}\frac{1}{\varrho}\frac{\mathrm{d}}{\mathrm{d}\varrho}\varrho\int_{t=0}^1 \chi(t)\int_{p=0}^\infty J_1(p\varrho)\cos(pt)\,\mathrm{d}p\,\mathrm{d}t\,,$$

and can then use the relation (A7) to get

$$I_1^0 = -\frac{Ea^4}{2(1-\nu^2)}\frac{1}{\varrho}\frac{\mathrm{d}}{\mathrm{d}\varrho}\int_0^1 \chi(t)\left[1-\frac{t}{\sqrt{t^2-\varrho^2}}H(t-\varrho)\right]\mathrm{d}t \tag{4.17}$$

and, since $\varrho < 1$,

$$\begin{aligned}I_1^0 &= -\frac{Ea^4}{2(1-\nu^2)}\frac{1}{\varrho}\frac{\mathrm{d}}{\mathrm{d}\varrho}\left[\int_0^1 \chi(t)\,\mathrm{d}t - \int_\varrho^1 \chi(t)\frac{t\,\mathrm{d}t}{\sqrt{t^2-\varrho^2}}\right]\\ &= \frac{Ea^4}{2(1-\nu^2)}\frac{1}{\varrho}\frac{\mathrm{d}}{\mathrm{d}\varrho}\int_\varrho^1 \chi(t)\frac{t\,\mathrm{d}t}{\sqrt{t^2-\varrho^2}}\,.\end{aligned}$$

Integrating by parts, we obtain

$$I_1^0 = \frac{Ea^4}{2(1-\nu^2)}\frac{1}{\varrho}\frac{\mathrm{d}}{\mathrm{d}\varrho}\left[\chi(1)\sqrt{1-\varrho^2} - \int_\varrho^1 \chi'(t)\sqrt{t^2-\varrho^2}\,\mathrm{d}t\right]\,,$$

then differentiating to give

$$I_1^0 = -\frac{Ea^4}{2(1-\nu^2)}\left[\frac{\chi(1)}{\sqrt{1-\varrho^2}} - \int_\varrho^1 \frac{\chi'(t)}{\sqrt{t^2-\varrho^2}}\,\mathrm{d}t\right]\,.$$

We obtain thus the required equation (4.15) for the stress.

Derivation of Equation for Displacement. Equations (4.11) and (4.9) lead to

$$u_z(r,0) = -\frac{2(1-\nu^2)}{Ea^4}\int_0^\infty \varphi(p)J_0(p\varrho)\,\mathrm{d}p\,.$$

Replacing $\varphi(p)$ by its expression, and changing the order of integration, we get

$$u_z(r,0) = \int_{t=0}^1 \chi(t)\int_{p=0}^\infty J_0(p\varrho)\cos(pt)\,\mathrm{d}p\,\mathrm{d}t\,,$$

and hence (4.16a) and then (4.16b) using the relation (A5) with $\varrho > 1$.

Way's Theorem. A theorem due to Way [21] in 1940 states that radial displacements and surface stresses, outside the area of contact, are independent of the (axisymmetric) distribution of the normal stresses inside the area of contact. In other words,

$$\boxed{u_r(r,0) = -\frac{(1-2\nu)(1+\nu)}{2\pi E}\frac{P}{r}, \qquad r > a}, \tag{4.18}$$

$$\sigma_r(r,0) = \frac{1-2\nu}{2\pi}\frac{P}{r^2}, \qquad r > a, \tag{4.19}$$

$$\sigma_\theta(r,0) = -\frac{1-2\nu}{2\pi}\frac{P}{r^2}, \qquad r > a. \tag{4.20}$$

The surface outside the area of contact is thus subject to pure shear.

We can derive the above formulae with the help of the above results. The radial displacement in surface is:

$$u_r(r,0) = \frac{(1+\nu)(1-2\nu)}{Ea^4} I_0^1,$$

with

$$I_0^1 = \int_0^\infty \varphi(p) J_1(p\varrho)\, dp.$$

Writing the expression for $\varphi(p)$, and changing the order of integration, we get

$$I_0^1 = -\frac{Ea^4}{2(1-\nu^2)} \int_{t=0}^1 \chi(t) \int_{p=0}^\infty J_1(p\varrho)\cos(pt)\, dp\, dt$$

and, using (A7) with $\varrho > t$ we get

$$\begin{aligned} I_0^1 &= -\frac{Ea^4}{2(1-\nu^2)}\frac{a}{r}\int_0^1 \chi(t)\, dt, \qquad r > a \\ &= -\frac{a^4}{2\pi}\frac{P}{r}. \end{aligned} \tag{4.21}$$

Hence we obtain (4.18).

The hoop stress in the surface, given by (4.12b), reduces to

$$\sigma_\theta(r,0) = \frac{1}{a^5}\left[2\nu I_1^0 + (1-2\nu)\frac{I_0^1}{\varrho}\right].$$

It can be seen easily, after (4.17), that $I_1^0 = 0$ for $\varrho > 1$, since $H(t-\varrho) = 0$. Equation (4.21) then gives directly (4.20). Regarding (4.19), it is deduced from (4.12e), which states that $\sigma_r + \sigma_\theta = 0$.

Concerning the radial displacement of (4.18), it must be pointed out that, first, it is zero at infinity; second, it is negative for positive P, i.e. the points of the free surface move radially towards a punch under compression; third, the points move away if the punch exerts a tensile adhesive force. More surprisingly, there is no dilation or contraction of the surface, as noted by Barber [22]. The surface dilation is defined by:

$$e_s = \varepsilon_r + \varepsilon_\theta = \frac{\partial u_r}{\partial r} + \frac{u_r}{r} ,$$

and from (4.18) we can see that it is zero for $r > a$. If we draw a circle of radius b around the punch, its area varies after application of the load P by the quantity δA given by

$$\delta A = 2\pi b u_r(b) = -\frac{(1-2\nu)(1+\nu)}{E} P ,$$

which is independent of its radius b. All the concentric circles drawn on the surface have an area which varies by the same quantity. These effects arise from the hypothesis of frictionless contact between the elastic half-space and the punch. When one presses a punch onto a surface, an area of free surface slides below it, which is independent of the profile of the punch and depends only on the applied load.

4.1.3 Influence of the Surface Energy

Let us study the stress $\sigma_z(r,0)$ and the displacement $u_z(r,0)$ at the vicinity of the edge of the contact. First let $r = a(1-\varepsilon)$. Using (4.15) we get

$$\sigma_z(r,0) \approx -\frac{E}{2a(1-\nu^2)} \left[\frac{\chi(1)}{\sqrt{2\varepsilon}} - \int_{1-\varepsilon}^{1} \frac{\chi'(t)\,\mathrm{d}t}{\sqrt{t^2-1+2\varepsilon}} \right] .$$

Changing the variable by $u = 1-t$, we then get

$$\sigma_z(r,0) \approx -\frac{E}{2a(1-\nu^2)} \left[\frac{\chi(1)}{\sqrt{2\varepsilon}} - \int_{0}^{\varepsilon} \frac{\chi'(1-u)\,\mathrm{d}u}{\sqrt{2(\varepsilon-u)}} \right] .$$

The integral in this expression tends towards zero as $\sqrt{\varepsilon}$. At the edge of the contact the stress thus tends towards infinity as $1/\sqrt{\varepsilon}$.

Figure 4.1 shows that the discontinuity of displacement $[u_z]$ at the edge of the punch is given by

$$[u_z] = f\left(\frac{r}{a}\right) - \delta + u_z(r,0) . \tag{4.22}$$

Making a power expansion at $r = a(1+\varepsilon)$ and using (4.7a) and (4.16a), we can see that

$$[u_z] = -\int_1^{\varrho} \frac{\chi(t)\,\mathrm{d}t}{\sqrt{\varrho^2 - t^2}} .$$

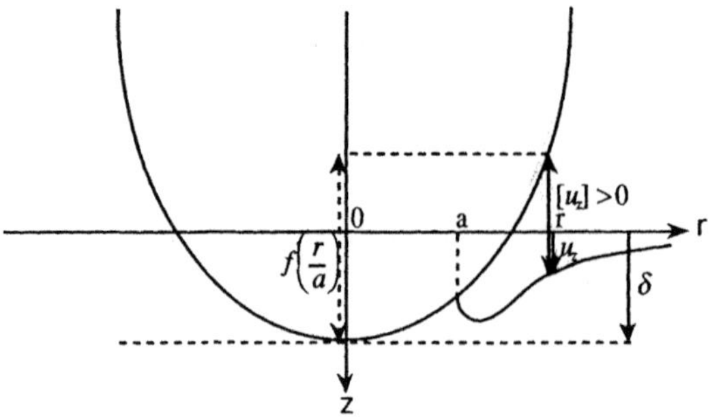

Fig. 4.1. Discontinuity of displacements at the edge of a punch

Near $t = 1$ we thus have

$$\begin{aligned}
[u_z] &= -\chi(1) \int_1^{\varrho} \frac{\mathrm{d}t}{\sqrt{\varrho^2 - t^2}} \\
&= -\chi(1) \left[\frac{\pi}{2} - \sin^{-1} \frac{1}{\varrho} \right] \\
&= -\chi(1) \cos^{-1} \frac{1}{1+\varepsilon} \\
&= -\chi(1) \sqrt{2\varepsilon}\,.
\end{aligned}$$

The connection is thus normal when $\chi(1) \neq 0$ since

$$\frac{\mathrm{d}\,[u_z]}{\mathrm{d}r} \approx -\frac{\chi(1)}{\sqrt{2\varepsilon}} \frac{1}{a}\,.$$

These infinite stresses and these vertical connections when $\chi(1) \neq 0$ led Sneddon to impose $\chi(1) = 0$ for punches with a smooth profile. (For a flat punch, we have $f(x) = f'(x) = 0$ and $\chi(1) = 2\delta/\pi$). In fact, these stresses and displacements are characteristic of fracture mechanics. With

$$K_{\mathrm{I}} = -\frac{E}{2(1-\nu^2)} \sqrt{\frac{\pi}{a}}\, \chi(1)\,, \tag{4.23}$$

we get:

$$\sigma_z(a-x) = \frac{K_{\mathrm{I}}}{\sqrt{2\pi x}}\,, \tag{4.24}$$

$$[u_z]\,(a+x) = 4\frac{1-\nu^2}{E} K_{\mathrm{I}} \sqrt{\frac{x}{2\pi}}\,, \tag{4.25}$$

characteristic[1] of mode I in plane strain, where K_{I} is the stress intensity factor. Note that $[u_z]$ and thus K_{I} are always positive (the solids are impe-

[1] The discontinuity of displacement is half that in fracture mechanics since the punch is not deformable.

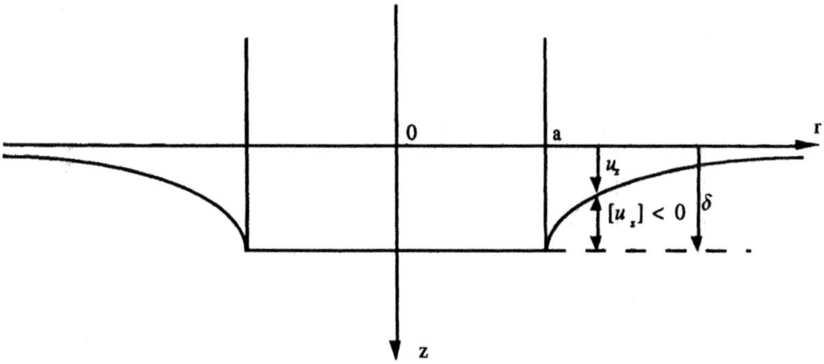

Fig. 4.2. Discontinuity of displacements at the edge of a flat punch

netrable): the stress σ_z is always tensile on the edge of the contact. (In the particular case of the flat punch in compression, Fig. 4.2, the discontinuity of u_z is not defined).

The analogy with fracture mechanics appears naturally if we consider the free edge of the contact as a crack which advances towards the interior of the contact when the load decreases, and recedes when it increases. Let us introduce the *energy release rate*, G. The strain being locally plane (i.e. there is a circular crack), and the punch not being deformable, we have

$$G = \frac{1}{2}\frac{1-\nu^2}{E}K_{\mathrm{I}}^2 .$$

Call P_1 the apparent load which, if $\chi(1)$ was null, would give the same contact radius as that observed under load P with $\chi(1) \neq 0$. Using (4.13) and (4.14), we get

$$P_1 = \frac{2aE}{1-\nu^2}\left[\int_0^1 \frac{f'(x)\,\mathrm{d}x}{\sqrt{1-x^2}} - \int_0^1 \frac{xf(x)\,\mathrm{d}x}{\sqrt{1-x^2}}\right] . \tag{4.26a}$$

(Note that for a flat punch we have $P_1 = 0$.) We then have from (4.26a)

$$\begin{aligned} P_1 - P &= -\frac{\pi a E}{1-\nu^2}\chi(1) \\ &= (4\pi a^3)^{1/2} K_{\mathrm{I}} , \end{aligned} \tag{4.26b}$$

i.e.

$$\boxed{G = \frac{1-\nu^2}{E}\frac{(P_1-P)^2}{8\pi a^3}} . \tag{4.27}$$

This very general formula is valid for any axisymmetric punch, at any point during loading or unloading. Equilibrium is defined by $G = w$ (Griffith's criterion), and we will have

$$P_1 - P = \left(8\pi a^3 \frac{Ew}{1-\nu^2}\right)^{1/2} , \tag{4.28}$$

$$\delta = \int_0^1 \frac{f'(x)\,\mathrm{d}x}{\sqrt{1-x^2}} - \left(\frac{1-\nu^2}{E} 2\pi a w\right)^{1/2} . \tag{4.29}$$

We will return later to these points.

4.2 Concentrated Force

The problem of stresses and displacements due to a concentrated force applied onto the surface of an elastic half-space was studied by Boussinesq in 1885. We will apply the method of Hankel transforms, following Little [23].

Let us consider a force P uniformly distributed inside a circle of radius a, where the radius tends towards zero. The boundary conditions are

$$\begin{aligned} \tau_{rz}(r,0) &= 0 , \\ \sigma_z(r<a,0) &= -\frac{P}{\pi a^2} , \\ \sigma_z(r>a,0) &= 0 . \end{aligned}$$

The minus sign in the middle expression arises from the fact that when the load P is positive, the stress is compressive and thus negative. As seen in Sect. 4.1.1, the first boundary condition leads to:

$$\Phi_0^{\mathrm{H}} = B(2\nu + \xi z)\mathrm{e}^{-\xi z} . \tag{4.30}$$

Let us take the Hankel transform of zeroth order of the two last boundary conditions. The Hankel transform of σ_z was given in Sect. 2.6 as

$$(\sigma_z)_0^{\mathrm{H}} = (1-\nu)\frac{\mathrm{d}^3\Phi_0^{\mathrm{H}}}{\mathrm{d}z^3} - (2-\nu)\xi^2 \frac{\mathrm{d}\Phi_0^{\mathrm{H}}}{\mathrm{d}z} .$$

We thus deduce that

$$[\sigma_z(r,0)]_0^{\mathrm{H}} = B\xi^3 ,$$

and the transformed boundary condition becomes

$$\begin{aligned} B\xi^3 &= -\int_0^a r\frac{P}{\pi a^2} J_0(\xi r)\,\mathrm{d}r \\ &= -\frac{P}{\pi a^2}\int_0^a rJ_0(\xi r)\,\mathrm{d}r . \end{aligned}$$

The integral is taken from 0 to a only, since $\sigma_z = 0$ for $r > a$. Let us use now the relation (A2) in the form

$$\xi r J_0(\xi r) = \frac{\mathrm{d}}{\mathrm{d}(\xi r)}[\xi r J_1(\xi r)] .$$

Using it,we get

$$B\xi^3 = -\frac{P}{\pi a^2}\frac{1}{\xi^2}\left[\xi r J_1(\xi r)\right]_{r=0}^{a}$$
$$= -\frac{P}{\pi}\frac{1}{\xi a}J_1(\xi a)\,.$$

Knowing that $J_1(x) \approx x/2$ when $x \to 0$, we have, at the limit when $a \to 0$,

$$B = -\frac{P}{2\pi\xi^3}\,. \tag{4.31}$$

We can return to displacements and stresses, (4.10–12) with I_n^m written in the form:

$$I_n^m = -\frac{Pa^{n+4}}{2\pi}Z_{n+1}^m\,. \tag{4.32}$$

Returning to Appendix A, we obtain:

$$I_0^0 = -\frac{Pa^4}{2\pi}\frac{1}{(r^2+z^2)^{1/2}}\,, \tag{4.33}$$

$$I_0^1 = -\frac{Pa^4}{2\pi}\frac{1}{r}\left(1-\frac{z}{\sqrt{r^2+z^2}}\right)\,, \tag{4.34}$$

$$I_1^0 = -\frac{Pa^5}{2\pi}\frac{z}{(r^2+z^2)^{3/2}}\,, \tag{4.35}$$

$$I_1^1 = -\frac{Pa^5}{2\pi}\frac{r}{(r^2+z^2)^{3/2}}\,, \tag{4.36}$$

$$I_2^0 = -\frac{Pa^6}{2\pi}\frac{2z^2-r^2}{(r^2+z^2)^{5/2}}\,,$$

$$I_2^1 = -\frac{Pa^6}{2\pi}\frac{3rz}{(r^2+z^2)^{5/2}}\,.$$

4.2.1 Displacements

In cylindrical coordinates (r, z), the displacements are given by:

$$2\mu u_r = \frac{P}{2\pi}\left[\frac{rz}{(r^2+z^2)^{3/2}} - \frac{(1-2\nu)r}{r^2+z^2+z(r^2+z^2)^{1/2}}\right]\,, \tag{4.37a}$$

$$2\mu u_z = \frac{P}{2\pi}\left[\frac{z^2}{(r^2+z^2)^{3/2}} + \frac{2(1-\nu)}{(r^2+z^2)^{1/2}}\right]\,, \tag{4.38a}$$

and in spherical coordinates (ϱ, φ) where φ is the angle with the z-axis, by

$$
\boxed{
\begin{aligned}
2\mu u_r &= \frac{P}{2\pi\varrho}\left(\cos^2\varphi + \cos\varphi - 1 + 2\nu\right)\tan\frac{\varphi}{2}\,, \qquad (4.37b)\\
2\mu u_z &= \frac{P}{2\pi\varrho}\left(\cos^2\varphi + 2 - 2\nu\right) \qquad (4.38b)
\end{aligned}
}\,.
$$

Note that in the bulk u_z is positive, whereas u_r changes its sign. Radial displacements are centrifugal ($u_r > 0$) inside a cone of angle φ defined by

$$\cos\varphi(1 + \cos\varphi) = 1 - 2\nu\,.$$

They are centripetal outside that cone, as shown in Fig. 4.3.

Surface displacements ($z = 0$) are given by

$$
\boxed{
\begin{aligned}
u_r(r,0) &= -\frac{(1-2\nu)(1+\nu)}{2\pi E}\frac{P}{r}\,, \qquad (4.39)\\
u_z(r,0) &= \frac{1-\nu^2}{\pi E}\frac{P}{r} \qquad (4.40)
\end{aligned}
}
$$

This is in agreement with Way's theorem, (4.18). Far enough from the origin to have $u_r \ll r$, any straight line of the surface passing through the origin takes the shape of an equilateral hyperbola. Note that at the surface

$$\frac{u_r}{u_z} = \frac{1-2\nu}{2(1-\nu)}\,,$$

and thus points move along directions making the same angle with the z-axis. The radial displacement at the surface is zero only for $\nu = 1/2$.

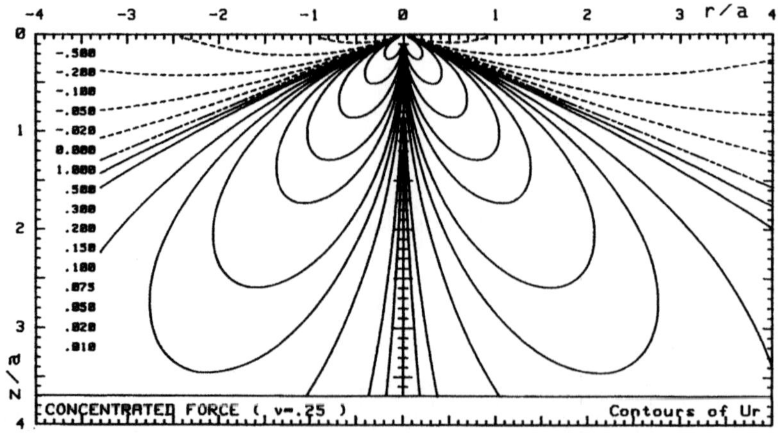

Fig. 4.3. Boussinesq's concentrated force: radial displacements

4.2.2 Stresses

In cylindrical coordinates the stresses are given by

$$\sigma_r = \frac{P}{2\pi}\left\{(1-2\nu)\left[\frac{1}{r^2} - \frac{z}{r^2(r^2+z^2)^{1/2}}\right] - \frac{3r^2 z}{(r^2+z^2)^{5/2}}\right\}, \tag{4.41a}$$

$$\sigma_\theta = \frac{P}{2\pi}(1-2\nu)\left[-\frac{1}{r^2} + \frac{z}{r^2(r^2+z^2)^{1/2}} + \frac{z}{(r^2+z^2)^{3/2}}\right], \tag{4.42a}$$

$$\sigma_z = -\frac{3P}{2\pi}\frac{z^3}{(r^2+z^2)^{5/2}}, \tag{4.43a}$$

$$\tau_{rz} = -\frac{3P}{2\pi}\frac{rz^2}{(r^2+z^2)^{5/2}}, \tag{4.44a}$$

$$\tau_{r\theta} = \tau_{\theta z} = 0,$$

and the hydrostatic stress is given by

$$\sigma_0 = -\frac{P}{\pi}\frac{1+\nu}{3}\frac{z}{(r^2+z^2)^{3/2}}.$$

In spherical coordinates, the stresses are:

$$\sigma_r = \frac{P}{2\pi\varrho^2}\left[\frac{1-2\nu}{1+\cos\varphi} - 3\sin^2\varphi\cos\varphi\right], \tag{4.41b}$$

$$\sigma_\theta = \frac{P}{2\pi\varrho^2}(1-2\nu)\left(\cos\varphi - \frac{1}{1+\cos\varphi}\right), \tag{4.42b}$$

$$\sigma_z = -\frac{P}{2\pi\varrho^2}3\cos^3\varphi, \tag{4.43b}$$

$$\tau_{rz} = -\frac{P}{2\pi\varrho^2}3\cos^2\varphi\sin\varphi. \tag{4.44b}$$

Note that (4.43b) and (4.44b) show that on any horizontal face the stress vector is on the radius vector. The stresses can be written in the generalized form

$$\sigma_{ij} = \frac{P}{2\pi\varrho^2}f_{ij}(\varphi,\nu). \tag{4.45}$$

They vary as the inverse square of the distance to the contact point, multiplied by a term depending only on φ and ν. They are thus independent of the Young's modulus.

The stresses are infinite at the origin, but it is possible to replace the concentrated force by forces statically equivalent and distributed on a small disk of radius a, centered on the origin. We can thus reintroduce normalized lengths r/a and z/a, and normalized stresses σ/p_{m} where $p_{\mathrm{m}} = P/\pi a^2$ is the mean pressure (positive). The advantage of this normalization is to allow the comparison with the stresses induced by the contact of an axisymmetric punch of any shape which, according to the Saint-Venant principle, must be identical at large r/a.

Surface stresses. Making $z = 0$ in (4.41a) to (4.44a), we can show that these stresses are

$$\sigma_r(r,0) = \frac{1-2\nu}{2\pi}\frac{P}{r^2}, \tag{4.46}$$

$$\sigma_\theta(r,0) = -\frac{1-2\nu}{2\pi}\frac{P}{r^2}, \tag{4.47}$$

$$\sigma_z(r,0) = \tau_{rz}(r,0) = 0 .$$

The state of stress is biaxial: the radial stress is still tensile, and the hoop stress is compressive. The surface is thus submitted to a pure shear, where

$$\tau_{\max}(r,0) = \frac{1-2\nu}{2\pi}\frac{P}{r^2} .$$

As for the radial displacement $u_r(r,0)$, the surface stresses cancel for $\nu = 1/2$.

Note that on any horizontal plane at level z, the stress σ_z can be written as

$$\sigma_z = -\frac{3P}{2\pi}\frac{\cos^5\varphi}{z^2},$$

a formula much used in soil mechanics.

Stresses on the z-axis. These stresses, normalized on Fig. 4.4, are obtained by making $\varphi = 0$ and letting $\varrho = z$, such that

$$\sigma_r(0,z) = \frac{1-2\nu}{4\pi}\frac{P}{z^2}, \tag{4.48}$$

$$\sigma_\theta(0,z) = \frac{1-2\nu}{4\pi}\frac{P}{z^2}, \tag{4.49}$$

$$\sigma_z(0,z) = -\frac{3}{2\pi}\frac{P}{z^2}, \tag{4.50}$$

$$\tau_{rz}(0,z) = 0 .$$

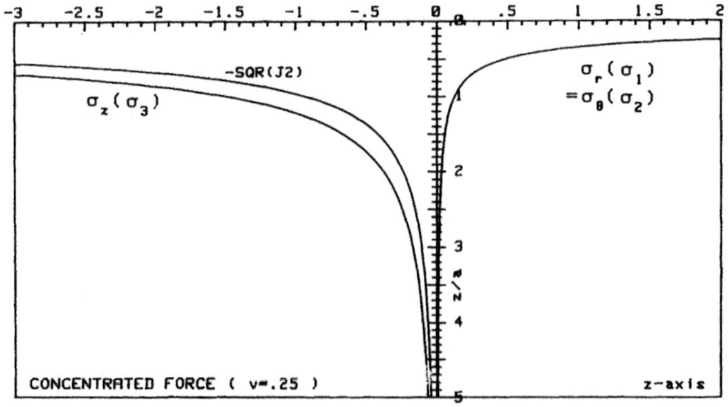

Fig. 4.4. Reduced stresses on the z-axis

The z-axis is a principal axis where the state of stress is triaxial. Its points are subjected to two non-zero maximum shear stresses:

$$\tau_{\max} = \frac{\sigma_\theta - \sigma_z}{2} = \frac{7-2\nu}{8\pi}\frac{P}{z^2} ,$$

$$\tau_{\max} = \frac{\sigma_r - \sigma_z}{2} = \frac{7-2\nu}{8\pi}\frac{P}{z^2} .$$

Study of principal stresses. By symmetry, σ_θ is a principal stress; it is the hoop stress σ_2. The two others are in the plane $\theta = \text{const}$. Let $\sigma_1 > \sigma_3$. In the surface and on the z-axis ($\tau_{rz} = 0, \sigma_r > 0$), we have $\sigma_1 \equiv \sigma_r$, $\sigma_3 \equiv \sigma_z$.

As all these stresses can be written in the form of (4.45), we can compare the stresses at a given point by comparing the terms f_{ij}. Figure 4.5 displays, in the case $\nu = 1/4$, the f_1, f_2, f_3 i.e. the angular variations of σ_1, σ_2, σ_3. We can see that the stress σ_1 is everywhere a traction, that the stress σ_3 is everywhere compressive. The hoop stress σ_2 is a traction in a cone of angle $51°50'$ defined by

$$\cos\varphi(1+\cos\varphi) = 1$$

and compressive outside (in the surface in particular). We should note that $\sigma_1 > \sigma_2 > \sigma_3$ almost everywhere, except in a shallow zone ($\varphi > 77°13'$ for $\nu = 1/4$) where $\sigma_3 > \sigma_2$.

Figure 4.6 displays contours of the principal stresses in reduced coordinates for $\nu = 1/4$. The hydrostatic stress, the von Mises criterion, and the stress trajectories are given in section in Fig. 4.7. In the surface, stress trajectories are circles (σ_2) and radial lines (σ_1). We can note that σ_3 starts perpendicularly to the surface (as does σ_z) and then bends; and that σ_1 can be identified as σ_r in the surface and on the z-axis. The dotted-dashed lines represent the isoclines at 45°.

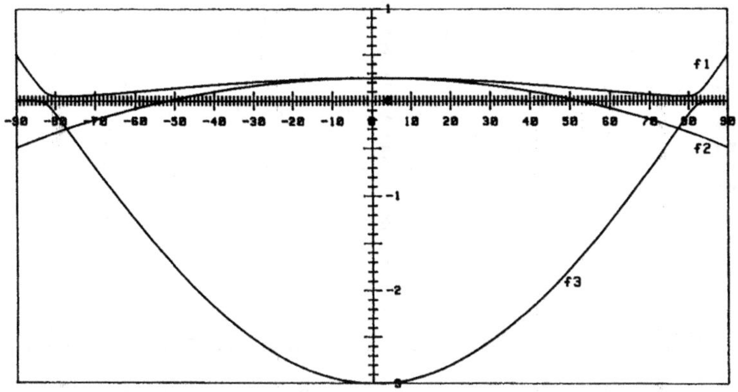

Fig. 4.5. Angular variations of σ_1, σ_2 and σ_3 for $\nu = 1/4$

(a)

(b)

(b)

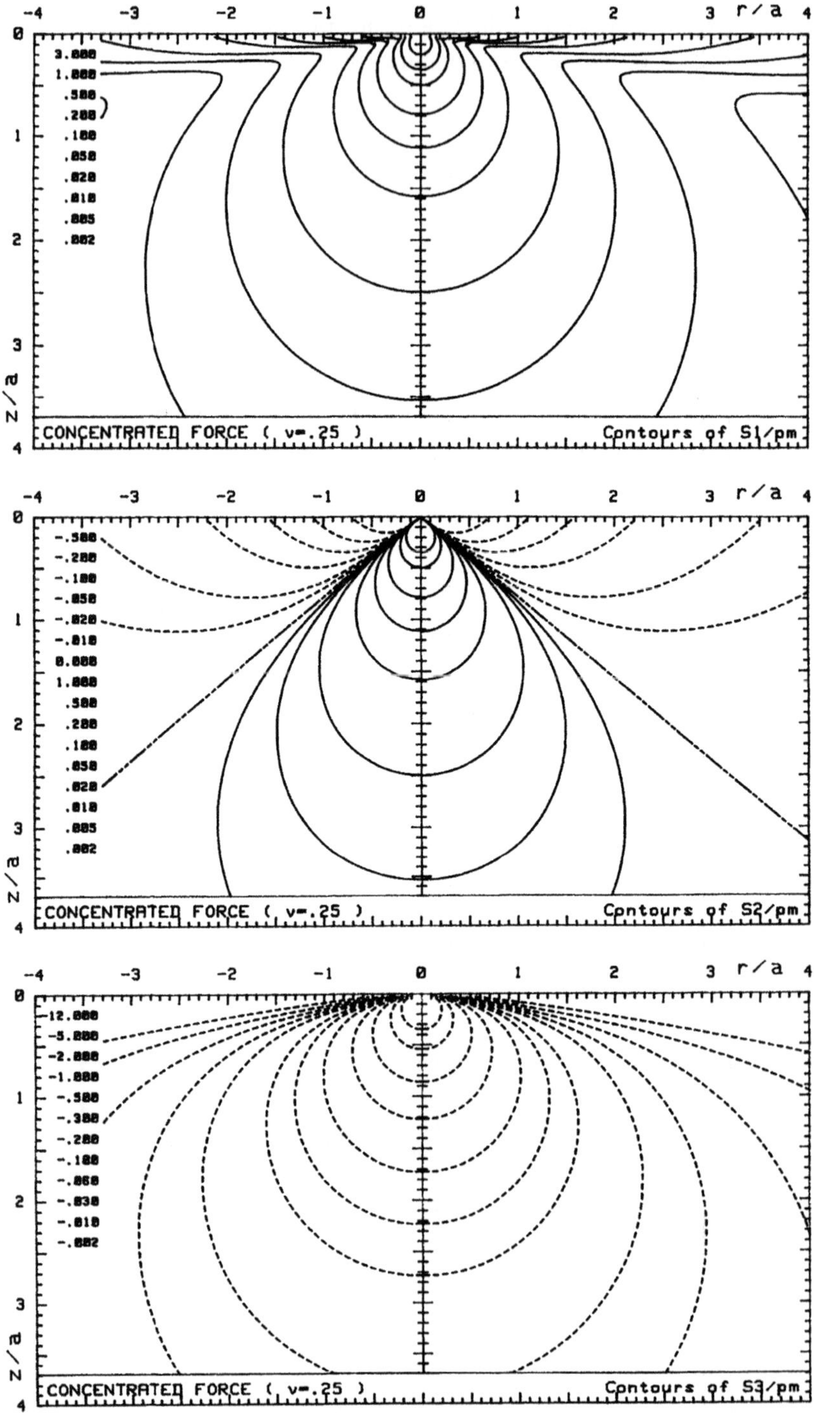

Fig. 4.6. Contours of the principal stresses $\sigma_1/p_{\rm m}$ **(a)**, $\sigma_2/p_{\rm m}$ **(b)** and $\sigma_3/p_{\rm m}$ **(c)** for $\nu = 1/4$ with a concentrated force

(a)
(b)
(c)

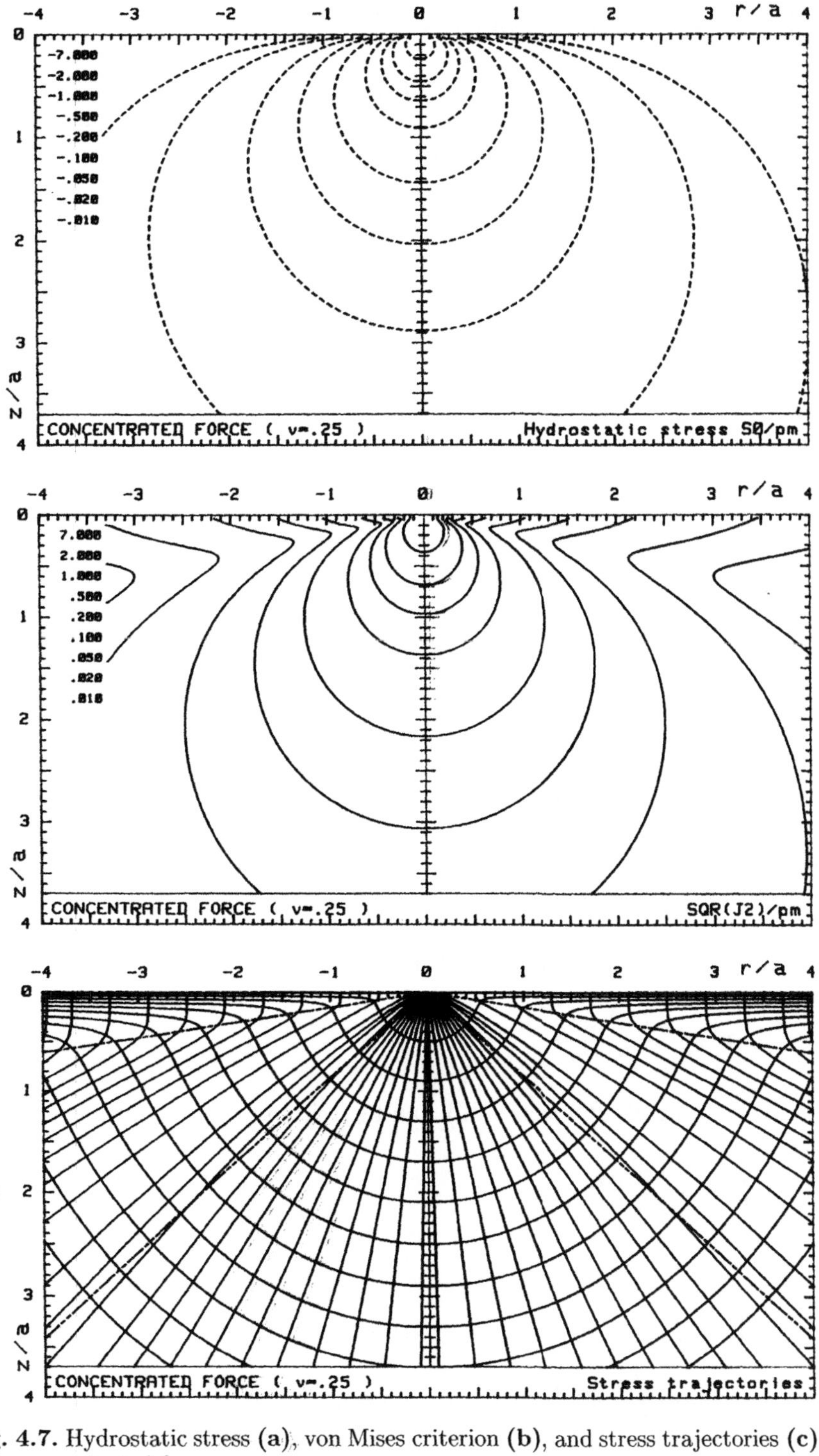

Fig. 4.7. Hydrostatic stress (**a**), von Mises criterion (**b**), and stress trajectories (**c**) for a concentrated force

4.3 The Circular Flat Punch

The problem of the rigid and circular flat punch of radius a pressing the surface of an elastic half-space under a load P was partially solved by Boussinesq in 1885, who gave the distribution of the stress $\sigma_z(r,0)$ in the area of contact, the penetration δ of the punch and the vertical displacement u_z of the surface outside the contact area. The full problem was solved by Sneddon [24] in 1946.[2] Let us use the reduced coordinates $\varrho = r/a$ and $\zeta = z/a$ and apply the general theorems of Sect. 4.1.2. The shape of the punch being given by

$$f(\varrho) = 0\,, \tag{4.51}$$

we have

$$\chi(t) = \chi(1) = \frac{2\delta}{\pi} \tag{4.52}$$

and we get

$$\delta = \frac{1-\nu^2}{2E}\frac{P}{a}\,, \tag{4.53}$$

$$\sigma_z(r,0) = -\frac{P}{2\pi a^2}\frac{1}{\sqrt{1-\varrho^2}}\,, \qquad r < a\,, \tag{4.54}$$

$$u_z(r,0) = \frac{1-\nu^2}{\pi E}\frac{P}{a}\sin^{-1}\frac{1}{\varrho}\,, \qquad r \geq a\,, \tag{4.55}$$

which are Boussinesq's results. The profile of the elastic half-space near the edge of the contact is displayed in Fig. 4.2.

Substituting (4.52) and (4.53) in (4.7), we get

$$\varphi(p) = -\frac{Pa^3}{2\pi}\frac{\sin p}{p}\,, \tag{4.56}$$

and the integrals I_n^m, from (4.9), can be written as

$$I_n^m = -\frac{Pa^3}{2\pi}S_n^m\,, \tag{4.57}$$

with

$$\begin{aligned} S_n^m &= \int_0^\infty p^{n-1}\mathrm{e}^{-p\zeta}\sin p J_m(p\varrho)\,\mathrm{d}p \\ &= \mathrm{Im}\int_0^\infty p^{n-1}\mathrm{e}^{-p(\zeta-i)}J_m(p\varrho)\,\mathrm{d}p\,. \end{aligned} \tag{4.58}$$

[2] Note the typographical errors in equations (3), (8) and (16) of [24].

Equations (4.10–12) then become

$$2\mu u_r = -\frac{P}{2\pi a}\left[(1-2\nu)S_0^1 - \zeta S_1^1\right] , \tag{4.59}$$

$$2\mu u_z = \frac{P}{2\pi a}\left[2(1-\nu)S_0^0 + \zeta S_1^0\right] , \tag{4.60}$$

$$\frac{\sigma_r}{p_\mathrm{m}} = -\frac{1}{2}\left[S_1^0 - \zeta S_2^0 - (1-2\nu)\frac{S_0^1}{\varrho} + \frac{\zeta S_1^1}{\varrho}\right] , \tag{4.61}$$

$$\frac{\sigma_\theta}{p_\mathrm{m}} = -\frac{1}{2}\left[2\nu S_1^0 + (1-2\nu)\frac{S_0^1}{\varrho} - \frac{\zeta S_1^1}{\varrho}\right] , \tag{4.62}$$

$$\frac{\sigma_z}{p_\mathrm{m}} = -\frac{1}{2}\left[S_1^0 + \zeta S_2^0\right] , \tag{4.63}$$

$$\frac{\tau_{rz}}{p_\mathrm{m}} = -\frac{1}{2}\left[\zeta S_2^1\right] , \tag{4.64}$$

where $p_\mathrm{m} = P/\pi a^2$ is the mean pressure. The hydrostatic pressure is simply

$$\frac{\sigma_0}{p_\mathrm{m}} = -\frac{1+\nu}{3} S_1^0 . \tag{4.65}$$

These integrals are computed in Appendix A as functions of the parameters

$$\begin{aligned} s &= \sqrt{\left(\varrho^2+\zeta^2-1\right)^2 + 4\zeta^2} \\ &= \sqrt{\left(\varrho^2+\zeta^2+1\right)^2 - 4\varrho^2} \\ &= \sqrt{[\zeta^2+(\varrho-1)^2]\,[\zeta^2+(\varrho+1)^2]} , \end{aligned} \tag{4.66}$$

$$\sin\varphi = \frac{2\zeta}{s} , \tag{4.67a}$$

$$\cos\varphi = \frac{\varrho^2+\zeta^2-1}{s} , \qquad 0 < \varphi < \pi , \tag{4.67b}$$

$$\tan\theta = \frac{1}{\zeta} , \tag{4.68}$$

$$u = \frac{1}{2}(\varrho^2+\zeta^2-1+s) . \tag{4.69}$$

These parameters have simple geometrical interpretations, as shown by Fig. 4.8. First, we can see immediately that s is given by

$$s = \frac{\mathrm{MA}\cdot\mathrm{MB}}{a^2} ,$$

and iso-s are Cassini's ovals; as regards φ, it is the angle AMB. The classical relation in triangles,

$$\mathrm{AB}^2 = \mathrm{MA}^2 + \mathrm{MB}^2 - 2\mathrm{MA}\cdot\mathrm{MB}\cos\varphi ,$$

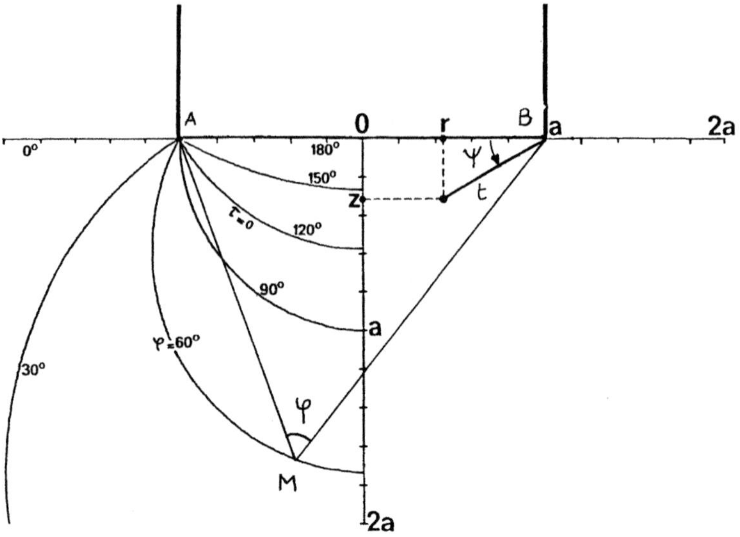

Fig. 4.8. Contours of φ, and polar coordinates at the edge of the contact

is in effect written as

$$4a^2 = [z^2 + (r-a)^2] + [z^2 + (r+a)^2] - 2a^2 s \cos\varphi \,.$$

Hence for φ the expression (4.67b) applies. The iso-φ are the subtended angles of the segment AB.

We can also verify that

$$u = \left(\frac{\mathrm{MA} + \mathrm{MB}}{2a}\right)^2 - 1 \,.$$

The iso-u lines are thus ellipses with foci at A and B.

Calculations give

$$S_0^0 = \tan^{-1}\left(\frac{1}{s^{1/2}\cos(\varphi/2)}\right) = \tan^{-1}\frac{1}{\sqrt{u}} \,,$$

$$S_1^0 = \frac{1}{s^{1/2}}\sin\frac{\varphi}{2} = \frac{\zeta}{s\sqrt{u}} \,,$$

$$S_2^0 = \frac{\sqrt{1+\zeta^2}}{s^{3/2}}\sin\left(\frac{3\varphi}{2} - \theta\right) = \frac{1}{s^3\sqrt{u}}\left[-s(\zeta^2+u) + 4\zeta^2(1+u)\right] \,,$$

$$S_0^1 = \frac{1}{\varrho}\left(1 - s^{1/2}\sin\frac{\varphi}{2}\right) = \frac{1}{\varrho}\left(1 - \frac{\zeta}{\sqrt{u}}\right) \,,$$

$$S_1^1 = \frac{\sqrt{1+\zeta^2}}{\varrho s^{1/2}}\sin\left(\theta - \frac{\varphi}{2}\right) = \frac{\varrho\sqrt{u}}{s(1+u)} \,,$$

$$S_2^1 = \frac{\varrho}{s^{3/2}}\sin\frac{3\varphi}{2} = \frac{\varrho}{s^2}\frac{\zeta}{\sqrt{u}}\left(3 - \frac{4\zeta^2}{us}\right) \,.$$

4.3.1 Study of the Displacements

In the surface and below the punch ($\varrho < 1$, $\varphi = \pi$, $u = 0$) the displacements are given by

$$u_r(r,0) = -\frac{(1-2\nu)(1+\nu)}{2\pi E}\frac{P}{r}(1-\sqrt{1-\varrho^2})\,, \tag{4.70}$$

$$u_z(r,0) = \delta = \frac{1-\nu^2}{2E}\frac{P}{a}\,. \tag{4.71}$$

In the surface and for $\varrho > 1$ ($\varphi = 0$, $u = \varrho^2 - 1$) we have

$$u_r(r,0) = -\frac{(1-2\nu)(1+\nu)}{2\pi E}\frac{P}{r}\,, \tag{4.72}$$

$$u_z(r,0) = \frac{1-\nu^2}{\pi E}\frac{P}{a}\sin^{-1}\frac{a}{r}\,. \tag{4.73}$$

According to Way's theorem, the displacement $u_r(r,0)$ for $r > a$ is that of the Boussinesq concentrated force given in (4.39). The radial (centripetal) displacements of the surface are at a maximum at $r = a$, as shown in Fig. 4.9. This calculation assumes, of course, that points move freely below the punch, i.e. that the friction coefficient is zero.

For $r \gg a$, the vertical displacement of the surface points is

$$u_z = \frac{1-\nu^2}{\pi E}\frac{P}{r}\,,$$

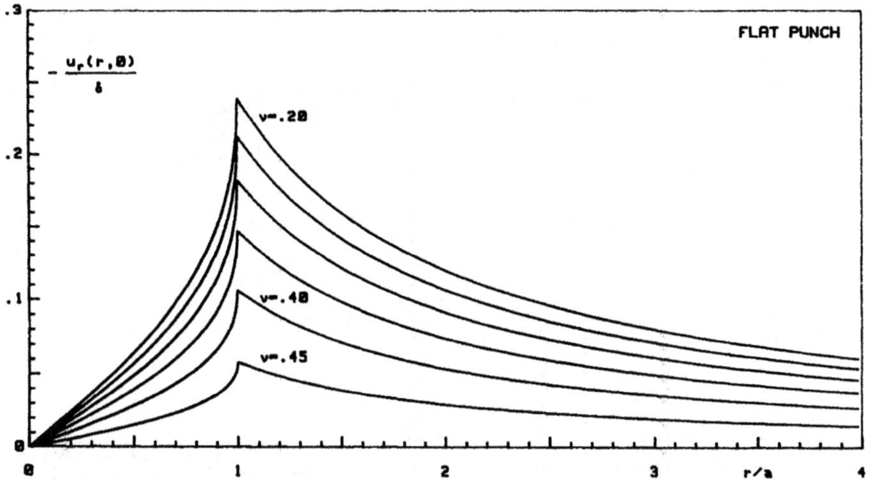

Fig. 4.9. Radial displacement in a surface

as for the the Boussinesq concentrated force, and we thus have

$$\frac{u_r(r,0)}{u_z(r,0)} \underset{r\to\infty}{\longrightarrow} \frac{1-2\nu}{2(1-\nu)} .$$

At the edge of the punch and outside, we have

$$u_z(r,0) \approx \frac{1-\nu^2}{\pi E}\frac{P}{a}\left(\frac{\pi}{2} - \sqrt{2(\varrho-1)}\right) ,$$

and hence $\mathrm{d}u_z/\mathrm{d}r$ is infinite at the edge of the punch. The connection of the elastic half-space to the punch is thus vertical, as shown in Fig. 4.2.

On the z-axis, we have

$$u_r(0,z) = 0 ,$$

$$u_z(0,z) = \frac{1+\nu}{2\pi E}\frac{P}{a}\left[2(1-\nu)\tan^{-1}\frac{1}{\zeta} + \frac{\zeta}{1+\zeta^2}\right] . \tag{4.74}$$

In the bulk, the displacements u_z are all positive, but the displacements u_r can change their sign:

$$2\mu u_r(r,z) = \frac{P}{2\pi a}\left[\frac{\varrho\zeta\sqrt{u}}{s(1+u)} - \frac{1-2\nu}{\varrho}\left(1-\frac{\zeta}{\sqrt{u}}\right)\right] , \tag{4.75}$$

$$2\mu u_z(r,z) = \frac{P}{2\pi a}\left[\frac{\zeta^2}{s\sqrt{u}} + 2(1-\nu)\tan^{-1}\frac{1}{\sqrt{u}}\right] , \tag{4.76}$$

Figure 4.10 displays the contours of u_r/δ (for $\nu = 1/4$), as functions of r/a and z/a. We can see that at large distances the curve $u_r = 0$ approaches the cone corresponding to the concentrated force ($\varphi = 68°32'$ for $\nu = 1/4$).

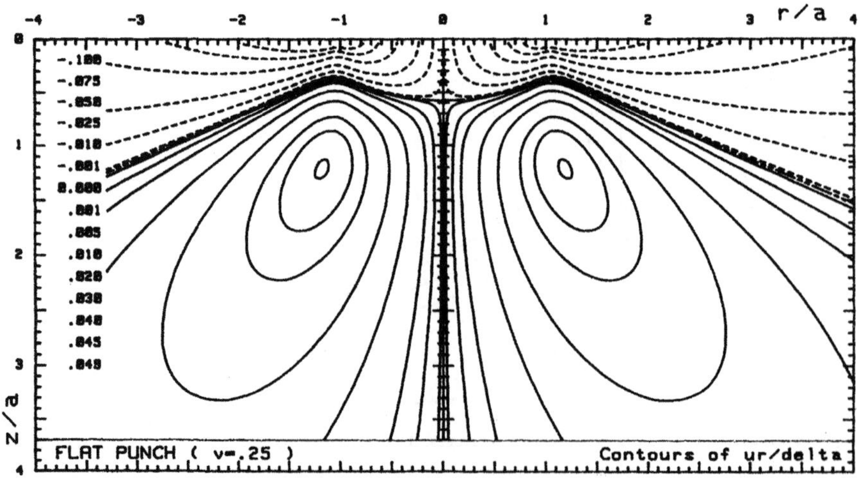

Fig. 4.10. Contours of the radial displacements given by u_r/δ

4.3.2 Study of the Stresses

In the Surface. Outside the contact ($\varrho > 1$, $\varphi = 0$), we have

$$\sigma_r(r,0) = \frac{1-2\nu}{2\pi}\frac{P}{r^2} ,$$
$$\sigma_\theta(r,0) = -\frac{1-2\nu}{2\pi}\frac{P}{r^2} ,$$
$$\sigma_z(r,0) = \tau_{rz}(r,0) = 0 .$$

According to Way's theorem, they are indeed the stresses due to a concentrated force, as per (4.46) and (4.47). The surface is thus subjected to pure shear.

In the area of contact ($\varrho < 1$, $\varphi = \pi$), we get

$$\frac{\sigma_r}{p_\mathrm{m}} = -\frac{1}{2}\left\{(1-\varrho^2)^{-1/2} - (1-2\nu)\frac{1}{\varrho^2}\left[1-(1-\varrho^2)^{1/2}\right]\right\} , \tag{4.77}$$

$$\frac{\sigma_\theta}{p_\mathrm{m}} = -\frac{1}{2}\left\{2\nu(1-\varrho^2)^{-1/2} + (1-2\nu)\frac{1}{\varrho^2}\left[1-(1-\varrho^2)^{1/2}\right]\right\} , \tag{4.78}$$

$$\frac{\sigma_z}{p_\mathrm{m}} = -\frac{1}{2}(1-\varrho^2)^{-1/2} ,$$
$$\tau_{rz} = 0 . \tag{4.79}$$

In particular, at the centre of the contact we have

$$\frac{\sigma_r}{p_\mathrm{m}} = \frac{\sigma_\theta}{p_\mathrm{m}} = -\frac{1+2\nu}{4} ,$$
$$\frac{\sigma_z}{p_\mathrm{m}} = -\frac{1}{2} .$$

These surface stresses are displayed in Fig. 4.11 for $\nu = 1/4$. Note the discontinuity of stresses at $r = a$.

On the z-axis. The stress tensor is:

$$\frac{\sigma_r}{p_\mathrm{m}} = \frac{\sigma_\theta}{p_\mathrm{m}} = \frac{1}{2(1+\zeta^2)}\left[\frac{1-2\nu}{2} - \frac{1}{1+\zeta^2}\right] , \tag{4.80}$$

$$\frac{\sigma_z}{p_\mathrm{m}} = -\frac{1}{2(1+\zeta^2)}\left[3 - \frac{2}{1+\zeta^2}\right] , \tag{4.81}$$
$$\tau_{rz} = 0 .$$

For $z = 0$, the stresses at the centre of the contact can be deduced, and for $z \to \infty$ those due to a concentrated force are obtained. These stresses are principal stresses. The stress σ_z is always negative (compression) and goes through a maximum at $z/a = 1/\sqrt{3}$. The two others stresses, first

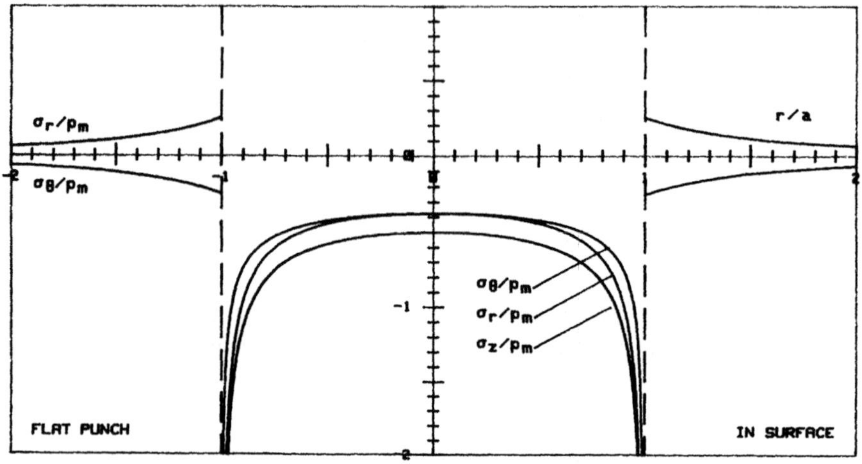

Fig. 4.11. Surface stresses

compressive, become tensile for $z/a > [(1+2\nu)/(1-2\nu)]^{1/2}$. The maximum shear stress, given by

$$\frac{\tau_{\max}}{p_m} = \frac{1}{2}\frac{\sigma_z - \sigma_\theta}{p_m} = -\frac{(7-2\nu)\zeta^2 + 1 - 2\nu}{8(1+\zeta^2)^2},$$

goes through a maximum at $z/a = [(5+2\nu)/(7-2\nu)]^{1/2}$.

These stresses are displayed in Fig. 4.12 as a function of z/a for $\nu = 1/4$. They approach those of the Boussinesq concentrated force for $z/a > 4$.

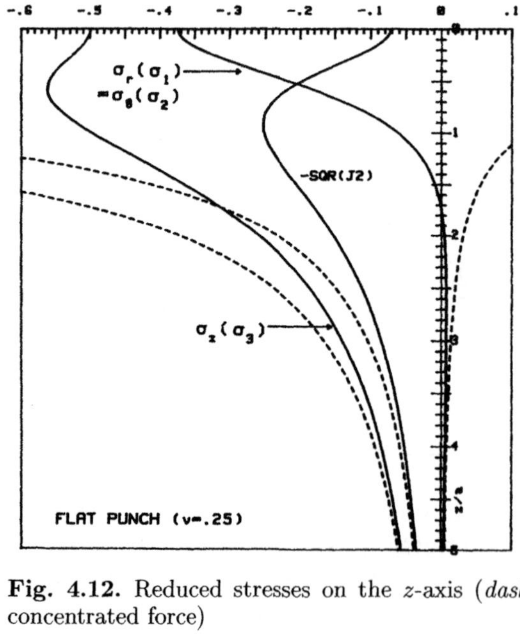

Fig. 4.12. Reduced stresses on the z-axis (*dashed line* showing the Boussinesq concentrated force)

In the Bulk. Figure 4.13 displays the contours of the principal stresses σ_1, $\sigma_2(\equiv \sigma_\theta)$ and σ_3 in the bulk. On the surface and along the z-axis, we have $\sigma_1 \equiv \sigma_r$ and $\sigma_3 \equiv \sigma_z$. Note that only the stress σ_3 is everywhere compressive. The stress σ_1 is locally compressive below the punch, and a traction elsewhere. The hoop stress σ_2 is compressive below the surface, and then tensile inside a truncated cone.

Figure 4.14 displays, for a flat punch, the hydrostatic stress, the von Mises criterion and the stress trajectories of σ_1 and σ_3 (tangential at all points to σ_1 and σ_3). We can see that plastic deformation takes place at the edge of the punch. We can see further that the stress σ_3 starts perpendicularly to the surface, where it is identical to σ_z (even if $\sigma_z = 0$ outside the contact). Let us recall that the stress trajectories of σ_2 are circles round the z-axis. One can note also the "regular" aspect of the stress trajectories, in contrast to the complicated aspect of the contours of the principal stresses. The curves in dashed-dotted lines, superimposed on the stress trajectories, are the isoclines at 0° ($\varphi = 2\pi/3$) and 45°.

4.3.3 Stresses and Displacements at the Edge of the Contact

Let us use the polar coordinates (t, ψ) near the edge of the contact (see Fig. 4.8), such that

$$\varrho = 1 - \frac{t}{a}\cos\psi \,,$$

$$\zeta = \frac{t}{a}\sin\psi \,,$$

with $t/a \ll 1$, and make a power expansion of (4.66–68). We get

$$s \approx \frac{2t}{a} \,,$$

$$\sin\varphi = \sin\psi \,,$$

$$\cos\varphi = -\cos\psi \,,$$

$$\theta = \frac{\pi}{2} \,.$$

The integrals S_n^m become

$$S_0^0 = \frac{\pi}{2} - \sqrt{\frac{2t}{a}}\sin\frac{\psi}{2} \,,$$

$$S_0^1 = 1 - \sqrt{\frac{2t}{a}}\cos\frac{\psi}{2} \,,$$

$$\zeta S_1^1 = \sqrt{\frac{2t}{a}}\cos\frac{\psi}{2}\sin^2\frac{\psi}{2} \,,$$

$$S_1^0 = \frac{1}{\sqrt{2t/a}}\cos\frac{\psi}{2} \,,$$

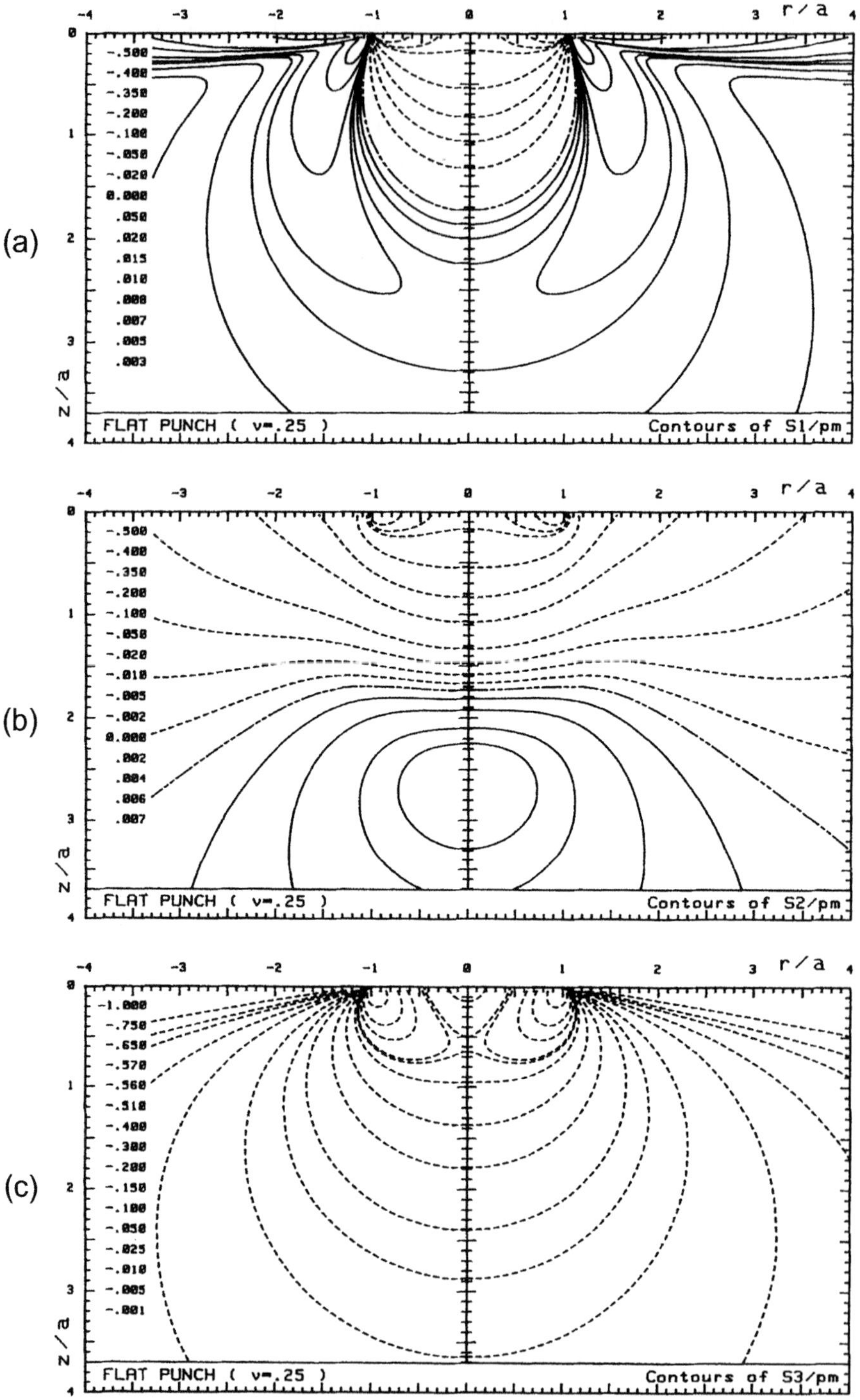

Fig. 4.13. Contours of the principal stresses under a flat punch: **a** σ_1; **b** σ_2; **c** σ_3

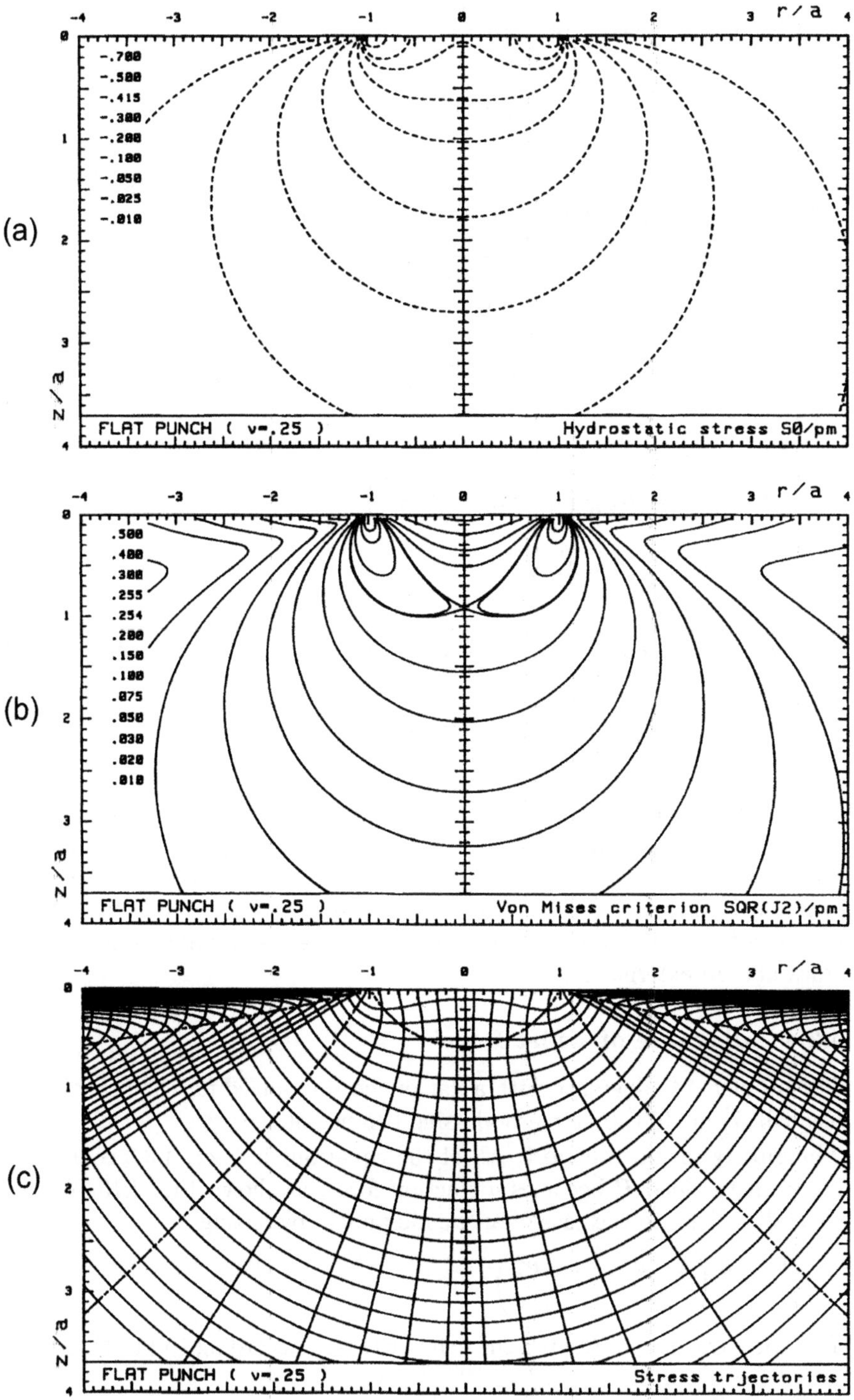

Fig. 4.14. Hydrostatic stress (**a**), von Mises criterion (**b**) and stress trajectories (**c**) for a flat punch

$$\zeta S_2^0 = \frac{1}{\sqrt{2t/a}} \cos\frac{\psi}{2} \sin\frac{\psi}{2} \sin\frac{3\psi}{2} ,$$

$$\zeta S_2^1 = -\frac{1}{\sqrt{2t/a}} \cos\frac{\psi}{2} \sin\frac{\psi}{2} \cos\frac{3\psi}{2} .$$

Letting $P = -K_\mathrm{I}\sqrt{4\pi a^3}$, which is in agreement with (4.26b), the power expansions of displacements and stresses can written as

$$u_r(r,z) - u_r(a,0) = -\frac{K_\mathrm{I}}{\mu}\sqrt{\frac{2t}{\pi}} \cos\frac{\psi}{2}\left(1 - 2\nu + \sin^2\frac{\psi}{2}\right) , \tag{4.82}$$

$$u_z(r,z) - \delta = +\frac{K_\mathrm{I}}{\mu}\sqrt{\frac{2t}{\pi}} \sin\frac{\psi}{2}\left(1 - 2\nu + \sin^2\frac{\psi}{2}\right) , \tag{4.83}$$

$$\sigma_r = \frac{K_\mathrm{I}}{\sqrt{2\pi t}} \cos\frac{\psi}{2}\left(1 - \sin\frac{\psi}{2} \sin\frac{3\psi}{2}\right) , \tag{4.84}$$

$$\sigma_\theta = 2\nu\frac{K_\mathrm{I}}{\sqrt{2\pi t}} \cos\frac{\psi}{2} , \tag{4.85}$$

$$\sigma_z = \frac{K_\mathrm{I}}{\sqrt{2\pi t}} \cos\frac{\psi}{2}\left(1 + \sin\frac{\psi}{2} \sin\frac{3\psi}{2}\right) , \tag{4.86}$$

$$\tau_{rz} = -\frac{K_\mathrm{I}}{\sqrt{2\pi t}} \cos\frac{\psi}{2} \sin\frac{\psi}{2} \cos\frac{3\psi}{2} . \tag{4.87}$$

Note first that $\sigma_\theta = \nu(\sigma_r + \sigma_z)$, which proves that a state of plane strain exists at the edge of the contact. Except for τ_{rz} and u_r for which the signs are reversed, the above formulae are the classic ones of fracture mechanics in mode I and plane strain. The reason for the change of sign for τ_{rz} and u_r arises from the change in the direction of the r-axis between an internal crack and an external crack.

4.3.4 Adherence of a Flat Punch

The problem of adherence with a flat punch was solved for the first time by Kendall [7] in 1971, by a method using balance of energy.

Thermodynamic method. All of the equations in Sect. 4.3.3 remain valid when the load P becomes a traction force, as shown Fig. 4.15. We will generalize them to the case of an elastic punch, by letting

$$\frac{1}{K} = \frac{3}{4}\left(\frac{1-\nu_1^2}{E_1} + \frac{1-\nu_2^2}{E_2}\right) . \tag{4.88}$$

The "penetration" of the punch is thus written, according to (4.71), as

$$\delta = \frac{2P}{3aK} . \tag{4.89}$$

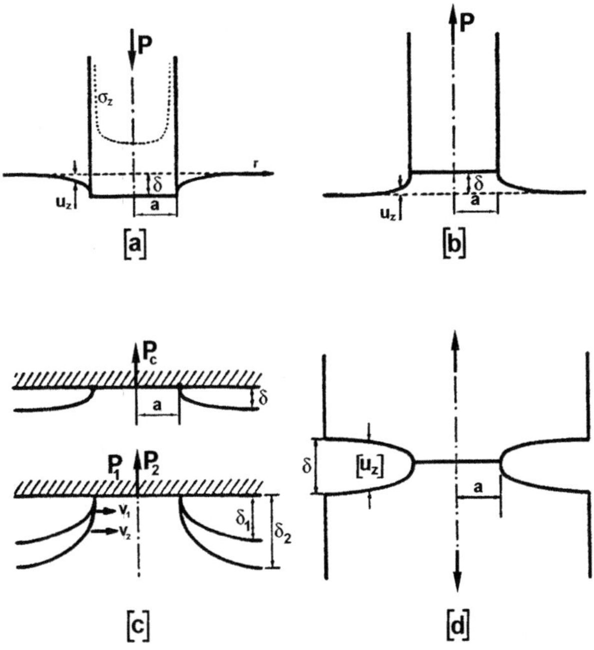

Fig. 4.15. Analogy between the adherence of a flat punch and the fracture of a deeply notched bar

To compute the energy release rate G, we must evaluate the mechanical energy (elastic plus potential) and its variation when the area of contact varies by $\mathrm{d}A$. To vary the radius of the punch in order to study the variations of mechanical energy may appear absurd, but examination of Fig. 4.15c shows that (4.89) applies to the case where the contact radius a is smaller than the radius of the punch. We can thus study variations of elastic and potential energies for fluctuations of contact area at constant load P. For a load P and a radius of contact a, the potential and elastic energies are, respectively,

$$U_{\mathrm{P}} = -P\delta = -\frac{2P^2}{3aK} ,$$

$$U_{\mathrm{E}} = \int P(\delta)\,\mathrm{d}\delta = \frac{1}{2}P\delta = \frac{P^2}{3aK} ,$$

and the energy release rate

$$G = \left(\frac{\partial U_{\mathrm{E}}}{\partial A} + \frac{\partial U_{\mathrm{P}}}{\partial A}\right)_{\mathrm{P}} = \frac{1}{2\pi a}\frac{\partial}{\partial a}\left(-\frac{P^2}{3aK}\right)$$

can be expressed as

$$\boxed{G = \frac{P^2}{6\pi a^3 K} = \frac{3K\delta^2}{8\pi a}} . \tag{4.90}$$

Note that (4.89) and (4.90) are the *equations of state* of the system, and that the *Maxwell relations* are:

$$\left(\frac{\partial G}{\partial \delta}\right)_A = \left(\frac{\partial P}{\partial A}\right)_\delta = \frac{P}{2\pi a^2} , \tag{4.91}$$

$$\left(\frac{\partial G}{\partial P}\right)_A = -\left(\frac{\partial \delta}{\partial A}\right)_P = \frac{P}{3\pi a^3 K} . \tag{4.92}$$

The equilibrium defined by $G = w$ is always unstable, since we have

$$\left(\frac{\partial G}{\partial A}\right)_P = 3\left(\frac{\partial G}{\partial A}\right)_\delta = -\frac{P^2}{4\pi^2 a^5 K} = -\frac{9K\delta^2}{16\pi^2 a^3} < 0 .$$

Let us consider the unstable equilibrium of Fig. 4.15c, under a constant load P. If a fluctuation momentarily increases the contact area, G decreases; then, because $G < w$, the crack recedes and the contact area increases, decreasing G still more. There is then more and more departure from the equilibrium, and the process stops when the edge of the punch is reached. If, on the other hand, a fluctuation decreases the contact area, G increases and, as $G > w$, the crack propagates and the contact area decreases, increasing G still more. The process stops only when complete separation is reached.

Let us now return to the adherence of the punch displayed in Fig. 4.15b. For $P = 0$, we have $G = 0$, but the contact area cannot increase further. In proportion to any increase in the traction force P, G increases at constant a, until the (unstable) equilibrium $G = w$ is reached under a load P_c. At this moment a fluctuation can only decrease the contact area, and this diminution will continue until a complete rupture has occured under this load P_c defined by $G = w$. Thus

$$\boxed{P_c = \sqrt{6\pi a^3 K w}} , \tag{4.93}$$

which is by definition the adherence force (the pull-off force) of a punch of radius a. Similarly, at fixed grips, the instability will occur at the critical displacement

$$\delta_c = \sqrt{\frac{8\pi a w}{3K}} . \tag{4.94}$$

Note with caution that this adherence force is not proportional to the area of the contact, nor to its perimeter, and is not proportional to the energy of adhesion!

Fracture Mechanics Method. Kendall [8] noted in 1973 that this geometry was similar to that of a deeply notched cylindrical bar. It happens that Paris and Sih [25] gave in 1965 the stress intensity factor for this case, namely

$$K_I = \frac{1}{2}\sigma\sqrt{\pi a} = \frac{P}{2\pi a^2}\sqrt{\pi a} = \frac{P}{2a\sqrt{\pi a}} ;$$

hence, since the strain is locally plane in axial symmetry,

$$G = \frac{1-\nu^2}{E} {K_\mathrm{I}}^2 = \frac{1-\nu^2}{E} \frac{P^2}{4\pi a^3} = \frac{P^2}{6\pi a^3 K} .$$

Recall that only knowledge of the stresses at the edge of the contact was needed to compute G. The power expansion as a function of $t = (a - r)$ of (4.54), with a positive sign since we have traction, gives

$$\sigma \approx \frac{P}{2a\sqrt{\pi a}} \frac{1}{\sqrt{2\pi t}} ,$$

and hence

$$K_\mathrm{I} = \frac{P}{2a\sqrt{\pi a}} .$$

Energies of the System and Berry's Representation. We have seen, in Chap. 3, that for transformations at fixed load the thermodynamic potential to use was the Gibbs free energy, given by

$$\begin{aligned} \mathcal{G} &= U_\mathrm{E} - P\delta + U_\mathrm{S} - TS , \\ \mathrm{d}\mathcal{G} &= -\delta \,\mathrm{d}P + (G - w)\,\mathrm{d}A - S\,\mathrm{d}T , \end{aligned}$$

or, if thermal effects are neglected, the enthalpy $H = U_\mathrm{E} - P\delta + U_\mathrm{S}$. Figure 4.16 displays the variation of this expression for energy along the curve of equilibrium points, as also for evolutions of the system at constant load P. Using (4.89) and (4.90), we have at any equilibrium point:

$$U_\mathrm{E} = 2Aw = -2U_\mathrm{S} ,$$

so that

$$H = -\frac{3}{2} U_\mathrm{E} = \frac{3}{4} U_\mathrm{P} = 3U_\mathrm{S} = -3Aw .$$

Hence, the curve $H(A)$ is a straight line with $(\partial H/\partial A)_\mathrm{P} = 0$ everywhere and $\mathrm{d}H/\,\mathrm{d}A = 0$ nowhere.

Similarly, for transformations at fixed grips, the thermodynamic potential to use is the Helmholtz free energy, in accordance with (3.4), namely

$$\begin{aligned} F &= U_\mathrm{E} + U_\mathrm{S} - TS , \\ \mathrm{d}F &= P\,\mathrm{d}\delta + (G - w)\,\mathrm{d}A - S\,\mathrm{d}T , \end{aligned}$$

or, if thermal effects are neglected, the internal energy $U = U_\mathrm{E} + U_\mathrm{S}$. Variations in potential along equilibrium points or for evolutions at fixed grips are displayed in Fig. 4.17. At every equilibrium point we have

$$U = \frac{1}{2} U_\mathrm{E} = -U_\mathrm{S} = Aw$$

and the curve $U(A)$ is a straight line.

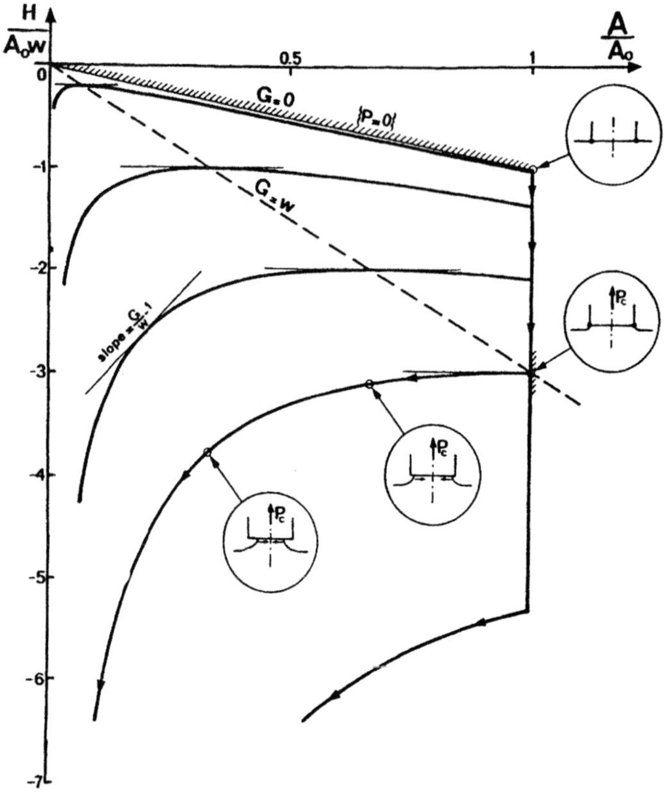

Fig. 4.16. Enthalpy of the system as a function of the contact area, in reduced coordinates. The *dashed straight line* corresponds to equilibrium points, and the *curves* represent the variation of enthalpy at fixed load. The curves at G = const are *straight lines* passing through the origin

The Berry representation of the system (see Sect. 3.2.2) is given in Fig. 4.18. The surface energy U_S is the area between the loading curve and the locus of the Griffith criterion. This locus is the cubic

$$P = \frac{9K^2}{16\pi w}\delta^3 \tag{4.95}$$

or, in reduced coordinates ($x = \delta K/w, y = PK/w^2$):

$$y = \frac{9}{16\pi}x^3 .$$

Points such as B or B_2 on this locus have the contact area A as a parameter. For a punch with an area A_0, the loading curve is the straight line OB (A_0 constant). At equilibrium point B, the elastic energy $U_E \equiv \frac{1}{2}P_c\delta_c$ is given by the area of the triangle OBI, and the surface energy $U_S = -\frac{1}{2}U_E$ by the cross-hatched negative area. (For a cubic $y = kx^3$, the area under the curve

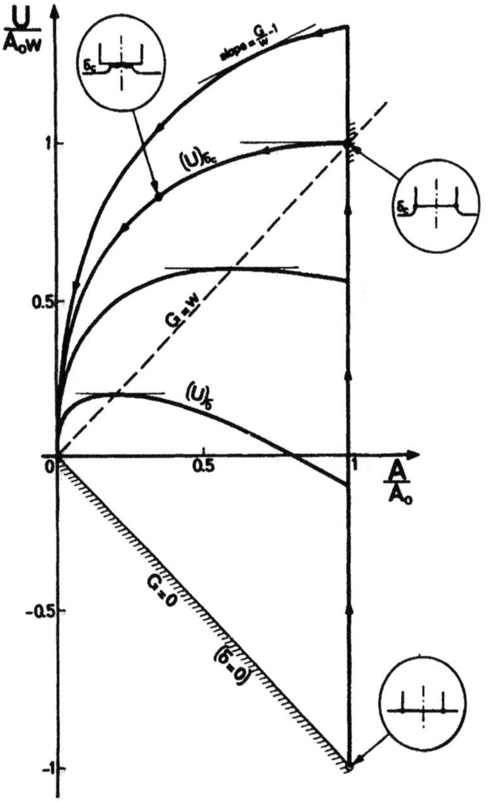

Fig. 4.17. Internal energy of the system as a function of the contact area, in reduced coordinates. The *dashed straight line* corresponds to equilibrium points, and the *curves* represent the variation of internal energy at fixed grips. The curves at $G = \text{const}$ are *straight lines* passing through the origin

is $\frac{1}{4}yx$.) During crack propagation at fixed load (Fig. 4.18b), or at fixed grips (Fig. 4.18c), the contact area and $|U_{\mathrm{S}}|$ decrease, and the elastic energy, given by the area of the triangle $\mathrm{OB_1I}$ tends respectively towards infinity or zero.

In Fig. 4.18b, the thermodynamic potential $H = U_{\mathrm{E}} - P\delta + U_{\mathrm{S}}$ at point $\mathrm{B_1}$ is the negative area $\mathrm{JB_1B_2O}$ (the equilibrium curve between $\mathrm{B_2}$ and O), and its variation $\mathrm{BB_1B_2}$ (the dotted area) during the crack propagation is the work

$$\int_{A_0}^{A} (G - w)\,\mathrm{d}A$$

supplied by the system: it corresponds to the kinetic energy of the crack and/or to the dissipated heat.

In Fig. 4.18c, the thermodynamic potential $U = U_{\mathrm{E}} + U_{\mathrm{S}}$ is the positive area $\mathrm{OB_2B_1I}$ (the equilibrium curve between O and $\mathrm{B_2}$), and the work supplied by the system is still $\mathrm{BB_1B_2}$.

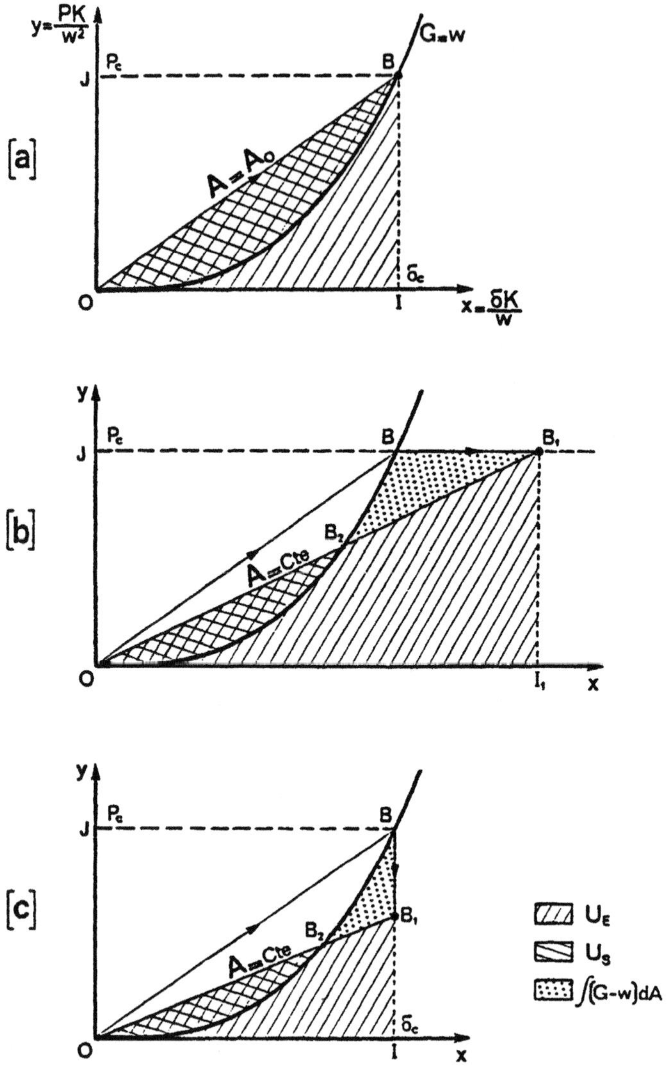

Fig. 4.18. Berry's representation

4.4 Contact of Spheres: the Hertz Theory

The problem of a spherical punch pressing an elastic half-space or of the contact of two elastic spheres was solved by Hertz [1] in 1882, who gave the stress distribution in the area of contact, the penetration δ of the punch (or the approach of the two spheres) and the radius of contact as a function of the load.[3] Hertz also gave the general formulae for the stress tensor, but did

[3] The same results have been given independently by Bousinessq [26] in January 1883

not pursue its calculation to the end. This work was done by Huber [27] in 1904, but with sign errors corrected by Love [28] in 1929. The first diagrams of principal stresses and stress trajactories were drawn by Fuchs [29] in 1913, by Morton and Close [30] in 1922, and then with precision by Lawn [31] in 1968.

We will give the results of the general theorems by Sneddon [17], but due to the importance of Hertzian contact in mechanics we will present first the elementary method, such as that found in Timoshenko and Goodier [32] for example. But beforehand let us give a mnemotechnic mean to recover easily the form of the relation between load P, contact radius a and sphere radius R. A representative value of the stress is $\sigma \approx P/\pi a^2$, whereas a representative value of the strain is $\varepsilon \approx a/R$. Applying Hooke's law $\sigma = \varepsilon E$, we get $a^3 \approx PR/E$.

4.4.1 Elementary Method

Let us consider, as shown in Fig. 4.19, two elastic spheres or radii R_1 and R_2. If no pressure is applied between the two spheres, the contact occurs only at the point O (in the absence, of course, of surface energy). If one sphere is pressed on the other with a force P, the centres of the two spheres approach by a quantity δ, and the contact is achieved on a circle of radius a. There are these two quantities δ and a that we will compute, starting from the following hypotheses:

1. the ratio a/R is small;
2. no friction occurs at the interface;
3. no tensile stress exists in area of contact (unilateral contact).

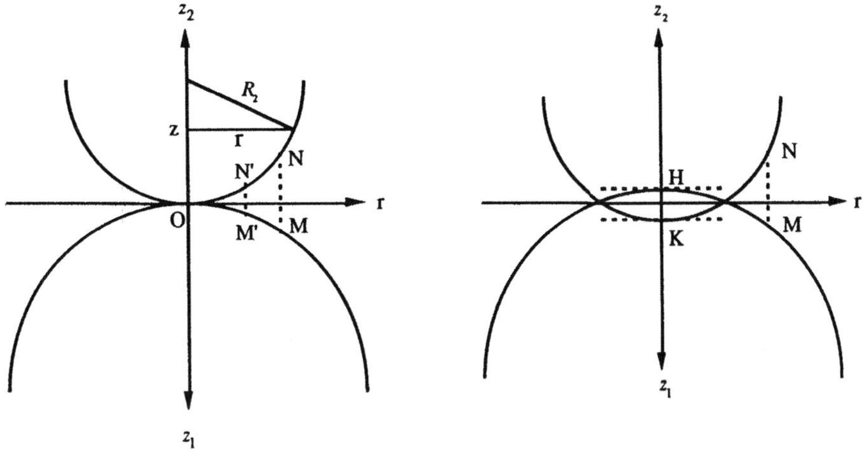

Fig. 4.19. Two spheres in contact: (**a**) at a single point; (**b**) by a circle of contact

The equations of the two circles displayed in Fig. 4.19a are

$$r^2 + (z_1 - R_1)^2 = R_1^2 \,,$$
$$r^2 + (z_2 - R_2)^2 = R_2^2 \,.$$

Near the z-axis, the distance between the two points M and N near O is

$$z_1 + z_2 = \frac{r^2}{2R} \,, \tag{4.96}$$

where

$$\frac{1}{R} = \frac{1}{R_1} + \frac{1}{R_2} \,.$$

If the deformation is taken to remain localized near the contact, we see that the centres of the two spheres are closer by the quantity $\delta = \mathrm{HK}$, as also are all pairs of points MN outside the contact circle (Fig. 4.19b).

On the other hand, due to the noninterpenetrability of the two spheres, the pairs of points such as M'N' (Fig. 4.20) are closer by a lesser quantity, equal to their initial separation $z_1 + z_2$. Call $u_z(1)$ and $u_z(2)$ the vertical elastic displacements due to the local deformation. For every point of the contact area, we thus have

$$u_z(1) + u_z(2) = \mathrm{HK} - (z_1 + z_2)$$
$$= \delta - \frac{r^2}{2R} \,.$$

In this calculation, we do not have to take into account the tangential displacements $u_r(r, 0)$, since the coordinates of the points are taken before deformation (Lagrangian coordinates). The points M and N are thus on the same vertical line.

Consider the local displacements $u_z(1)$ and $u_z(2)$. Taking the same notation as in the problem of the Boussinesq concentrated force, described by

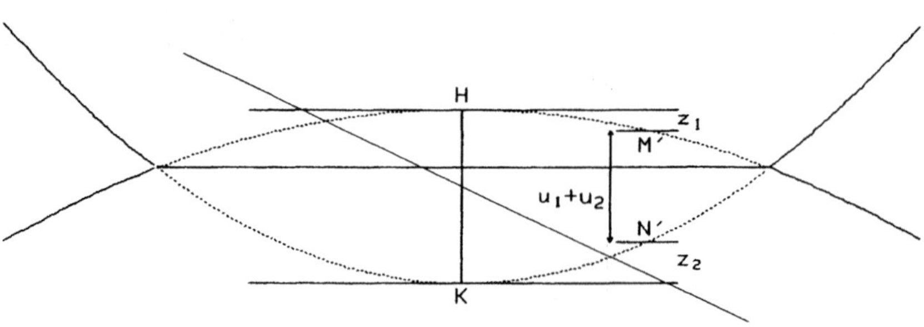

Fig. 4.20. Two spheres in contact by a circle (close-up)

(4.40), the displacement $\mathrm{d}^2u_z(1)$ of a point M of the sphere 1 situated on the surface of contact, under the action of the force $ps\,\mathrm{d}\psi\,\mathrm{d}s$ exerted by an element of the sphere 2, (see Fig. 4.22 for notations), is given by

$$\mathrm{d}^2u_z(1) = \frac{1-\nu_1^2}{\pi E_1}\frac{ps\,\mathrm{d}\psi\,\mathrm{d}s}{s}$$

and hence, by integration, we have

$$u_z(1) = \frac{1-\nu_1^2}{\pi E_1}\iint p\,\mathrm{d}\psi\,\mathrm{d}s\,.$$

Likewise for $u_z(2)$. And with the notation

$$k_i = \frac{1-\nu_i^2}{\pi E_i}\,,$$

we get

$$\delta - \frac{r^2}{2R} = (k_1+k_2)\iint p\,\mathrm{d}\psi\,\mathrm{d}s\,.$$

We have thus to find an expression of the pressure distribution which satisfies this relation for any point of the contact.

We can show that this condition is fulfilled by assuming that the pressure distribution on the contact area is represented by the ordinate of a half-sphere of radius a constructed on the surface of contact, as shown in Fig. 4.21, so that, if p_0/a is the scale factor, the total force is the volume of the half-sphere:

$$P = \frac{p_0}{a}\frac{2}{3}\pi a^3\,,$$

where p_0 is the maximum pressure, equal to 1.5 times the mean pressure given by $p_\mathrm{m} = P/\pi a^2$. The "hemispherical" pressure distribution can be written as

$$p = p_0\sqrt{1-\frac{r^2}{a^2}}\,. \tag{4.97}$$

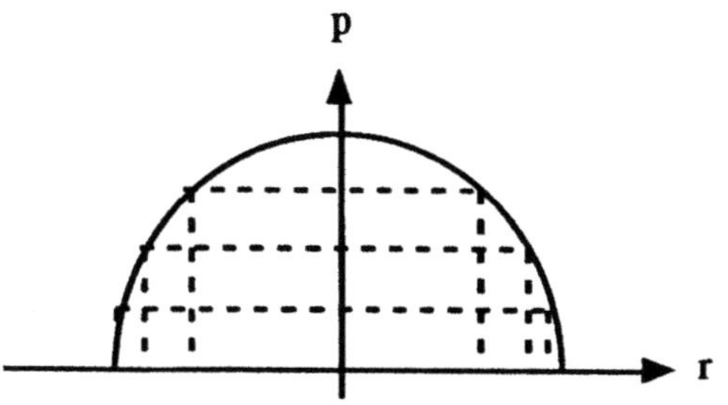

Fig. 4.21. Half-sphere on a contact surface

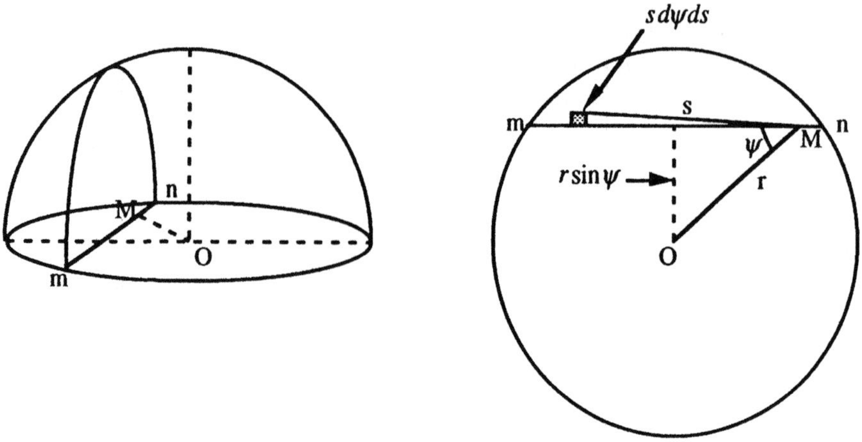

Fig. 4.22. Circle of contact within a half-sphere

We have first to integrate $\int p\, \mathrm{d}s$ along a chord mn of the circle of contact (Fig. 4.22). Apart from the factor p_0/a, the integral $\int_m^n p\, \mathrm{d}s$ is represented by the area

$$A = \frac{\pi}{2}(a^2 - r^2 \sin^2 \psi)$$

of the half-circle of diameter mn.

We then get

$$\begin{aligned}(k_1 + k_2) \iint p\, \mathrm{d}\psi\, \mathrm{d}s &= (k_1 + k_2)\frac{p_0}{a}\pi \int_0^{\pi/2} (a^2 - r^2 \sin^2 \psi)\, \mathrm{d}\psi \\ &= (k_1 + k_2) p_0 \frac{\pi^2}{4a}(2a^2 - r^2)\,.\end{aligned}$$

The fact that this distribution is in the form $\delta - r^2/2R$ shows that the chosen pressure distribution was correct, and by identification we get

$$\begin{aligned}\delta &= (k_1 + k_2) p_0 \frac{\pi^2 a}{2}\,,\\ \frac{1}{2R} &= (k_1 + k_2) p_0 \frac{\pi^2}{4a}\,.\end{aligned}$$

Letting

$$\frac{1}{K} = \frac{3}{4}\left(\frac{1 - \nu_1^2}{E_1} + \frac{1 - \nu_2^2}{E_2}\right)\,, \tag{4.98}$$

we have $p_0 = (3K/2\pi)\sqrt{\delta/R}$, $a = \sqrt{R\delta}$ and thus $P = K\sqrt{R\delta^3}$. In other form, these can be expressed as

$$a^3 = \frac{PR}{K}, \tag{4.99}$$

$$\delta = \frac{a^2}{R} = \left(\frac{P^2}{K^2R}\right)^{1/3}. \tag{4.100}$$

The penetration δ is twice the height of the portion of sphere of chord $2a$ (Fig. 4.23). If we compare this value to the penetration of a flat punch of same radius a, we see that it is 1.5 times larger.

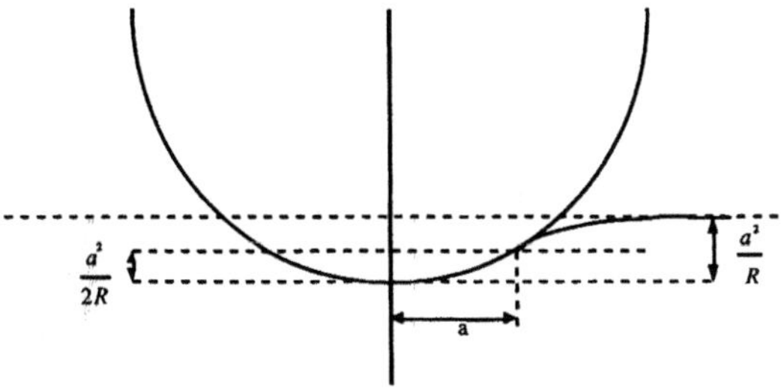

Fig. 4.23. Penetration of a sphere (close-up)

The force needed to have a penetration δ is thus:

$$P = KR^{1/2}\delta^{3/2}. \tag{4.101}$$

The area of contact πa^2 increases as $P^{2/3}$; the mean pressure p_m increases as $P^{1/3}$.

4.4.2 Stored Elastic Energy

Using (4.99) and (4.100) the stored elastic energy can be expressed as

$$U_E = \int P\,d\delta = \int \frac{2}{3}\frac{P^{2/3}}{K^{2/3}R^{1/3}}\,dP,$$

i.e.

$$U_E = \frac{2}{5}\frac{P^{5/3}}{K^{2/3}R^{1/3}} = \frac{2}{5}KR^{1/2}\delta^{5/2} = \frac{2}{5}P\delta. \tag{4.102}$$

This energy is less than the energy which would be stored if δ varied linearly with P. The relative difference from $\frac{1}{2}P\delta$ is 20%. It is because the

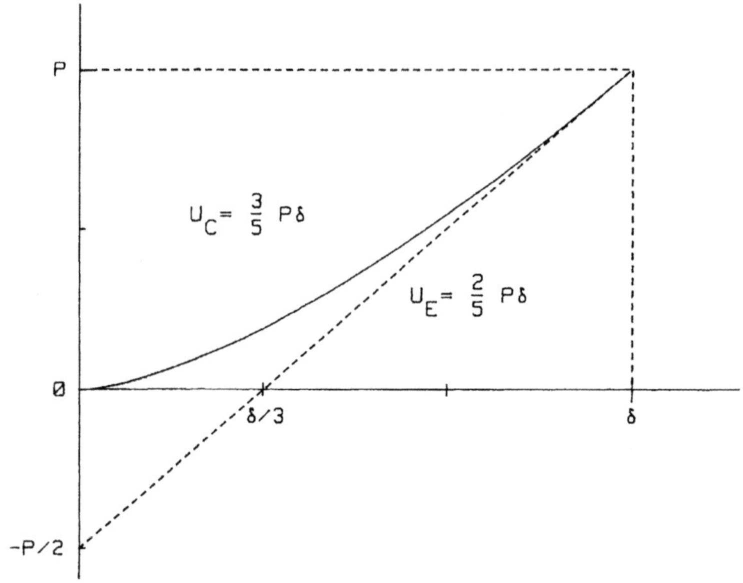

Fig. 4.24. Elastic and complementary energies in the Hertz problem

geometry changes with the load that this relation is not linear (see Fig. 4.24). We can establish that the Castigliano theorem does not apply, since

$$\frac{\mathrm{d}U_{\mathrm{E}}}{\mathrm{d}P} = \frac{2}{3}\frac{a^2}{R} \neq \delta \, ,$$

but that we have still

$$\frac{\mathrm{d}U_{\mathrm{E}}}{\mathrm{d}\delta} = P$$

Introducing the complementary energy $U_{\mathrm{C}} = P\delta - U_{\mathrm{E}} = \frac{3}{5}P\delta$, we can see that the *generalized Castigliano theorem* correctly gives the displacement:

$$\frac{\mathrm{d}U_{\mathrm{C}}}{\mathrm{d}P} = \delta \, .$$

The tangent to the curve $P(\delta)$, at a point where the contact radius is a, has a slope (the stiffness of the contact) given by

$$k = \frac{\mathrm{d}P}{\mathrm{d}\delta} = \frac{3aK}{2}$$

and cuts the axis of δ at $\delta/3$ and the axis of P at $-P/2$. Note that the compliance and the stiffness are those of a flat punch of radius a.

4.4.3 Use of the General Theorems

Let us now directly apply the general theorems of Sneddon [17] for a spherical punch of radius R, whose profile is given by

$$f(\varrho) = f\left(\frac{r}{a}\right) = R - \sqrt{R^2 - a^2\varrho^2}\,, \tag{4.103}$$

where a is the radius of contact. For the moment we do not make the hypothesis $a \ll R$. We assume the contact to be frictionless ($\tau_{rz} = 0$ in $z = 0$), and we neglect the influence of surface energy.

Let us compute

$$\chi(t) = \frac{2}{\pi}\left(\delta - t\int_0^t \frac{f'(\varrho)\,\mathrm{d}\varrho}{\sqrt{t^2 - \varrho^2}}\right).$$

The integral is written as

$$\begin{aligned}\int_0^t \frac{f'(\varrho)\,\mathrm{d}\varrho}{\sqrt{t^2 - \varrho^2}} &= a^2\int_0^t \frac{\varrho\,\mathrm{d}\varrho}{\sqrt{(t^2-\varrho^2)(R^2 - a^2\varrho^2)}} \\ &= \frac{a}{2}\int_0^1 \frac{\mathrm{d}u}{\sqrt{1-u}\sqrt{R^2/a^2t^2 - u}}\,,\end{aligned}$$

where $u = \varrho^2/t^2$. We then get

$$\chi(t) = \frac{2}{\pi}\left(\delta - \frac{at}{2}\ln\frac{R+at}{R-at}\right), \tag{4.104}$$

$$\chi'(t) = -\frac{a}{\pi}\left(\ln\frac{R+at}{R-at} + \frac{2aRt}{R^2 - a^2t^2}\right).$$

From (4.14), which states that

$$P = \frac{\pi a E}{1-\nu^2}\int_0^1 \chi(t)\,\mathrm{d}t\,,$$

we deduce that

$$P = \frac{2aE}{1-\nu^2}\left(\delta + \frac{R^2-a^2}{4a}\ln\frac{R+a}{R-a} - \frac{R}{2}\right). \tag{4.105}$$

By making $\chi(1) = 0$, i.e. neglecting the influence of surface energy, we get

$$\delta = \frac{a}{2}\ln\frac{R+a}{R-a} \tag{4.106a}$$

$$P = \frac{E}{1-\nu^2}\left[\frac{R^2+a^2}{2}\ln\frac{R+a}{R-a} - aR\right], \tag{4.107a}$$

equations which were given by Sneddon [17] and Ting [33]. For $a/R \to 0$ we have

$$\ln\frac{R+a}{R-a} = 2\tanh^{-1}\frac{a}{R} \approx 2\left(\frac{a}{R} + \frac{a^3}{3R^3}\right)$$

and hence

$$\boxed{\delta = \frac{a^2}{R}\,,} \tag{4.106b}$$

$$\boxed{P = \frac{4}{3}\frac{E}{(1-\nu^2)}\frac{a^3}{R}\,.} \tag{4.107b}$$

We obtain again (4.99) and (4.100).

Figure 4.25 shows that the approximation $a \ll R$ remains valid up to $a/R = 0.4$. In fact, even (4.106a) and (4.107a) are not valid for large a/R since boundary displacements of the order of $(a/R)^2$ are comparable with the strain neglected in linear elasticity theory.

If we directly make the hypothesis $a \ll R$, calculations are considerably simplified. We then get

$$f(\varrho) = \frac{a^2}{2R}\varrho^2\,, \tag{4.108}$$

$$\chi(t) = \frac{2}{\pi}\left(\delta - \frac{a^2}{R}t^2\right)\,, \tag{4.109}$$

$$P = \frac{2aE}{1-\nu^2}\left(\delta - \frac{a^2}{3R}\right)\,. \tag{4.110}$$

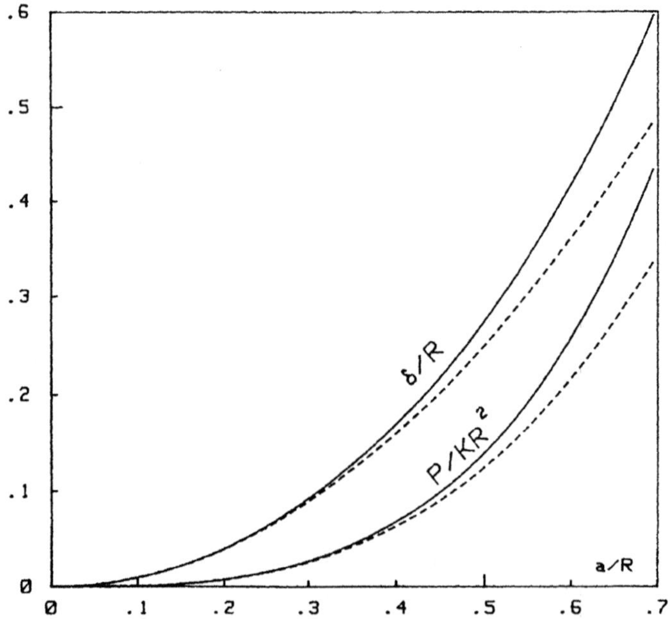

Fig. 4.25. Curves δ/R and P/KR^2 (*solid lines*) and their approximations $(a/R)^2$ and $(a/R)^3$ (*dashed lines*)

If we make $\chi(1) = 0$ as above, we immediately obtain the expressions for δ and P, as given in (4.106b) and (4.107b), and

$$\chi(t) = \frac{2a^2}{\pi R}(1 - t^2) \, . \tag{4.111}$$

We deduce that

$$\sigma_z = -\frac{E}{1-\nu^2}\frac{2a}{\pi R}\sqrt{1-\varrho^2} \, ,$$

i.e.

$$\boxed{\frac{\sigma_z(r,0)}{p_{\mathrm{m}}} = -\frac{3}{2}\sqrt{1-\varrho^2} \, , \qquad \varrho < 1} \, , \tag{4.112}$$

where $p_{\mathrm{m}} = P/\pi a^2$ is the mean pressure. We also get

$$\boxed{u_z(r,0) = \frac{a^2}{\pi R}\left[\sqrt{\varrho^2 - 1} + (2-\varrho^2)\sin^{-1}\frac{1}{\varrho}\right] \, , \qquad \varrho > 1} \, , \tag{4.113a}$$

a result initially given by Boussinesq [2]. Regarding the discontinuity of displacement, defined by (4.22), it is

$$[u_z] = \frac{a^2}{\pi R}\left[\sqrt{\varrho^2 - 1} + (\varrho^2 - 2)\cos^{-1}\frac{1}{\varrho}\right] \, . \tag{4.113b}$$

4.4.4 Calculation of Stresses and Displacements

Putting the expression for $\chi(t)$ given in (4.111), into that of $\varphi(p)$ from (4.7), we get

$$\begin{aligned} \varphi(p) &= -\frac{3a^6 K}{4\pi R}\int_0^1 (1-t^2)\cos pt \, \mathrm{d}t \\ &= p_0 a^5 \left(\frac{\cos p}{p^2} - \frac{\sin p}{p^3}\right) \, , \end{aligned} \tag{4.114}$$

where $p_0 = \frac{3}{2}\frac{P}{\pi a^2}$ is the pressure at the centre of the contact. The integrals I_n^m given in (4.9) become

$$\begin{aligned} I_n^m &= \int_0^\infty \mathrm{e}^{-p\zeta} p^n \varphi(p) J_m(p\varrho) \, \mathrm{d}p \\ &= p_0 a^5 \int_0^\infty \mathrm{e}^{-p\zeta}\left(p^{n-2}\cos p - p^{n-3}\sin p\right) J_m(p\varrho) \, \mathrm{d}p \\ &= p_0 a^5 \left(C_{n-1}^m - S_{n-2}^m\right) \, , \end{aligned}$$

with

$$C_n^m = \int_0^\infty p^{n-1} e^{-p\zeta} \cos p \, J_m(p\varrho) \, dp \,, \tag{4.115}$$

$$S_n^m = \int_0^\infty p^{n-1} e^{-p\zeta} \sin p \, J_m(p\varrho) \, dp \,. \tag{4.116}$$

These integrals are computed in Appendix A, and we have, with the parameters s and u defined in Chapt. 3, the following:

$$\begin{aligned}\frac{I_0^0}{p_0 a^5} &= C_{-1}^0 - S_{-2}^0 \\ &= \frac{1}{4}\left[\frac{3\zeta^2}{\sqrt{u}} - \sqrt{u} + (\varrho^2 - 2\zeta^2 - 2)\tan^{-1}\frac{1}{\sqrt{u}}\right] ,\end{aligned} \tag{4.117}$$

$$\begin{aligned}\frac{I_0^1}{p_0 a^5} &= C_{-1}^1 - S_{-2}^1 \\ &= -\frac{1}{3\varrho}\left[1 - \left(\frac{\zeta}{\sqrt{u}}\right)^3\right] + \frac{\varrho\zeta}{2}\left(\tan^{-1}\frac{1}{\sqrt{u}} - \frac{\sqrt{u}}{1+u}\right) ,\end{aligned} \tag{4.118}$$

$$\frac{I_1^0}{p_0 a^5} = C_0^0 - S_{-1}^0 = \zeta\left(\tan^{-1}\frac{1}{\sqrt{u}} - \frac{1}{\sqrt{u}}\right) , \tag{4.119}$$

$$\frac{I_1^1}{p_0 a^5} = C_0^1 - S_{-1}^1 = \frac{\varrho}{2}\left(\frac{\sqrt{u}}{1+u} - \tan^{-1}\frac{1}{\sqrt{u}}\right) , \tag{4.120}$$

$$\frac{I_2^0}{p_0 a^5} = C_1^0 - S_0^0 = \frac{\sqrt{u}}{s} - \tan^{-1}\frac{1}{\sqrt{u}} , \tag{4.121}$$

$$\frac{I_2^1}{p_0 a^5} = C_1^1 - S_0^1 = -\frac{\zeta}{\sqrt{u}}\frac{\varrho}{s(1+u)} . \tag{4.122}$$

The displacements and stresses are thus written as

$$\begin{aligned}u_r = &-\frac{2\delta}{\pi}\left\{\frac{1-2\nu}{3(1-\nu)}\frac{1}{\varrho}\left[1 - \left(\frac{\zeta}{\sqrt{u}}\right)^3\right]\right. \\ &\left. -\varrho\frac{\zeta}{\sqrt{u}}\left(\sqrt{u}\tan^{-1}\frac{1}{\sqrt{u}} - \frac{u}{1+u}\right)\right\} ,\end{aligned} \tag{4.123}$$

$$\begin{aligned}u_z = &\frac{\delta}{\pi(1-\nu)}\left\{(1-\nu)\sqrt{u} - (1-3\nu)\frac{\zeta^2}{\sqrt{u}}\right. \\ &\left. + \left[(2-\varrho^2) - \nu(2\zeta^2 + 2 - \varrho^2)\right]\tan^{-1}\frac{1}{\sqrt{u}}\right\} ,\end{aligned} \tag{4.124}$$

$$\frac{\sigma_r}{p_0} = \frac{1-2\nu}{3\varrho^2}\left[1-\left(\frac{\zeta}{\sqrt{u}}\right)^3\right]$$
$$+\frac{\zeta}{\sqrt{u}}\left[(1+\nu)\sqrt{u}\tan^{-1}\frac{1}{\sqrt{u}}+(1-\nu)\frac{u}{1+u}\right.$$
$$\left.+\frac{\zeta^2}{u^2+\zeta^2}-2\right], \tag{4.125}$$
$$\frac{\sigma_\theta}{p_0} = -\frac{1-2\nu}{3\varrho^2}\left[1-\left(\frac{\zeta}{\sqrt{u}}\right)^3\right]$$
$$+\frac{\zeta}{\sqrt{u}}\left[(1+\nu)\sqrt{u}\tan^{-1}\frac{1}{\sqrt{u}}-(1-\nu)\frac{u}{1+u}-2\nu\right], \tag{4.126}$$
$$\frac{\sigma_z}{p_0} = -\frac{1}{s}\left(\frac{\zeta}{\sqrt{u}}\right)^3, \tag{4.127}$$
$$\frac{\tau_{rz}}{p_0} = -\frac{\zeta}{\sqrt{u}}\frac{\varrho\zeta}{s(1+u)}, \tag{4.128}$$
$$\tau_{r\theta} = \tau_{\theta z} = 0\,.$$

Note that the hydrostatic stress is

$$\frac{\sigma_0}{p_0} = -\frac{2(1+\nu)}{3}\frac{\zeta}{\sqrt{u}}\left[1-\sqrt{u}\tan^{-1}\frac{1}{\sqrt{u}}\right].$$

Equations (4.125–128) form the Huber [27] tensor, corrected by Love [28].

4.4.5 Study of the Stresses

Stresses in the Surface. We must make $\zeta = 0$ in the stress tensor, with

$$u = 0 \quad \text{and} \quad \zeta/\sqrt{u} = \sqrt{1-\varrho^2} \quad \text{for } \varrho < 1\,,$$
$$u = \varrho^2 - 1 \quad \text{and} \quad \zeta/\sqrt{u} = 0 \quad \text{for } \varrho > 1\,.$$

In the Hertz circle we get

$$\frac{\sigma_r}{p_0} = \frac{1-2\nu}{3}\frac{1}{\varrho^2}\left[1-(1-\varrho^2)^{3/2}\right]-(1-\varrho^2)^{1/2}, \tag{4.129}$$
$$\frac{\sigma_\theta}{p_0} = -\frac{1-2\nu}{3}\frac{1}{\varrho^2}\left[1-(1-\varrho^2)^{3/2}\right]-2\nu(1-\varrho^2)^{1/2}, \tag{4.130}$$
$$\frac{\sigma_z}{p_0} = -(1-\varrho^2)^{1/2}, \tag{4.131}$$
$$\tau_{rz} = \tau_{z\theta} = \tau_{r\theta} = 0\,.$$

The stress σ_z is the "hemispherical" distribution of pressure.

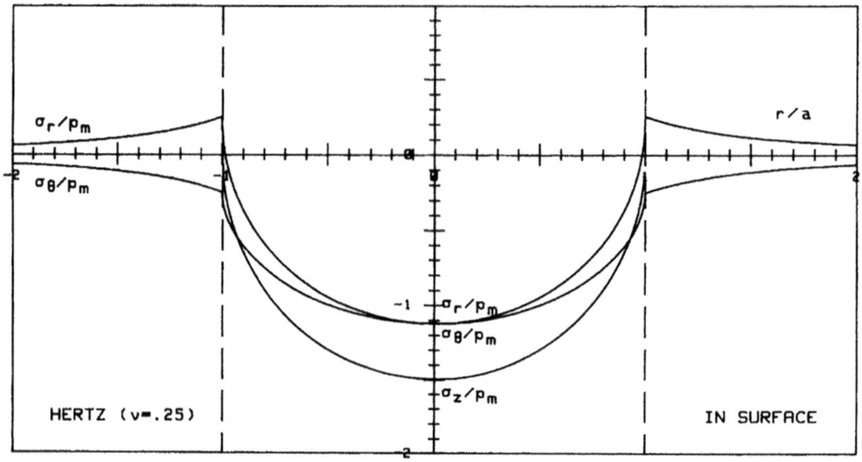

Fig. 4.26. Stresses in the surface

Outside the Hertz circle, we have

$$\sigma_r = \frac{1-2\nu}{2\pi}\frac{P}{r^2}\,,$$
$$\sigma_\theta = -\frac{1-2\nu}{2\pi}\frac{P}{r^2}\,,$$
$$\sigma_z = 0\,,$$
$$\tau_{rz} = \tau_{z\theta} = \tau_{r\theta} = 0\,.$$

We thus obtain again the stresses due to a concentrated force.

In the surface, the stresses σ_r, σ_θ and σ_z are principal stresses. They are displayed in Fig. 4.26. The hoop stress σ_θ is always a compression. The radial stress σ_r is compressive in most of the area of contact; outside the area of contact it is tensile, and reaches its maximum on the Hertz circle. For brittle materials, a crack in mode I will start on the edge of the circle and then penetrate below the surface: it is a *Hertzian fracture*.

Stresses on the z-Axis. For $\varrho = 0$, we find that

$$u = \zeta^2\,,$$
$$\frac{1}{\varrho^2}\left[1-\left(\frac{\zeta}{\sqrt{u}}\right)^3\right] = \frac{3}{2}\frac{1}{1+\zeta^2}\,.$$

Hence the stresses along the z-axis are given by

$$\boxed{\begin{aligned} \frac{\sigma_r}{p_0} &= -(1+\nu)\left(1-\zeta\tan^{-1}\frac{1}{\zeta}\right)+\frac{1}{2(1+\zeta^2)}\,, && (4.132)\\ \frac{\sigma_\theta}{p_0} &= \frac{\sigma_r}{p_0}\,, && (4.133)\end{aligned}}$$

$$\boxed{\begin{aligned} &\frac{\sigma_z}{p_0} = -\frac{1}{1+\zeta^2}\,, \\ &\tau_{rz} = \tau_{z\theta} = \tau_{r\theta} = 0\,. \end{aligned}} \tag{4.134}$$

The z-axis is a principal axis. The stresses along it are displayed in Fig. 4.27 as a function of the reduced depth z/a. The higher stress is the compressive stress σ_z, at its maximum at the surface. The two other principal stresses are weaker: starting compressive, they decrease rapidly and become weakly tensile for $z/a > 1.26$ if $\nu = 1/4$ and for $z/a > 1.48$ if $\nu = 0.3$; but they are always compression for $\nu = 1/2$.

The maximum shear stress $\tau_{\max} = (\sigma_r - \sigma_z)/2$ goes through a maximum as follows:

$$\begin{aligned} |\tau_{\max}| &= 0.48 p_{\mathrm{m}} \quad \text{for } z/a = 0.46 \quad \text{if } \nu = \frac{1}{4}\,, \\ |\tau_{\max}| &= 0.45 p_{\mathrm{m}} \quad \text{for } z/a = 0.49 \quad \text{if } \nu = \frac{1}{3}\,, \\ |\tau_{\max}| &= 0.40 p_{\mathrm{m}} \quad \text{for } z/a = 0.55 \quad \text{if } \nu = \frac{1}{2}\,. \end{aligned}$$

In an elasto-plastic material, it is at that point, termed the *Hertz point*, that plastic deformation begins, since $\sqrt{J_2}$ is also at a maximum.

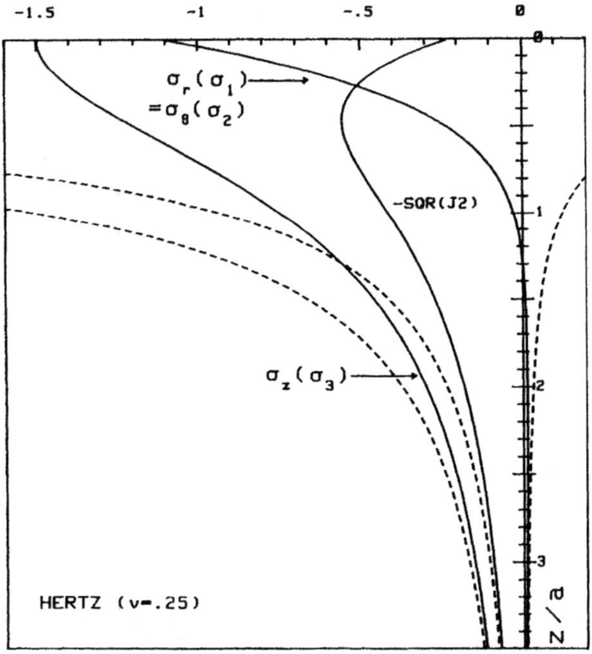

Fig. 4.27. Stresses normalized by p_{m} on the z-axis (The *dashed line* is the Boussinesq concentrated force)

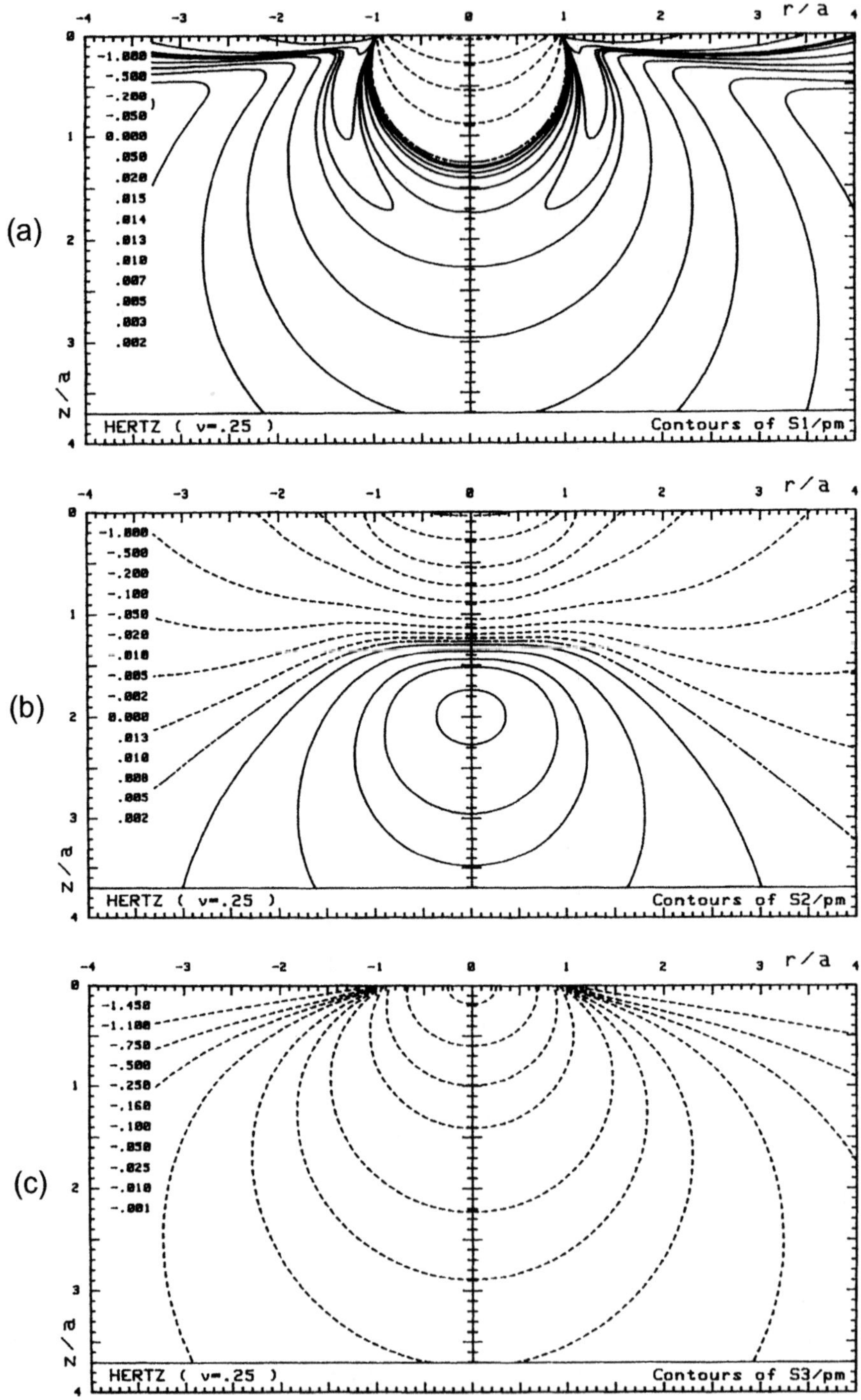

Fig. 4.28. Contours of the principal stresses: (**a**) σ_1; (**b**) σ_2; (**c**) σ_3

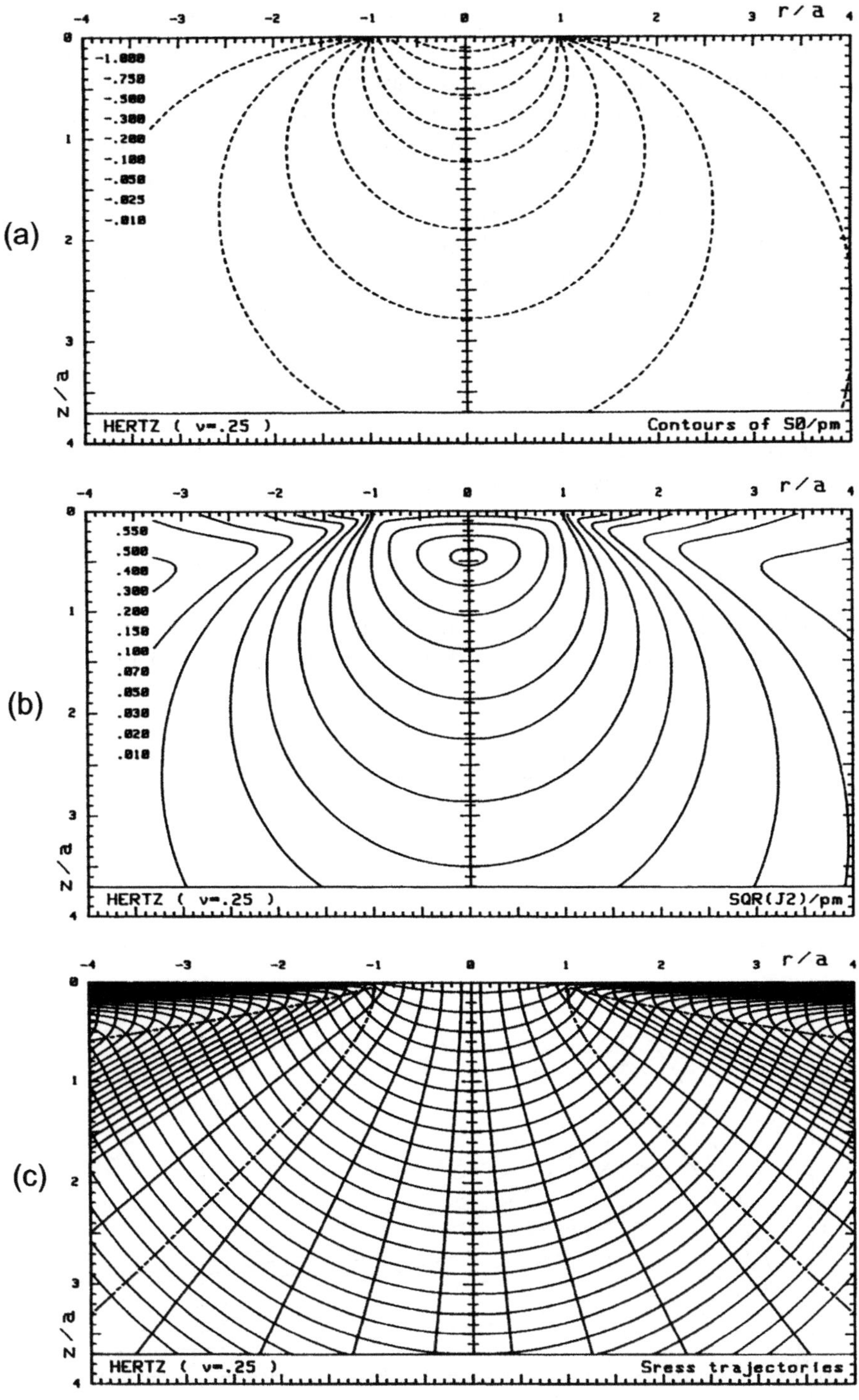

Fig. 4.29. Hydrostatic stress **(a)**, von Mises criterion **(b)**, and stress trajectories **(c)** (*dotted-dashed lines*: isoclines at 45°)

Bulk stresses. Figure 4.28 displays, for $\nu = 1/4$, the contours of principal stresses σ_1, σ_2 ($\equiv \sigma_\theta$) and σ_3 in a plane passing through the symmetry axis. On the surface and on the z-axis we have $\sigma_1 = \sigma_r$, $\sigma_3 = \sigma_z$. The stress σ_1, outside a compressive pear-shaped zone below the surface, is a traction, whose maximum value is reached in a thin sheet just below the surface. The hoop stress σ_2, becomes a traction at a depth of about $1.3a$, and never exceeds σ_1, although they are equal along the z-axis.

Figure 4.29 displays the hydrostatic stress, the von Mises yield criterion and the stress trajectories. At large distances, the results are evidently similar to the contours and stress trajectories of the Boussinesq concentrated force.

4.4.6 Study of Displacements

The vertical displacement $u_z(r, 0)$ of the surface points is given by (4.113). The asymptotic behaviour for $r \to \infty$ gives

$$u_z(r, 0) \approx \frac{1 - \nu^2}{\pi E} \frac{P}{r} ,$$

which is, indeed, that of a concentrated force.

The radial displacement of surface points, according to (4.123), is given by

$$u_r(r, 0) = -\frac{(1 + \nu)(1 - 2\nu)}{E} \frac{P}{2\pi r} \left[1 - \left(1 - \frac{r^2}{a^2} \right)^{3/2} \right] , \qquad r < a ,$$

$$u_r(r, 0) = -\frac{(1 + \nu)(1 - 2\nu)}{E} \frac{P}{2\pi r} , \qquad r > a .$$

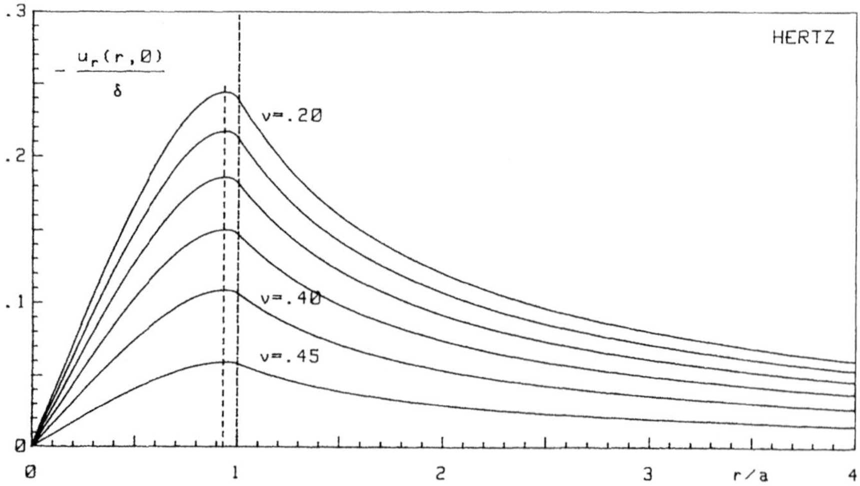

Fig. 4.30. Radial displacements at the surface

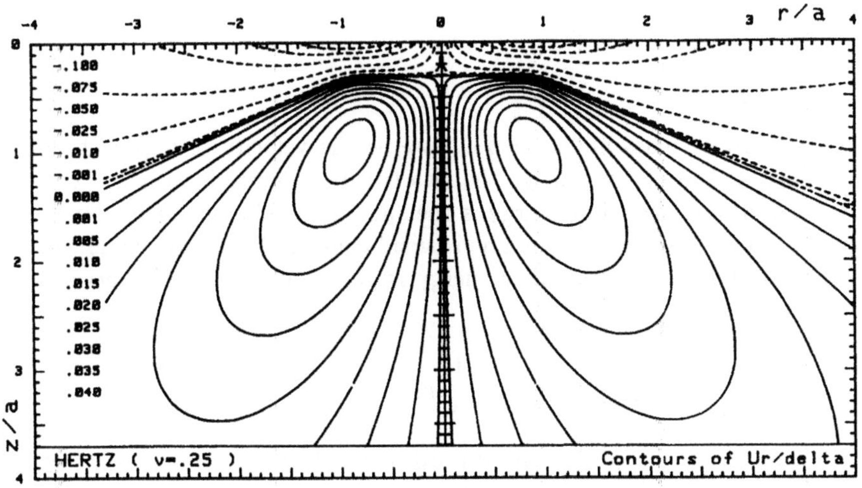

Fig. 4.31. Contours of radial displacements

Figure 4.30 displays the reduced displacements u_r/δ of surface points. They reach a maximum not on the edge of the contact but at $\varrho = (3/4)^{1/4} = 0.930$. The radial displacements in the bulk are given in Fig. 4.31. At large distances, they approach those of the concentrated force.

4.4.7 Impact Between Two Spheres

The problem of the impact between two spheres was also solved by Hertz [1], but only for velocities slow enough that interactions with elastic waves could be neglected and the deformations treated as quasi-static. Consider the normal impact between two spheres of (respectively) radius R_1 and R_2, of mass m_1 and m_2, of modulus and Poisson's ratio E_1, ν_1 and E_2, ν_2, and initial velocity v_{10} and v_{20} (positive scalar quantities). Let us use coordinates where the velocity $\boldsymbol{V}$ of the centre of inertia of the system is zero. We have thus:

$$M\boldsymbol{V} = \sum_i m_i \boldsymbol{v}_{i0} = 0 \,,$$

i.e.

$$m_1 v_{10} = m_2 v_{20} \,.$$

The kinetic energy of the system before impact is thus

$$\begin{aligned} U_{\mathrm{K}} &= \frac{1}{2}(m_1 v_{10}^2 + m_2 v_{20}^2) = \frac{1}{2}\frac{m_1 m_2 v^2 + (m_1 v_{10} - m_2 v_{20})^2}{m_1 + m_2} \\ &= \frac{1}{2} m v^2 \,, \end{aligned}$$

where

$$m = \frac{m_1 m_2}{m_1 + m_2}$$

is the reduced mass and $v = v_{10} + v_{20}$ is the velocity of approach before an impact. During an impact the two spheres approach with the velocity $\dot{\delta} = \mathrm{d}\delta/\mathrm{d}t$ and the total energy is the sum of the kinetic energy $\frac{1}{2}m\dot{\delta}^2$ (we assume that the deformed volume is small compared with the volume of the spheres) and the stored elastic energy is given by (4.102). Conservation of energy gives

$$\frac{1}{2}m\left(\frac{\mathrm{d}\delta}{\mathrm{d}t}\right)^2 + \frac{2}{5}KR^{1/2}\delta^{5/2} = \frac{1}{2}mv^2 . \tag{4.135}$$

This equation can also be established by starting from the laws of dynamics. If P is the compressive force at time t, the retardations of the two spheres are given by

$$m_1\frac{\mathrm{d}v_1}{\mathrm{d}t} = -P ,$$
$$m_2\frac{\mathrm{d}v_2}{\mathrm{d}t} = -P .$$

The velocity of approach being $\mathrm{d}\delta/\mathrm{d}t = v_1 + v_2$, we obtain

$$\frac{\mathrm{d}^2\delta}{\mathrm{d}t^2} = -\frac{P}{m} ,$$

where m is the reduced mass. Replacing P by its expression as a function of δ, (4.101), we get

$$\ddot{\delta} = -\frac{KR^{1/2}}{m}\delta^{3/2} . \tag{4.136}$$

Multiplying both sides by $m\dot{\delta}$ and integrating we obtain (4.135) again.

The maximum approach, the maximum radius of contact and the maximum force are all reached when the velocity $\dot{\delta}$ vanishes, and are given by

$$\delta_{\max} = \left(\frac{5}{4}\frac{mv^2}{KR^{1/2}}\right)^{2/5} , \tag{4.137}$$

$$a_{\max} = \left(\frac{5}{4}\frac{mv^2R^2}{K}\right)^{1/5} , \tag{4.138}$$

$$P_{\max} = (K^2R)^{1/5}\left(\frac{5}{4}mv^2\right)^{3/5} . \tag{4.139}$$

The duration of the impact is obtained by integration of (4.135) written in the form

$$dt = \frac{d\delta}{\sqrt{v^2 - \frac{4}{5}\left(\frac{KR^{1/2}}{m}\right)\delta^{5/2}}} . \tag{4.140}$$

Letting $x = \delta/\delta_{\max}$, the duration of the compression is given by the integral

$$t_c = \frac{\delta_{\max}}{v} \int_0^1 \frac{dx}{\sqrt{1 - x^{5/2}}} , \tag{4.141}$$

which is of the type

$$\int_0^1 \frac{dx}{\sqrt{1 - x^n}} = \frac{\sqrt{\pi}}{n} \frac{\Gamma(1/n)}{\Gamma(1/n + 1/2)} . \tag{4.142}$$

The duration of the impact is twice the duration of compression and is thus given by

$$T_c = \frac{4\sqrt{\pi}}{5} \frac{\Gamma(2/5)}{\Gamma(9/10)} \frac{\delta_{\max}}{v} = 2.943 \frac{\delta_{\max}}{v} ,$$

i.e.

$$T_c = 3.218 \left(\frac{m^2}{K^2 R v}\right)^{1/5} . \tag{4.143}$$

The duration of the impact is thus inversely proportional to $v^{1/5}$. The evolution of $\delta/\delta_{\max}$ with time can be obtained only by numerical integration of (4.140), which can also be written (letting $\tau = t/T_c$) as

$$\frac{dx}{d\tau} = \pm 2.943 \sqrt{1 - x^{5/2}} . \tag{4.144}$$

Hunter [34] and Deresiewicz [35] have shown that the sinusoidal approximation

$$\frac{\delta}{\delta_{\max}} = \sin\left(\frac{\pi t}{T_c}\right) \tag{4.145}$$

provides an excellent approximation. Figure 4.32 displays the variation of approach during impact. The variations of the force and of the contact area, given by

$$\frac{P}{P_{\max}} = \left(\frac{\delta}{\delta_{\max}}\right)^{3/2} ,$$

$$\frac{a}{a_{\max}} = \left(\frac{\delta}{\delta_{\max}}\right)^{1/2} ,$$

are also displayed. The sinusoidal approximations are shown by the dashed lines.

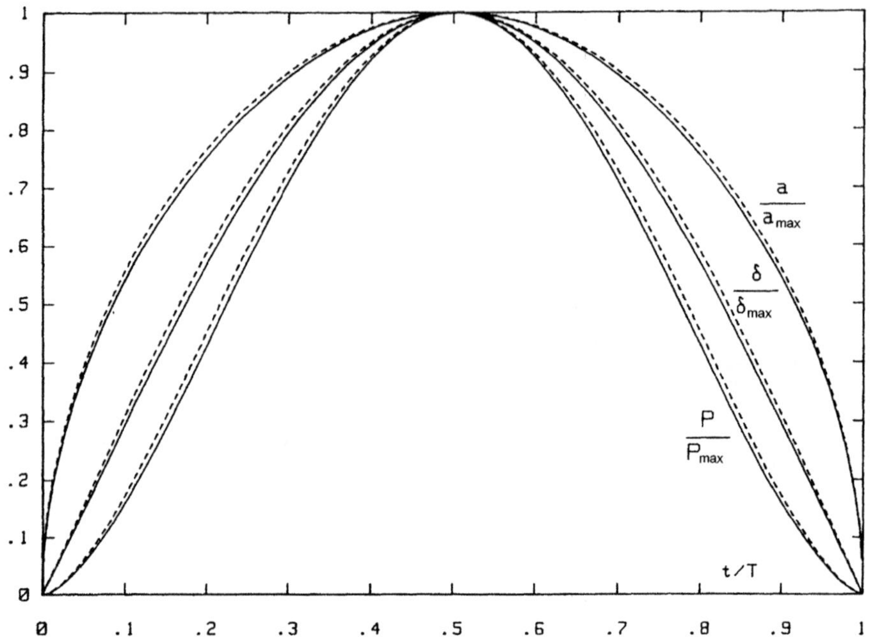

Fig. 4.32. Variation of the force, of the penetration and of the contact radius during a Hertzian impact. The dashed line is a sinusoidal approximation

For two identical spheres of radius R_1 and density ϱ, we have $R = R_1/2$ and $M = m_1/2 = 2\pi\varrho R_1^3/3$, and hence

$$\delta_{\text{max}} = \left(\frac{5\pi\sqrt{2}}{6}\frac{\varrho}{K}\right)^{2/5} R_1 v^{4/5} ,$$

$$a_{\text{max}} = \left(\frac{5\pi}{24}\frac{\varrho}{K}\right)^{1/5} R_1 v^{2/5} ,$$

$$P_{\text{max}} = \frac{1}{2^{1/5}}\left(\frac{5\pi}{6}\varrho\right)^{3/5} K^{2/5} R_1^2 v^{6/5} ,$$

$$T_{\text{c}} = 4.97\left(\frac{\varrho}{K}\right)^{2/5}\frac{R_1}{v^{1/5}} .$$

Alternatively, for a sphere of radius R and a half-space ($R = R_1$, $m = m_1$), we get

$$\delta_{\text{max}} = \left(\frac{5\pi}{3}\frac{\varrho}{K}\right)^{2/5} R v^{4/5} ,$$

$$a_{\text{max}} = \left(\frac{5\pi}{3}\frac{\varrho}{K}\right)^{1/5} R v^{2/5} ,$$

$$P_{\max} = \left(\frac{5\pi}{3}\varrho\right)^{3/5} K^{2/5} R^2 v^{6/5} ,$$

$$T_c = 5.71 \left(\frac{\varrho}{K}\right)^{2/5} \frac{R}{v^{1/5}} .$$

In both cases the maximum approach, the maximum radius of contact and the duration of the impact are all proportional to the sphere radius. The maximum contact force increases as the square of the sphere radius. We can transform these impact velocities in height of fall h by the relation $v = \sqrt{2gh}$.

This theory is in principle restricted to velocities which are slow compared with the velocity of sound. The criterion proposed by Love [36] is that the duration of the contact is long enough to permit stress waves to traverse many times. So let us consider the case of two identical metallic spheres of radius R_1, with $\nu = 1/3$. The time for a wave to go there and back is

$$t_1 = \frac{4R_1}{c_0} ,$$

where

$$c_0 = \sqrt{\frac{\lambda + 2\mu}{\varrho}} = \sqrt{\frac{3E}{2\varrho}}$$

is the velocity of sound. Let us express the duration of the contact as a function of c_0. We get, with $K = 3E/4$,

$$T_c = 6.56 \frac{R_1}{c_0^{4/5} v^{1/5}} ,$$

and we have

$$\frac{t_1}{T_c} \approx \left(\frac{v}{c_0}\right)^{1/5} .$$

This ratio must be much less than unity. If we take this ratio equal to be 0.1, we obtain $v = 10^{-5} c_0$ as a limit of validity of the Hertz theory.

In fact the behaviour ceases to be elastic before this limit, with the occurence of fractures with brittle materials or plastic deformation with metals. Take the example of a metal with $\nu = 1/3$. We have seen that the first plastic deformation appears for $\tau_{\max} = 0.5\sigma_0 \approx 0.45 p_m$, i.e. $p_m = 1.10\sigma_0$, where σ_0 is the elastic limit. In the case of the impact of a sphere on a plane, the maximum mean pressure, independent of the spherical radius, is

$$p_m = \left(\frac{P}{\pi a^2}\right)_{\max} = \frac{a_{\max} K}{\pi R} = \left(\frac{5}{3\pi^4}\varrho K^4\right)^{1/5} v^{2/5} ,$$

and plastic deformation appears for

$$v_c = 10 \left(\frac{\sigma_0^5}{\varrho K^4}\right)^{1/2} .$$

Introducing the velocity of sound, we can express this last equation as

$$v_c = 14.15 \left(\frac{\sigma_0}{E}\right)^{5/2} c_0 \, .$$

In fcc metals the ratio σ_0/E is about 10^{-3} (the proportionality between σ_0 and E arises from the fact that interactions between dislocations are elastic). Plastic deformation will thus appear for $v > 5 \times 10^{-7} c_0$, which is very small.

4.5 Contacts of Spheres: the JKR Theory

The Hertzian contact of a sphere with a plane is what is called a *unilateral problem*; only compressive stresses can exist in the contact area. If the solids have a surface energy, tensile stresses exist inside and outside the contact area, which becomes larger than predicted by the Hertz theory. Let us neglect, for the moment, the adhesion forces outside the contact area.

4.5.1 Thermodynamic Method

Let P_1 be the apparent Hertzian load which would give the same radius of contact a as that observed in the presence of surface energy. By definition, we have

$$\boxed{P_1 = \frac{a^3 K}{R}} \, . \tag{4.146}$$

To determine the penetration δ and the elastic energy U_E stored under the load P, Johnson *et al.* (JKR) [6] used the loading path shown in Fig. 4.33.

Assuming the surface energy to be zero, we apply the apparent Hertzian load P_1 (at point L). The radius of contact is thus a, the penetration $\delta_1 = a^2/R$, as per (4.99), and the stored elastic energy, given by (4.102), is

$$\frac{2}{5}\frac{P_1 a^2}{R} = \frac{2}{5}\frac{a^5 K}{R^2} \, .$$

If the load is then decreased from P_1 to P, while maintaining the radius of contact constant at a and increasing the energy of adhesion from zero up to its value w (at point B), the unloading curve is a straight line (LB) since the area remains constant. It is equivalent to the unloading of a flat punch since the displacements of points in the contact area are uniform. The variation of penetration is thus given by (4.89):

$$\Delta\delta = \frac{2(P_1 - P)}{3aK} \, ,$$

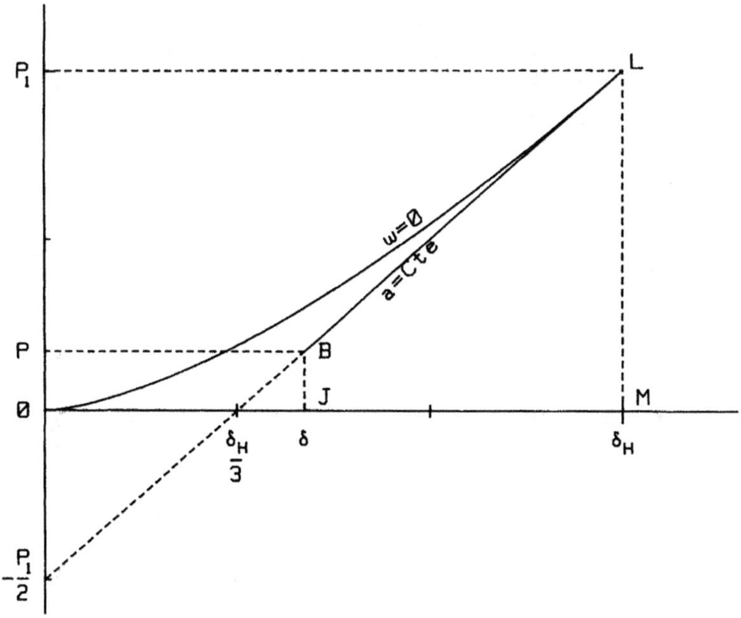

Fig. 4.33. JKR loading path. $G = 0$ from O to L, then $\mathrm{d}a = 0$ from L to B

and the variation of elastic energy is represented by the area of the trapezium LBJM:

$$\Delta U_{\mathrm{E}} = \frac{2(P_1 - P)}{3aK}\frac{P_1 + P}{2} = \frac{P_1^2}{3aK} - \frac{P^2}{3aK} \, .$$

So, at point B, under the load P and in presence of adhesion, we have:[4]

$$\boxed{\delta = \frac{a^2}{3R} + \frac{2P}{3aK}} \, , \tag{4.147}$$

and

$$\begin{aligned} U_{\mathrm{E}} &= \frac{P^2}{3aK} + \frac{a^5 K}{15R^2} \\ &= \frac{3aK}{4}\left(\delta - \frac{a^2}{3R}\right)^2 + \frac{a^5 K}{15R^2} \, . \end{aligned}$$

Let us recall the general formula, (3.4a), established in Chap. 3:

$$\mathrm{d}U_{\mathrm{E}} = P\,\mathrm{d}\delta + G\,\mathrm{d}A$$

and remark that the loading path used by JKR to compute the elastic energy stored at a point of the diagram $P(\delta)$ is precisely that discussed in Sect. 3.2.2

[4] Note that (4.147) was given for the first time by Shield [37] in 1967.

concerning the Berry representation: a curve starting from the origin and having a branch at $G = 0$ and a branch at A constant, so that the elastic energy is simply the area OLBJ below this loading path.

Differentiating U_{E} at constant displacement δ, and introducing the potential energy $U_{\mathrm{P}} = -P\delta$, the energy release rate is given by

$$G = \frac{(P_1 - P)^2}{6\pi a^3 K} \tag{4.148a}$$

$$= \frac{\left(a^3 K/R - P\right)^2}{6\pi a^3 K} \tag{4.148b}$$

$$= \frac{3K}{8\pi a}\left(\delta - \frac{a^2}{R}\right)^2 . \tag{4.148c}$$

Note that the elastic energy can also be written as a function of G, in the form

$$U_{\mathrm{E}} = \frac{2}{5}(P\delta + GA) . \tag{4.149}$$

Equations (4.147) and (4.148) are the equations of state of the system. (Note that (4.148a) is in the general form discussed in Sect. 4.1.3.) At this stage the radius of contact a under a load P is still unknown. It will be determined by the Griffith equilibrium condition $G = w$, which is the Boussinesq supplementary condition needed to solve the problem. For $w = 0$ we evidently have the Hertz results given in (4.99), (4.100) and (4.102).

The stability of the system is given by the sign of the derivatives of G:

$$\left(\frac{\partial G}{\partial A}\right)_P = \frac{P_1^2 - P^2}{4\pi^2 a^5 K} , \tag{4.150a}$$

$$\left(\frac{\partial G}{\partial A}\right)_\delta = \frac{3K}{16\pi^2 a^3}\left(\frac{a^2}{R} - \delta\right)\left(\frac{3a^2}{R} + \delta\right) , \tag{4.150b}$$

and the Maxwell relations, (3.8) and (3.9) become

$$\left(\frac{\partial G}{\partial \delta}\right)_A = \left(\frac{\partial P}{\partial A}\right)_\delta = -\frac{P_1 - P}{2\pi a^2} , \tag{4.151}$$

$$\left(\frac{\partial G}{\partial P}\right)_A = -\left(\frac{\partial \delta}{\partial A}\right)_P = -\frac{P_1 - P}{3\pi a^3 K} . \tag{4.152}$$

The distribution of normal stresses in the area of contact, shown in Fig. 4.34, is obtained by adding a Hertzian distribution, (4.112), under the

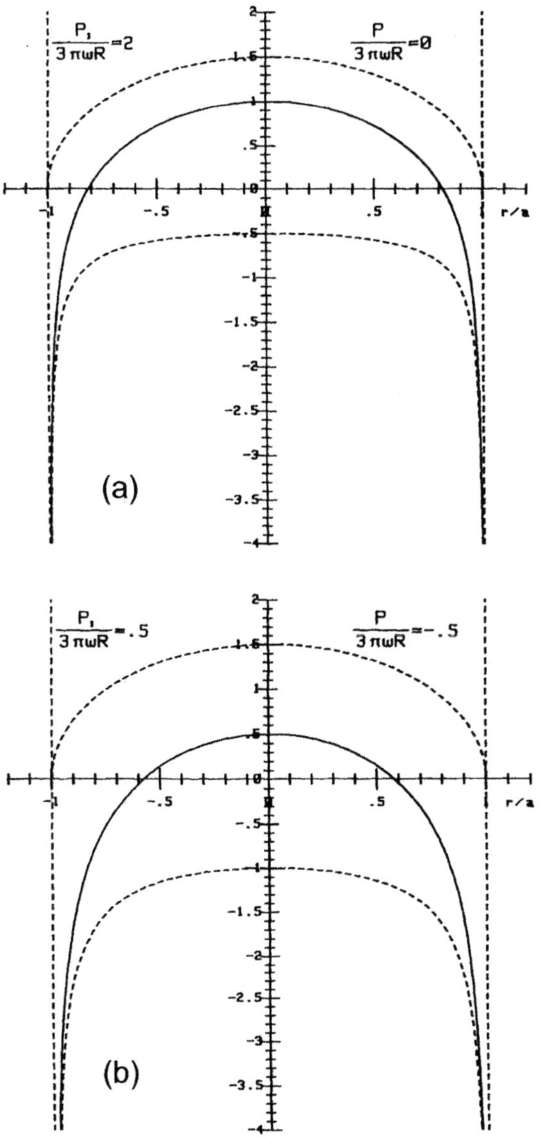

Fig. 4.34. Distribution of pressure in the area of contact for $P = 0$ and $P = -3\pi wR/2$. The pressure P is normalized to the pressure $P_1/\pi a^2$. The Hertz and Boussinesq distributions are shown by the *dashed lines*

load P_1, and a distribution for a flat punch, (4.54), under a tensile load $P_1 - P$. So we have

$$\boxed{\sigma_z(r,0) = \frac{P_1 - P}{2\pi a^2}\frac{1}{\sqrt{1-\varrho^2}} - \frac{3}{2}\frac{P_1}{\pi a^2}\sqrt{1-\varrho^2}\,, \qquad \varrho < 1} \,. \tag{4.153a}$$

The area of contact is divided into two zones: a compressed central zone of radius

$$r^* = a\sqrt{1 - \frac{P_1 - P}{3P_1}} , \tag{4.153b}$$

and an annular zone in which the stress is tensile, becoming infinite on the edge. Of course, the integral of the stresses in the area of contact remains equal to P.

The displacements of the surface points outside the area of contact are obtained by adding to the Hertzian displacements under load P_1, as per (4.113a), those of a flat punch, as per (4.55), and under a tensile force $P_1 - P$ where

$$\boxed{\begin{aligned} u_z(r,0) = &-\frac{1-\nu^2}{E}\frac{P_1 - P}{\pi a}\sin^{-1}\frac{1}{\varrho} \\ &+\frac{a^2}{\pi R}\left[\sqrt{\varrho^2 - 1} + (2 - \varrho^2)\sin^{-1}\frac{1}{\varrho}\right] , \qquad \varrho > 1 \end{aligned}} \; . \tag{4.154}$$

To obtain the profile of an adhesive sphere in contact with a rigid plane, it suffices to add the quantity

$$z - \delta = \frac{r^2}{2R} - \left(\frac{a^2}{3R} + \frac{2P}{3aK}\right)$$

to the displacement u_z given by (4.154).

4.5.2 Use of the Sneddon General Theorems

Let us return to the general equations of Sects. 4.1.2 and 4.1.3, taking for the profile of the sphere the approximation

$$f(x) = \frac{a^2}{2R}x^2 , \qquad x = \frac{r}{a} , \tag{4.155}$$

and letting

$$\frac{1}{K} = \frac{3}{4}\frac{1-\nu^2}{E} ,$$

which can be generalized in the form of (4.98). We get

$$\chi(t) = \frac{2}{\pi}\left(\delta - \frac{a^2}{R}t^2\right) ,$$

$$\chi(1) = \frac{2}{\pi}\left(\delta - \frac{a^2}{R}\right) ,$$

and (4.14) gives

$$\delta = \frac{a^2}{3R} + \frac{2P}{3aK} .$$

Let us calculate the apparent load P_1 by (4.26a). We get

$$\begin{aligned} P_1 &= \frac{3a^3K}{2R}\left[\int_0^1 \frac{x\,\mathrm{d}x}{\sqrt{1-x^2}} - \int_0^1 \frac{x^3\,\mathrm{d}x}{2\sqrt{1-x^2}}\right] \\ &= \frac{a^3K}{R} . \end{aligned}$$

Hence, using (4.27), we obtain directly the energy release rate of (4.148b).

Introducing the stress intensity factor K_{I}, defined by (4.23), (4.15) and (4.16b) we get

$$\sigma_z(r,0) = \frac{K_{\mathrm{I}}}{\sqrt{\pi a}}\frac{1}{\sqrt{1-r^2/a^2}} - \frac{3aK}{2\pi R}\sqrt{1-\frac{r^2}{a^2}} , \tag{4.156}$$

$$\begin{aligned} u_z(r,0) &= -\frac{2(1-\nu^2)}{\pi E}K_{\mathrm{I}}\sqrt{\pi a}\sin^{-1}\frac{a}{r} \\ &\quad + \frac{a^2}{\pi R}\left[\sqrt{\frac{r^2}{a^2}-1} + \left(2-\frac{r^2}{a^2}\right)\sin^{-1}\frac{a}{r}\right] , \end{aligned} \tag{4.157}$$

which are, indeed, (4.153a) and (4.154). Regarding the discontinuity of displacement defined by (4.22), it is written as

$$\begin{aligned} [u_z] &= \frac{2(1-\nu^2)}{\pi E}K_{\mathrm{I}}\sqrt{\pi a}\cos^{-1}\frac{a}{r} \\ &\quad + \frac{a^2}{\pi R}\left[\sqrt{\frac{r^2}{a^2}-1} - \left(2-\frac{r^2}{a^2}\right)\cos^{-1}\frac{a}{r}\right] , \end{aligned} \tag{4.158}$$

and is valid both for a rigid sphere on an elastic half-space as well as for an elastic sphere on a rigid plane. In the latter case, $[u_z]$ represents simply the profile of the deformed sphere.

4.5.3 Equilibrium Relations and Adherence Force

The equilibrium equation $G = w$ permits the linking of two of the three variables a, P and δ of the equation of state (4.147), and the drawing of the equilibrium curves $a(P)$, $\delta(a)$, $P(\delta)$. Figure 4.35 displays the relation between P_1 and P (i.e. between a^3 and P) in reduced coordinates $P/3\pi wR$ and $a^3K/3\pi wR^2$, which is written, using (4.148a) in the form

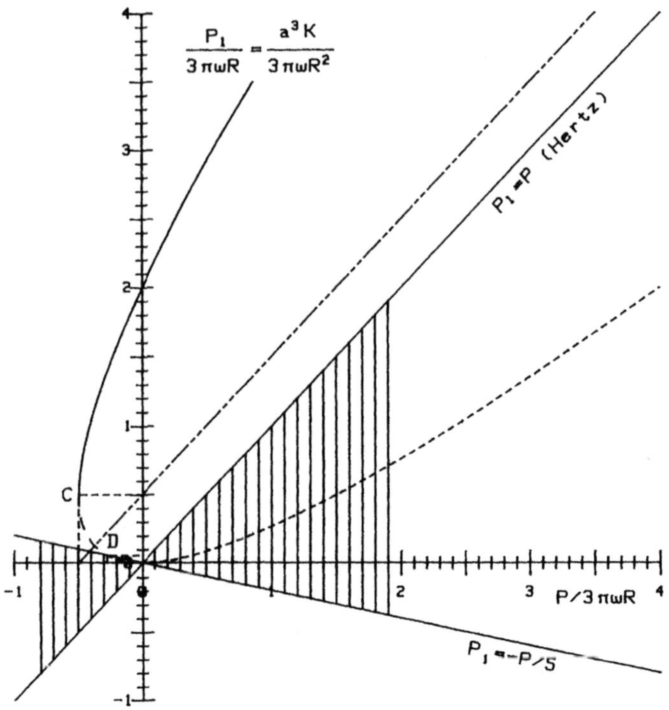

Fig. 4.35. Equilibrium relation between the apparent Hertzian load P_1 and the applied load P, in reduced coordinates. The positive determinant is shown in the solid line

$$\left(\frac{P_1}{3\pi wR} - \frac{P}{3\pi wR}\right)^2 = 2\frac{P_1}{3\pi wR}, \tag{4.159}$$

and whose branches correspond to the two determinants

$$P_1 = P + 3\pi wR \pm \sqrt{6\pi wRP + (3\pi wR)^2}. \tag{4.160}$$

This last equation represents a parabola inclined by 45° on the axes, whose slope at any point is

$$\frac{\mathrm{d}P_1}{\mathrm{d}P} = \frac{P_1 - P}{P_1 - P - 3\pi wR} = \frac{2P_1}{P_1 + P}.$$

Figure 4.36 displays the equilibrium relation between a and P deduced from (4.148b). We have superimposed their variations at fixed displacement δ, given by (4.147). These last curves have their slope defined by (4.151) and have thus a vertical tangent on the Hertz curve. The cycle LMN represents the evolution of the system when a sudden variation of displacement is imposed, from $z = 8$ to $z = 5$, whereas the part LN of the JKR curve represents a quasistatic transformation.

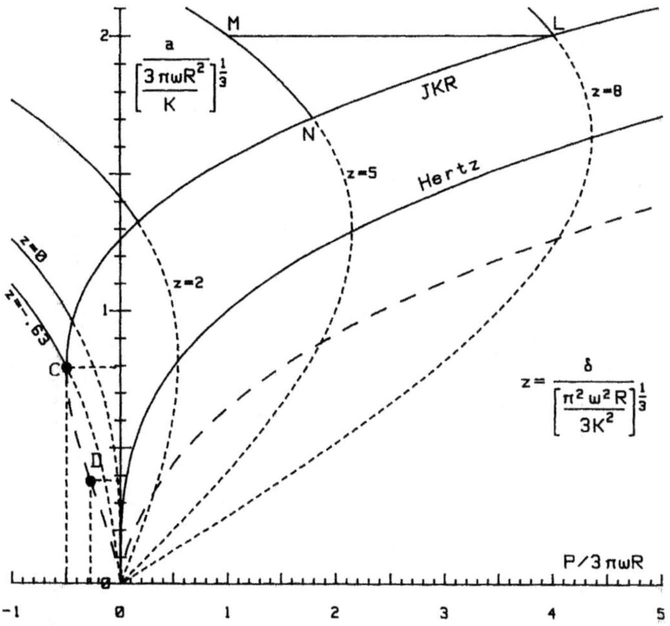

Fig. 4.36. Equilibrium relation between the radius of contact a and the applied load P, in reduced coordinates. Superimposed are the curves at constant δ, independent of w

Figure 4.37 displays the equilibrium relation between δ and a, deduced from (4.148c). The equation of the branch $\delta < \delta_H$ is

$$\boxed{\delta = \frac{a^2}{R} - \sqrt{\frac{8\pi a w}{3K}}} \,. \tag{4.161}$$

Its slope is given by

$$\frac{d\delta}{da} = \frac{5P_1 + P}{3a^2 K} \,. \tag{4.162}$$

On this figure are superimposed the variations of δ and a at fixed load P deduced from (4.147). These curves have their slope defined by (4.152), and thus have their horizontal tangent on the Hertz curve. Note that on Figs. 4.36 and 4.37 the curves $(P)_\delta$ and $(\delta)_P$ are independent of w.

Finally, the equilibrium relation between P and δ is given in Fig. 4.38. The slope of this curve is given by

$$\frac{dP}{d\delta} = \frac{9aK}{2} \frac{P_1 + P}{5P_1 + P} \,,$$

and the curves at constant a, as seen above, are straight lines such as BL, tangent to the Hertz curve and independent of w. Their slope, given by

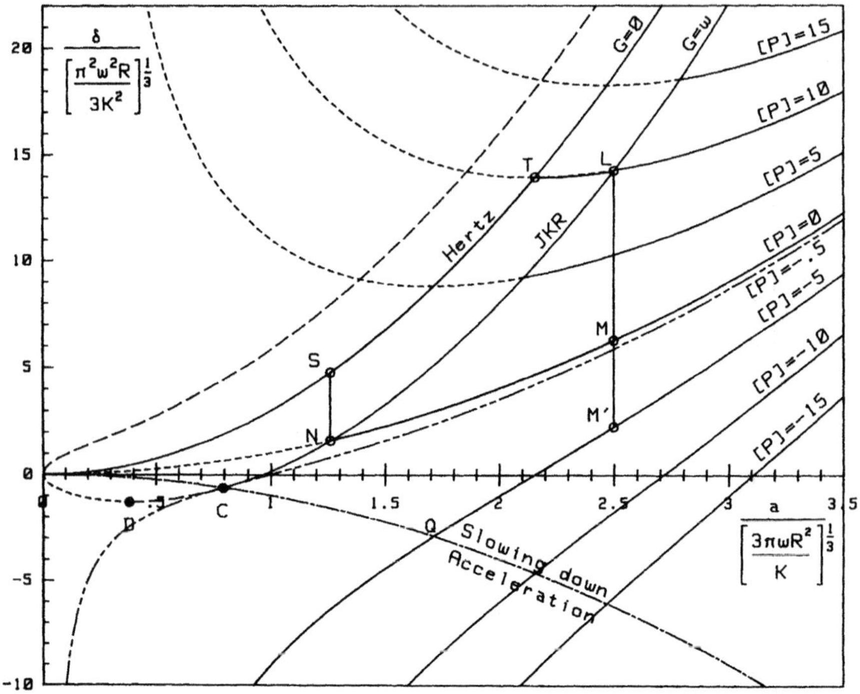

Fig. 4.37. Equilibrium relation between penetration δ and radius of contact a, in reduced coordinates. Superimposed are the curves at constant P, independent of w

$$\left(\frac{\partial P}{\partial \delta}\right)_a = \frac{3aK}{2} = k \,,$$

is the stiffness of the contact.

Stability of Equilibrium at Fixed Load. The stability of equilibrium at fixed load is defined by $(\partial G/\partial A)_P > 0$, as per (4.150a), which is also written, by extracting P^2 from (4.148a) with $G = w$, as

$$\left(\frac{\partial G}{\partial A}\right)_P = \frac{P_1 - P - 3\pi w R}{2\pi^2 a^2 R} > 0 \,.$$

This shows that the positive determinant must be taken in (4.160), which is thus written as

$$\boxed{a^3 = \frac{PR}{K}\left[1 + \frac{3\pi w R}{P} + \sqrt{2\frac{3\pi w R}{P} + \left(\frac{3\pi w R}{P}\right)^2}\right]} \,, \tag{4.163}$$

a result given by JKR [6] in 1971. In particular, under zero load, the radius of contact and the penetration are, respectively,

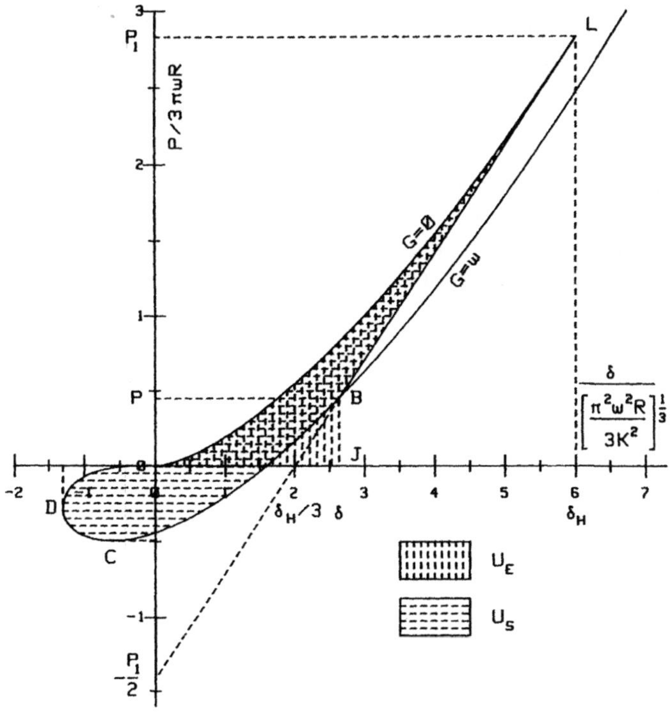

Fig. 4.38. Equilibrium relation between load P and penetration δ, in reduced coordinates. Berry's representation of elastic and surface energies at point B

$$a_0 = \left(\frac{6\pi w R^2}{K}\right)^{1/3} , \tag{4.164}$$

$$\delta_0 = \frac{a_0^2}{3R} = \left(\frac{4\pi^2 w^2 R}{3K^2}\right)^{1/3} . \tag{4.165}$$

At the instability point (point C on the figures) we have

$$P_c = -P_1 = -\frac{3}{2}\pi w R , \tag{4.166}$$

$$a_{\min} = \left(\frac{3}{2}\frac{\pi w R^2}{K}\right)^{1/3} = 0.63 a_0 , \tag{4.167}$$

$$\delta_{\min} = -\frac{a^2}{3R} = -\left(\frac{\pi^2 w^2 R}{12K^2}\right)^{1/3} . \tag{4.168}$$

P_c is the *adherence force at fixed load.* At this point da/dP is infinite, $dP/d\delta = 0$ and, using (4.162) or (4.152),

$$\frac{d\delta}{dA} = \left(\frac{\partial \delta}{\partial A}\right)_P = \frac{2}{3\pi R} .$$

The branches of stable equilibrium at fixed load are given in full line in Figs. 4.35–4.37. Note that at this unstable point we still have a large compressed zone in the area of contact, since in accordance with (4.153b) the stresses are compressive on $r < a/\sqrt{3} = 0.58a$ (see Fig. 4.34b).

Stability of Equilibrium at Fixed Grips. Using (4.147), we can see in (4.150b) that $(\partial G/\partial A)_\delta$ vanishes for $P + 5P_1 = 0$. The hatched zone of Fig. 4.35 corresponds to unstable equilibrium. We can see that at fixed grips, P_1 can decrease beyond the summit of the parabola (point D). At this point we have

$$\delta_{\min} = -\frac{3a^2}{R} = -\left(\frac{3}{4}\frac{\pi^2 w^2 R}{K^2}\right)^{1/3}, \tag{4.169}$$

$$a_{\min} = \left(\frac{\pi w R^2}{6K}\right)^{1/3} = 0.30 a_0 , \tag{4.170}$$

$$P_c = -5P_1 = -\frac{5}{6}\pi w R . \tag{4.171}$$

P_c is the *adherence force at fixed grips*. At this point we have $d\delta/da = 0$ and $d\delta/dP = 0$. These results are in agreement with predictions of Sect. 3.2.1, and are also clear from Fig. 4.38. At fixed load (apparatus of zero stiffness), we can move on the equilibrium curve from top to bottom, but we cannot pass beyond the point C where the tangent has a slope of zero. At fixed grips (apparatus of infinite stiffness), we can move on the equilibrium curve from right to left, but we cannot pass beyond the point D where the tangent has a infinite slope. According to (3.19), for an apparatus of stiffness k_m, the instability will appear at the point where the tangent has a slope $-k_m$. Between points C and D, (4.163) is no longer valid, since the sign must be changed in front of the square root. On the other hand, the determinant

$$P = \frac{a^3 K}{R} - \sqrt{6\pi a^3 K w} \tag{4.172}$$

remains valid all along the branch of stable equilibrium at fixed grips. Note that at the point D the area of contact is entirely under traction (Fig. 4.39), any compression having disppeared, as given by (4.153b), for $P < -2P_1$, i.e. $a^3 < 2\pi w R^2/3K$.

Behaviour out of Equilibrium. Let us return to Fig. 4.37 and consider the equilibrium point L under load P with the radius of contact a and the elastic displacement δ. If a load $P' < P$ is suddenly applied, instantaneously G takes a new value, higher than w, and the penetration δ decreases at constant radius of contact. Then the radius of contact, the penetration and G change at load

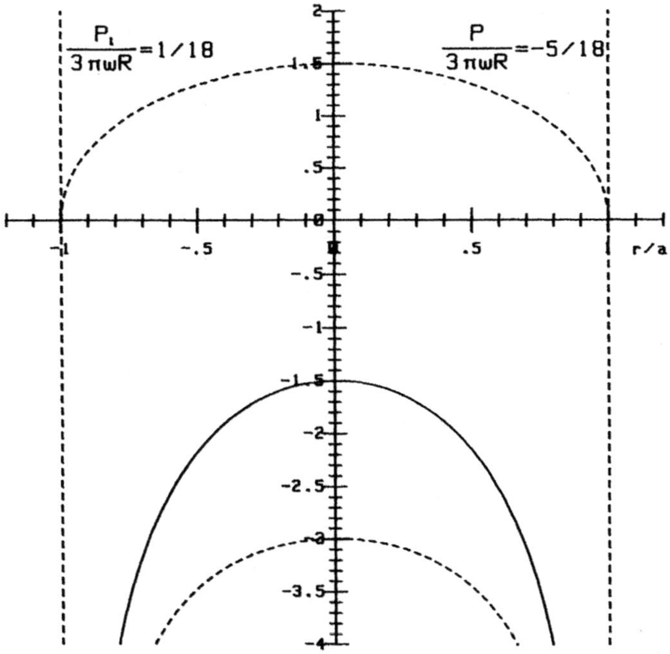

Fig. 4.39. Pressure distribution in the area of contact, at the limit of stability at fixed δ

P' constantly along the curve $(\delta)_{P'}$. The instantaneous displacement (branch LM or LM') is that of a flat punch under the load $P - P'$:

$$\Delta\delta = \frac{2(P - P')}{3aK} .$$

The evolution along the curve $(\delta)_{P'}$ (for which the kinetics can be studied, and for viscoelastic solids we can show that the crack velocity is a function of the force $G - w$ per unit length) can lead either to a new equilibrium (branch MN) if the new load P' is higher than $P_c = -\frac{3}{2}\pi wR$, or to a rupture (branch M'Q) if $P' < P_c$. In the first case the crack continuously slows down from M to N; in the second case it slows down from M' to Q and then accelerates. An inflexion point is observed in the curve $a(t)$ when $(\partial G/\partial a)_{P'} = 0$, i.e. when the radius of contact a reaches the value defined by $a^3K/R = -P'$. The variation of G as a function of the radius of contact is displayed in Fig. 4.40 for various unloadings P'. We can see that what was called *controlled rupture* in Sect. 3.2.1, here along the equilibrium curve LNC, corresponds to a sequence of small triangles such as LMN, with at each time a wait for the equilibrium to be reached. It is a *quasistatic transformation.*

If now, instead of a sudden unloading of the sphere, we apply a sudden loading starting from an equilibrium point such as N, the behaviour does not exactly mirror the previous results. In effect, instantaneously G becomes lower than w, and the penetration increases at constant radius of contact. But

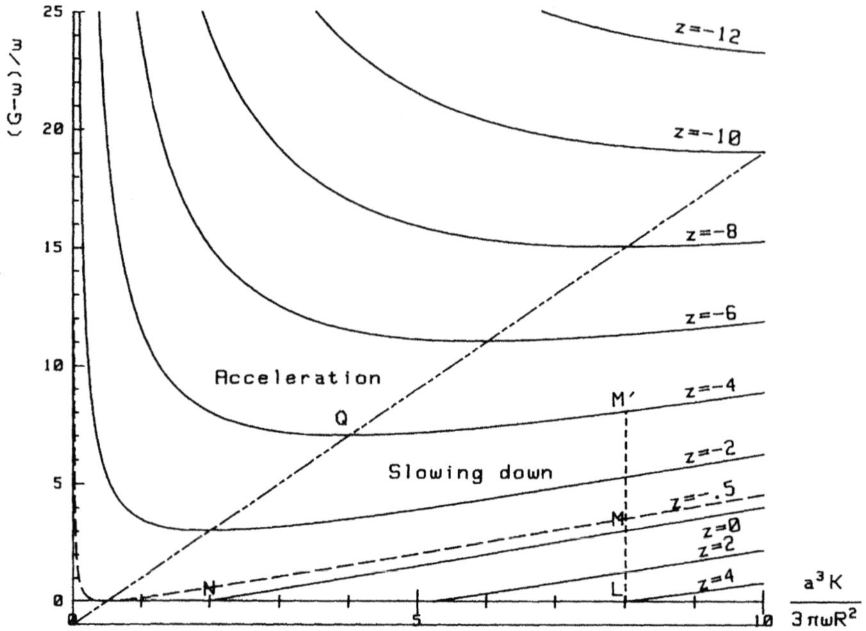

Fig. 4.40. Evolution of the energy release rate G with the cube of the radius of contact for various unloadings P', in reduced coordinates ($z = P'/3\pi wR$). The straight line $P_1 = -P'$, the locus of minimums of $G - w$, separates the slowing down and acceleration zones of the crack

G cannot be negative (or more exactly, K_I cannot be negative, for otherwise interpenetration would occur), and it is necessary to follow the curve $G = 0$, i.e. the Hertz curve ST, with a tangential connection of the sphere to the half-space, up to the point T where the radius of contact and the penetration will increase at constant load with a vertical connection. This behaviour is observed experimentally.

An Apparent Contradiction. We have seen that the adherence of a sphere to a plane is proportional to the energy of adhesion w. However, we have found previously that the adherence of a flat punch on a plane is proportional to $\sqrt{w}$, and we will see below that the adherence of a cone on a plane is proportional to w^2. Adherence thus depends on adhesion, but also on the geometry of the system. This is why these two quantities must be distinguished by distinct words.

The apparent Hertzian load P_1 is simply the load which, in the theory of the unilateral Hertz contact, would give the same radius of contact a (but not the same penetration δ). To have the same penetration as in the JKR theory, namely

$$\delta = \frac{a^2}{3R}\left(1 + 2\frac{P}{P_1}\right) = \frac{P_1^{2/3}}{3R^{1/3}K^{2/3}}\left(1 + 2\frac{P}{P_1}\right) ,$$

an apparent Hertzian load

$$P_2 = P_1 \left(\frac{P_1 + 2P}{3P_1} \right)^{3/2} ,$$

with $P < P_2 < P_1$, should be applied – but then the radius of contact would be less than a.

4.5.4 Energies of the System and Berry's Representation

Recall from Chap. 3 that elastic, potential and surface energies of the system are given by

$$U_{\mathrm{E}} = \frac{a^5 K}{3R^2} \left(\frac{1}{5} + \frac{P^2}{P_1^2} \right) ,$$

$$-P\delta = -\frac{Pa^2}{3R} \left(1 + 2\frac{P}{P_1} \right) ,$$

$$U_{\mathrm{S}} = -\pi a^2 w ,$$

and that the thermodynamic potentials H (at fixed load) and U (at fixed grips) are given by

$$H = U_{\mathrm{E}} + U_{\mathrm{S}} - P\delta ,$$

$$U = U_{\mathrm{E}} + U_{\mathrm{S}} = \int_{\mathrm{equ.}} P \, \mathrm{d}\delta .$$

At equilibrium, P is related to P_1 by (4.172) such that

$$P = P_1 - \sqrt{6\pi w R P_1} .$$

Figure 4.41 displays these various energies as a function of the radius of contact a, at equilibrium. At rupture at fixed load, the enthalpy H displays a horizontal tangent, whereas for the rupture at fixed grips it is the internal energy U.

In Fig. 4.42, to this variation of enthalpy along the equilibrium curve are superimposed the variations of enthalpy with the area of contact A, at constant load P. The slope of these last curves is then $(G-w)/w$. Instability appears at the point where the minima disappear (the inflexion point at C). The circuit LMN represents the return to equilibrium after a sudden unloading.

As seen above in Fig. 4.38, the elastic energy U_{E} is the area OLBJ below the loading path. The surface energy of the system U_{S} is thus, except for the sign, the area ODCBLO contained between the equilibrium curve and the loading path (which can be verified by computing the curvilinear integral, as occurs in [38]). The internal energy U of the system is shown by the algebraic area ODCBJ below the equilibrium curve, and its enthalpy H is represented by the area ODCBP (negative).

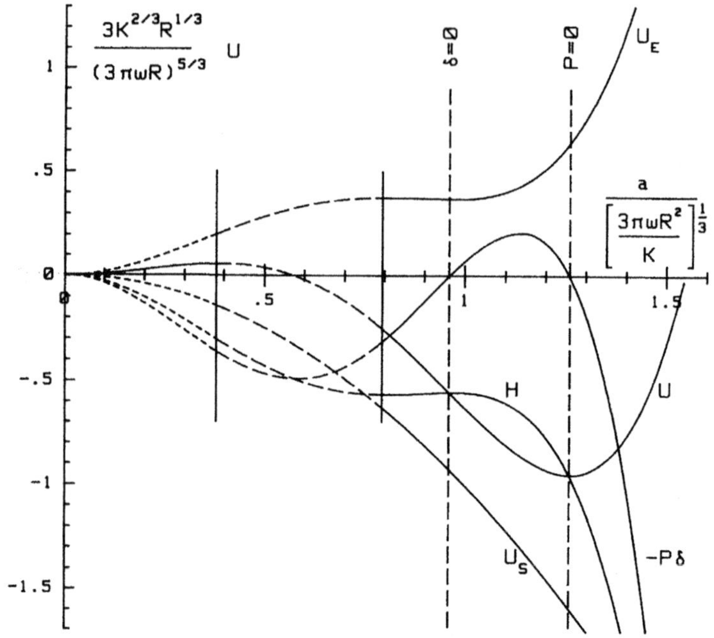

Fig. 4.41. Variation of elastic, potential and surface energies, and of thermodynamic potentials H and U with the radius of contact a, along the equilibrium curve

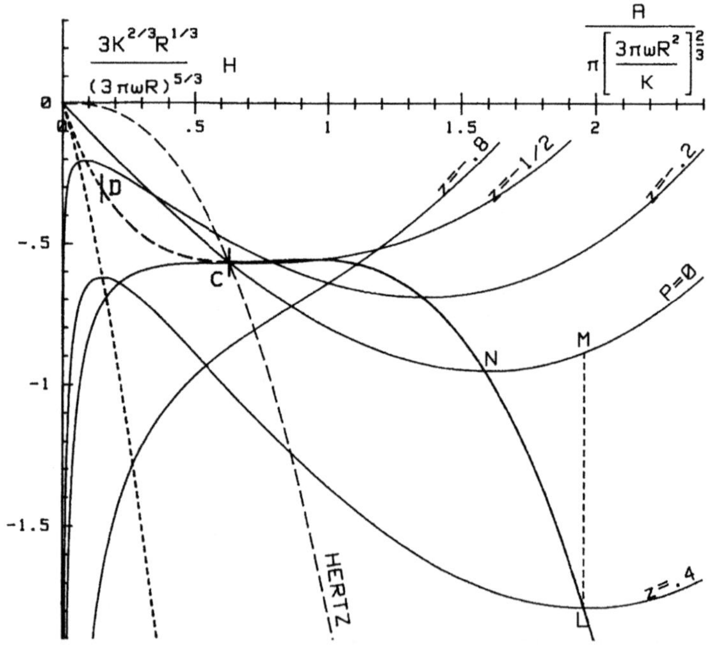

Fig. 4.42. Variations of enthalpy H of the system with the area of contact A, along the equilibrium curve and at constant load P ($z = P/3\pi wR$)

4.5.5 Profile of a Surface

A Hertzian contact (with a tangential connection) and a JKR contact under the same load P are compared schematically in Fig. 4.43. In the second case, the connection is perpendicular to the surface, and the area of contact and the penetration are larger.

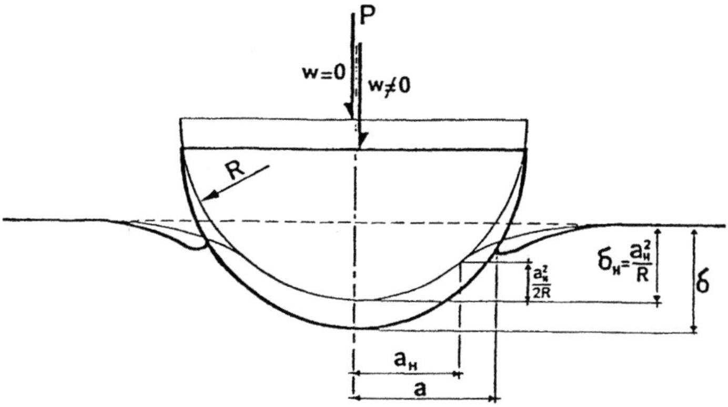

Fig. 4.43. Comparison between a Hertzian contact ($w = 0$) and a JKR contact ($w \neq 0$) under the same load P

The profile of the surface, given by (4.154), is displayed in Fig. 4.44, where the ratio $u_z/(a^2/R)$, independent of ν, is plotted as a function of r/a for various values of $P/3\pi wR$ such that

$$\frac{u_z}{a^2/R} = -\frac{4}{3\pi}\left(1 - \frac{P}{P_1}\right)\sin^{-1}\frac{1}{\varrho} + \frac{1}{\pi}\left[\sqrt{\varrho^2 - 1} + (2 - \varrho^2)\sin^{-1}\frac{1}{\varrho}\right] ,$$

the ratio P/P_1 being given by (4.160). The smaller the normalized force $P/3\pi wR$, the greater the departure from the Hertzian profile. The dashed lines in Fig. 4.44 represent the solution with the minus sign in front of the square root (points between C and D in Figs. 4.35–4.37).

The power expansion of $[u_z]$ in the vicinity of the edge of the contact, in $r = a(1 + \varepsilon)$ is:

$$[u_z] = -\chi(1)\sqrt{2\varepsilon} + \left(5\chi(1) + 32\frac{a^2}{\pi R}\right)\frac{(2\varepsilon)^{3/2}}{24} . \tag{4.173}$$

We can establish that the contact with the sphere is perpendicular if $\chi(1) \neq 0$ and tangential if $\chi(1) = 0$.

The asymptotic development of u_z for $r \to \infty$ is

$$u_z(r, 0) = \frac{1 - \nu^2}{\pi E}\left[\frac{P}{r} + \left(\frac{P}{2} - \frac{a^3 K}{5R}\right)\frac{a^2}{3r^3}\right] , \tag{4.174}$$

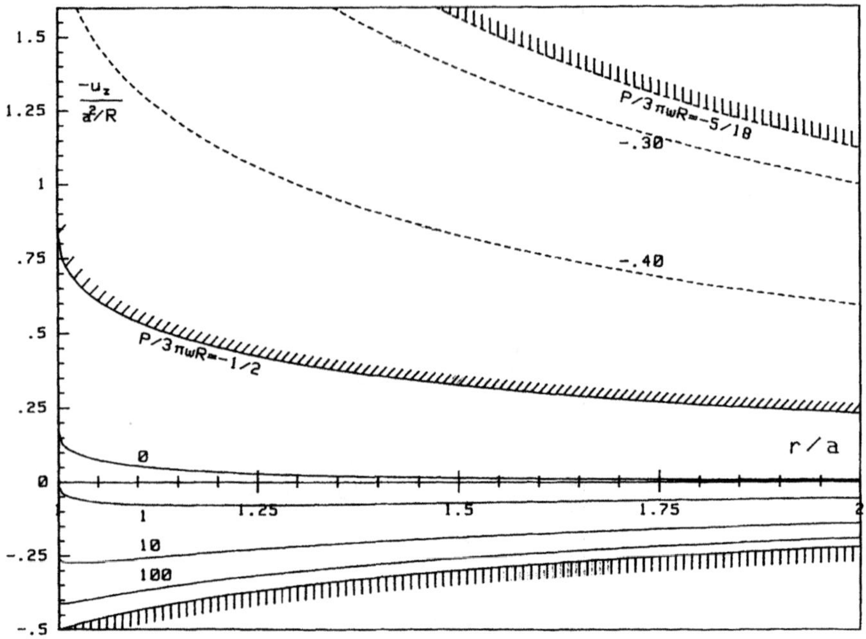

Fig. 4.44. Profile of a deformed surface

which can be written as

$$u_z(r,0) = \frac{1-\nu^2}{\pi E}\left[\frac{P}{r} - \frac{\int_0^a 2\pi r^3 \sigma_z \, dr}{4r^3}\right] ,$$

a general formula given by Boussinesq [2] in 1885. Under zero load, u_z decreases as a^5/r^3 at large distances.

4.5.6 The Stress Tensor

The stress tensor was studied by Barquins and Maugis [16]. It is obtained by superposition of Hertzian stresses $(\sigma)_{\rm H}$, for a compressive load P_1 and a mean pressure $(p_{\rm m})_{\rm H} = P_1/\pi a^2$, onto the stresses of a flat punch under a tensile load $P_1 - P$ and a mean pressure $(p_{\rm m})_P = (P, -P)/\pi a^2$. If we reduce these stresses to the mean pressure $p_{\rm m} = P/\pi a^2$, we get

$$\frac{\sigma}{p_{\rm m}} = \frac{P_1}{P}\left(\frac{\sigma}{p_{\rm m}}\right)_{\rm H} - \left(\frac{P_1}{P} - 1\right)\left(\frac{\sigma}{p_{\rm m}}\right)_P . \tag{4.175}$$

The case $P = 0$ must be treated separately. The stresses are reduced to the mean pressure $p'_{\rm m} = P_1/\pi a^2$, and we then have

$$\frac{\sigma}{p'_{\rm m}} = \left(\frac{\sigma}{p_{\rm m}}\right)_{\rm H} - \left(\frac{\sigma}{p_{\rm m}}\right)_P . \tag{4.176}$$

(a)

(b)

(c)

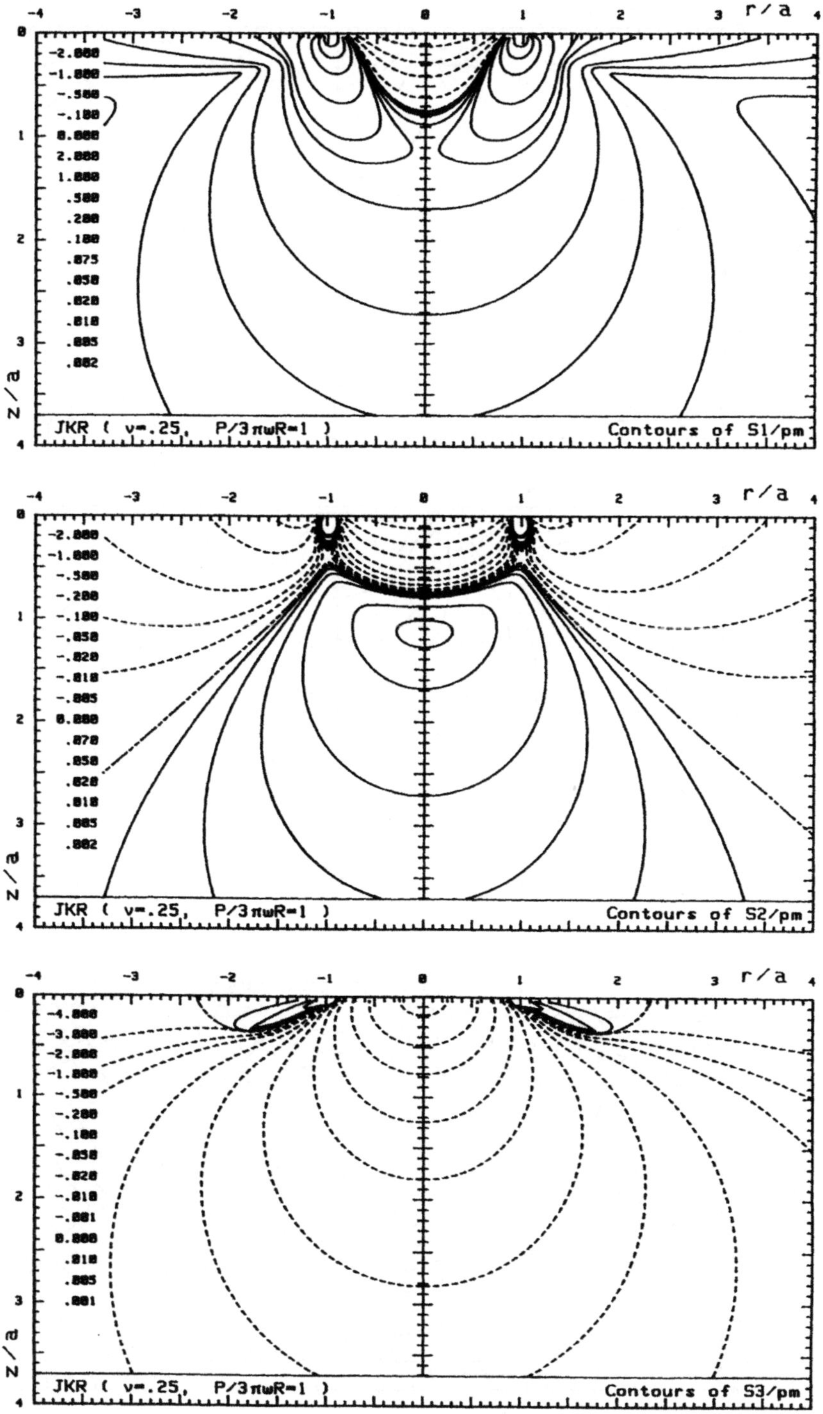

Fig. 4.45. Contours of principal stresses: **(a)** σ_1; **(b)** σ_2; **(c)** σ_3

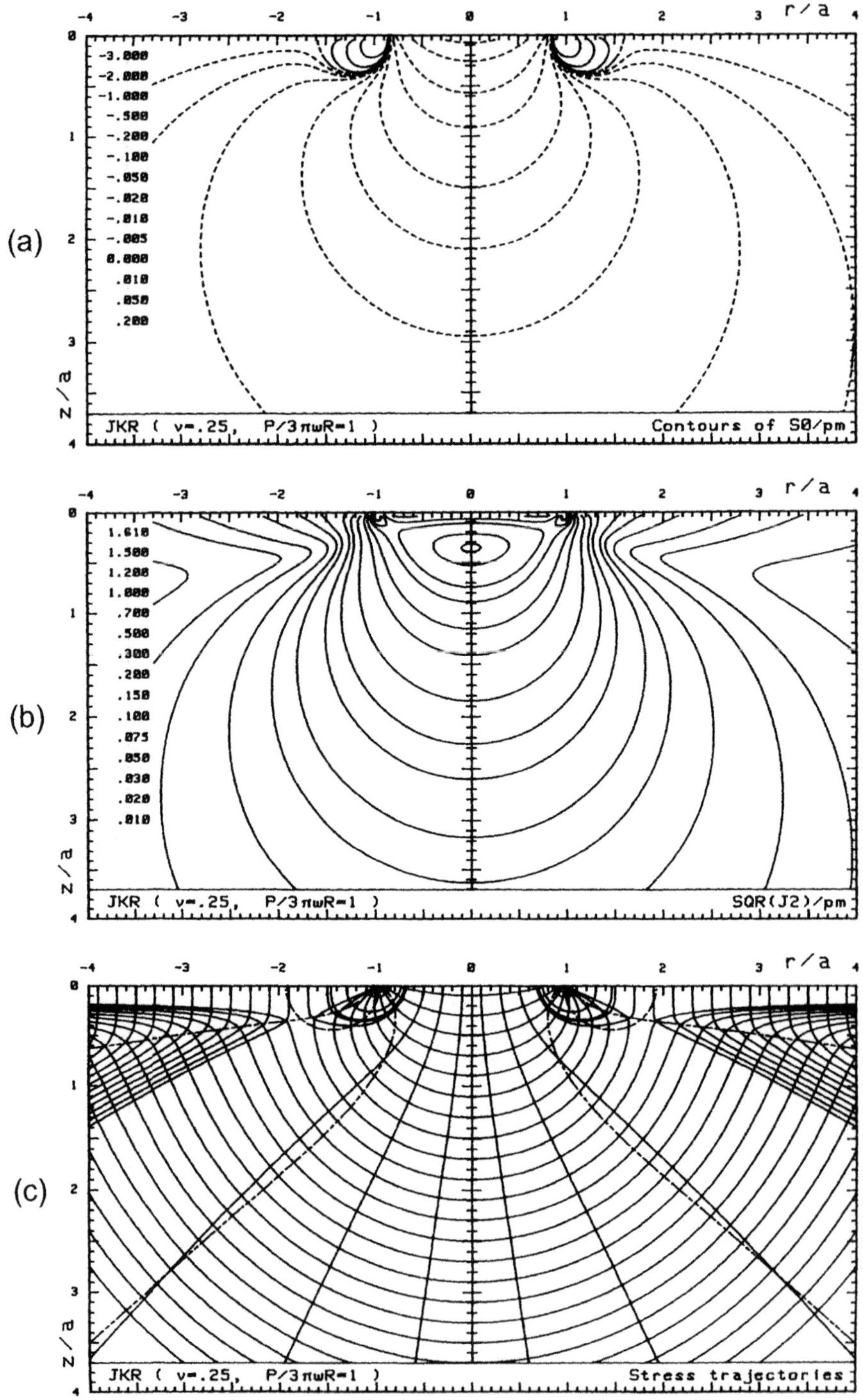

Fig. 4.46. Hydrostatic stress **(a)**, von Mises criterion **(b)**, and stress trajectories **(c)** (The *dashed-dotted lines* are isoclines at 0° and 45°)

(a)

(b)

(c)

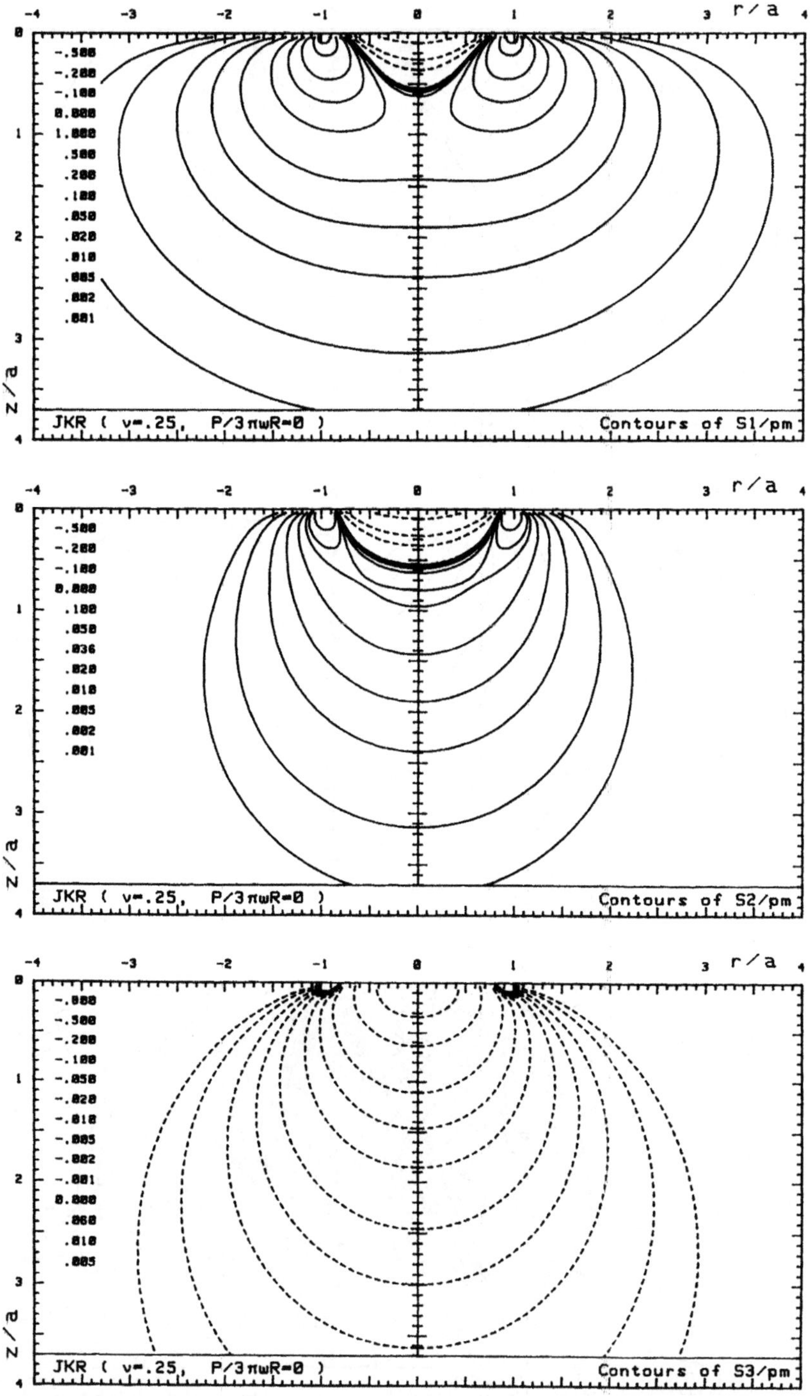

Fig. 4.47. Contours of principal stresses for $P = 0$ (reduced to the mean pressure $p'_{\mathrm{m}} = P_1/\pi a^2$): **(a)** σ_1; **(b)** σ_2; **(c)** σ_3

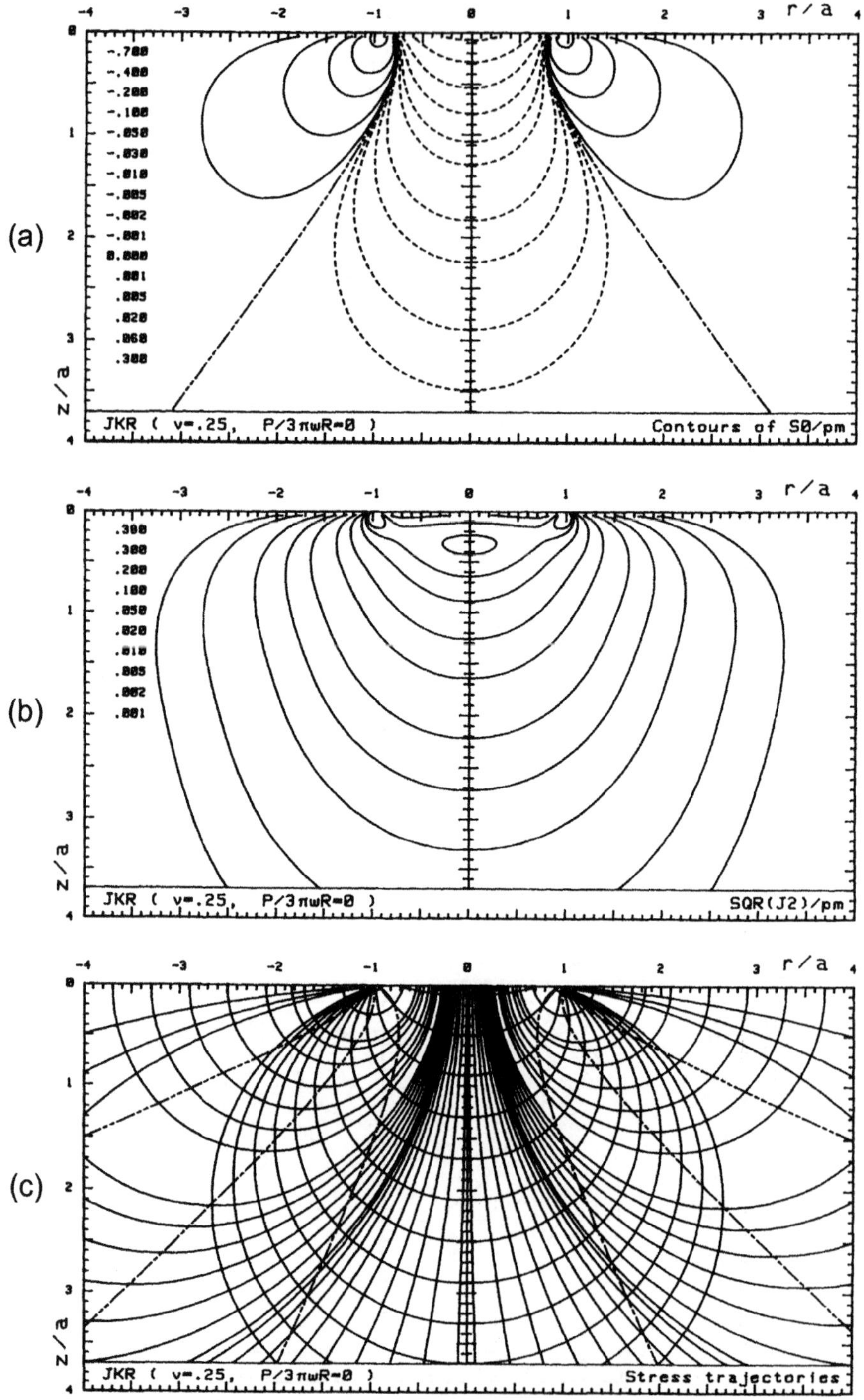

Fig. 4.48. Hydrostatic stress **(a)**, von Mises criterion **(b)**, and stress trajectories **(c)** for $P = 0$ (The *dashed-dotted lines* are isoclines at 0° and 45°)

It suffices to combine the stresses $(\sigma/p_{\mathrm{m}})_{\mathrm{H}}$ given in Sect. 4.4.4 and the stresses $(\sigma/p_{\mathrm{m}})_{\varrho}$ given in Sect. 4.3.

Figures 4.45 and 4.46 display in a plane of symmetry, for $P/3\pi wR = 1$ and for $\nu = 0.25$, the contours of principal stresses σ_1, $\sigma_2 \, (\equiv \sigma_\theta)$ and σ_3, of the hydrostatic pressure, of the von Mises criterion and the stress trajectories. At the surface and on the z-axis $\sigma_1 \equiv \sigma_r$ and $\sigma_3 \equiv \sigma_z$.

Figures 4.47 and 4.48 are similar graphical representations, but for contact under zero load.

4.6 Contacts of Spheres: the DMT Theory

As mentioned in the introduction, the adherence force of the JKR theory being independent of Young's modulus, it is incompatible with the adherence force

$$P_{\mathrm{c}} = -2\pi wR \tag{4.177}$$

computed by Bradley and Deryagin for a rigid sphere and a rigid plane (see Sect. 1.3.3). In 1975 Derjaguin, Muller and Toporov [12] gave their theory (the DMT theory) in which adhesion forces act in an annular zone around the contact but without deforming the profile, which remains Hertzian (Young's modulus is thus assumed large). The main results of their theory were that the adherence force $2\pi wR$ is reached when the area of contact vanishes (no instability), that the radius of contact under zero load is

$$a_0^3 = \frac{\pi wR^2}{K} ,$$

and that the adhesion forces around the contact add to the applied load a force which decreases from $2\pi wR$ to πwR when the penetration increases.

In a later paper [39], and still assuming a Hertzian profile but using a different method, Derjaguin *et al.* found again an adherence force $2\pi wR$, but this time the overload due to adhesion forces increased from $2\pi wR$ when the penetration increased, the radius of contact under zero load being, within a few percentage points:

$$a_0^3 = \frac{2\pi wR^2}{K} . \tag{4.178}$$

Note immediately that if the overload remains equal to $2\pi wR$, instead of varying with penetration, the results would be

$$\boxed{\begin{aligned} \frac{a^3K}{R} &= P + 2\pi wR , && (4.179a) \\ \delta &= \frac{a^2}{R} , && (4.179b) \end{aligned}}$$

and

$$G = \frac{a^3 K/R - P}{2\pi R} , \tag{4.180}$$

which is, indeed, the result obtained in [18].

4.7 Contacts of Spheres: the JKR–DMT Transition

At a meeting in Porto Rico in 1976 Tabor [13], comparing the two theories, pointed out that the main defect of the DMT theory was to neglect the deformations due to adhesion forces around the contact, whereas that of the JKR theory was to neglect the adhesion forces. He called attention to the fact that, as in the JKR theory, the stress distribution has a discontinuity in the DMT theory. Stresses are compressive in the contact and vanish at $r = a^-$, whereas they are tractions outside, reaching the theoretical stress at $r = a^+$. Considering the expression for the height h of the discontinuity of displacement[5] in the JKR theory, given by

$$h \approx \left(\frac{w^2 R}{E^2}\right)^{1/3} ,$$

Tabor noted that if h becomes of the order of magnitude of the equilibrium interatomic distance Z_0, the JKR theory must be corrected to take into account attractive forces outside the contact. The behaviour should vary continuously from one limit to the other, depending on a parameter

$$\left(\frac{w^2 R}{E^2 Z_0^3}\right)^{1/3} \approx \frac{h}{Z_0} . \tag{4.181}$$

In 1980 Muller *et al.* [14] gave a self-consistent numerical calculation using a Lennard–Jones potential, and they renounced the hypothesis that adhesion forces do not modify the Hertzian profile. The result was a smooth transition from the JKR theory towards the DMT theory when a parameter μ proportional to the parameter (introduced by Tabor) decreases. The DMT theory applies for $\mu \ll 1$ (hard solids, small radius of curvature, weak energy

---

[5] This discontinuity is the displacement $\Delta\delta$ due to the flat punch movement seen in Sect. 4.5.1:

$$\Delta\delta = \frac{2(P_1 - P)}{3aK} .$$

If $P = 0$, (4.164) for a_0 leads to

$$\Delta\delta^3 = \frac{6\pi^2 w^2 R(1-\nu^2)^2}{E^2} .$$

of adhesion) and that of JKR for $\mu \gg 1$ (soft solids, large radius of curvature, large energy of adhesion). DMT showed later [15] that these two limit theories do not depend on the exact shape of the interaction potential between the surfaces.

We have seen in Sect. 4.5.2 that, by applying the principles of fracture mechanics, the outcome of the JKR theory could be obtained, but not that for the DMT theory. This is because we used the classical relation between G and K_{I}^2, which is only valid in linear elastic fracture mechanics when the interaction zone is small compared to a characteristic length of the system—here, the radius of contact. By applying a Dugdale model where the adhesion forces are assumed constant and equal to σ on an annulus of width $d = c - a$ around the contact, we obtain the JKR–DMT transition, but this time with the advantage of literal formulae [18]. The method is to compute two stress-intensity factors: one, K_{I}, due to external loading alone, is that of the JKR theory; the other, K_m, corresponds to constant stresses acting in an exterior circular crack (the problem studied in Sect. 3.6.3). The cancellation of the sum $K_{\mathrm{I}} + K_m$ suppresses the singular stresses, insures the continuity of stresses, and determines the length d. The energy release rate is computed by the J integral, which reduces here to $\sigma_0 \delta_t$, and the problem is solved by writing the equilibrium equation $G = w$.

4.7.1 Study of Stresses

The stress distribution in the area of contact, due to the external loading (under the JKR theory) is (see (4.156)):

$$\sigma_z(r,0) = \frac{K_{\mathrm{I}}}{\sqrt{\pi a}} \frac{1}{\sqrt{1-\varrho^2}} - \frac{3aK}{2\pi R}\sqrt{1-\varrho^2}\,, \qquad \varrho < 1\,, \tag{4.182}$$

with

$$K_{\mathrm{I}} = \frac{a^3 K/R - P}{2a\sqrt{\pi a}}\,, \tag{4.183}$$

where $\varrho = r/a$. The stress distribution due to the internal loading[6] is given by (3.144) and (3.145), in which we must change the internal pressure p in $-\sigma_0$, so that

$$\sigma_z(r,0) = \frac{K_m}{\sqrt{\pi a}} \frac{1}{\sqrt{1-\varrho^2}} + \frac{2\sigma_0}{\pi} \tan^{-1}\sqrt{\frac{m^2-1}{1-\varrho^2}}\,, \qquad \varrho < 1\,, \tag{4.184}$$

with

$$K_m = -\frac{\sigma_0 a}{\sqrt{\pi a}} \left[\sqrt{m^2-1} + m^2 \tan^{-1}\sqrt{m^2-1}\right]\,, \tag{4.185}$$

[6] Recall that $\cos^{-1}(1/x) = \tan^{-1}\sqrt{x^2-1}$ if $x > 0$.

where $m = c/a > 1$. These stresses are additive and we can see that the singular stresses disappear if

$$\begin{aligned} K_{\mathrm{I}} &= -K_m \\ &= \frac{\sigma_0 a}{\sqrt{\pi a}} \left[\sqrt{m^2-1} + m^2 \tan^{-1} \sqrt{m^2-1} \right] . \end{aligned} \tag{4.186}$$

This relation determines the value of m as a function of σ_0 and K_{I} (which is a function of a and P). The stress distribution in the area of contact thus reduces to

$$\boxed{\sigma_z(\varrho,0) = -\frac{3aK}{2\pi R}\sqrt{1-\varrho^2} + \frac{2\sigma_0}{\pi} \tan^{-1} \sqrt{\frac{m^2-1}{1-\varrho^2}} , \qquad \varrho < 1} , \tag{4.187}$$

with $\sigma_z(1,0) = \sigma_0$, so that the condition $K_{\mathrm{I}} + K_m = 0$ ensures also the continuity of stresses.

In the limit case, when $m \to 1$, (4.186) shows that

$$\frac{K_{\mathrm{I}}\sqrt{\pi a}}{\sigma_0 a} \approx 2\sqrt{m^2-1} , \tag{4.188a}$$

which can be written, with $d = c - a$, as

$$d \approx \frac{\pi K_{\mathrm{I}}{}^2}{8\sigma_0^2} . \tag{4.188b}$$

We can recognize the classical (SSY) approximation of fracture mechanics (see (3.177)). The stress distribution, given by (4.187), thus reduces to the JKR distribution, as per (4.182).

In the other limit case, when $m \to \infty$, (4.186) leads to

$$\frac{K_{\mathrm{I}}}{\sigma_0\sqrt{\pi a}} \approx \frac{m^2}{2} , \tag{4.189}$$

which shows that $\sigma_0 \to 0$. In these conditions the stress distribution, given by (4.187), becomes

$$\sigma_z(\varrho,0) \approx -\frac{3aK}{2\pi R}\sqrt{1-\varrho^2} ,$$

which is the Hertzian distribution, (4.131), as in the DMT theory.

4.7.2 Study of Displacements

The discontinuity in displacements of the surface, outside the area of contact, due to the external loading is that of the JKR theory (see (4.158)), so that

$$\begin{aligned} [u_z] = &\frac{2(1-\nu^2)}{\pi E} K_{\mathrm{I}}\sqrt{\pi a} \cos^{-1}\frac{1}{\varrho} \\ &+ \frac{a^2}{\pi R}\left[\sqrt{\varrho^2-1} + (\varrho^2-2)\cos^{-1}\frac{1}{\varrho} \right] , \qquad \varrho > 1 . \end{aligned} \tag{4.190}$$

Under the internal loading it varies by

$$u_T(\varrho,0) = \frac{2(1-\nu^2)}{\pi E} K_m \sqrt{\pi a} \cos^{-1}\frac{1}{\varrho} \tag{4.191}$$
$$- \frac{4(1-\nu^2)}{\pi E}\sigma_0 a \left[\sqrt{m^2-1}\left(\sqrt{\varrho^2-1} - \cos^{-1}\frac{1}{\varrho}\right) - m^2 \int_1^{\min(\varrho,m)} \frac{\sqrt{\varrho^2-t^2}}{t^2\sqrt{m^2-t^2}}\,\mathrm{d}t\right],$$

given by (3.150). Adding the two expressions, with $K_\mathrm{I} + K_m = 0$ we get:

$$[u_z] = \frac{a^2}{\pi R}\left[\sqrt{\varrho^2-1} + (\varrho^2-2)\cos^{-1}\frac{1}{\varrho}\right] - \frac{4(1-\nu^2)}{\pi E}\sigma_0 a \left[\sqrt{m^2-1}\left(\sqrt{\varrho^2-1} - \cos^{-1}\frac{1}{\varrho}\right) - m^2 \int_1^{\min(\varrho,m)} \frac{\sqrt{\varrho^2-t^2}}{t^2\sqrt{m^2-t^2}}\,\mathrm{d}t\right]. \tag{4.192}$$

The opening δ_t of the crack (COD) is the discontinuity of displacement at $r = c$. Making $\varrho = m$, we get:

$$\delta_t = \frac{a^2}{\pi R}\left[\sqrt{m^2-1} + (m^2-2)\tan^{-1}\sqrt{m^2-1}\right] + \frac{4(1-\nu^2)}{\pi E}\sigma_0 a \left[\sqrt{m^2-1}\tan^{-1}\sqrt{m^2-1} - m + 1\right]. \tag{4.193}$$

We can establish that the curve representating (4.192) has a vertical tangent at $r = c$. This behaviour arises from the elliptical integral. By differentiating under the integral for $\varrho < m$ or for $\varrho > m$, we get

$$\frac{\mathrm{d}I(\varrho)}{\mathrm{d}\varrho} = \varrho \int_1^{\min(\varrho,m)} \frac{\mathrm{d}t}{t^2\sqrt{\varrho^2-t^2}\sqrt{m^2-t^2}},$$

and for $\varrho = m$

$$\left(\frac{\mathrm{d}I}{\mathrm{d}\varrho}\right)_{\varrho=m} = \left[-\frac{1}{mt} + \frac{1}{2m^2}\ln\left|\frac{m+t}{m-t}\right|\right]_1^m,$$

which is infinite.

In the limit case, when $m \to \infty$, (4.192) reduces to

$$[u_z] = \frac{a^2}{\pi R}\left[\sqrt{\varrho^2-1} + (\varrho^2-2)\cos^{-1}\frac{1}{\varrho}\right],$$

which is the discontinuity of displacement in the Hertzian case (cf (4.113b)).

In the other limit case, when $m \to 1$, we can see that the limit of (4.192) is the JKR profile. In effect, by using the second theorem of the mean with $1 < \xi < m$, we can write

$$m^2 \int_1^m \frac{\sqrt{\varrho^2 - t^2}}{t^2\sqrt{m^2 - t^2}}\, dt = m^2 \sqrt{\varrho^2 - \xi^2} \int_1^m \frac{dt}{t^2\sqrt{m^2 - t^2}}$$
$$= \sqrt{\varrho^2 - \xi^2}\sqrt{m^2 - 1}\,,$$

equivalent to $\sqrt{\varrho^2 - 1}\sqrt{m^2 - 1}$ when $m \to 1$. Using (4.188a) we get

$$[u_z] = \frac{2(1-\nu^2)}{\pi E} K_{\mathrm{I}} \sqrt{\pi a} \cos^{-1}\frac{1}{\varrho}$$
$$+ \frac{a^2}{\pi R}\left[\sqrt{\varrho^2 - 1} + (\varrho^2 - 2)\cos^{-1}\frac{1}{\varrho}\right],$$

which is (4.158).

4.7.3 Penetration

The penetration δ of the sphere is the sum of that of JKR, from (4.147), and of the approach due to the adhesive forces outside the contact, from (3.148) and (3.141) where p is replaced by $-\sigma_0$. Then we have

$$\delta = \frac{a^2}{3R} + \frac{2P}{3aK} + \frac{1-\nu^2}{E}\sigma_0 a \left[\left(m^2 \tan^{-1}\sqrt{m^2 - 1}\right) - \sqrt{m^2 - 1}\right]. \tag{4.194}$$

If we put this relation into (4.186), we can see that the condition of cancellation of singular stresses reduces to a condition on penetration, whereby

$$\boxed{\delta = \frac{a^2}{R} - \frac{8\sigma_0}{3K}\sqrt{c^2 - a^2}}\,. \tag{4.195}$$

In contrast, if we replace σ_0 by its value given by (4.186), then (4.194) becomes

$$\delta = \left(\frac{a^2}{3R} + \frac{2P}{3aK}\right) + \left(\frac{2a^2}{3R} - \frac{2P}{3aK}\right)\frac{(m^2 \tan^{-1}\sqrt{m^2 - 1}) - \sqrt{m^2 - 1}}{(m^2 \tan^{-1}\sqrt{m^2 - 1}) + \sqrt{m^2 - 1}}\,.$$

When $m \to \infty$ we obtain the Hertz result again, and when $m \to 1$ we obtain that of JKR.

At this stage, the ratio $m = c/a$ is given as a function of P and a or δ and a, but the equilibrium curves $\delta(a)$, $a(P)$ and $P(\delta)$ cannot be plotted, for a relation linking c to σ_0 and a is still missing. This relation is the Griffith equilibrium equation $G = w$.

4.7.4 Energy Release Rate

As seen in Sect. 3.7.3, the J integral in a Dugdale model reduces to

$$G = J = \sigma_0 \delta_t ,$$

and hence, using (4.193), to

$$G = \frac{\sigma_0 a^2}{\pi R}\left[\sqrt{m^2-1} + (m^2-2)\tan^{-1}\sqrt{m^2-1}\right] + \frac{4(1-\nu^2)}{\pi E}\sigma_0^2 a\left[\left(\sqrt{m^2-1}\tan^{-1}\sqrt{m^2-1}\right) - m + 1\right] , \qquad (4.196)$$

which becomes, using (4.183) and (4.186),

$$G = \frac{P_1 - P}{2\pi R}\left(1 - \frac{2\tan^{-1}\sqrt{m^2-1}}{\sqrt{m^2-1} + m^2\tan^{-1}\sqrt{m^2-1}}\right) + \frac{4(P_1-P)^2\left[\sqrt{m^2-1}\tan^{-1}\sqrt{m^2-1} - m + 1\right]}{3\pi R P_1\left[\sqrt{m^2-1} + \left(m^2\tan^{-1}\sqrt{m^2-1}\right)\right]^2} . \qquad (4.197)$$

In the limit case, when $m \to 1$, let $m = 1+\varepsilon$. Recalling the approximation

$$\tan^{-1}\sqrt{m^2-1} \approx \sqrt{2\varepsilon}\left(1 - \frac{5\varepsilon}{12}\right) ,$$

we get

$$G \approx \frac{(P_1-P)^2}{6\pi R P_1}\left(1 - \frac{13\varepsilon}{6}\right) + \frac{P_1 - P}{2\pi R}\frac{4\varepsilon}{3} , \qquad (4.198)$$

which gives the JKR result when $\varepsilon \to 0$.

In the limit case when $m \to \infty$ we have, for m large, (4.197) reducing to

$$G \approx \frac{P_1 - P}{2\pi R}\left(1 - \frac{2}{m^2}\right) + \frac{4(P_1-P)^2}{3\pi R P_1}\frac{2\pi - 4}{\pi^2 m^3} ,$$

which gives

$$\boxed{G = \frac{P_1 - P}{2\pi R}} \qquad (4.199)$$

when $m \to \infty$, and which is thus the energy release rate in the DMT approximation.

4.7.5 Equilibrium Curves

Let us introduce the adimensional parameters

$$A = \frac{a}{(\pi w R^2/K)^{1/3}} , \tag{4.200}$$

$$\overline{P} = \frac{P}{\pi w R} , \tag{4.201}$$

$$\Delta = \frac{\delta}{(\pi^2 w^2 R/K^2)^{1/3}} , \tag{4.202}$$

$$\lambda = \frac{2\sigma_0}{(\pi w K^2/R)^{1/3}} . \tag{4.203}$$

With this notation, the Hertz theory becomes

$$A^3 = \overline{P} ,$$
$$\Delta = A^2 .$$

The DMT becomes

$$A^3 = \overline{P} + 2 ,$$
$$\Delta = A^2 .$$

And that of JKR becomes

$$A^3 = \overline{P} + A\sqrt{6A} ,$$
$$\Delta = \frac{A^3 + 2\overline{P}}{3A} = A^2 - \frac{2}{3}\sqrt{6A} .$$

Equations (4.186) and (4.195), for the cancellation of singularities, become

$$\overline{P} = A^3 - \lambda A^2 \left(\sqrt{m^2-1} + m^2 \tan^{-1}\sqrt{m^2-1}\right) , \tag{4.204}$$

$$\Delta = A^2 - \frac{4}{3} A\lambda\sqrt{m^2-1} , \tag{4.205}$$

whereas the Griffith equilibrium equation (4.196) becomes

$$\frac{\lambda A^2}{2}\left[\sqrt{m^2-1} + (m^2-2)\tan^{-1}\sqrt{m^2-1}\right] + \frac{4\lambda^2 A}{3}\left[\left(\sqrt{m^2-1}\tan^{-1}\sqrt{m^2-1}\right) - m + 1\right] = 1 . \tag{4.206}$$

Note that the condition of no singularity, $K_{\mathrm{I}} + K_m = 0$, and the equilibrium equation, $G = w$, are two independent relations, so that the nonsingularity of stresses is maintained, even in nonequilibrium conditions (such as during crack propagation).

For a given parameter λ and a given radius of contact a, we can extract m from (4.206) and put it in both (4.204) and (4.205) to obtain the equilibrium curves $a(P)$, as shown in Fig. 4.49, $a(\delta)$ (Fig. 4.50) and $P(\delta)$ (Fig. 4.51).

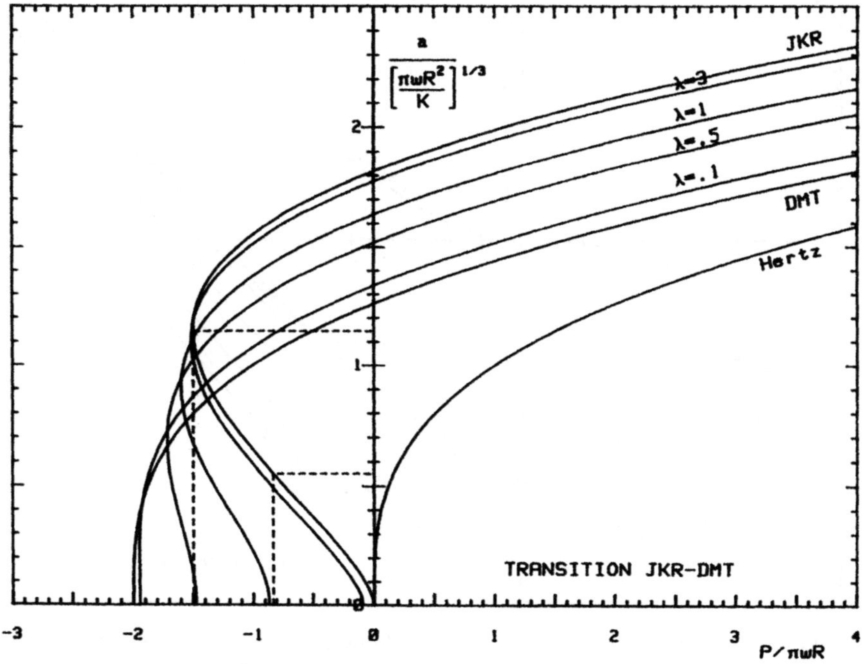

Fig. 4.49. Relation between radius of contact a and applied load P for various λ

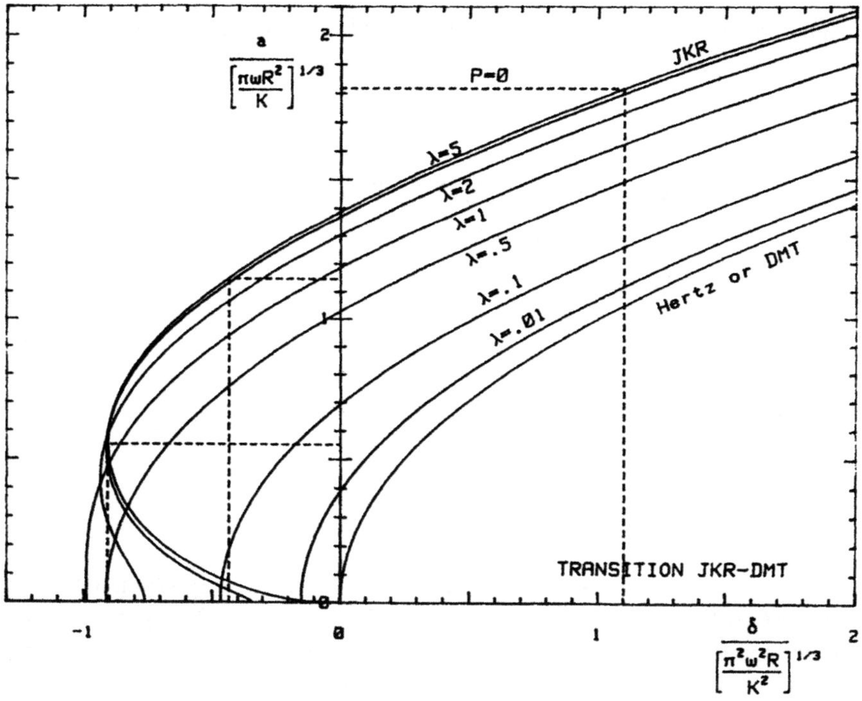

Fig. 4.50. Relation between radius of contact a and penetration δ for various λ

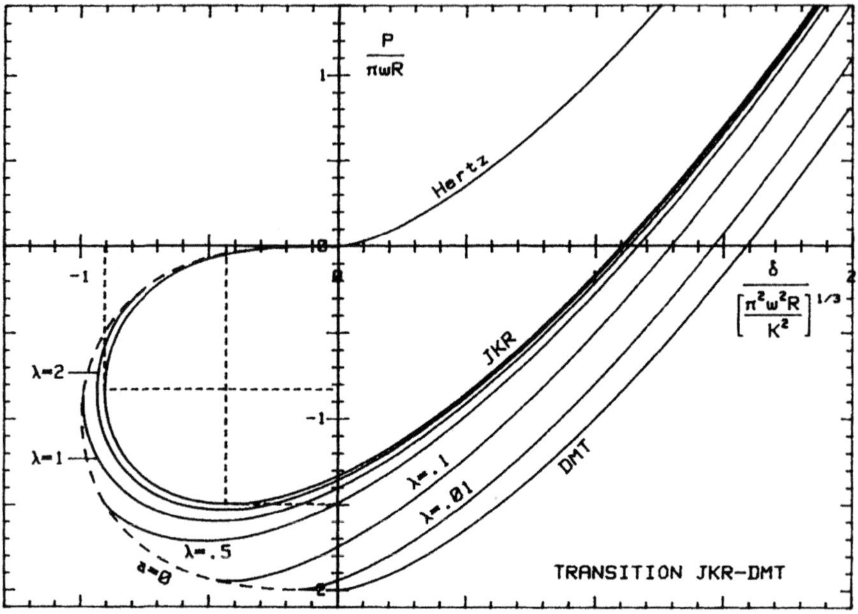

Fig. 4.51. Relation between applied load P and penetration δ. Adherence at fixed load or at fixed grips corresponds to a horizontal or vertical tangent, when it exists, in which case separation occurs with a jump. Otherwise separation occurs at $a = 0$

When a tends toward zero, m tends toward infinity, and (4.206) reduces to

$$A = \frac{\alpha(\lambda)}{m} ,$$

with

$$\alpha(\lambda) = \frac{-2\lambda^2(\pi - 2) + \sqrt{4\lambda^4(\pi-2)^2 + 9\lambda\pi}}{3\lambda\pi/2} .$$

We get thus, when $a \to 0$,

$$\overline{P} = -\frac{\pi}{2}\lambda[\alpha(\lambda)]^2 , \tag{4.207}$$

$$\Delta = -\frac{4}{3}\lambda\alpha(\lambda) , \tag{4.208}$$

or else

$$\overline{P}(\overline{P} + 2) = \frac{9\pi(\pi - 2)}{32}\Delta^3 .$$

This curve is drawn as the dashed line in Fig. 4.51. It is symmetrical with respect to the straight line of ordinate -1. Equation (4.207) gives the adherence force when separation occurs at radius of contact zero. (It is $-2\pi wR$ when $\lambda = 0$). The adherence force at fixed load or at fixed grips is displayed as a function of λ in Fig. 4.52.

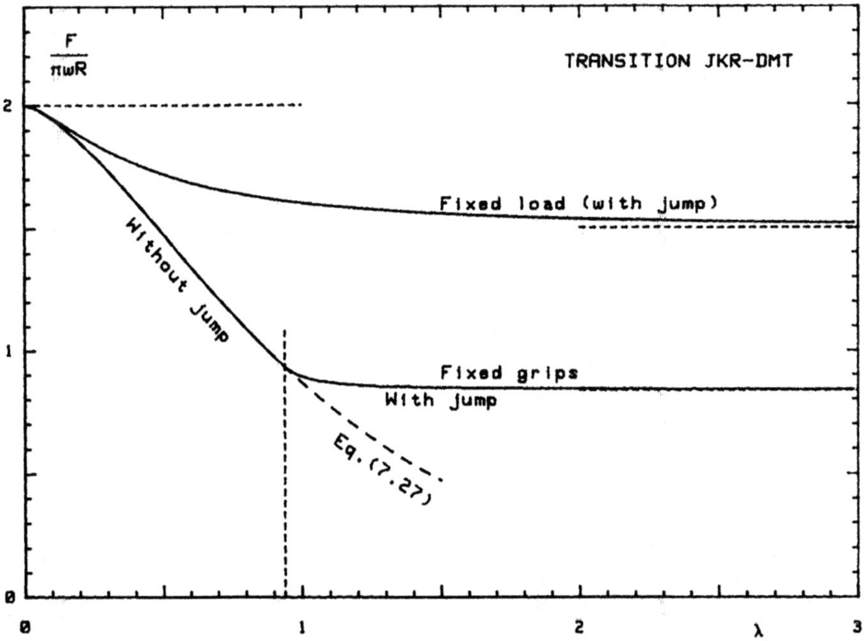

Fig. 4.52. Adherence at fixed load and at fixed grips as a function of λ. At fixed load there is always a jump; at fixed grips the jump appears at $\lambda > 0.938$

The discontinuity of displacement outside the circle of contact is given by (4.192), which can be written as a function of elliptical integrals of the first and the second kind, i.e. $E(\varphi, k)$ and $F(\varphi, k)$, and with the same normalization as for δ. For $r < c$ this is given by

$$\overline{[u_z]} = \frac{A^2}{\pi}\left[\sqrt{\varrho^2-1} + (\varrho^2-2)\cos^{-1}\frac{1}{\varrho}\right] + \frac{8\lambda A}{3\pi}\sqrt{m^2-1}\cos^{-1}\frac{1}{\varrho} + \frac{8\lambda A}{3\pi}\left[\sqrt{\frac{\varrho^2-1}{m^2-1}} - mE(\varphi,k)\right], \tag{4.209}$$

with

$$\varphi = \sin^{-1}\left(\frac{m}{\varrho}\sqrt{\frac{\varrho^2-1}{m^2-1}}\right), \qquad k = \frac{\varrho}{m} = \frac{r}{c}.$$

For $r > c$ it is given by

$$\overline{[u_z]} = \frac{A^2}{\pi}\left[\sqrt{\varrho^2-1} + (\varrho^2-2)\cos^{-1}\frac{1}{\varrho}\right] + \frac{8\lambda A}{3\pi}\sqrt{m^2-1}\cos^{-1}\frac{1}{\varrho} + \frac{8\lambda A}{3\pi}\left[\sqrt{\frac{m^2-1}{\varrho^2-1}} - \varrho E(\varphi,k) + \left(1-\frac{m^2}{\varrho^2}\right)\varrho F(\varphi,k)\right], \tag{4.210}$$

with

$$\varphi = \sin^{-1}\left(\frac{\varrho}{m}\sqrt{\frac{m^2-1}{\varrho^2-1}}\right) ,$$

$$k = \frac{m}{\varrho} = \frac{c}{r} .$$

Equations (4.209) and (4.210) describe the distance between a rigid sphere and an elastic plane, or the profile of a deformable sphere on a rigid plane, or else the distance between two elastic cylinders crossed at right angle. They are displayed in Fig. 4.53 for the case $P = 0$. At equilibrium the opening of the crack at $r = c$ (COD) is $\delta_t = w/\sigma_0$, i.e. in normalized coordinates:

$$\Delta_t = \frac{2}{\pi\lambda} . \tag{4.211}$$

This curve is plotted as the dashed line in Fig. 4.53.

For a given load P and a given parameter λ, we can compute m and A by (4.204) and (4.206) and draw the stress distribution in the area of contact using (4.187), which is written in normalized coordinates as

$$\frac{\pi\sigma_z}{(\pi w K^2/R)^{1/3}} = -\frac{3A}{2}\sqrt{1-\varrho^2} + \lambda\tan^{-1}\sqrt{\frac{m^2-1}{1-\varrho^2}} . \tag{4.212}$$

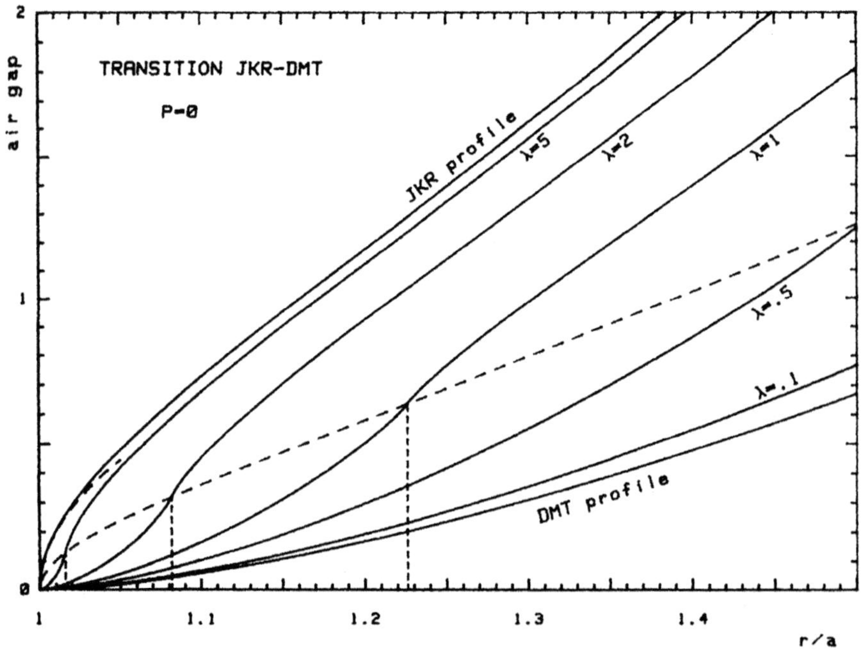

Fig. 4.53. Profile of the "crack" for various λ and for $P = 0$. The COD and the parabolic approximation for the JKR profile are shown by the *dashed line*

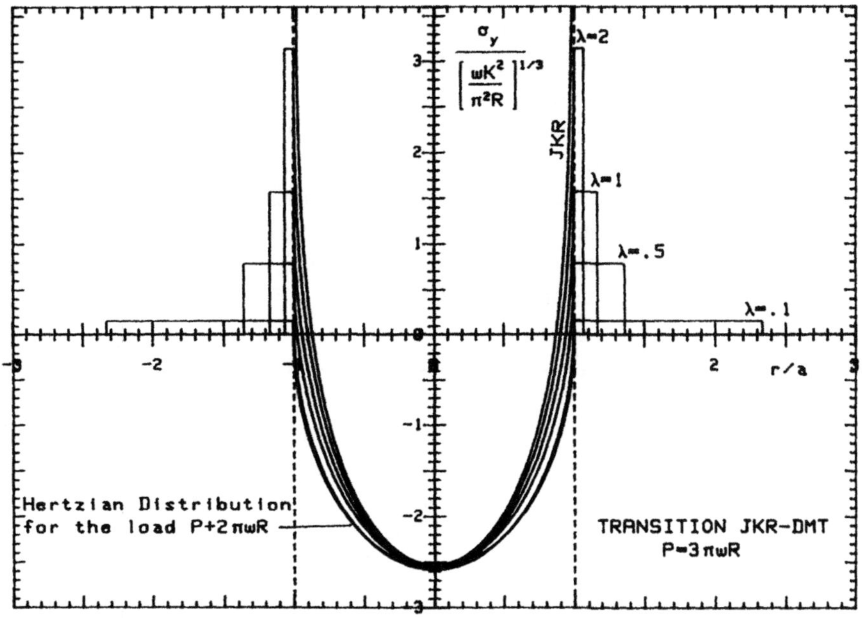

Fig. 4.54. Stress distribution for $P = 3\pi w R$

Figure 4.54 displays this distribution for $P = 3\pi w R$. We can verify, by integration, that such a distribution equilibrates the sum of the applied load P with the overload $\pi(c^2 - a^2)\sigma_0$ due to adhesion forces around the contact.

Note that there is always a continuity of stress through the boundary $r = a$ where the stress is σ_0 (in principle the theoretical stress σ_{th}). In the DMT approximation this stress must tend to zero in order to have continuity with the stress at $r = a_-$, but its integral is finite and the total attractive force outside the contact is $2\pi w R$. In the area of contact the stress distribution is Hertzian and its integral is $P + 2\pi w R$. Any negative applied loads are thus sustained by the adhesion forces outside the contact. In the JKR approximation the distribution of adhesion stresses reduces to a singular stress at $r = a_+$. The total attractive force outside the contact is zero, and the integral of the stresses in the area of contact is equal to the applied load P. The negative applied loads are sustained by tensile stresses in the area of contact.

4.7.6 Application to a Liquid Meniscus

Experiments on crossed cylinders of mica have shown [40–42] a change in the profile of the JKR–DMT type when the size of the meniscus has increased. As a liquid meniscus on the edge of a contact is a good example of a Dugdale zone, with a constant stress σ_0 equal to the Laplace pressure $\sigma_0 = \gamma_{\mathrm{L}}/\varrho_k$, and a crack opening displacement $\delta_t = 2\varrho_k \cos\theta$ (where γ_{L} is the surface energy of the liquid, ϱ_k is the radius of curvature of the meniscus, and θ is the contact angle), the above theory can be applied with $J = \sigma_0 \delta_t = 2\gamma_L \cos\theta$.[7] (In fact, the contour integral round the crack tip leads rather to $J = w' + 2\gamma_L \cos\theta$, where w' is the energy of adhesion through the liquid, but the present theory does not permit us to distinguish between σ_{th} and the Laplace pressure.)

Experiments have been repeated with more precision by Gauthier-Manuel *et al.* [44], on crossed mica cylinders, and compared with the theory. Figures 4.55–4.57 display the observed profile in dry air, in humid air and in the presence of a large meniscus, respectively.

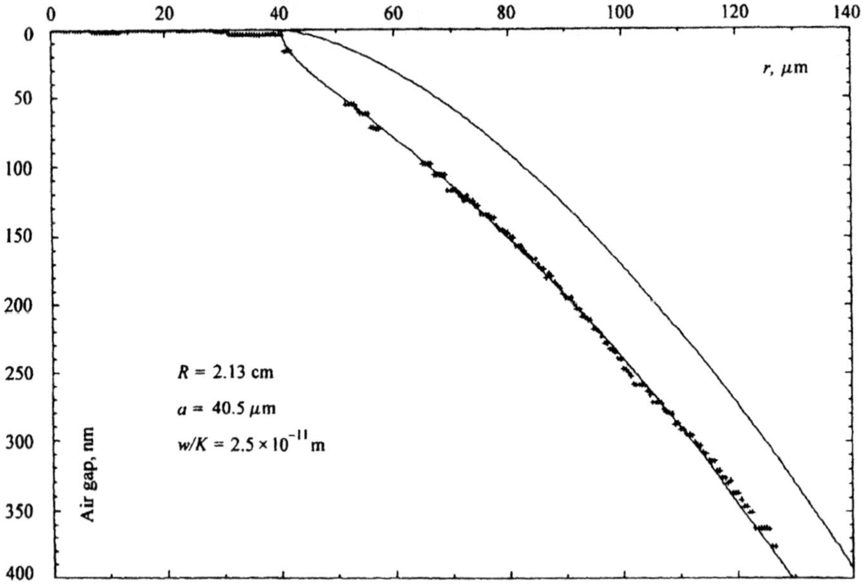

Fig. 4.55. Contact of mica–mica in dry air. A JKR profile is observed, in agreement with (4.158) fitted with $w/K = 2.5 \times 10^{-11}$ m. The Hertzian profile is given for comparison

4.8 Conical Punch

The problem of the conical punch was studied by Love [45] in 1939, then by Harding and Sneddon [19] in 1945, and also by Sneddon on his own [17, 46].

4.8.1 General Theorems

Let us consider a cone of semi-angle α and let $\beta = \pi/2 - \alpha$ and apply the general theorems of Sect. 4.1.2. The profile of the cone is defined by

[7] The problem of the contact of elastic spheres in the presence of a liquid meniscus has also been studied by Fogden and White [43] by a completely different method. The results can be shown to be identical.

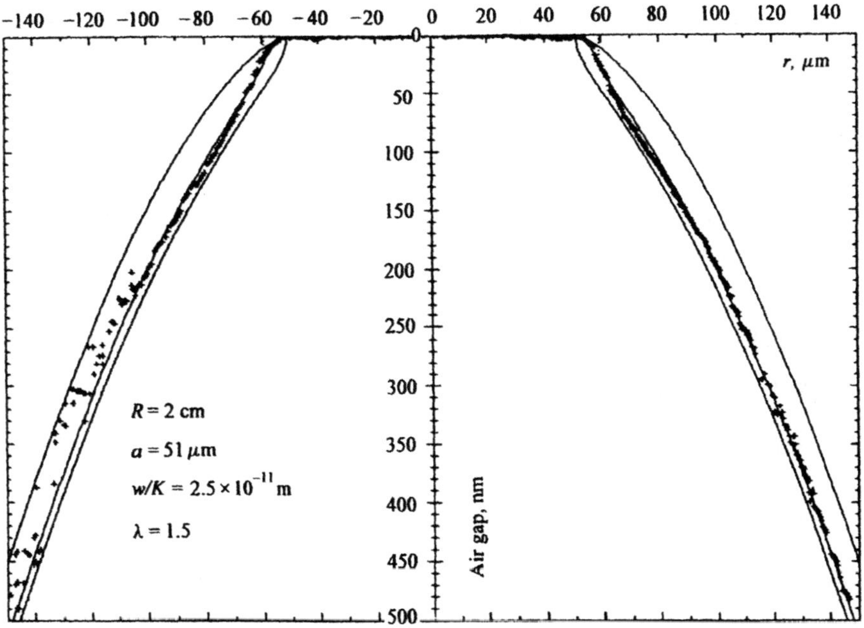

Fig. 4.56. Contact of mica–mica in humid air. The Hertzian profile and the JKR profile (with $w/K = 2.5 \times 10^{-11}$ m) are drawn for comparison. The observed profile is in agreement with (4.209) and (4.210) with $\lambda = 1.5$ and $w/K = 2.5 \times 10^{-11}$ m

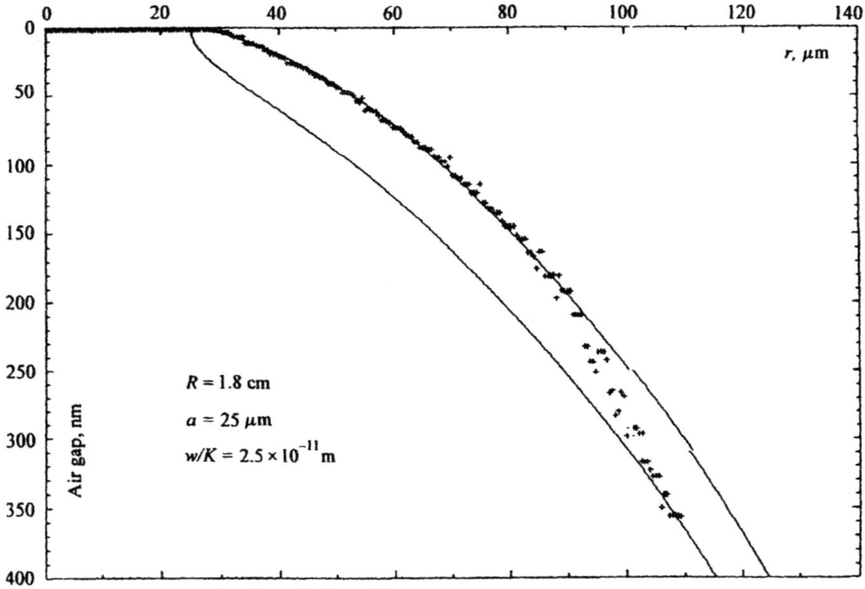

Fig. 4.57. Contact of mica–mica with a large water meniscus. A Hertzian profile is observed, in agreement with (4.113b). The JKR profile (with $w/K = 2.5 \times 10^{-11}$ m) is given for comparison

$$f(x) = f\left(\frac{r}{a}\right) = xa\tan\beta\,, \tag{4.213}$$
$$f'(x) = a\tan\beta\,,$$

and we have

$$\chi(t) = \frac{2}{\pi}\left(\delta - t\int_0^t \frac{f'(x)}{\sqrt{t^2-x^2}}\,\mathrm{d}x\right)$$
$$= \frac{2\delta}{\pi} - ta\tan\beta\,, \tag{4.214}$$
$$\chi(1) = \frac{2\delta}{\pi} - a\tan\beta\,, \tag{4.215}$$
$$\chi'(t) = -a\tan\beta\,.$$

Letting

$$E' = \frac{E}{1-\nu^2}\,,$$

we get

$$\boxed{P = 2aE'\left(\delta - \frac{\pi a}{4}\tan\beta\right)}\,, \tag{4.216}$$

which is the first equation of state of the system and

$$\sigma_z(r,0) = -\frac{E'}{2}\left[\frac{\chi(1)}{\sqrt{a^2-r^2}} + \tan\beta\cosh^{-1}\frac{a}{r}\right], \quad 0 < r < a\,, \tag{4.217}$$
$$u_z(r,0) = \frac{2\delta}{\pi}\sin^{-1}\frac{a}{r} + (\sqrt{r^2-a^2} - r)\tan\beta, \quad r \geq a\,,$$
$$= \chi(1)\sin^{-1}\frac{a}{r} + \left(\sin^{-1}\frac{a}{r} - \frac{r}{a} + \sqrt{\frac{r^2}{a^2}-1}\right)a\tan\beta\,. \tag{4.218}$$

For the moment let $\chi(1) = 0$, and we have for the non-adhesive cone

$$\delta = \frac{\pi a}{2}\tan\beta\,, \tag{4.219a}$$
$$\frac{P}{\pi a^2} = p_{\mathrm{m}} = \frac{E'\tan\beta}{2}\,. \tag{4.220a}$$

$$\frac{\sigma_z(r,0)}{p_{\mathrm{m}}} = -\cosh^{-1}\frac{a}{r}\,, \quad 0 < r \leq a\,, \tag{4.221}$$
$$u_z(r,0) = \frac{2\delta}{\pi}\left(\sin^{-1}\frac{a}{r} - \frac{r}{a} + \sqrt{\frac{r^2}{a^2}-1}\right), \quad r > a\,. \tag{4.222}$$

Equations (4.219a) and (4.220a) can be written as

$$\delta = \frac{P}{aE'} , \tag{4.219b}$$

$$P = \frac{2E'}{\pi \tan \beta} \delta^2 . \tag{4.220b}$$

We note that the mean pressure p_{m} is independent of the load and depends only on the angle of the cone, that the penetration δ is proportional to the radius of contact (it is twice that of the flat punch of same radius under the same load), and that the relation between the force and displacement is parabolic. The stored elastic energy is

$$U_{\mathrm{E}} = \frac{2E'}{3\pi \tan \beta} \delta^3 = \frac{\pi^2 E' \tan^2 \beta}{12} a^3 = \frac{1}{3} P\delta . \tag{4.223}$$

The displacement at $r = a$ is

$$u_z(a, 0) = \delta \left(1 - \frac{2}{\pi}\right)$$

and the geometry of the contact is displayed in Fig. 4.58.

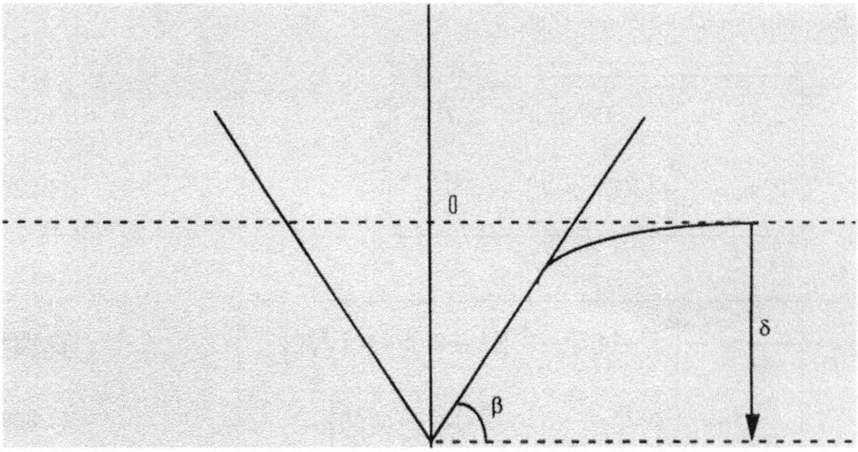

Fig. 4.58. Contact of a conical punch

The derivative of u_z is

$$\frac{\mathrm{d}u_z}{\mathrm{d}r} = -\tan \beta \left(1 - \frac{a}{r}\sqrt{\frac{r^2}{a^2} - 1}\right) \tag{4.224}$$

and the contact is indeed tangential at $r = a$. (Recall that the z-axis is directed downwards.)

4.8.2 Calculation of Displacements and Stresses

Using (4.219a), $\chi(t)$ becomes

$$\chi(t) = (1-t)a\tan\beta\,. \tag{4.225}$$

Putting this expression in that of $\varphi(p)$ in (4.7), we get

$$\begin{aligned}\varphi(p) &= -\frac{Ea^5}{2(1-\nu^2)}\tan\beta\int_0^1 (1-t)\cos(pt)\,\mathrm{d}t\\ &= -p_{\mathrm{m}}a^5\frac{1-\cos p}{p^2}\,,\end{aligned}$$

and the integrals I_n^m from (4.9) become

$$I_n^m = -p_{\mathrm{m}}a^5 J_n^m\,,$$

with

$$\begin{aligned}J_n^m &= \int_0^\infty p^{n-2}(1-\cos p)\mathrm{e}^{-p\zeta}J_m(p\varrho)\,\mathrm{d}p\\ &= Z_{n-1}^m - C_{n-1}^m\,.\end{aligned} \tag{4.226}$$

Equations (4.10–12) are thus

$$2\mu u_r = -\frac{P}{\pi a}\left[(1-2\nu)J_0^1 - \zeta J_1^1\right]\,, \tag{4.227}$$

$$2\mu u_z = \frac{P}{\pi a}\left[2(1-\nu)J_0^0 + \zeta J_1^0\right]\,, \tag{4.228}$$

$$\frac{\sigma_r}{p_{\mathrm{m}}} = -\left\{J_1^0 - \zeta J_2^0 - \frac{1}{\varrho}\left[(1-2\nu)J_0^1 - \zeta J_1^1\right]\right\}\,, \tag{4.229}$$

$$\frac{\sigma_\theta}{p_{\mathrm{m}}} = -\left\{2\nu J_1^0 + \frac{1}{\varrho}\left[(1-2\nu)J_0^1 - \zeta J_1^1\right]\right\}\,, \tag{4.230}$$

$$\frac{\sigma_z}{p_{\mathrm{m}}} = -\left[J_1^0 + \zeta J_2^0\right]\,, \tag{4.231}$$

$$\frac{\tau_{rz}}{p_{\mathrm{m}}} = -\zeta J_2^1\,, \tag{4.232}$$

and the hydrostatic stress is simply

$$\frac{\sigma_0}{p_{\mathrm{m}}} = -\frac{2(1+\nu)}{3}J_1^0\,. \tag{4.233}$$

The integrals Z_n^m and C_n^m are computed in Appendix A, and we find, with the parameters s, u, φ, θ, defined in Sect. 4.3, that

$$J_0^0 = -\zeta J_1^0 - \sqrt{\varrho^2+\zeta^2} + \sqrt{u} + \tan^{-1}\frac{1}{\sqrt{u}}\,, \tag{4.234}$$

$$\begin{aligned} J_1^0 &= \frac{1}{2}\ln\frac{s+1+\zeta^2+2\sqrt{s(1+\zeta^2)}\cos(\theta-\varphi/2)}{\left(\zeta+\sqrt{\varrho^2+\zeta^2}\right)^2} \\ &= \ln\frac{(1+\zeta/\sqrt{u})\sqrt{1+u}}{\zeta+\sqrt{\varrho^2+\zeta^2}}\,, \end{aligned} \tag{4.235}$$

$$\begin{aligned} J_2^0 &= \frac{1}{\sqrt{\varrho^2+\zeta^2}} - \frac{1}{s^{1/2}}\cos\frac{\varphi}{2} \\ &= \frac{1}{\sqrt{\varrho^2+\zeta^2}} - \frac{\sqrt{u}}{s}\,, \end{aligned} \tag{4.236}$$

$$\begin{aligned} J_0^1 &= \frac{1}{2}\left[\varrho J_1^0 - \zeta J_1^1 + \frac{1}{\varrho}\left(1 - s^{1/2}\sin\frac{\varphi}{2}\right)\right] \\ &= \frac{1}{2}\left[\varrho J_1^0 - \zeta J_1^1 + \frac{1}{\varrho}\left(1 - \frac{\zeta}{\sqrt{u}}\right)\right]\,, \end{aligned} \tag{4.237}$$

$$\begin{aligned} J_1^1 &= \frac{1}{\varrho}\left(\sqrt{\varrho^2+\zeta^2} - s^{1/2}\cos\frac{\varphi}{2}\right) \\ &= \frac{1}{\varrho}\left(\sqrt{\varrho^2+\zeta^2} - \sqrt{u}\right)\,, \end{aligned} \tag{4.238}$$

$$\begin{aligned} J_2^1 &= \frac{1}{\varrho}\left[\frac{\sqrt{1+\zeta^2}}{s^{1/2}}\cos\left(\theta-\frac{\varphi}{2}\right) - \frac{\zeta}{\sqrt{\varrho^2+\zeta^2}}\right] \\ &= \frac{\zeta}{\varrho}\left(\frac{1+u}{s\sqrt{u}} - \frac{1}{\sqrt{\varrho^2+\zeta^2}}\right)\,. \end{aligned} \tag{4.239}$$

4.8.3 Study of Stresses

At the Surface. We put $\zeta = 0$ in the stress tensor, with

$$\varphi = \pi\,,\quad s = 1-\varrho^2\,,\quad u = 0\,,\quad \frac{\zeta}{\sqrt{u}} = \sqrt{1-\varrho^2} \quad \text{for}\quad \varrho \le 1\,,$$

$$\varphi = 0\,,\quad s = \varrho^2-1\,,\quad u = s\,,\quad \frac{\zeta}{\sqrt{u}} = 0 \quad \text{for}\quad \varrho \ge 1\,.$$

In the area of contact, we get

$$J_1^0(\varrho,0) = \ln\frac{1+\sqrt{1-\varrho^2}}{\varrho} = \cosh^{-1}\frac{1}{\varrho}\,,$$

$$J_0^1(\varrho,0) = \frac{1}{2}\left[\varrho J_1^0 + \frac{1-\sqrt{1-\varrho^2}}{\varrho}\right]\,,$$

$$\frac{\sigma_r(\varrho,0)}{p_m} = -\frac{1+2\nu}{2}\cosh^{-1}\frac{1}{\varrho} + \frac{1-2\nu}{2}\frac{1-\sqrt{1-\varrho^2}}{\varrho^2}\,, \tag{4.240}$$

$$\frac{\sigma_\theta(\varrho,0)}{p_m} = -\frac{1+2\nu}{2}\cosh^{-1}\frac{1}{\varrho} - \frac{1-2\nu}{2}\frac{1-\sqrt{1-\varrho^2}}{\varrho^2}\,, \tag{4.241}$$

$$\frac{\sigma_z(\varrho,0)}{p_m} = -\cosh^{-1}\frac{1}{\varrho}\,, \tag{4.242}$$

$$\tau_{rz} = 0\,.$$

Outside the area of contact, we get, with $J_1^0 = 0$ and $J_0^1 = 1/2\varrho$,

$$\sigma_r(r,0) = \frac{1-2\nu}{2\pi}\frac{P}{r^2}\,,$$

$$\sigma_\theta(r,0) = -\frac{1-2\nu}{2\pi}\frac{P}{r^2}\,,$$

$$\sigma_z = \tau_{rz} = 0\,.$$

In agreement with Way's theorem, these are the stresses due to a concentrated load, and they are displayed in Fig. 4.59.

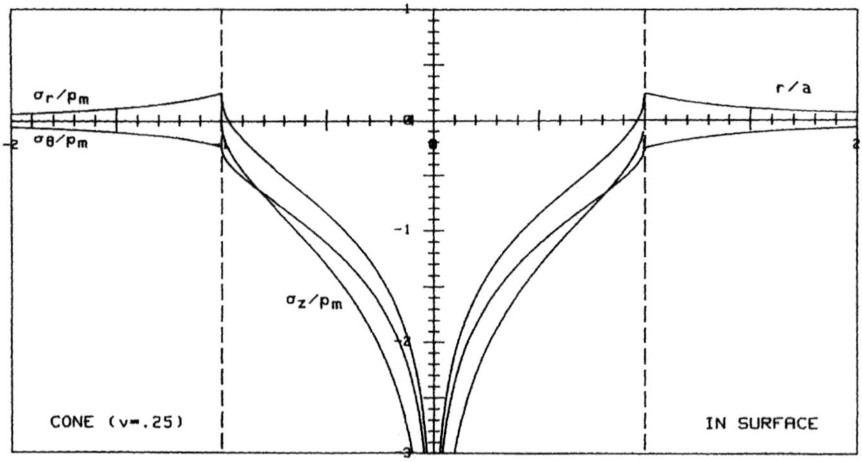

Fig. 4.59. Surface stresses

On the z-Axis. For $\varrho = 0$ we get $s = 1 + \zeta^2$ and $u = \zeta^2$; hence

$$J_1^0(0,\zeta) = \frac{1}{2}\ln\left(1+\frac{1}{\zeta^2}\right)\,,$$

$$J_2^0(0,\zeta) = \frac{1}{\zeta(1+\zeta^2)}\,.$$

For J_0^1/ϱ and J_1^1/ϱ we must remove the indetermination or return to (4.226) with $J_1(p\varrho) \approx p\varrho/2$ when $\varrho \to 0$. Keeping the terms in ϱ^2 we get

$$s = 1 + \zeta^2 - \varrho^2 \frac{1-\zeta^2}{1+\zeta^2} \,,$$

$$\sqrt{u} = \zeta \left(1 + \frac{\varrho^2}{2(1+\zeta^2)}\right) \,,$$

$$1 - \frac{\zeta}{\sqrt{u}} = \frac{\varrho^2}{2(1+\zeta^2)} \,;$$

hence

$$\frac{\zeta}{\varrho} J_1^1(0,\zeta) = \frac{1}{2(1+\zeta^2)} \,,$$

$$\frac{1}{\varrho} J_0^1(0,\zeta) = \frac{1}{4} \ln \left(1 + \frac{1}{\zeta^2}\right) .$$

The stresses on z-axis are thus

$$\frac{\sigma_r(0,\zeta)}{p_\mathrm{m}} = -\frac{1}{2}\left[\frac{1+2\nu}{2} \ln\left(1+\frac{1}{\zeta^2}\right) - \frac{1}{1+\zeta^2}\right] \,, \tag{4.243}$$

$$\frac{\sigma_\theta(0,\zeta)}{p_\mathrm{m}} = \frac{\sigma_r(0,\zeta)}{p_\mathrm{m}} \,,$$

$$\frac{\sigma_z(0,\zeta)}{p_\mathrm{m}} = -\frac{1}{2}\left[\ln\left(1+\frac{1}{\zeta^2}\right) + \frac{2}{1+\zeta^2}\right] \,, \tag{4.244}$$

$$\tau_{rz} = 0 \,.$$

The z-axis is a principal axis where the state of stress is triaxial ($\sigma_1 \equiv \sigma_r$, $\sigma_2 \equiv \sigma_\theta$, $\sigma_3 \equiv \sigma_z$) with two equal principal stresses. The stresses σ_r and σ_θ, compressive at the surface, become weak tensions at depth.

The hydrostatic stress is

$$\frac{\sigma_0(0,\zeta)}{p_\mathrm{m}} = -\frac{1+\nu}{3} \ln\left(1+\frac{1}{\zeta^2}\right) \,, \tag{4.245}$$

and the equivalent stress, for the von Mises criterion, is

$$\frac{\sigma_{eq}(0,\zeta)}{p_\mathrm{m}} = \frac{1}{2}\left[\frac{1-2\nu}{2} \ln\left(1+\frac{1}{\zeta^2}\right) + \frac{3}{1+\zeta^2}\right] . \tag{4.246}$$

These stresses are displayed in Fig. 4.60, and compared with those due to a concentrated force.

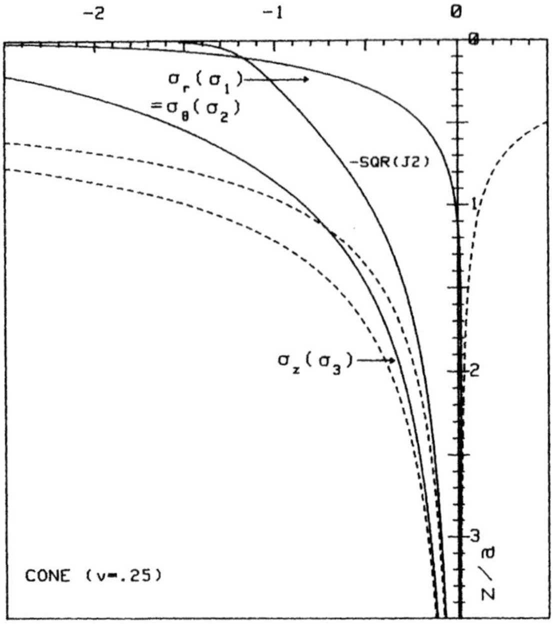

Fig. 4.60. Reduced stresses on the z-axis. (The *dashed line* is the Boussinesq concentrated load)

In the bulk. Figure 4.61 displays, for $\nu = 1/4$, the contours of principal stresses σ_1, σ_2 ($\equiv \sigma_\theta$) and σ_3 in a plane passing through the axis of symmetry. At the surface and on the axis, $\sigma_1 \equiv \sigma_r$ and $\sigma_3 \equiv \sigma_z$. The stress σ_3 is everywhere compressive. The stress σ_1, outside a compressive bulb-shaped zone below the area of contact, is a tension whose maximum value is reached in a thin layer just below the surface outside the contact. The hoop stress σ_2 becomes tensile at a depth of about $1.1a$, and never exceeds σ_1, although they are equal along the z-axis.

Figure 4.62 displays the hydrostatic stress σ_0, the von Mises criterion, and the stress trajectories of σ_1 and σ_3. At a large distance, the contours and stress trajectories for a Boussinesq concentrated force can of course be obtained.

4.8.4 Study of displacements

The vertical displacement $u_z(r,0)$ of the surface points is given by (4.222). The asymptotic behaviour for $r \to \infty$ is

$$u_z(r,0) \approx \frac{\delta}{\pi}\frac{a}{r} = \frac{1-\nu^2}{\pi E}\frac{P}{r},$$

which is, indeed, that of a concentrated force.

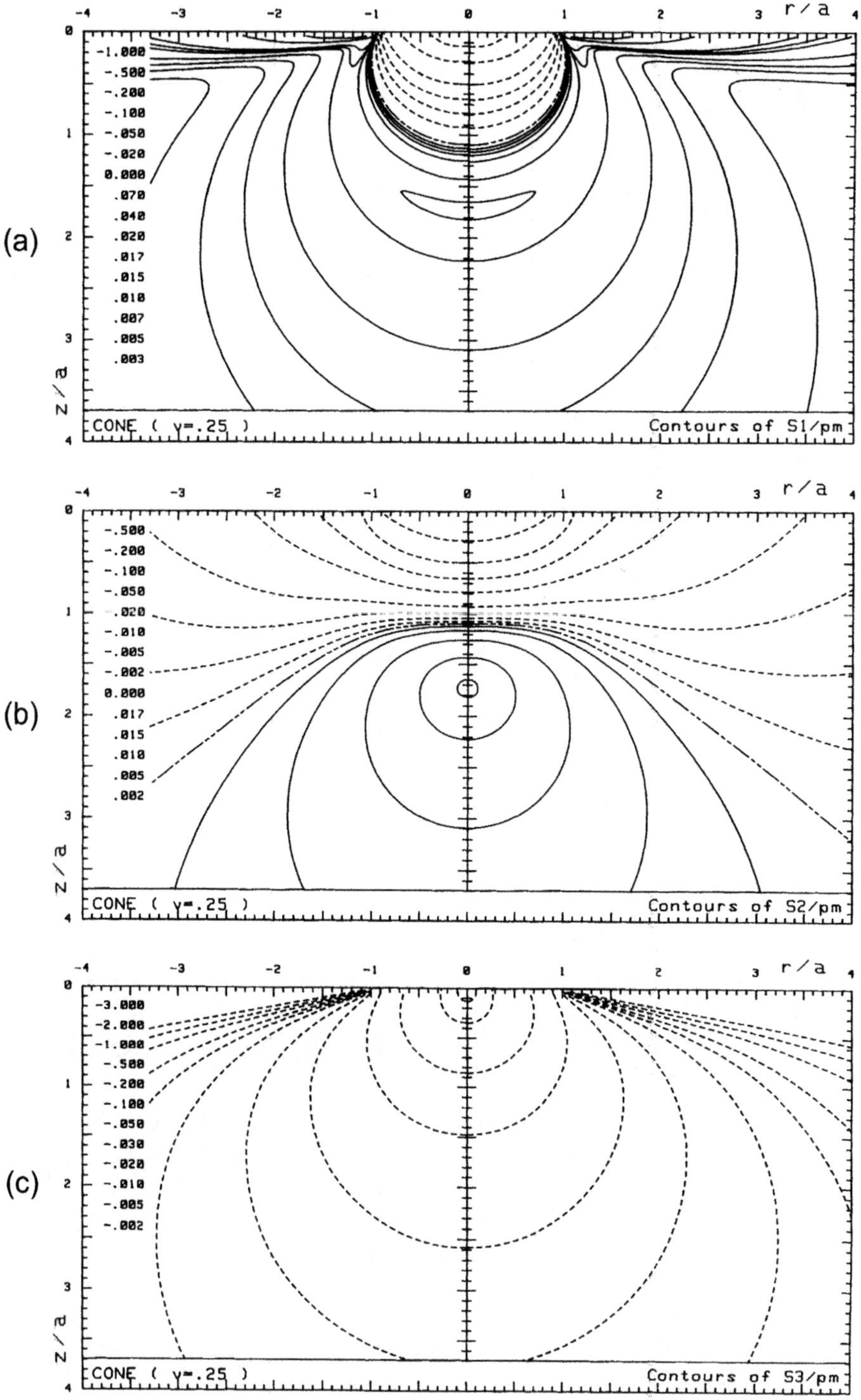

Fig. 4.61. Contours of principal stresses in the bulk: **(a)** σ_1; **(b)** σ_2; **(c)** σ_3

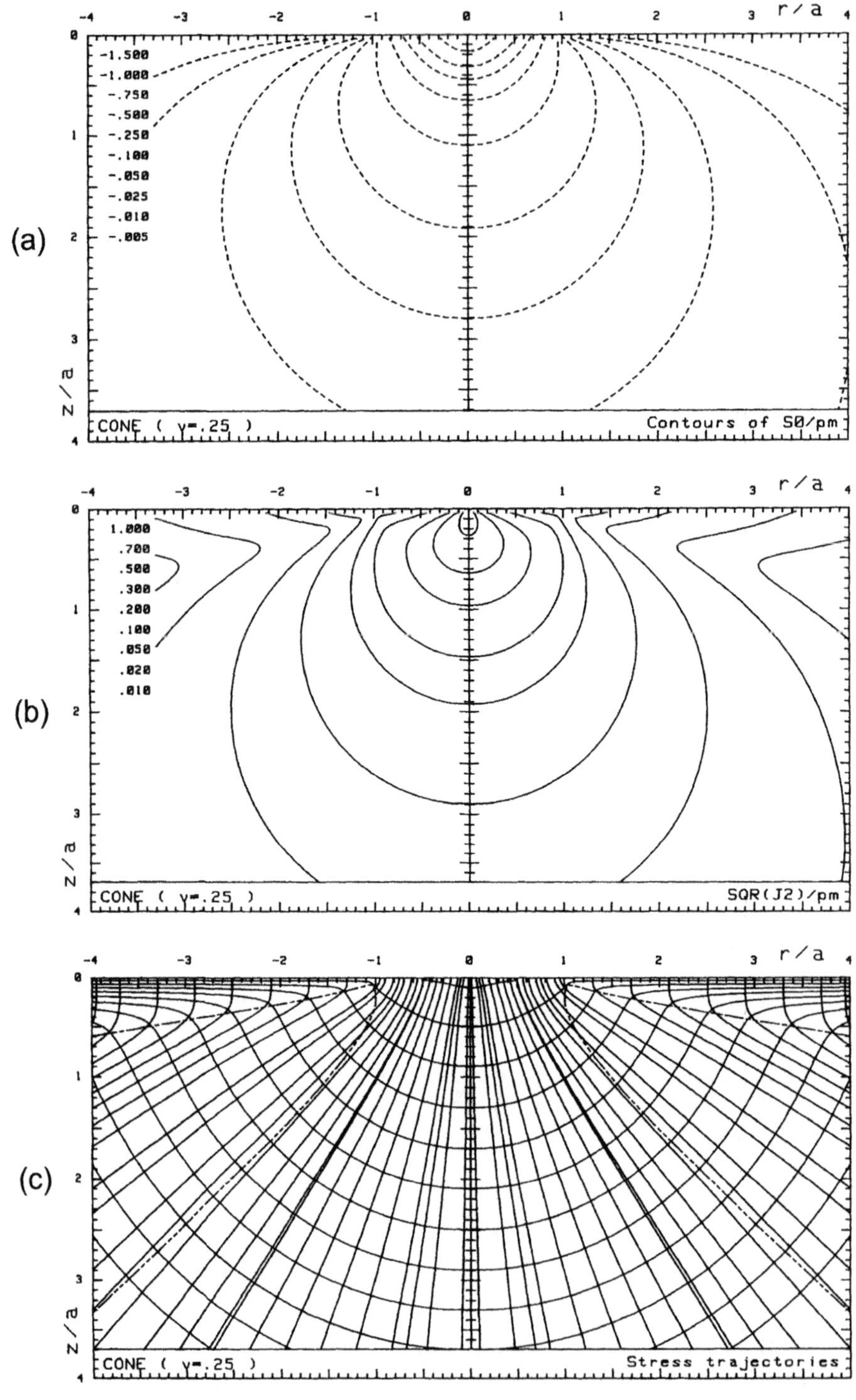

Fig. 4.62. Hydrostatic stress **(a)**, von Mises criterion **(b)**, and stress trajectories **(c)** in the bulk (the *dashed-dotted lines* being isoclines at 45°)

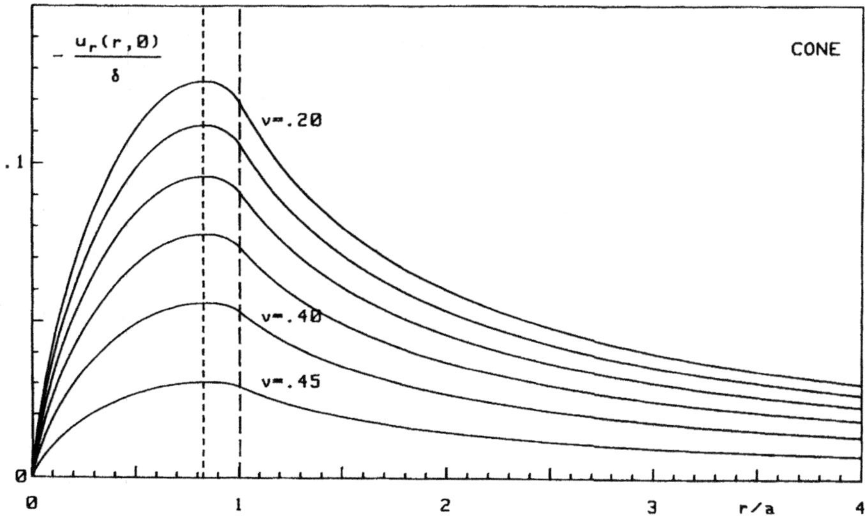

Fig. 4.63. Radial displacements in a surface for various values of ν

The radial displacement of surface points is given by (4.227), so that

$$u_r(r,0) = -\frac{(1+\nu)(1-2\nu)}{E}\frac{P}{2\pi r} \times \left[1 - \sqrt{1-\frac{r^2}{a^2}} + \frac{r^2}{a^2}\cosh^{-1}\frac{a}{r}\right], \qquad r \le a ,$$

$$u_r(r,0) = -\frac{(1+\nu)(1-2\nu)}{E}\frac{P}{2\pi r}, \qquad r \ge a . \tag{4.247}$$

The radial displacements outside the circle of contact are those of a concentrated force, in agreement with Way's theorem. Figure 4.63 displays the reduced displacements u_r/δ of the surface points for various values of ν. They exhibit a maximum, not on the edge of the contact, but at $r/a = 0.825$. The radial displacements in surface vanish for $\nu = 1/2$.

Figure 4.64 displays the contours of u_r/δ, for $\nu = 1/4$, as a function of r/a and z/a.

4.8.5 Adherence of a Conical Punch

As in the JKR theory for a sphere, it is possible to make a balance of elastic, potential and surface energies. The elastic energy is evaluated by first applying an equivalent load (see (4.220a)) given by

$$\boxed{P_1 = \frac{\pi E' \tan\beta}{2} a^2}, \tag{4.248}$$

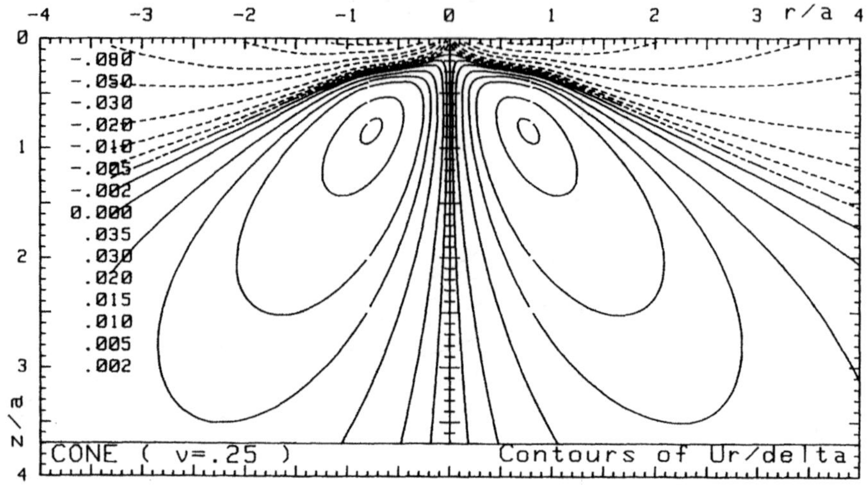

Fig. 4.64. Contours of reduced radial displacements

followed by unloading at constant radius of contact a. The elastic displacement during this unloading is given by (4.89) and the stored elastic energy becomes

$$U_{\mathrm{E}} = \frac{1}{4}\left(\frac{P^2}{aE'} + \frac{\pi^2 E' \tan^2\beta}{12} a^3\right) . \tag{4.249}$$

It is easier to make use of the general theorems [47].

The expression for P_1 given above corresponds, of course, to (4.26a) and we can deduce directly, from the general theorems, (4.23) and (4.27):

$$K_{\mathrm{I}} = -\frac{E'}{2}\sqrt{\frac{\pi}{a}}\chi(1) = \frac{P_1 - P}{\sqrt{4\pi a^3}} ,$$
$$G = \frac{K_{\mathrm{I}}^2}{2E'} ,$$

or, using (4.248) and the equation of state (4.216):

$$G = \frac{\left(\frac{\pi E'}{2}a^2 \tan\beta - P\right)^2}{8\pi E' a^3} \tag{4.250a}$$
$$= \frac{E'}{2\pi a}\left(\delta - \frac{\pi a}{2}\tan\beta\right)^2 . \tag{4.250b}$$

Note that the elastic energy given by (4.249), can be written in the form

$$U_{\mathrm{E}} = \frac{1}{3}(P\delta + 2GA) . \tag{4.251}$$

Equilibrium Equations. Equilibrium is given by $G = w$, from which the radius of contact and the penetration under zero load are:

$$a_0 = 32\frac{w}{\pi E' \tan^2 \beta} , \tag{4.252}$$

$$\delta_0 = 24\frac{w}{E' \tan \beta} , \tag{4.253}$$

which suggests the use of reduced parameters $x = P/P^*$, $y = a/a^*$, $z = \delta/\delta^*$, with

$$P^* = \frac{w^2}{\pi E' \tan^3 \beta} ,$$

$$a^* = \frac{w}{\pi E' \tan^2 \beta} ,$$

$$\delta^* = \frac{w}{E' \tan \beta} .$$

The equilibrium equations become

$$x = \frac{1}{2}\left(y^2 \pm \sqrt{32y^3}\right) , \tag{4.254}$$

$$z = \frac{1}{2}\left(y \pm \sqrt{8y}\right) . \tag{4.255}$$

Note that with these parameters, (4.219a) and (4.220a) reduce to

$$x = y^2/2 ,$$

$$z = y/2 .$$

These curves are displayed in Figs. 4.65 and 4.66. In Fig. 4.66 are superimposed the curves giving the variation of δ with a at constant load P, given by (4.216) which, in reduced coordinates, is

$$z = \frac{y}{4} + \frac{x}{2y} . \tag{4.256}$$

The equilibrium relation between load and penetration is displayed in Fig. 4.67, the minus sign being taken in (4.254) and (4.255).

Stability of Equilibrium and Adherence. The stability of equilibrium is given by the sign of the derivatives of G:

$$\left(\frac{\partial G}{\partial A}\right)_P = \frac{(P_1 - P)(P_1 + 3P)}{16\pi^2 E' a^5} , \tag{4.257}$$

$$\left(\frac{\partial G}{\partial A}\right)_\delta = -\frac{E'}{4\pi^2 a^3}\left(\delta - \frac{\pi a}{2}\tan\beta\right)\left(\delta + \frac{\pi a}{2}\tan\beta\right) . \tag{4.258}$$

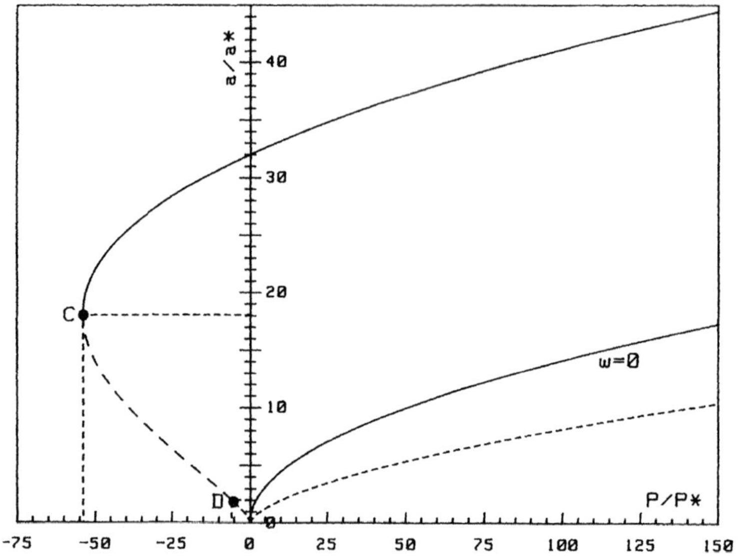

Fig. 4.65. Equilibrium radius of contact as a function of the applied load, in reduced coordinates. Point C corresponds to adherence at fixed load, and point D to adherence at fixed grips

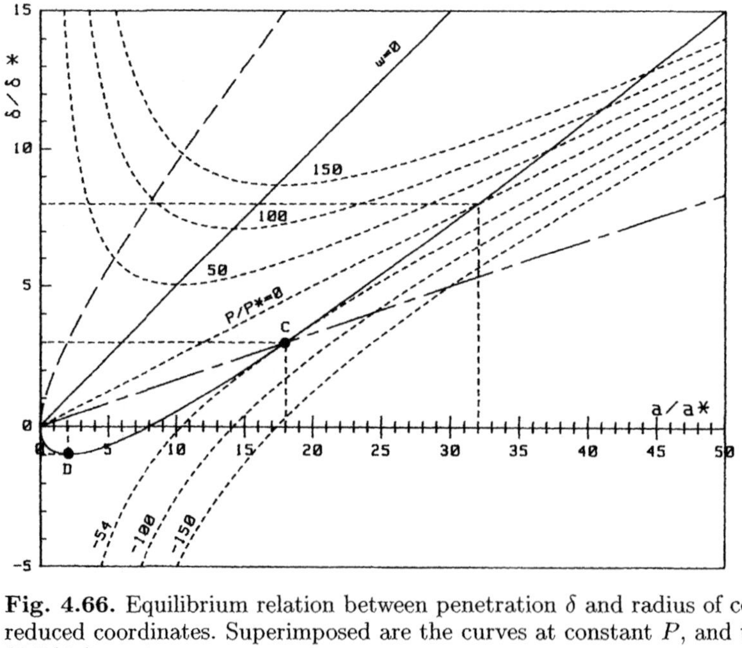

Fig. 4.66. Equilibrium relation between penetration δ and radius of contact a, in reduced coordinates. Superimposed are the curves at constant P, and the locus of $(\partial G/\partial a)_P = 0$

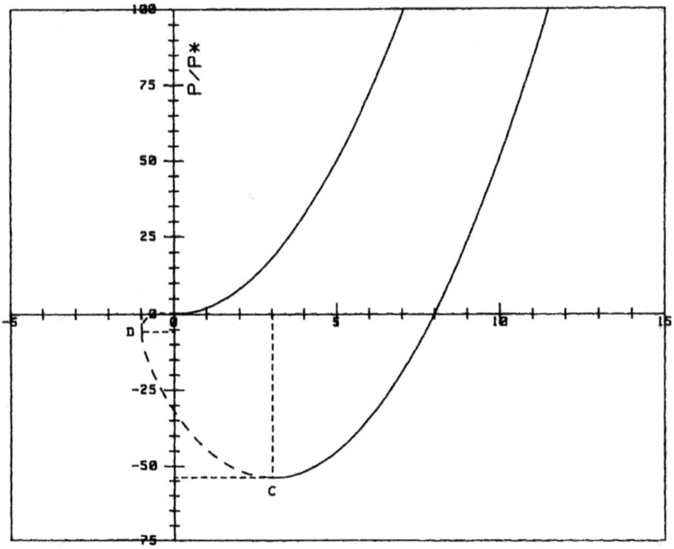

Fig. 4.67. Equilibrium relation between load P and penetration δ

Stable zones are on the branches with the minus sign chosen in (4.254) and (4.255).

The adherence force at fixed load is given by $(\partial G/\partial a)_P = 0$. It is the point C with a vertical tangent in Fig. 4.65 and a horizontal tangent in Fig. 4.67. At this point we have

$$P_c/P^* = -54\,, \tag{4.259a}$$
$$a_c/a^* = 18\,, \tag{4.259b}$$
$$\delta_c/\delta^* = 3\,. \tag{4.259c}$$

The adherence force at fixed grips is given by $(\partial G/\partial a)_\delta = 0$. It is the point D with a horizontal tangent in Fig. 4.66 and a vertical tangent in Fig. 4.67. At this point we have

$$P_c/P^* = -6\,, \tag{4.260a}$$
$$a_c/a^* = 2\,, \tag{4.260b}$$
$$\delta_c/\delta^* = -1\,. \tag{4.260c}$$

4.8.6 Impact and Rebound of a Cone

The problem of the impact of a rigid cone of mass m and velocity V on an elastic half-space (modulus E and ν) was studied by Graham [48] in 1973.

It is easily modelled upon that of the Hertzian impact (Sect. 4.4.7), with an elastic energy U_E given by (4.223). The law of energy conservation is thus written:

$$\frac{1}{2}m\left(\frac{\mathrm{d}\delta}{\mathrm{d}t}\right)^2+\frac{2E'}{3\pi\tan\beta}\delta^3=\frac{1}{2}mV^2\,. \tag{4.261}$$

The maximum approach, the maximum radius of contact and the maximum force are reached when the velocity $\dot{\delta}$ vanishes. They are given by

$$\delta_\mathrm{max}=\left[\frac{3\pi\tan\beta}{4E'}mV^2\right]^{1/3}, \tag{4.262}$$

$$a_\mathrm{max}=\left[\frac{6}{\pi^2E'\tan^2\beta}mV^2\right]^{1/3}, \tag{4.263}$$

$$P_\mathrm{max}=\left[\frac{9E'}{2\pi\tan\beta}m^2V^4\right]^{1/3}. \tag{4.264}$$

The duration of the impact is obtained by integration of (4.261), written in the form

$$\mathrm{d}t=\frac{\mathrm{d}\delta}{\sqrt{V^2-\frac{4E'\delta^3}{3\pi m\tan\beta}}}\,. \tag{4.265}$$

Letting

$$x=\frac{\delta}{\delta_\mathrm{max}}=\frac{a}{a_\mathrm{max}}\,,$$

the duration of the compression is given by

$$t_\mathrm{c}=\frac{\delta_\mathrm{max}}{V}\int_0^1\frac{\mathrm{d}x}{\sqrt{1-x^3}}\,. \tag{4.266}$$

The duration of the impact is twice the duration of compression and is

$$T_\mathrm{c}=\frac{2\sqrt{\pi}\,\Gamma(1/3)}{3\;\;\Gamma(5/6)}\frac{\delta_\mathrm{max}}{V}=2.804\frac{\delta_\mathrm{max}}{V}\,,$$

i.e.

$$T_\mathrm{c}=3.732\left(\frac{m\tan\beta}{E'}\frac{1}{V}\right)^{1/3}. \tag{4.267}$$

The duration of the impact is thus inversely proportional to $V^{1/3}$. The evolution of δ with time is obtained by numerical integration of (4.265), which can also be written, with $\tau=t/T_\mathrm{c}$, as

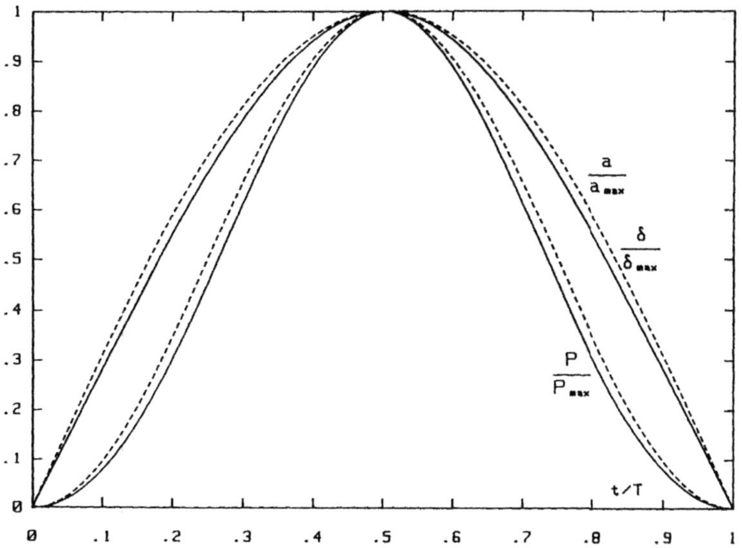

Fig. 4.68. Evolution of the force, of the penetration and of the radius of contact during the rebound of a cone. The *dashed lines* are the sinusoidal approximation

$$\frac{\mathrm{d}x}{\mathrm{d}\tau} = \pm 2.804\sqrt{1 - x^3}\,. \tag{4.268}$$

Graham [48] has shown that the sinusoidal approximation

$$\frac{a}{a_{\max}} = \frac{\delta}{\delta_{\max}} = \sin\left(\frac{\pi t}{T_{\mathrm{c}}}\right)$$

is as good as for the sphere. Figure 4.68 displays the evolution of the penetration and of the radius of contact with time. The evolution of the force, given by

$$\frac{P}{P_{\max}} = \left(\frac{\delta}{\delta_{\max}}\right)^2$$

is also displayed. The sinusoidal approximations are shown as the dashed line.

4.9 Liquid Bridges

Adherence between elastic solids can be masked by capillary forces which can be of the same order of magnitude. The exact calculation of these capillary forces can be complex, but is simplified for small volumes of liquid, in which case it is possible to use a circular arc approximation for the meniscus. This force is due to both the Laplace pressure and to the tension exerted by the liquid [49, 50], but as the second term is negligible for small volumes of liquid it is interesting to evaluate this adherence by fracture mechanics, using the Dugdale model with the Laplace pressure $\sigma_0 = \gamma/\varrho$ (where ϱ is the radius of curvature of the meniscus) as pressure acting at the tip of a circular external crack of opening $\delta_t = 2\varrho\cos\theta$ where θ is the contact angle. From (3.163) we have thus

$$J = G = 2\gamma\cos\theta\,. \tag{4.269}$$

Let us take a rigid axisymmetric punch of profile $f(r)$, and let b be the wetted radius and z the separation between the punch and the rigid plane. We can compute the energy release rate G from the potential energy $U_\mathrm{P} = -Pz$ as

$$G = \left(\frac{\partial U_\mathrm{P}}{\partial A}\right)_P = -P\frac{\mathrm{d}z}{2\pi b\,\mathrm{d}b}\,.$$

As a first approximation let us consider the liquid bridge as a cylinder of constant volume V, so that

$$V = \pi z b^2(z) + \int_0^{b(z)} 2\pi r f(r)\,\mathrm{d}r\,.$$

The condition $\mathrm{d}V/\mathrm{d}z = 0$ gives $\mathrm{d}z/\mathrm{d}b$, and hence the result

$$\boxed{G = \frac{P}{\pi b^2}[z + f(b)]}\,. \tag{4.270}$$

Knowing the profile $f(r)$ of the punch, we can obtain the adherence force by equating (4.269) and (4.270). So, for a flat punch, with $f(r) = 0$, we obtain the classical result

$$P = \frac{2\pi b^2\gamma\cos\theta}{z}\,. \tag{4.271}$$

For a sphere, with $f(r) = r^2/2R$,

$$P = \frac{4\pi\gamma R\cos\theta}{1 + 2zR/b^2}\,, \tag{4.272}$$

a result given by Israelachvili [51].

For a cone of semi-angle β, with $f(r) = r/\tan\beta$,

$$P = \frac{2\pi b\gamma\cos\theta\tan\beta}{1 + z\tan\beta/b}\,. \tag{4.273}$$

This last equation is an approximation for large β of an expression given by Coughlin *et al.* [52] for the case $z = 0$.

For a meniscus between two spheres of radius R, evaluation of the volume leads to

$$P = \frac{2\pi\gamma R\cos\theta}{1 + zR/b^2} \,. \tag{4.274}$$

As experiments are generally made at constant volume of the meniscus, it is desirable to introduce the volume V into above equations, with $V = \pi b^2 z$ for two parallel plates, $V = \pi b^2 z + \pi b^4/4R$ for a sphere and a plate, and $V = \pi b^2 z + \pi b^4/2R$ for two spheres. Equations (4.271), (4.272) and (4.274) then become

$$P = \frac{2V\gamma\cos\theta}{z^2} \qquad \text{(two plates)}\,, \tag{4.275}$$

$$P = 4\pi R\gamma\cos\theta\left[1 - \frac{1}{\sqrt{1+\frac{V}{\pi R z^2}}}\right] \qquad \text{(sphere/plane)}\,, \tag{4.276}$$

$$P = 2\pi R\gamma\cos\theta\left[1 - \frac{1}{\sqrt{1+\frac{2V}{\pi R z^2}}}\right] \qquad \text{(sphere/sphere)}\,, \tag{4.277}$$

which are decreasing functions of z. Figure 4.69 displays the equilibrium curves $P(z)$ for a sphere and a plane.

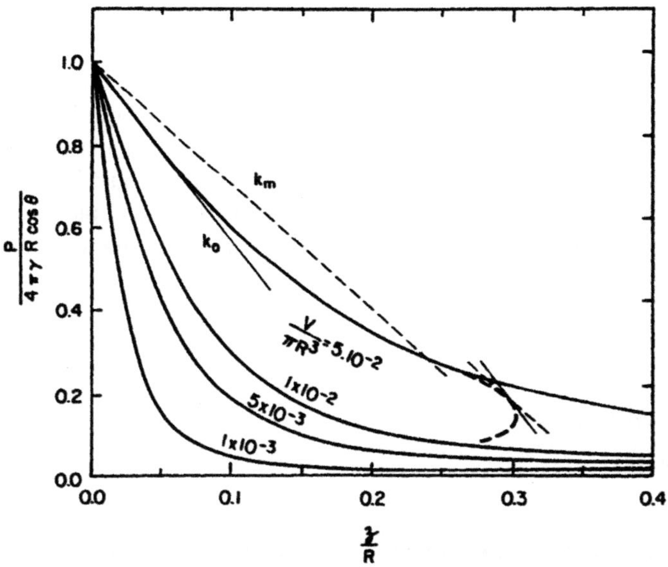

Fig. 4.69. Equilibrium curves for force–displacement for a liquid bridge between a sphere and a plane, and for various values of the reduced volume $V/\pi R^3$ of the liquid bridge. Stable equilibrium can only be observed for a stiffness k_m of the measuring apparatus larger than the initial slope $-\mathrm{d}P/\mathrm{d}z$

The stability of equilibrium depends on the stiffness k_m of the measuring apparatus, and can been evaluated either from the sign of $(\partial G/\partial A)_\Delta$, or from (3.19). This last method is more direct and shows that equilibrium is stable if $k_m > -\mathrm{d}P/\mathrm{d}z$. For a liquid bridge between a sphere and a plane, for example, the initial slope is:

$$\left(\frac{\mathrm{d}P}{\mathrm{d}z}\right)_{z=0} = -k_0 = -4\pi\gamma\cos\theta\sqrt{\frac{\pi R^3}{V}}\,.$$

It is only for $k_m > k_0$ that the equilibrium curve can be followed, eventually up to the point where again $\mathrm{d}P/\mathrm{d}z = -k_m$, as for the curve drawn with a dashed line. This shape arises from the fact that, for a given volume, there are two solutions for the gorge radius [53–56]. The criterion for instability at fixed grips is the point where the two solutions converge [56]: the point with a vertical tangent on the curve $P(z)$. If $k_m < k_0$ the equilibrium at fixed Δ is unstable for $z = 0$, and a jump occurs along the straight line

$$P = 4\pi\gamma R\cos\theta - k_m\delta$$

until the point where it again intersects the equilibrium curve, which is then followed until rupture. If this straight line does not intersect the equilibrium curve, the adherence force is simply $P = 4\pi\gamma R\cos\theta$. Some experimenters have observed this instability at $z = 0$ [57–59], others have found extended equilibrium curves [60, 61]. When one studies the contact of spheres with a meniscus, one can expect two jumps: the first at the limit of stability of the solid/solid contact, and the second at the limit of stability of the liquid bridge.

We can note that in this cylindrical approximation for the liquid bridge, the adherence force at $z = 0$ is independent of the volume of the liquid bridge. In fact, a circular approximation for the meniscus or an exact calculation [62] shows that it decreases with volume. In such experiments, care must be taken to avoid viscosity effects. For a liquid bridge between two parallel plates, for example, Stefan's law modified by Healey [63] tells us that the time to remove two plates from z_1 to z_2 with a force F is

$$t = \frac{3\pi\eta b^4}{8F}\left(\frac{1}{z_1^2} - \frac{1}{z_2^2}\right),$$

i.e. that the viscous force is

$$F = \frac{3\eta V^2}{2\pi z^5}\frac{\mathrm{d}z}{\mathrm{d}t}\,.$$

For a liquid bridge between two spheres, the viscous force was given by Matthewson [64] in the form

$$F = 6\pi\eta R^2\left(1 - \frac{2zR}{b^2}\right)^2\frac{\mathrm{d}z}{z\,\mathrm{d}t}\,.$$

The influence of the viscosity on the adherence due to liquid bridges has been clearly shown by McFarlane and Tabor [57] and can be suspected in experiments by Mason and Clark [60] on oil bridges.

4.10 Contact and Adherence of Rough Elastic Solids

4.10.1 Description of a Rough Surface

Let us consider in Fig. 4.70 the surface of a solid defined by contour lines. We are principally interested by its *summits*, the points of its surface higher than its immediate neighbours and at which or near which contacts will establish. To characterize this surface, its profile[8] $f(x)$ is plotted with respect to a reference plane, along parallel lines of length L. A series of *peaks* and *valleys* is obtained (see Fig. 4.71).

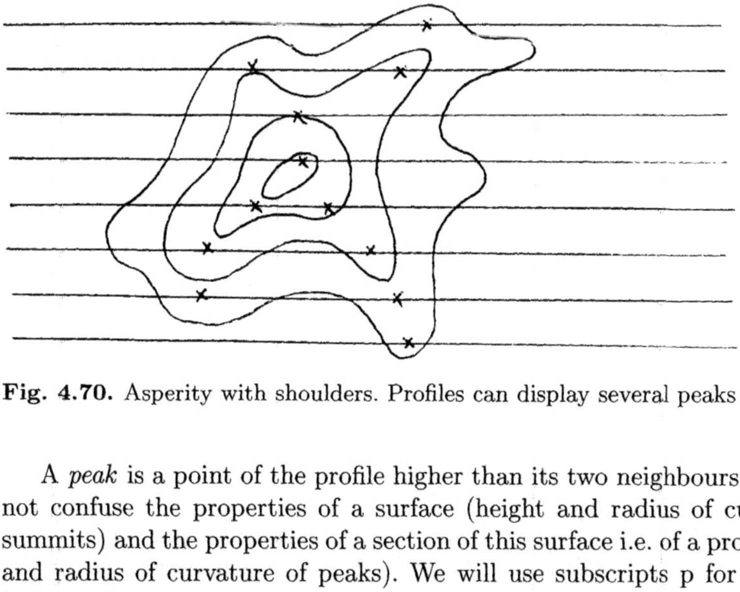

Fig. 4.70. Asperity with shoulders. Profiles can display several peaks

A *peak* is a point of the profile higher than its two neighbours. One must not confuse the properties of a surface (height and radius of curvature of summits) and the properties of a section of this surface i.e. of a profile (height and radius of curvature of peaks). We will use subscripts p for peaks and indices s for summits.

A section has little chance to pass through a summit, and Fig. 4.70 shows that the number of detected peaks near an asperity depends on the shape of this asperity, and corresponds to the various shoulders encountered. This obvious difference, neglected in the early work, was clearly emphasized by Nayak [65] in 1971. Note furthermore that characterization of a rough surface poses the same problems as the measurement of the length of the Brittany

[8] These profiles are often displayed with a ratio of about 100 between the vertical and horizontal scales. In reality, mean slopes of profiles and surfaces are weak, of the order of 10°

coast: this length increases when the length of the standard of measure (the sampling interval) decreases. A rough surface can be seen as a fractal object.

Vertical Distribution of a Profile. A profile $f(x)$ being drawn, a datum line or mean line can be defined as $\overline{f}(x)$ (a straight line if the solid is a plane, or a circular arc if the solid is rounded), where the mean square deviation with respect to that line, given by

$$\sigma^2 = \frac{1}{L}\int_0^L (f - \overline{f})^2 \, \mathrm{d}x \,, \tag{4.278}$$

is a minimum. We then have

$$\int_0^L (f - \overline{f}) \, \mathrm{d}x = 0 \,, \tag{4.279}$$

and the areas situated above and below this line are equal. Let us prove this result for a straight line of ordinate $\overline{f}$. Let

$$g = \overline{f} + \lambda \,,$$

and consider the integral

$$F(\lambda) = \int_0^L [f(x) - g]^2 \, \mathrm{d}x \,.$$

We want this integral to be at a minimum for $\lambda = 0$, i.e. $F'(0) = 0$. So we have

$$F'(\lambda) = -2\int_0^L [f(x) - \overline{f} - \lambda] \, \mathrm{d}x \,,$$

$$F'(0) = -2\int_0^L [f(x) - \overline{f}] \, \mathrm{d}x \,.$$

The condition $F'(0) = 0$ is thus (4.279).

The heights will be measured with respect to this line and denoted $z(x)$. The *average roughness* R_a, also named CLA (centre line average), is defined by

$$R_a = \frac{1}{L}\int_0^L |z| \, \mathrm{d}x \,, \tag{4.280}$$

and the *root mean square roughness* σ is defined by

$$\sigma^2 = \frac{1}{L}\int_0^L z^2 \, \mathrm{d}x \,, \tag{4.281}$$

where σ^2 is named the *variance* and σ the *standard deviation.*

If we denote by $\varphi(z)\,\mathrm{d}z$ the probability for a point of the surface to have a height between z and $z+\mathrm{d}z$, (φ being a density of probability), the average roughness and the r.m.s. roughness are given by

$$R_{\mathrm{a}} = \int_{-\infty}^{+\infty} |z|\,\varphi(z)\,\mathrm{d}z = 2\int_{0}^{+\infty} z\varphi(z)\,\mathrm{d}z\ , \tag{4.282}$$

$$\sigma^2 = \int_{-\infty}^{+\infty} z^2\varphi(z)\,\mathrm{d}z\ . \tag{4.283}$$

If we refer to the linear distribution of a unit of mass, $\varphi(z)$ is the linear density, $z = 0$ is the position of the centre of gravity, and σ is the radius of gyration. One is led to define central moments of order r as

$$\overline{z^r} = \int_{-\infty}^{+\infty} z^r\,\varphi(z)\,\mathrm{d}z \tag{4.284}$$

and reduced (non-dimensional) central moments as $\overline{z^r}/\sigma^r$. The first central moment gives the mean and is zero, the second central moment is the variance, the third reduced central moment, denoted S_q (for skewness), describes the asymmetry and is zero for a symmetrical distribution (Gaussian, for example), and the fourth reduced central moment, denoted E_{k}, describes the shape (the kurtosis), either flattened or sharp, and has the value 3 for a Gaussian shape (cf Peklenik [66]).

The probability for a point of the profile having a height larger than z is given by the *cumulative height distribution*

$$\Phi(z) = \int_{z}^{\infty} \varphi(u)\,\mathrm{d}u\ , \tag{4.285}$$

which varies from 0 to 1 when z varies from $+\infty$ to $-\infty$. This S-shaped curve, seen in Fig. 4.71, is named the *Abbott bearing area curve* (after Abbott and Firestone, 1933). If the profile is cut in slices parallel to the mean line, the

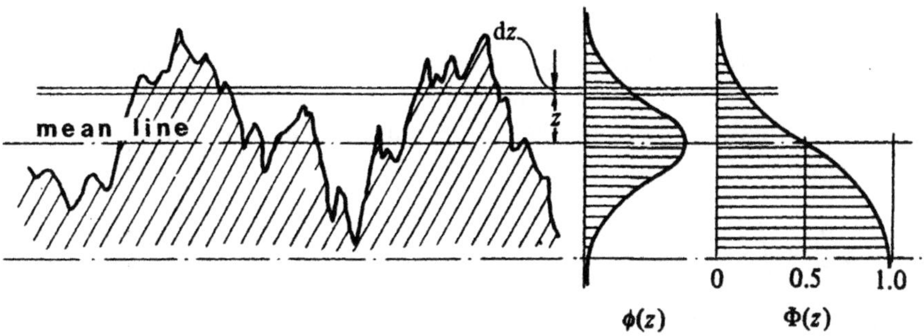

Fig. 4.71. Heights density of probability $\phi(z)$ and cumulative height distribution $\Phi(z)$

length of plateaux which appear as the slices are removed, follows this curve. Abbott thought that wear follows this pattern and that it slows down as the area of plateaux increases, implying that the variation of plateau lengths on the profile and the variation of plateau areas on the surface follow the same relation, which is not evident.

Besides the cumulative height distribution, the cumulative peak height distribution $\Phi_{\mathrm{p}}(z_{\mathrm{p}})$, of density $\varphi_{\mathrm{p}}(z_{\mathrm{p}})$ can also be studied. The mean peak height will be

$$\overline{z}_{\mathrm{p}} = \int_{-\infty}^{+\infty} z_{\mathrm{p}}\varphi_{\mathrm{p}}(z_{\mathrm{p}})\,\mathrm{d}z_{\mathrm{p}} \tag{4.286}$$

and the variance

$$\sigma_{\mathrm{p}}^2 = \int_{-\infty}^{+\infty} (z_{\mathrm{p}} - \overline{z}_{\mathrm{p}})^2\,\varphi_{\mathrm{p}}(z_{\mathrm{p}})\,\mathrm{d}z_{\mathrm{p}}\ . \tag{4.287}$$

This mean peak height $\overline{z}_{\mathrm{p}}$ is generally positive with respect to the mean line.

Greenwood and Williamson [67] have shown experimentally that the height distribution of a number of technical surfaces was near to Gaussian, and that the peak height distribution was also Gaussian but with a different mean value and a different standard deviation. By microtopography [68] it was also shown that the distribution of summit heights is also Gaussian.

Horizontal Distribution of a Profile. Two profiles can have the same height and peak height distributions but differ in the horizontal extension. A simple means of characterizing this horizontal distribution is to compute the root mean square slope σ_{m} and the root mean square curvature σ_{κ}. However, these two measures are first and second derivatives of the profile and are very sensitive to the details of the profile, i.e. to the precision with which the profile was measured, and increase when this precision increases. At the limit one would obtain a continuous curve having derivative nowhere!

The autocorrelation function and the power spectral density borrowed from the theory of random processes can be used in preferences. Consider a pair of points $z(x)$ and $z(x + \lambda)$ situated on the profile and separated by a distance λ. Then form the product $z(x)z(x + \lambda)$. Treat λ as the length of a rule which can slide along the profile. Let it move by $\mathrm{d}x$; then we define a new pair of points $z(x+ \mathrm{d}x)$ and $z(x+ \mathrm{d}x+\lambda)$, giving another product. By repeating this process along the whole available length $L - \lambda$, we can define a function

$$\frac{1}{L-\lambda}\int_0^{L-\lambda} z(x)z(x+\lambda)\,\mathrm{d}x\ ,$$

which for $L \gg \lambda$ becomes

$$R_z(\lambda) = \frac{1}{L}\int_0^{L} z(x)z(x+\lambda)\,\mathrm{d}x\ . \tag{4.288}$$

This function is termed the *covariance* or the *autocorrelation function*, and it reduces to the variance σ^2 when λ is zero.

Sometimes the name "autocorrelation function" is reserved for the normalized covariance

$$\varrho_z(\lambda) = \frac{R_z(\lambda)}{\sigma^2} , \tag{4.289}$$

which is thus 1 at $\lambda = 0$. When λ is small compared with the width of a peak the product $z(x)z(x+\lambda)$ is finite and positive. If λ is large compared with the mean width of peaks, this product can be positive or negative and the net contribution of these terms to the integral is negligible. Plotted as a function of λ, the autocorrelation function varies from 1 to zero (but not necessarily in a monotonic manner). Its mean rate of decrease characterizes the horizontal spreading of the roughness. If the profile contains a periodic signal of period λ_0 more or less masked by noise, it will be clearly revealed by oscillations of the same period in the autocorrelation function.

It is often interesting to represent a profile as a range of frequencies. The *power spectral density* is the Fourier transform of the autocorrelation function:

$$G_z(f) = \frac{2}{\pi} \int_0^\infty R_z(\lambda) \cos(2\pi f \lambda) \, \mathrm{d}\lambda . \tag{4.290}$$

In optics this function represents the energy carried at a frequency f. A profile having a periodicity hidden in the noise will display a peak in power spectral density at this frequency (measured in cycles per unit length).

Peaks and Summits: Theory of Random Processes. It is not easy to pass from the characteristics of a profile to the characteristics of a surface. Nayak [65] has introduced into surface topography the theory of random processes. He showed that for isotropic random surfaces whose height distribution is Gaussian (with standard deviation σ) all the characteristics of the surface are contained in the power spectral density of the profile. The main results of this theory (see also Greenwood [69, 70]) are the following:

1. If $\eta_\mathrm{p}$is the density of peaks per unit length, the density of summits per unit area is given by

$$\eta_\mathrm{s} = \frac{2\pi}{3\sqrt{3}} {\eta_\mathrm{p}}^2 = 1.209\eta_\mathrm{p}^2 . \tag{4.291}$$

2. If $\overline{z}_\mathrm{p}$ is the mean height of peaks, the mean height of summits is given by

$$\overline{z}_\mathrm{s} = \frac{4\sqrt{2}}{\pi} \overline{z}_\mathrm{p} = 1.8\overline{z}_\mathrm{p} , \tag{4.292}$$

with a standard deviation nearly equal to σ.

3. If $\overline{\kappa}_{\mathrm{p}}$ is the mean curvature of peaks, the mean curvature of summits is given by

$$\overline{\kappa}_{\mathrm{s}} = \frac{8\sqrt{2}}{3\pi}\overline{\kappa}_{\mathrm{p}} = 1.20\overline{\kappa}_{\mathrm{p}} \,. \tag{4.293}$$

4. The higher the peaks and summits, the smaller their radius of curvature.

4.10.2 Models of Rough Surfaces

Model of Greenwood and Williamson. In the Greenwood and Williamson [67] model, the surface is uniformly covered by asperities (density η_{s}) whose summits have all the same radius of curvature R, but with a random height. Summits have a mean height $\overline{z}_{\mathrm{s}}$ above the mean line previously defined. We will take as reference the mean plane of summit height. The distribution of the height z_{s} of the summits, *measured with respect to this mean plane of summits*, is Gaussian, with a standard deviation σ_{s}, given by

$$\varphi_{\mathrm{s}}(z_{\mathrm{s}}) = \frac{1}{\sigma_{\mathrm{s}}\sqrt{2\pi}} \exp\left(-\frac{z_{\mathrm{s}}^2}{2{\sigma_{\mathrm{s}}}^2}\right) \,. \tag{4.294}$$

The expression $\varphi_{\mathrm{s}}(z_{\mathrm{s}})\,\mathrm{d}z_{\mathrm{s}}$ is thus the probability for a summit to have a height between z_{s} and $z_{\mathrm{s}} + \mathrm{d}z_{\mathrm{s}}$ above the summits mean plane (Fig. 4.72).

In fact parameters σ_{s}, R and η_{s} are not independent, and Archard [71] has shown that they are linked by the relation

$$0.03 < \sigma_{\mathrm{s}} R \eta_{\mathrm{s}} < 0.05 \,. \tag{4.295}$$

A more precise relation has been given by McCool [72] in the context of random processes.

Model of Whitehouse and Archard. Whitehouse and Archard [73] have proposed a model of *profile* where the height distribution is Gaussian with a standard deviation σ and where the horizontal spread is defined by an exponential autocorrelation function

$$R_z(\lambda) = \sigma^2 \mathrm{e}^{-\lambda/\lambda^*} \,, \tag{4.296}$$

where λ^* is the correlation length. When $\lambda > 2.3\lambda^*$ the (normalized) autocorrelation function is less than 0.1, and one can consider that peaks separated by this distance are no longer correlated. A sampling interval $\lambda = 2.3\lambda^*$ will thus reveal the main structure of the profile. In this model the radii of curvature of peaks have a distribution which depends of their height: the higher the peaks, the lower their radius of curvature.

Defining, for a given mesh, a summit as a point higher that its four neighbours, Whitehouse and Archard [73] showed that the number of summits per unit area, η_{s}, their mean radius of curvature R, and the standard deviation of their height distribution σ^* are given by:

$$\eta_s = \frac{1}{5}\left(\frac{1}{2.3\lambda^*}\right)^2 , \tag{4.297}$$

$$R \propto \frac{\lambda^{*2}}{\sigma} , \tag{4.298}$$

$$\sigma^* \propto \sigma . \tag{4.299}$$

Two lengths are thus sufficient to define a surface: the standard deviation of height σ and the correlation length λ^*. These three equations give rise to the relation $\sigma^* R\eta_s =$ const. between the three quantities used by Greenwood and Williamson [67], in accordance with (4.295).

4.10.3 Contact and Adherence of Two Rough Planes

The problem of contact of rough surfaces is fundamental in the study of electrical contacts, thermal conductance, adhesion and friction. When two rough surfaces are in contact, they touch on higher asperities only, and the real area of contact is made through an archipelago of contact islands on which the pressure is very high. It was Holm in 1946 who introduced the idea that, despite weak nominal pressures, asperities could undergo large stresses exceeding the elastic limit. If asperities are plastically deformed as in a Brinell hardness test, they are subjected to a constant pressure equal to the hardness H of the material. The real area of contact is thus proportional to the applied load, i.e. $A = P/H$, and independent of the roughness. However, Archard [74] pointed out in 1957 that plastic deformation cannot be indefinitely repeated, for the geometry of the contact progressively changes until the load is elastically supported by asperities. Using a model where every roughness is covered with microroughnesses and every microroughness covered with nanoroughnesses (in other words, a fractal surface), Archard showed that for purely elastic contacts the real area of contact could also increase linearly with the load, a result due to the fact that the number of contacts increases with the load. Later on, more realistic surface models were used, with the same result that, for rough surfaces in elastic contact, the real area of contact increases more or less linearly with the load.

Hertzian Contacts: the Theory of Greenwood and Williamson. Let us consider the contact of two rough planes of nominal area A_0 and of rms roughness σ_1 and σ_2. The system can be replaced by a rigid smooth plane in contact with a rough plane (Fig. 4.72) of rms roughness $\sigma = \sqrt{\sigma_1^2 + \sigma_2^2}$, and of effective Young's modulus E^* defined by

$$\frac{1}{E^*} = \frac{1-\nu_1^2}{E_1} + \frac{1-\nu_2^2}{E_2} .$$

Let us take as reference the mean plane of summits situated at a height $\bar{z}_s$ above the mean line and let $\varphi_s(z_s)\,dz_s$ be the probability for a summit to

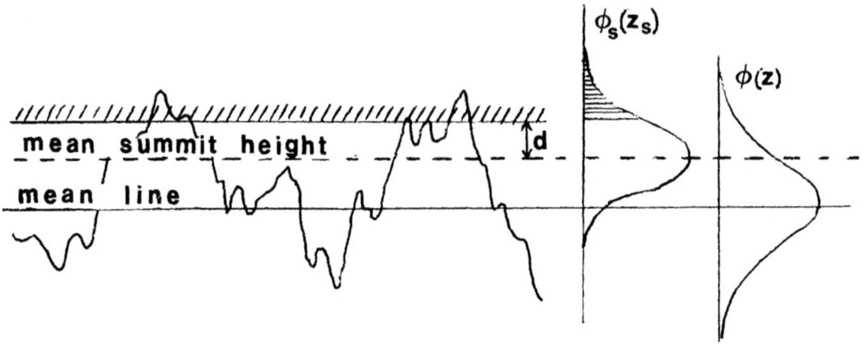

Fig. 4.72. Contact of a rough surface and a smooth flat surface. $\varphi(z)$ is the heights probability density, and $\varphi_s(z_s)$ is the height of the summits probability density

have its height between z_s and $z_s + \mathrm{d}z_s$ above this mean line. (In the following discussion we will write, for simplicity, z instead of z_s.) The distance d between the planes will be the distance between this mean plane of summits and the smooth plane. When this distance decreases, Hertzian contacts occur on the higher summits. The probability for a summit to be in contact is the probability for its height z be larger than d, so that

$$\Phi_s(d) = \int_d^\infty \varphi_s(z)\,\mathrm{d}z\ . \tag{4.300}$$

If there are N summits in a nominal area A_0 (so that $\eta_s = N/A_0$ is the number of summits per unit area), the number of summits in contact when the separation is d is

$$n = N \int_d^\infty \varphi_s(z)\,\mathrm{d}z\ . \tag{4.301}$$

Summits whose height is larger than d are flattened elastically by a quantity $\delta = z - d$. If the contact is Hertzian, the area of the ith contact and the load it carries are

$$A_i = \pi a_i^2 = \pi\delta_i R_i = \pi R_i(z_i - d)\ , \tag{4.302}$$

$$P_i = K R_i^{1/2}\delta_i^{3/2} = K R_i^{1/2}(z_i - d)^{3/2}\ , \tag{4.303}$$

where $K = 4E^*/3$. The total real area and the total load are obtained by summing (4.302) and (4.303) for all the asperities in contact:

$$A = N \int_d^\infty A_i \varphi_s(z)\,\mathrm{d}z\ , \tag{4.304}$$

$$P = N \int_d^\infty P_i \varphi_s(z)\,\mathrm{d}z\ . \tag{4.305}$$

Gaussian Distribution of Summit Heights. For a Gaussian distribution of summits, as per (4.294) and assuming that all summits have the radius of curvature R, we get

$$n = \frac{N}{\sqrt{2\pi}} \int_{d/\sigma_s}^{\infty} \exp\left(-\frac{z^2}{2{\sigma_s}^2}\right) \mathrm{d}\left(\frac{z}{\sigma_s}\right) ,$$

$$A = \frac{N\pi R\sigma_s}{\sqrt{2\pi}} \int_{d/\sigma_s}^{\infty} \left(\frac{z-d}{\sigma_s}\right) \exp\left(-\frac{z^2}{2{\sigma_s}^2}\right) \mathrm{d}\left(\frac{z}{\sigma_s}\right) ,$$

$$P = \frac{NKR^{1/2}{\sigma_s}^{3/2}}{\sqrt{2\pi}} \int_{d/\sigma_s}^{\infty} \left(\frac{z-d}{\sigma_s}\right)^{3/2} \exp\left(-\frac{z^2}{2{\sigma_s}^2}\right) \mathrm{d}\left(\frac{z}{\sigma_s}\right) ,$$

which can also be written

$$n/N = F_0(d^*) , \tag{4.306}$$

$$A^* = \frac{A}{NR\sigma_s} = \pi F_1(d^*) , \tag{4.307}$$

$$P^* = \frac{P}{NE^*R^{1/2}{\sigma_s}^{3/2}} = \frac{4}{3} F_{3/2}(d^*) , \tag{4.308}$$

with $d^* = d/\sigma_s$ and

$$\begin{aligned} F_n(d^*) &= \frac{1}{\sqrt{2\pi}} \int_{d^*}^{\infty} (x-d^*)^n \exp\left(-\frac{x^2}{2}\right) \mathrm{d}x , \\ &= \frac{1}{\sqrt{2\pi}} \int_0^{\infty} u^n \exp\left[-\frac{1}{2}(u+d^*)^2\right] \mathrm{d}u . \end{aligned} \tag{4.309}$$

The mean pressure is given by

$$ {p_m}^* = \frac{P^*}{A^*} = \frac{p_m}{E^*(\sigma_s/R)^{1/2}} = \frac{4}{3\pi} \frac{F_{3/2}(d^*)}{F_1(d^*)} . \tag{4.310}$$

Note that $F_{n+1}(h) = nF_{n-1}(h) - hF_n(h)$, that $F_n' = -nF_{n-1}'$ and that only the $F_n(0)$ are obtained in closed form:

$$F_0(0) = 1/2 ,$$

$$F_1(0) = 1/\sqrt{2\pi} ,$$

$$F_{3/2}(0) = \frac{2^{1/4}\Gamma(5/4)}{\sqrt{2\pi}} = 0.430 .$$

These integrals are numerally computed (with an upper limit equal to 6) and Fig. 4.73 displays variations of n/N, A^*, P^* and p_m^* with d^*. A negative

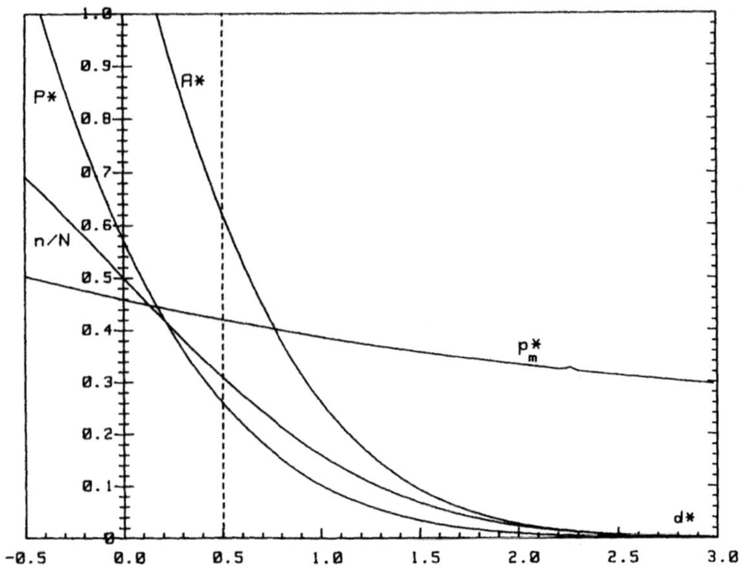

Fig. 4.73. Contact of a rough surface and a smooth flat surface. Real area of contact, load, mean pressure, and the proportion of summits in contact are plotted as a function of the separation (normalized coordinates)

value of d^* corresponds to a smooth plane pressed below the mean plane of summits. In fact the calculation assumes that deformations of asperities are independent of one another, i.e. that the real area of contact A remains small compared with the nominal area A_0. We can restrict to $d^* > 0.5$ for there are no more than 30% of summits in contact ($F_0(0.5) = 0.309$). Figure 4.74 displays the relation between A and P in logarithmic coordinates as well as the variation of mean pressure. The relation between A and P is almost linear (since the slope is 0.95).

Exponential Distribution of Summit Heights. As shown by Greenwood and Williamson, the calculation is more instructive if one uses an exponential distribution of summit heights:

$$\varphi_s(z) = \frac{C}{\sigma_s} \exp\left(-\frac{|z|}{\sigma_s}\right) \tag{4.311}$$

(where C is an arbitrary constant). We will follow Johnson [75] by writing only

$$A_i = f(\delta_i) \; ,$$
$$P_i = g(\delta_i) \; ,$$

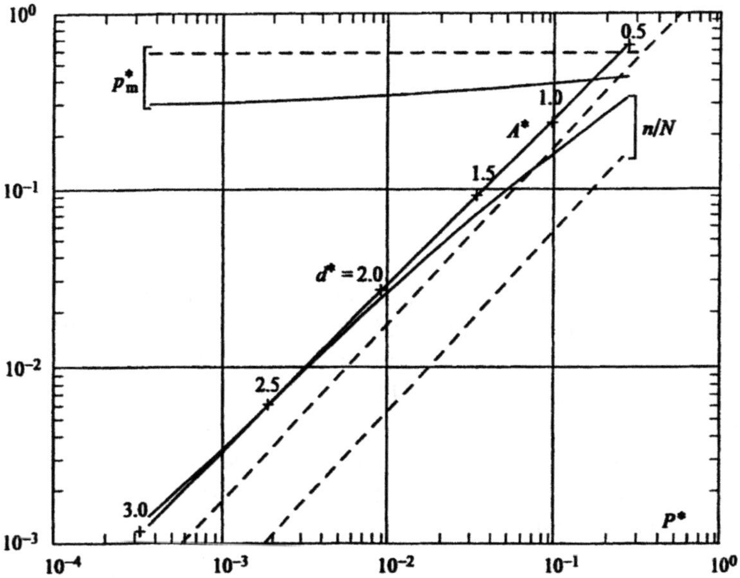

Fig. 4.74. Contact of a rough surface and a smooth flat surface. Real area of contact, mean pressure, and the proportion of summits in contact are plotted as a function of the total load (normalized coordinates). The *solid line* represents a Gaussian distribution of summit heights, the *dashed line* represents an exponential distribution

with $\delta_i = z_i - d$. The number of summits in contact, given by (4.301), becomes

$$\begin{aligned} n &= \frac{CN}{\sigma_s} \int_d^\infty \exp(-z/\sigma_s)\, \mathrm{d}z \\ &= CN \exp(-d^*) \int_0^\infty \exp(-\delta/\sigma_s)\, \mathrm{d}(\delta/\sigma_s) \\ &= CN \exp(-d^*) \,. \end{aligned} \tag{4.312}$$

Similarly, the real contact area, given by (4.304), and the total load, given by (4.305), are written:

$$\begin{aligned} A &= \frac{CN}{\sigma_s} \int_d^\infty f(\delta) \exp(-z/\sigma_s)\, \mathrm{d}z \\ &= CN \exp(-d^*) \int_0^\infty f(\delta) \exp(-\delta/\sigma_s)\, \mathrm{d}(\delta/\sigma_s) \\ &= nI_f \,, \end{aligned} \tag{4.313}$$

$$\begin{aligned} P &= \frac{CN}{\sigma_s} \int_d^\infty g(\delta) \exp(-z/\sigma_s)\, \mathrm{d}z \\ &= CN \exp(-d^*) \int_0^\infty g(\delta) \exp(-\delta/\sigma_s)\, \mathrm{d}(\delta/\sigma_s) \\ &= nI_g \,. \end{aligned} \tag{4.314}$$

We thus have

$$n = \frac{P}{I_g} ,$$

$$A = \frac{I_f}{I_g} P , \tag{4.315}$$

$$p_{\mathrm{m}} = \frac{P}{A} = \frac{I_g}{I_f} . \tag{4.316}$$

The definite integrals, denoted I_f and I_g, being independent of the separation d, the number of asperities in contact and the total area of contact are thus proportional to the load, and that independently of the law of deformation of asperities, elastic or otherwise. I_f and I_g are the mean size and the mean load of the contacts, independent of load. Of course, the size of an individual contact increases with the load, but at the same time new small contacts appear, so that the mean area per contact remains constant.

In the purely elastic case, we get

$$\begin{aligned} I_f &= \pi R \sigma_{\mathrm{s}} \int_0^\infty x \mathrm{e}^{-x}\, \mathrm{d}x \\ &= \pi R \sigma_{\mathrm{s}} , \\ I_g &= K R^{1/2} {\sigma_{\mathrm{s}}}^{3/2} \int_0^\infty x^{3/2} \mathrm{e}^{-x}\, \mathrm{d}x \\ &= \sqrt{\pi} E^* R^{1/2} {\sigma_{\mathrm{s}}}^{3/2} . \end{aligned}$$

In normalized coordinates, we have $A^* = P^* \sqrt{\pi}$, $p^*_{\mathrm{m}} = 1/\sqrt{\pi}$ and $n/N = P^*/\sqrt{\pi}$. These results are compared with the Gaussian distribution in Fig. 4.74.

Plastic Deformation and Plasticity Index. For a Hertzian contact, plastic deformation begins for a mean pressure

$$p_{\mathrm{m}} = 1.1 \sigma_0 \approx 0.37 H \tag{4.317}$$

(where σ_0 is the elastic limit and $H \equiv 3\sigma_0$ is the hardness). For a rough surface, Fig. 4.73 shows that the normalized mean pressure varies little with the load and remains near 0.35. We thus have approximately

$$p_{\mathrm{m}} \approx 0.35 E^* \sqrt{\frac{\sigma_{\mathrm{s}}}{R}} .$$

When this value exceeds $0.37H$, i.e. when approximately $E^* \sqrt{\sigma_{\mathrm{s}}/R} > H$, asperities are plastically deformed.

Greenwood and Williamson [67] as a result introduced a *plasticity index* ψ, given by

$$\boxed{\psi = \frac{E^*}{H} \sqrt{\frac{\sigma_{\mathrm{s}}}{R}}} . \tag{4.318}$$

When ψ is small, deformation of asperities in contact with a flat is essentially elastic. When ψ is large, plastic deformations become important, even under low load. Let us compute more precisely the proportion of plastically deformed contacts. For a Hertzian contact, the mean pressure can also be written as

$$p_{\mathrm{m}} = \frac{4E^*}{3\pi}\sqrt{\frac{\delta}{R}}\,,$$

and the criterion of plastic deformation, given in (4.317), becomes $\delta_{\mathrm{p}} = 0.76R(H/E^*)^2$ that we will write simply as

$$\delta_{\mathrm{p}} = R\left(\frac{H}{E^*}\right)^2 \tag{4.319}$$

since elastic/elastoplastic transition is very smooth. Figure 4.75 shows that all summits of height $z > d$ are in contact and flattened by a quantity $\delta = z - d$, and that all summits of height $z > d + \delta_{\mathrm{p}}$ are plastically deformed.

Just as the probability for a summit to have its height z larger than d was

$$\Phi(d) = \int_d^{\infty} \varphi(z)\,\mathrm{d}z\,,$$

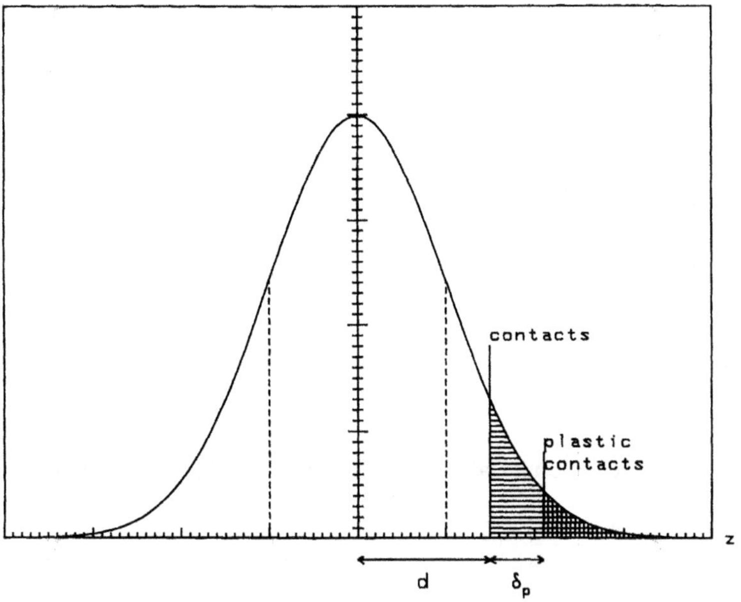

Fig. 4.75. All summits of height z larger than d are in contact and are flattened by a quantity $\delta = z - d$. Summits of height z larger than $d + \delta_{\mathrm{p}}$ are plastically deformed

so the probability for a summit to have its height larger than $d + \delta_\mathrm{p}$ and be plastically deformed is

$$\Phi(d + \delta_\mathrm{p}) = \int_{d+\delta_\mathrm{p}}^{\infty} \varphi(z)\,\mathrm{d}z\,.$$

The number of these plastically deformed summits when the separation is d is thus

$$n_\mathrm{p} = N \int_{d+\delta_\mathrm{p}}^{\infty} \varphi(z)\,\mathrm{d}z\,.$$

If we continue to use the Hertz law (which assumes that one remains in the vicinity of the criterion of plasticity, then for ψ slightly greater than unity) the area of the ith contact remains given by (4.302). The plastically deformed area is thus

$$A_\mathrm{p} = \pi N R \int_{d+\delta_\mathrm{p}}^{\infty} (z - d)\varphi(z)\,\mathrm{d}z\,.$$

For a Gaussian distribution and introducing, as above, nondimensional quantities, we get

$$\begin{aligned} n_\mathrm{p} &= N F_0\left(d^* + \delta_\mathrm{p}^*\right) \\ &= \frac{N}{\sqrt{2\pi}} \int_{\delta_\mathrm{p}^*}^{\infty} \exp\left[-\frac{1}{2}(\delta^* + d^*)^2\right] \mathrm{d}\delta^*\,, \end{aligned} \tag{4.320}$$

$$\begin{aligned} {A_\mathrm{p}}^* &= \frac{A_\mathrm{p}}{N R \sigma_\mathrm{s}} = \sqrt{\frac{\pi}{2}} \int_{d^*+\delta_\mathrm{p}^*}^{\infty} (x - d^*) \exp\left(-\frac{x^2}{2}\right) \mathrm{d}x \\ &= \sqrt{\frac{\pi}{2}} \int_{\delta_\mathrm{p}^*}^{\infty} \delta^* \exp\left[-\frac{1}{2}(\delta^* + d^*)^2\right] \mathrm{d}\delta^*\,. \end{aligned} \tag{4.321}$$

Note that ${A_\mathrm{p}}^*$ is not equal to $\pi F_1(d^* + \delta_\mathrm{p}^*)$.

The proportion of plasticized contacts is simply

$$\frac{n_\mathrm{p}}{n} = \frac{F_0(d^* + \delta_\mathrm{p}^*)}{F_0(d^*)}\,. \tag{4.322}$$

If we return to the exponential distribution, given by (4.311), we find that

$$n_\mathrm{p} = C N \exp(-d^*) \exp(-\delta_\mathrm{p}^*)$$

or, using (4.312),

$$\frac{n_\mathrm{p}}{n} = \exp(-\delta_\mathrm{p}^*)\,.$$

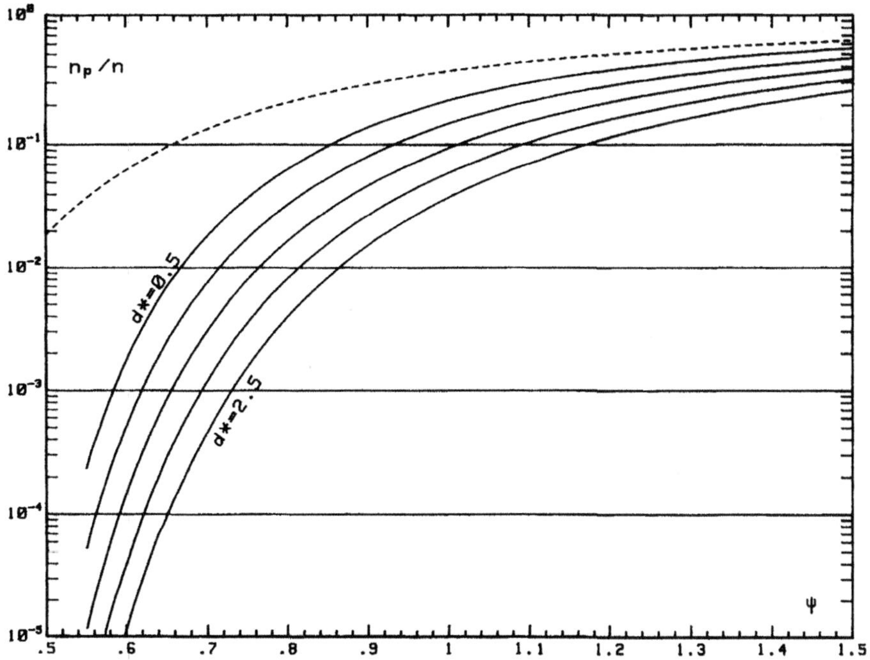

Fig. 4.76. Proportion of plasticized contacts, as a function of the plasticity index ψ and for a Gaussian distribution of summit heights. The dashed line represents an exponential distribution

The proportion of plastic contacts depends only on δ_p^*, i.e. of the plasticity index ψ, since

$$\delta_\mathrm{p}^* = \frac{R}{\sigma_\mathrm{s}}\left(\frac{H}{E^*}\right)^2 = \frac{1}{\psi^2} \, .$$

Figure 4.76 shows the proportion of plastic contacts as a function of the plasticity index, for Gaussian and exponential distributions. It is possible to consider that for $\psi < 0.6$ contacts are essentially elastic, and that for $\psi > 1$ plastic deformation becomes important.

Numerical Example. Let us take a numerical example for steel on steel (after McCool [72]), with values as follows: equivalent Young's modulus $E^* = 1.14 \times 10^5\,\mathrm{N\,mm^{-2}}$; elastic limit $\sigma_0 = 2070\,\mathrm{N\,mm^{-2}}$; density of summits $\eta_\mathrm{s} = 3900\,\mathrm{summits\,mm^{-2}}$; radius of curvature of summits $R = 6.6 \times 10^{-2}\,\mathrm{mm}$; standard deviation of summit heights $\sigma_\mathrm{s} = 2.4 \times 10^{-4}\,\mathrm{mm}$. We thus have $\psi = 1.11$ and a critical flattening, from (4.319), where

$$\delta_\mathrm{p}^* = \frac{1}{\psi^2} = 0.82 \, .$$

For a relative approach $d^* = d/\sigma_\mathrm{s} = 2.5$, we have: $F_0(2.5) = 6.21 \times 10^{-3}$, $F_1(2.5) = 2.00 \times 10^{-3}$ and $F_{3/2}(2.5) = 1.47 \times 10^{-3}$. The number of contacts, from (4.306), is

$$n = 3900 \times 6.21 \times 10^{-3} = 24.2 \, \text{contacts} \, \text{mm}^{-2} \, ,$$

so that 0.6% of the summits are in contact. The ratio of the real area of contact to the apparent area of contact, from (4.307), is given by

$$\frac{A}{A_0} = 3900 \times 6.6 \times 10^{-2} \times 2.4 \times 10^{-4} \times \pi \times 2 \times 10^{-3} = 3.9 \times 10^{-4} \, .$$

The nominal pressure, from (4.308), is given by

$$\begin{aligned} \frac{P}{A_0} &= 3900 \times 1.14 \times 10^5 \times \sqrt{6.6 \times 10^{-2}} \\ &\quad \times (2.4 \times 10^{-4})^{3/2} \times \frac{4}{3} \times 1.47 \times 10^{-3} \\ &= 0.83 \, \text{N} \, \text{mm}^{-2} \, . \end{aligned}$$

The mean pressure per contact is given by

$$p_{\text{m}} = \frac{P}{A} = \frac{0.83}{3.9 \times 10^{-4}} = 2.1 \times 10^3 \, \text{N} \, \text{mm}^{-2} \, ,$$

and the ratios are

$$\frac{n_{\text{p}}}{n} = \frac{F_0(2.5 + 0.82)}{F_0(2.5)} = \frac{4.5 \times 10^{-4}}{6.2 \times 10^{-3}} = 7.3\% \, ,$$

$$\frac{A_{\text{p}}}{A} = \frac{4.86 \times 10^{-4}}{F_1(2.5)} = 24\% \, ,$$

the value of A_{p} being obtained by (4.321).

Let us increase the load until the relative approach is $d^* = 0.5$. One has then $F_0(0.5) = 0.309$ (i.e. 31% of the summits are in contact), $F_1(0.5) = 0.198$, $F_{3/2}(0.5) = 0.195$. The above results become

$$n = 1200 \, \text{contacts per mm}^2 \, ,$$

$$\frac{A}{A_0} = 3.8 \times 10^{-2} \, ,$$

$$\frac{P}{A_0} = 110 \, \text{N} \, \text{mm}^{-2} \, ,$$

$$p_{\text{m}} = 2895 \, \text{N} \, \text{mm}^{-2} \, .$$

We can see that when the load is multiplied by about 130, the number of contacts and the real area of contact are multiplied respectively by about 50 and by about 100, whereas the mean pressure is only multiplied by 1.36. Concerning the plastic deformation, we have:

$$\frac{n_{\text{p}}}{n} = \frac{F_0(0.5 + 0.82)}{F_0(0.5)} = \frac{9.34 \times 10^{-2}}{3.085 \times 10^{-1}} = 30\% \, ,$$

$$\frac{A_{\text{p}}}{A} = \frac{0.120}{F_1(0.5)} = 61\% \, .$$

Contacts of the JKR Type: the Theory of Fuller and Tabor. The JKR theory of sphere/plane contact has been very well verified on soft solids such as gelatin or rubber, but experiments on more rigid solids, such as metals, have revealed only insignifiant adherence. This effect is due to roughness, and particularly to the dispersion of summit heights, some asperities (the higher) being compressed while others are stretched. This problem was solved by Johnson [9] for an exponential distribution of height, and by Fuller and Tabor [10] in 1975 for a Gaussian distribution.

Let us consider the contact of a rough surface and a smooth rigid plane over a nominal area A_0. The rough surface has N asperities uniformly distributed whose summits each have the same radius of curvature R ($\eta_s = N/A_0$ is the number of summits per unit area). The heights z of the summits are measured with respect to the summit mean plane. The distance d between the two solids is the distance between the smooth plane and this mean plane. The negative values of d correspond to a smooth plane pressed beyond this mean plane. The height distribution is Gaussian, with a standard deviation σ. The theory is based on that of Greenwood and Williamson, but the Hertz relations are changed into JKR relations, as per (4.147) and (4.172) so that

$$\delta = \frac{a^2}{3R} + \frac{2P}{3aK} = \frac{1}{3K^{2/3}R^{1/3}} \frac{P_1 + 2P}{P_1^{1/3}} , \tag{4.323}$$

$$P = \frac{a^3 K}{R} - \sqrt{6\pi a^3 K w} = P_1 - \sqrt{6\pi w R P_1} . \tag{4.324}$$

Let

$$P_c = \frac{3}{2}\pi w R , \tag{4.325}$$

$$\delta_c = \left(\frac{3}{4} \frac{\pi^2 w^2 R}{K^2} \right)^{1/3} = \left(\frac{P_c{}^2}{3K^2 R} \right)^{1/3} , \tag{4.326}$$

so that the limits of stability at fixed load and fixed grips are respectively $P/P_c = -1$ and $\delta/\delta_c = -1$. Eliminating P_1 between the two equations (4.323) and (4.324), we obtain the two results

$$\frac{\delta}{\delta_c} = (3\chi - 1) \left(\frac{1+\chi}{9} \right)^{1/3} , \quad \chi \geq 0 , \qquad \delta/\delta_c \geq -3^{-2/3} , \tag{4.327a}$$

$$\frac{\delta}{\delta_c} = -(3\chi + 1) \left(\frac{1-\chi}{9} \right)^{1/3} , \quad 0 \leq \chi \leq \frac{1}{3} , \quad \delta/\delta_c \leq -3^{-2/3} , \tag{4.327b}$$

with

$$\chi = \sqrt{1 + \frac{P}{P_c}} . \tag{4.328}$$

Equations (4.327a) and (4.327b) give δ/δ_c as a function of P/P_c. Let

$$\frac{P_i}{P_c} = g\left(\frac{\delta_i}{\delta_c}\right) \tag{4.329}$$

be the inverse relation (numerically evaluated) which expresses univocally the force P_i exerted on an asperity as a function of its approach δ_i, and which replaces (4.303). This function is displayed in Fig. 4.77, the branch CD corresponding to (4.327b). The Hertz relation is also given in the form

$$\frac{P}{P_c} = \frac{1}{\sqrt{3}}\left(\frac{\delta}{\delta_c}\right)^{3/2}, \tag{4.330}$$

and the DMT relation, in accordance with (4.179), becomes

$$\frac{P}{P_c} = \frac{1}{\sqrt{3}}\left(\frac{\delta}{\delta_c}\right)^{3/2} - \frac{4}{3}. \tag{4.331}$$

The probability for a summit to be in contact with the smooth plane is the probability for its height z to be larger than d

$$\Phi_s(z) = \int_d^\infty \varphi_s(z)\,dz\,.$$

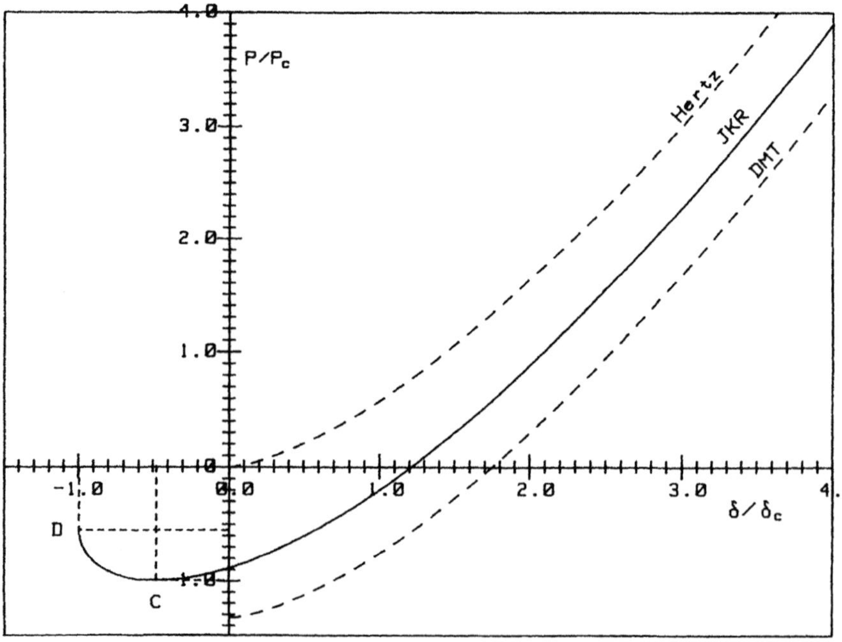

Fig. 4.77. Relation between load and penetration in the JKR theory, in reduced coordinates. The *dashed lines* represent the theories of Hertz and DMT

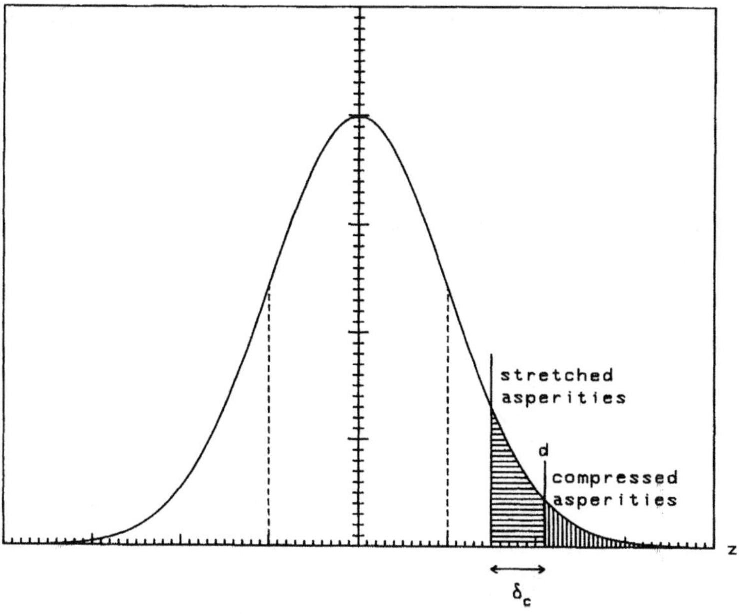

Fig. 4.78. JKR contacts. After an initial compression and return to the separation d, all summits of height z larger than $d - \delta_c$ are in contact and deformed by a quantity $z - d$. Those of height z larger than d are compressed; those of height z between d and $d - \delta_c$ are stretched

If there are N summits in the nominal area, the number n of summits in contact when the separation is d should be

$$n = N \int_d^{\infty} \varphi_s(z)\,dz .$$

There is, however, a difficulty. The above formula is valid only when the smooth plane and the rough plane approach progressively: asperities then come into contact one after another beginning with the highest. If the solids are first compressed and then the separation is increased up to the value d, a large number of asperities come into contact, and some of these (the less high) are stretched and break when their elongation exceeds δ_c. Figure 4.78, analogous to Fig. 4.75 for plastic deformation, shows that the lower limit of integration, in this case, is $d - \delta_c$. Since we are interested in adherence during separation, we will use this limit of integration.

The number of summits in contact when the separation is d is thus

$$n = N \int_{d-\delta_c}^{\infty} \varphi(z)\,dz , \tag{4.332}$$

i.e. for a Gaussian distribution

$$\frac{n}{N} = \frac{1}{\sigma\sqrt{2\pi}} \int_{d-\delta_c}^{\infty} \exp\left(-\frac{z^2}{2\sigma^2}\right) dz .$$

Changing the variable from z to δ $(= z - d)$ and using the nondimensional variables $\delta^* = \delta/\sigma$, $\delta_c^* = \delta_c/\sigma$, $d^* = d/\sigma$, we get

$$\frac{n}{N} = \frac{1}{\sqrt{2\pi}} \int_{-\delta_c^*}^{\infty} \exp\left[-\frac{1}{2}(\delta^* + d^*)^2\right] \mathrm{d}\delta^* \,. \tag{4.333}$$

The total force exerted between the two solids is the sum of the forces exerted by asperities in contact, so that

$$\begin{aligned} P &= N \int_{d-\delta_c}^{\infty} P_i \varphi(z)\, \mathrm{d}z \\ &= \frac{N P_c}{\sigma\sqrt{2\pi}} \int_{d-\delta_c}^{\infty} g\left(\frac{\delta_i}{\delta_c}\right) \exp\left(-\frac{z^2}{2\sigma^2}\right) \mathrm{d}z \,, \end{aligned}$$

i.e.

$$\frac{P(d^*, \delta_c^*)}{N P_c} = \frac{1}{\sqrt{2\pi}} \int_{-\delta_c^*}^{\infty} g\left(\frac{\delta^*}{\delta_c^*}\right) \exp\left[-\frac{1}{2}(\delta^* + d^*)^2\right] \mathrm{d}\delta^* \,. \tag{4.334}$$

Curves giving the normalized force P/NP_c as a function of the normalized distance d/σ are displayed in Fig. 4.79 for various values of the parameter

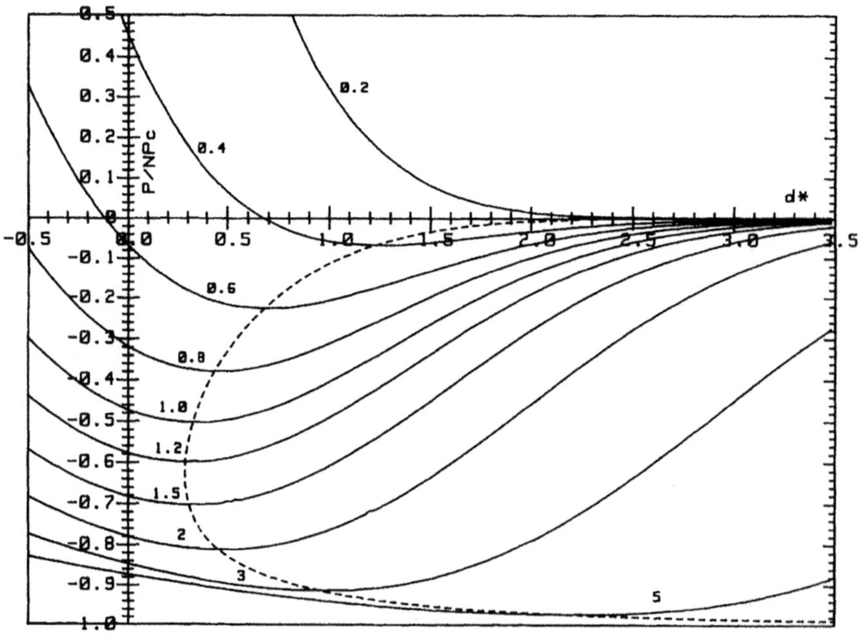

Fig. 4.79. JKR contacts. Normalized force as a function of the normalized distance $d^* = d/\sigma$ for various values of the parameter δ_c/σ

δ_c/σ. The locus of minima is obtained by differentiating (4.334) under the integral and is given by

$$\int_{-\delta_c^*}^{\infty} g\left(\frac{\delta^*}{\delta_c^*}\right)(d^* + \delta^*)\exp\left[-\frac{1}{2}(d^* + \delta^*)^2\right] d\delta^* = 0 . \tag{4.335}$$

If we operate at fixed grips, increasing the distance between the planes with a micrometer for example, asperities will break contact one after another and the adherence force will correspond to the adherence of the last asperity (the highest). On the other hand, if one operates at fixed load, an instability will occur at the minimum P_{min} of the curve and the adherence force will thus be $F = -P_{\text{min}}$. This adherence force is reported in Fig. 4.80, as a function of $1/\delta_c^*$, a quantity proportional to σ, and termed the *adhesion parameter*[9] by Fuller and Tabor [10].

If we return to the definition of δ_c^*, from (4.326), we can note that

$$\left(\frac{\sigma}{\delta_c}\right)^{3/2} = \sqrt{3}\frac{KR^{1/2}\sigma^{3/2}}{P_c} .$$

This value is proportional to the ratio of the Hertzian force to flatten an asperity of radius R by σ to the adherence force of this asperity. Fuller and Tabor [10] have termed the *modified adhesion parameter* the nondimensional quantity

$$\boxed{\theta = \frac{E^* R^{1/2}\sigma^{3/2}}{wR} = \frac{9\pi}{8\sqrt{3}}\left(\frac{\sigma}{\delta_c}\right)^{3/2}} . \tag{4.336}$$

Note that when the standard deviation σ of asperity heights tends towards zero, the adherence of a flat plane is not obtained, but N times the adherence at fixed load of a sphere of radius R instead.

Furthermore, the theory is established with the assumption of a small number of independent contacts. In Fig. 4.80 there is displayed the proportion n/N of summits still in contact at the moment of separation, by putting in (4.333) the values of d^* and δ_c^* taken at the minima P_{min}. If the ratios n/N are limited to less than 0.3, we can note that the range of validity of the theory is rather small.

Contacts of the DMT Type [76]. There is no difficulty in repeating the above calculations for contacts of the DMT type (see (4.179)). Equations (4.302) and (4.303) become

$$A_i = \pi a_i^2 = \pi R\delta_i = \pi R(z_i - d) ,$$

$$P_i = KR^{1/2}\delta_i^{3/2} - 2\pi wR = KR^{1/2}(z_i - d)^{3/2} - 2\pi wR ,$$

[9] This term seems ill-chosen. It would be expected for an adhesion parameter that it would increase with adhesion, i.e. with w. We are, rather, concerned with a roughness parameter.

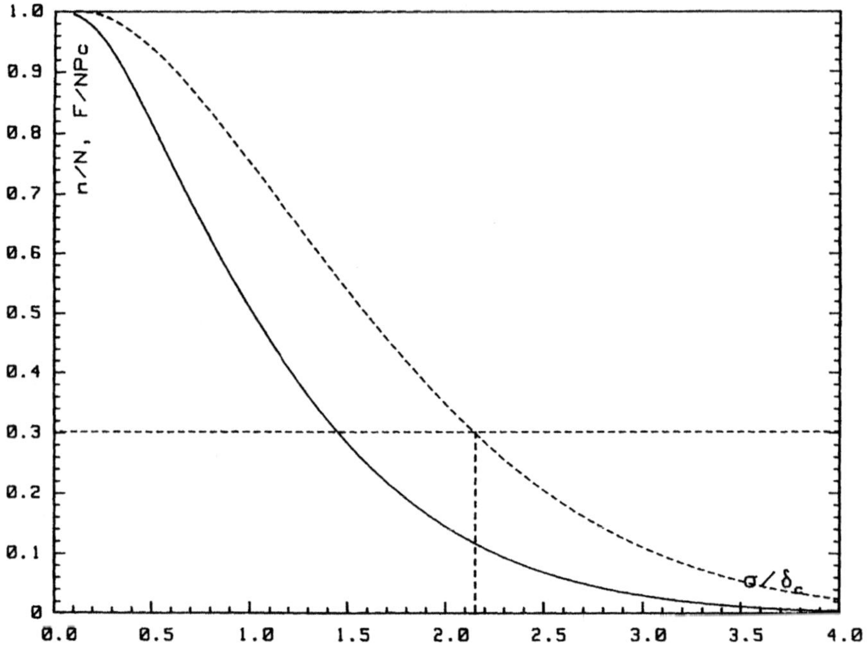

Fig. 4.80. JKR contacts. Normalized adherence force of a flat plane on a rough plane as a function of the parameter $\sigma/\delta_{\mathrm{C}}$. The dotted line represents the proportion of summits in contact at separation

but, contrary to the JKR model, asperities are never stretched and the radius of contact of an asperity vanishes when it is no longer compressed, as in the Hertzian case. However, an asperity can be compressed even when a traction force is applied, on condition that this force is less than $2\pi wR$. In effect, adhesion forces act around the contact and add an overload $2\pi wR$ to the applied load P. (Recall that in the JKR theory there is no overload but a modification in the distribution of stresses in an enlarged area of contact.)

The proportion of summits in contact, the real area of contact and the load are still given by (4.301), (4.304) and (4.305) where d is indeed the lower limit of integration. In the last equation, P is the load applied by the experimenter. The calculations of n and A are not modified, but for P we have

$$\begin{aligned} P &= N \int_d^\infty \left(K R^{1/2} \delta^{3/2} - 2\pi w R \right) \varphi(z)\,\mathrm{d}z \\ &= N K R^{1/2} \int_d^\infty \delta^{3/2} \varphi(z)\,\mathrm{d}z - 2\pi w R n \,. \end{aligned}$$

For a Gaussian distribution we find:

$$n/N = F_0(d^*) , \tag{4.337}$$

$$A^* = \frac{A}{NR\sigma_s} = \pi F_1(d^*) , \tag{4.338}$$

$$P^* = \frac{P}{NE^* R^{1/2} \sigma_s^{3/2}} = \frac{4}{3} F_{3/2}(d^*) - \frac{2\pi}{\theta} F_0(d^*) , \tag{4.339}$$

where θ is the adhesion parameter, from (4.336), P^* is the load applied by the experimenter, and the term in θ is the overload due to adhesion forces. For infinite $\theta (w = 0)$ we obtain the results of Greenwood and Williamson. The variations of n/N and A^* with d^* have been displayed in Fig. 4.73. To have $n/N < 0.3$, we must have $d^* > 0.5$.

To make the comparison with the case of JKR contacts, we can rewrite (4.339) in the form

$$\frac{P}{NP_c} = \frac{1}{\sqrt{3}} \left(\frac{\sigma}{\delta_c} \right)^{3/2} F_{3/2}(d^*) - \frac{4}{3} F_0(d^*) , \tag{4.340}$$

in which P_c and δ_c preserve their original significance, as per (4.325) and (4.326). The curves describing the normalized force P/NP_c as a function of the normalized distance d/σ are displayed in Fig. 4.81 for various values of the

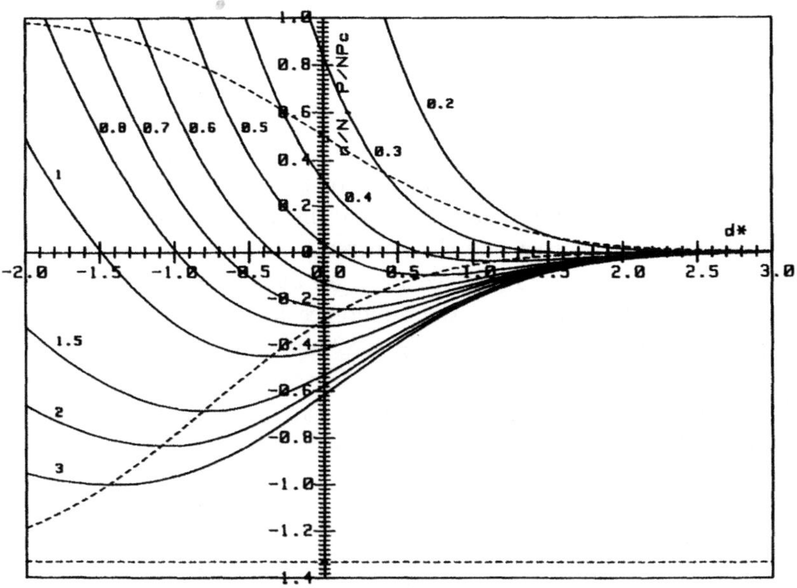

Fig. 4.81. DMT contacts. Normalized force as a function of the normalized separation $d^* = d/\sigma$ for various values of the parameter δ_c/σ. The *dashed line* represents the proportion n/N of summits in contact, and the locus of minima

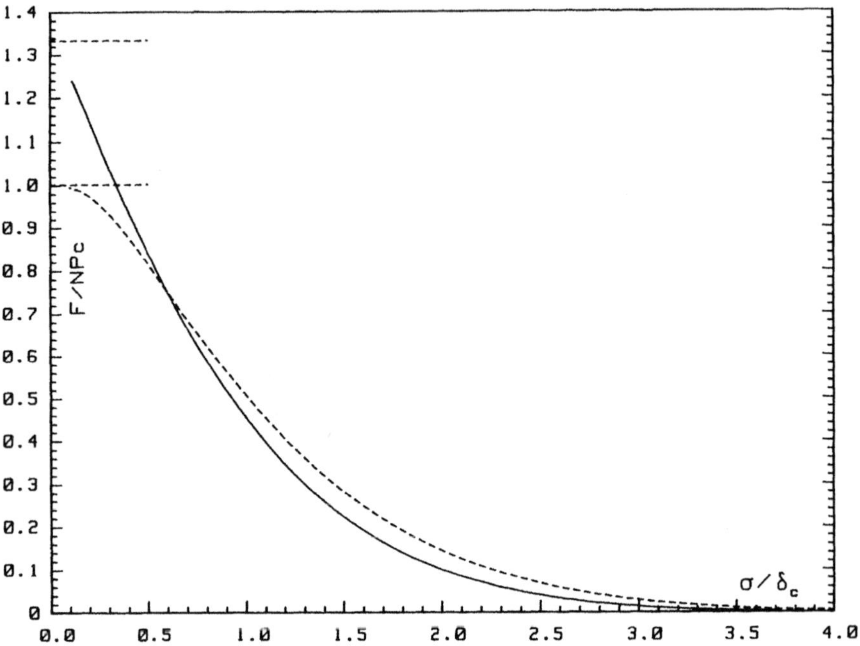

Fig. 4.82. DMT contacts. Normalized adherence force of a flat plane on a rough plane, as a function of the parameter σ/δ_c. The *dotted line* represents the Fuller and Tabor theory for JKR contacts

parameter δ_c/σ. The locus of minima is obtained by differentiating (4.340) under the integral and is given by

$$F_{1/2}(d^*) - \frac{8}{3\sqrt{6\pi}} \left(\frac{\delta_c}{\sigma}\right)^{3/2} \exp\left(-\frac{d^{*2}}{2}\right) = 0\,, \tag{4.341}$$

an implicit equation giving d^* as a function of δ_c/σ.

As for JKR contacts, if one works at fixed displacement, asperities break the contact one after another and the adherence force will be negligible. On the other hand, if one works at fixed load, an instability will occur at the minimum P_{min} of the curve and the adherence force will be $F = -P_{min}$. This adherence force is displayed in Fig. 4.82 and compared with the result of Fuller and Tabor [10] for JKR contacts. The two curves are rather close. Recall that both theories assume that contacts are independent, i.e. that the proportion n/N of summits in contact remains small.

Figure 4.83 displays the overload

$$Z^* = \frac{2\pi}{\theta} F_0(d^*) = \frac{2\pi}{\theta}\frac{n}{N} \tag{4.342}$$

due to adhesion forces which must be added to the load P^* applied by the experimenter. Starting from the limit of stability (*solid black circles*), it increases with the applied load. The variation of the total load $Z^* + P^*$

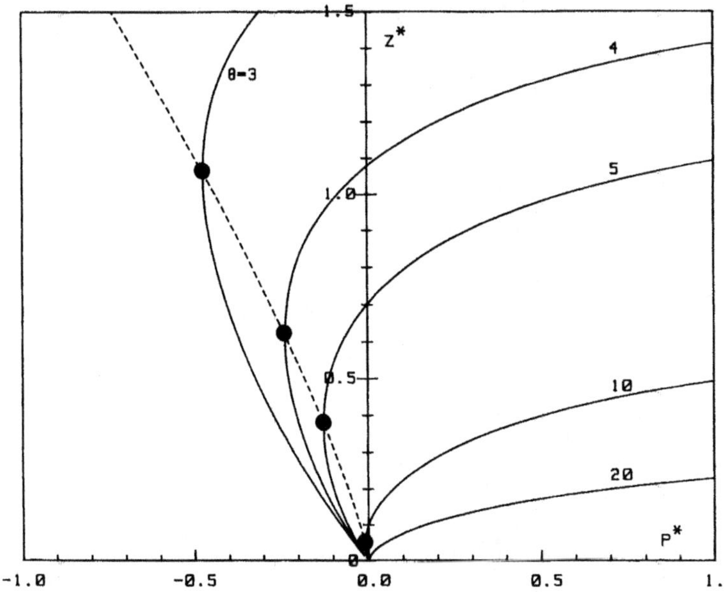

Fig. 4.83. DMT contacts. Overload Z^* due to adhesion forces as a function of the applied load P^*, for various values of the adhesion parameter θ (normalized coordinates). The *dashed line* represents the limit of stability (adherence force)

with the applied load P^* is displayed in Fig. 4.84. It must be emphasized that the value of Z at the limit of stability (*solid black circles*) is not the adherence force, which has value $-P$ at this point. (It is only for a single contact that the adherence force $F = 2\pi wR$ is equal to the overload Z.)

To illustrate this effect of overloading, let us take an example. Consider a polished steel surface ($E^* = 110\,\text{GPa}$) in contact over $1\,\text{cm}^2$ with a flat surface. Let $R = 0.5\,\text{mm}$ and $\sigma = 200\,\text{Å}$ be respectively the radius of curvature of summits and the standard deviation of their heights. We will take $N = 5 \times 10^5$ to satisfy (4.295). Apply a load P such that $P^* = 0.020$ ($P = 69.5\,\text{N}$). Then:

1. Assume that $w = 0$ and apply the Greenwood and Williamson theory. If the hardness is $H = 4\,\text{GPa}$, we obtain a plasticity index $\psi = 0.17$, which shows that contacts are mainly elastic. The load P^* is reached for a distance $d^* = 1.698$ and the proportion of summits in contact is 4.5%.
2. In air, the adhesion is weak and is due to van der Waals interactions. Let us take $w = 0.1\,\text{J}\,\text{m}^{-2}$. This gives $\theta = 139$ and $\sigma/\delta_c = 16.7$. The load P^* is reached at a distance $d^* = 1.656$ and the proportion of summits in contact is 4.9%. The overload due to adhesion forces is $Z^* = 0.002$, i.e. $Z = 7.7\,\text{N}$. The relative increase in load on asperities is thus 11%.
3. For clean surfaces in an ultrahigh vacuum, the adhesion forces are higher and of the order of J/m^2. Let us take $w = 1\,\text{J}\,\text{m}^{-2}$. This gives $\theta = 14$ and $\sigma/\delta_c = 3.60$. The load P^* is reached at a distance $d^* = 1.106$, and the

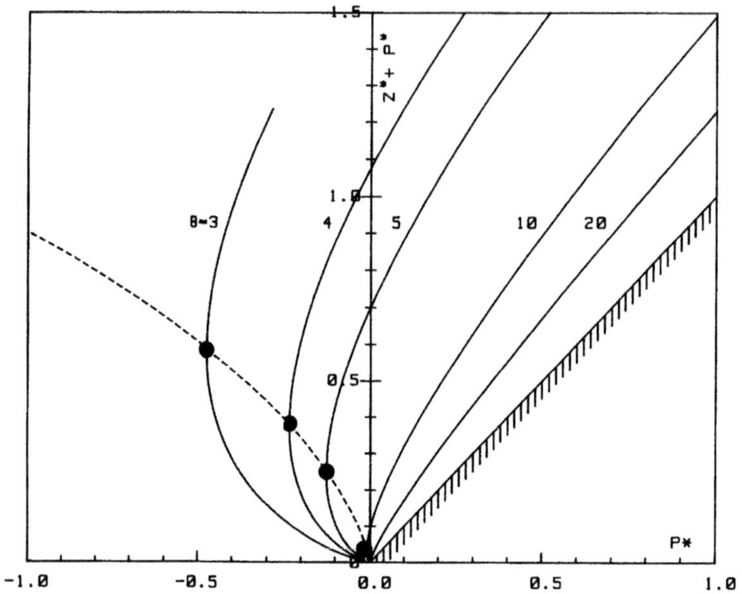

Fig. 4.84. DMT contacts. Total load $Z^* + P^*$ as a function of the applied load P^*, for various values of the adhesion parameter θ (normalized coordinates). The *dashed line* represents the limit of stability (adherence force).

proportion of summits in contact is 13.4%. The overload due to adhesion force is now $Z^* = 0.060$, i.e. $Z = 211$ N. The total load on the asperities has been multiplied by 4 (and clearly this ratio tends towards infinity when P tends towards zero).

This overload due to adhesion certainly plays an important role in the friction of solids. An analogy can be made with magnetic forces. The friction coefficient $f = T/P$ of steel on steel is about 0.2, but if the rider is magnetized, the friction coefficient increases enormously. The mechanisms of dissipation have not changed, but the load on asperities is now the applied load plus the overload due to magnetic forces. If T is the friction force and P the applied load, the friction coefficient μ should be defined by

$$T = \mu(P + Z) , \tag{4.343}$$

and the shape of the friction curve for T versus P should be that of Fig. 4.84. A similar equation has been repeatedly proposed in the past with Z a constant adherence force, so that the equation looks like the two-term Coulomb law

$$T = k + \mu P \tag{4.344}$$

(cf Tabor [77] and Derjaguin [78]). Such equations account for friction under negative load. In the present theory, Z is not the adherence force and it increases with the applied load P since the number of contacts increases and thus the overload.

References

1. H. Hertz: J. Reine Angew. Math. **92**, 156 (1882)
2. J. Boussinesq: *Application des potentiels à l'étude de l'équilibre et du mouvement des solides élastiques* (Gauthier-Villars, Paris 1885, nouveau tirage Blanchard, Paris 1969)
3. R.S. Bradley: Philos. Mag. **13**, 853 (1932)
4. B.V. Deryagin: Kolloid Z. **69**, 163 (1934); cf also B.V. Deryagin, N.A. Krotova, V.P. Smilga: *Adhesion of solids* (Consultant Bureau, New York 1978)
5. K.L. Johnson: Brit. J. Appl. Phys. **9**, 199 (1958)
6. K.L. Johnson, K. Kendall, A.D. Roberts: Proc. Roy. Soc. Lond. A **324**, 301 (1971)
7. K. Kendall: J. Phys. D: Appl. Phys. **4**, 1186 (1971)
8. K. Kendall: J. Adhesion **5**, 77 (1973)
9. K.L. Johnson: 'Non-Hertzian contact of elastic spheres'. In: *The mechanics of the contact between deformable bodies*, ed. by de Pater, Kalker (University Press, Delft 1975) pp. 26–40
10. K.N.G. Fuller, D. Tabor: Proc. Roy. Soc. Lond. A **345**, 327 (1975)
11. D. Maugis, M. Barquins, R. Courtel: Métaux, Corrosion, Industrie **605**, 1 (1976)
12. B.V. Derjaguin, V.M. Muller, Yu. P. Toporov: J. Colloid Interface Sci. **53**, 314 (1975)
13. D. Tabor: J. Colloid Interface Sci. **58**, 2 (1977)
14. V.M. Muller, V.S. Yushchenko, B.V. Derjaguin: J. Colloid Interface Sci. **77**, 91 (1980)
15. V.M. Muller, V.S. Yushchenko, B.V. Derjaguin: J. Colloid Interface Sci. **92**, 92 (1983)
16. M. Barquins, D. Maugis: J. Méc. Théor. Appl. **1**, 331 (1982)
17. I.N. Sneddon: Int. J. Eng. Sci. **3**, 47 (1965)
18. D. Maugis: J. Colloid Interface Sci. **150**, 243 (1992)
19. J.W. Harding, I.N. Sneddon: Proc. Camb. Phil. Soc. **41**, 16 (1945)
20. I.N. Sneddon: Proc. Glasgow Math. Assoc. **4**, 108 (1960)
21. S. Way: J. Appl. Mech. ASME **7** A147 (1940)
22. J.R. Barber: *Elasticity* (Kluwer Acad. Publ., 1992)
23. R.W. Little: *Elasticity* (Prentice Hall, 1973)
24. I.N. Sneddon: Proc. Camb. Phil. Soc. **42**, 29 (1946)
25. P.C. Paris, G.C. Sih: 'Stress analysis of cracks'. In: *Fracture toughness and testing applications*, ASTM, STP 381. (Philadelphia 1965) pp. 30–83
26. J. Boussinesq: C.R. Acad. Sci. **XCVI**, 245 (1883)
27. M.T. Huber: Ann. Phys. **14**, 153 (1904)
28. A.E.H. Love: Phil. Trans. **288**, 377 (1929)
29. S. Fuchs: Phys. Zeitschr. **14**, 1282 (1913)
30. W.B. Morton, L.J. Close: Phil. Mag. **43**, 320 (1922)
31. B.R. Lawn: J. Appl. Phys. **39**, 4828 (1968)
32. S. Timoshenko, J.N. Goodier: *Théorie de l'élasticité* (Béranger, Paris 1961)
33. T.C.T. Ting: Trans. ASME, J. Appl. Mech. **33**, 845 (1966)
34. S.C. Hunter: J. Mech. Phys. Solids **5**, 162 (1957)
35. H. Deresiewicz: Acta Mech. **6**, 110 (1968)
36. A.E.H. Love: *A treatise on mathematical theory of elasticity*, 4th edn. (Cambridge University Press, Cambridge 1952)
37. R.T. Shield: Z. Angew. Math. Phys. **18**, 682 (1967)
38. D. Maugis, M. Barquins: 'Fracture mechanics and adherence of viscoelastic solids'. In: *Adhesion and adsorption of polymers*, ed. by L.H. Lee (Plenum Press 1980) pp. 203–277

39. V.M. Muller, B.V. Derjaguin, Yu. P. Toporov: Colloids Surf. **7**, 251 (1983)
40. L.R. Fisher, J.N. Israelachvili: Colloids Surf. **3**, 303 (1981)
41. H.K. Christenson: J. Colloid Interface Sci. **104**, 234 (1985)
42. H.K. Christenson: J. Colloid Interface Sci. **121**, 170 (1988)
43. A. Fogden, L.R. White: J. Colloid Interface Sci. **138**, 414 (1990)
44. D. Maugis, B. Gauthier-Manuel: J. Adhesion Sci. Technol. **8**, 1311 (1994)
45. A.E.H. Love: Quat. J. Math. (Oxford) **10**, 161 (1939)
46. I.N. Sneddon: Proc. Camb. Phil. Soc. **44**, 492 (1948)
47. D. Maugis, M. Barquins: C.R. Acad. Sci. Paris B **291**, 161 (1980)
48. G.A.C. Graham: Int. J. Eng. Sci. **11**, 409 (1973)
49. R.A. Fisher: J. Agric. Sci. **16**, 492 (1926)
50. R.B. Heady, J.W. Cahn: Metall. Trans. **1**, 185 (1970)
51. J.N. Israelachvili: *Intermolecular and surface forces*, 2nd edn. (Academic Press 1992)
52. R.W. Coughlin, B. Elbirli, L. Vergara-Edwards: J. Colloid Interface Sci. **87**, 18 (1982)
53. M.A. Fortes: J. Colloid Interface Sci. **88**, 338 (1982)
54. E.A. Boucher, M.J.B. Evans, S. McGarry: J. Colloid Interface Sci. **89**, 154 (1982)
55. M.A. Erle, D.C. Dyson, N.R. Morrow: A. I. Ch. E. J. **17**, 115 (1971)
56. G. Lian, C. Thornton, M.J. Adams: J. Colloid Interface Sci. **161**, 138 (1993)
57. J.S. McFarlane, D. Tabor: Proc. Roy. Soc. London A **202**, 224 (1950)
58. W.J. O'Brien, J.J. Herman: J. Adhesion **5**, 91 (1973)
59. L.R. Fischer, J.N. Israelachvili: Chem. Phys. Lett. **76**, 325 (1980)
60. C. Mason, W.C. Clark: Chem. Eng. Sci. **20**, 859 (1965)
61. K. Hotta, K. Takeda, K. Ilnoya: Powder Technol. **10**, 231 (1974)
62. D.N. Mazzone, G.I. Tardos, R. Pfeffer: J. Colloid Interface Sci. **113**, 544 (1986)
63. A. Healey: Trans. Inst. Rubber Ind. **1**, 334 (1926)
64. M.J. Matthewson: Phil. Mag. **57**, 207 (1988)
65. P.R. Nayak: Trans ASME, J. Lubr. Technol. **93**, 398 (1971)
66. J. Peklenik: Proc. Inst. Mech. Eng. **182**, 108 (1967-1968)
67. J.A. Greenwood, J.B.P. Williamson: Proc. Roy. Soc. A **295**, 300 (1966)
68. J.B.P. Williamson: Proc. Inst. Mech. Eng. **182**, 21 (1967-1968)
69. J.A. Greenwood: Proc. Roy. Soc. A **393**, 133 (1984)
70. J.A. Greenwood: 'Contact of rough surfaces'. In: *Fundamentals of Friction: Macroscopic and Microscopic Processes*, ed. by I.L. Singer, H.M. Pollock (Kluwer Academic Publishers 1992) pp. 37–56
71. J.F. Archard: Tribology Int. Oct., 213–220 (1974)
72. J.I. McCool: Wear **107**, 37 (1986)
73. D.J. Whitehouse, J.F. Archard: Proc. Roy. Soc. A **316**, 97 (1970)
74. J.F. Archard: Proc. Roy. Soc. A **243**, 190 (1957)
75. K.L. Johnson: *Contact mechanics* (Cambridge University Press, Cambridge 1985)
76. D. Maugis: J. Adhesion Sci. Technol. **10**, 161 (1996)
77. D. Tabor: 'Interactions between surfaces: adhesion and friction'. In: *Surface physics of materials vol. II*, ed. by J.M. Blakely (Academic Press, New York 1975) pp. 475–529
78. B.V. Derjaguin: Wear **128**, 19 (1988)

5. Study of Some Geometries

5.1 Background

5.1.1 Bending of a Cantilever Beam

Let us consider (see Fig. 5.1) a beam of rectangular section (width b, height h) built in at $x = L$ and supporting a load P at $x = 0$. Angles and moments will be positive if they are in the direction of rotation of Ox to Oy. At a point of abcissa x the bending moment is thus

$$\mathcal{M}_f = -Px$$

and the shear force $T = P$. At a point $0 < x < L$ of the beam axis displayed in Fig. 5.1 the shear force is positive, the bending moment is negative and the curvature is positive. Consider a single element of the beam, of length $\mathrm{d}x$. If $p\,\mathrm{d}x$ is the load acting on the element $\mathrm{d}x$ (p is the weight per unit length in N/m), the shear force and the bending moment at $x + \mathrm{d}x$ are given respectively by

$$T_1 = T + \mathrm{d}T = T + p\,\mathrm{d}x \ ,$$

$$\mathcal{M}_1 = \mathcal{M} + \mathrm{d}\mathcal{M} = \mathcal{M} - T\,\mathrm{d}x - p\,\mathrm{d}x\frac{\mathrm{d}x}{2} \ .$$

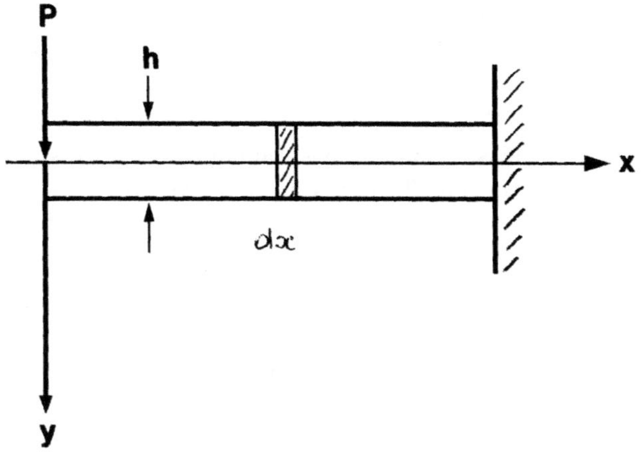

Fig. 5.1. Rectangular beam, of height h, built in at $x = L$

With limitation to the first-order terms, we get the relations

$$\frac{\mathrm{d}T}{\mathrm{d}x} = p \,, \tag{5.1}$$

$$\frac{\mathrm{d}\mathcal{M}_f}{\mathrm{d}x} = -T \tag{5.2}$$

between bending moment, shear force and lineic weight. Bending with only a bending moment (and thus constant) is said to be *pure bending* and bending with shear force (thus always accompanied by a bending moment) is said to be *simple bending.* We can see on a longitudinal section of the beam that fibres on the convex side are stretched, whereas they are compressed on the concave side.

The fibre which is not stretched nor compressed is called the *neutral fibre.* We can show that it is the centroidal axis of the cross-section. The deflection δ is the maximum vertical displacement of that fibre. If $b \ll h$, the beam is in a state of plane stress, which we shall assume implicitly in the following. If $b \gg h$ the beam is in a state of plane strain and all moduli E in formulae must be replaced by $E/(1-\nu^2)$ (see Chap. 2)

The equilibrium of forces acting on a thin slice of a bended beam shows that the angle of rotation $d\theta$ between two sections of curvilinear abcissae s and $s+\mathrm{d}s$ on the neutral axis is given by

$$\boxed{\frac{\mathrm{d}\theta}{\mathrm{d}s} = \frac{1}{R} = -\frac{\mathcal{M}_f}{EI}} \,, \tag{5.3}$$

where

$$I \equiv \frac{bh^3}{12}$$

is the moment of inertia of the cross-sectional area about the neutral axis and R is the radius of curvature. A fibre at a distance y from the neutral axis takes a radius of curvature $R-y$. An element of initial length $R\,\mathrm{d}\theta$ takes a length $(R-y)\,\mathrm{d}\theta$ and thus undergoes a strain $\varepsilon_x = -y/R$ (for an elastic deformation as well as for a nonelastic one). Application of Hooke's law shows that the stress at a height y (known as flexural stress) is given by

$$\boxed{\sigma_x = -\frac{y}{R}E = \frac{\mathcal{M}_f y}{I}} \,. \tag{5.4}$$

This flexural stress is maximum on the outermost fibres ($y = \pm h/2$):

$$\sigma_{\max} = \frac{6\mathcal{M}_f}{bh^2} \,.$$

With

$$\theta = \tan^{-1}\left(\frac{\mathrm{d}y}{\mathrm{d}x}\right),$$
$$\mathrm{d}s = \sqrt{\mathrm{d}x^2 + \mathrm{d}y^2},$$

we get

$$\frac{1}{R} = \frac{\mathrm{d}\theta}{\mathrm{d}s} = \frac{\mathrm{d}}{\mathrm{d}x}\left[\tan^{-1}\left(\frac{\mathrm{d}y}{\mathrm{d}x}\right)\right]\frac{\mathrm{d}x}{\mathrm{d}s}$$
$$= \frac{y''}{(1+y'^2)^{3/2}}, \tag{5.5}$$

and the equation of the elastica is given by

$$\boxed{\frac{y''}{(1+y'^2)^{3/2}} = -\frac{\mathcal{M}_f}{EI}}. \tag{5.6}$$

We will return on this point about peeling.

If deformations are small, we can make the following approximations:

$$\mathrm{d}s \approx \mathrm{d}x,$$
$$\theta \approx \tan\theta = \frac{\mathrm{d}y}{\mathrm{d}x},$$

and (5.3) is written as

$$\frac{\mathrm{d}^2 y}{\mathrm{d}x^2} = -\frac{\mathcal{M}_f}{EI}. \tag{5.7}$$

Integration of $y'' = Px/EI$ with $y = 0$ and $y' = 0$ at $x = L$ leads to

$$y = \frac{P}{6EI}(x^3 - 3L^2x + 2L^3).$$

The deflection at $x = 0$ is thus given by

$$\boxed{\delta = \frac{PL^3}{3EI} = \frac{4PL^3}{Ebh^3}}, \tag{5.8}$$

and the rotation of the free end is given by

$$\alpha = -\frac{PL^2}{2EI}. \tag{5.9}$$

The existence of a shear force provokes a translation of sections parallel to this shear force, and thus increases the deflection. Let us compute it with the help of the Castigliano theorem (Sect. 3.2), which states that

$$\delta = \frac{\mathrm{d}U_{\mathrm{E}}}{\mathrm{d}P}.$$

The elastic energy contained in the bended beam is given by

$$\begin{aligned} U_{\mathrm{E}} &= \int_0^L \left(\frac{\mathcal{M}_f^2}{2EI} + \chi \frac{T^2}{2\mu S} \right) \mathrm{d}x \\ &= \frac{2P^2L^3}{Ebh^3} + \chi(1+\nu)\frac{P^2L}{Ebh} , \end{aligned} \tag{5.10}$$

where χ is the *shear factor* which takes into account the distribution of shear stresses in the section (S/χ is the reduced section). For a rectangular section, this distribution is parabolic (shear stress zero at the surface and maximum on the neutral axis) and the shear factor is 6/5. We immediately deduce that

$$\delta = \frac{4PL^3}{Ebh^3} \left[1 + \frac{3}{5}(1+\nu)\frac{h^2}{L^2} \right] . \tag{5.11}$$

The correction term is less than 5% for $L > 4h$. The radius of curvature remains given by (5.3).

5.1.2 Beam on Elastic Foundation

Let us consider a beam that rests over its length on an elastic foundation, such that the reaction at any point is proportional to the displacement y undergone by the beam. This problem was examined for the first time by Winkler in 1867.

Using (5.1) and (5.2) between $\mathcal{M}_f$, T and p, (5.7) leads to

$$EI\frac{\mathrm{d}^4 y}{\mathrm{d}x^4} = p . \tag{5.12}$$

This equation implies small curvatures. In the part free of the load, the only pressure exerted on the beam is the elastic reaction that we will write as $p = -ky$, where k (in N/m^2) is the *foundation modulus*. Letting

$$\boxed{\lambda^4 = \frac{k}{4EI} = \frac{3k}{Ebh^3}} , \tag{5.13}$$

(5.12) becomes

$$\boxed{\frac{\mathrm{d}^4 y}{\mathrm{d}x^4} + 4\lambda^4 y = 0} , \tag{5.14a}$$

a differential equation of the fourth order with constant coefficients. With $\sigma_y = ky/b$, this equation is also written

$$\frac{\mathrm{d}^4 \sigma_y}{\mathrm{d}x^4} + 4\lambda^4 \sigma_y = 0 . \tag{5.14b}$$

Let $y = \mathrm{e}^{qx}$. We get

$$(q^4 + 4\lambda^4)\mathrm{e}^{qx} = 0 ,$$

which is satisfyied if q is a root of the characteristic equation

$$q^4 + 4\lambda^4 = 0 ,$$

i.e. $q = \sqrt[4]{-4}\lambda$. There are thus four distinct roots

$$\begin{aligned} q_1 &= (1+\mathrm{i})\lambda , \\ q_2 &= (-1+\mathrm{i})\lambda , \\ q_3 &= -(1+\mathrm{i})\lambda , \\ q_4 &= (1-\mathrm{i})\lambda . \end{aligned}$$

The general solution is thus

$$y = c_1\mathrm{e}^{q_1 x} + c_2\mathrm{e}^{q_2 x} + c_3\mathrm{e}^{q_3 x} + c_4\mathrm{e}^{q_4 x} ,$$

with $c_1 = \overline{c}_4$ and $c_2 = \overline{c}_3$, which can also be written as

$$\begin{aligned} y = {} & C_1 \sin\lambda x \sinh\lambda x + C_2 \sin\lambda x \cosh\lambda x \\ & + C_3 \cos\lambda x \sinh\lambda x + C_4 \cos\lambda x \cosh\lambda x , \end{aligned} \tag{5.15a}$$

or else

$$y = \mathrm{e}^{-\lambda x}(A\sin\lambda x + B\cos\lambda x) + \mathrm{e}^{\lambda x}(C\sin\lambda x + D\cos\lambda x) . \tag{5.15b}$$

5.1.3 Elastoplastic Bending

Let us consider a perfect elastoplastic material defined by the stress–strain diagram of Fig. 5.2 where σ_0 is the elastic limit (we assume that it is the same in compression as in traction). From O to A we have an elastic loading, from A to B a plastic loading, from B to D an elastic unloading, from D to E an inverse plastic loading.

As long as the bending moment is small, the bending stresses remain in the elastic domain and their distribution with y is given by (5.4) and displayed in Fig. 5.3a. Then, when the bending moment increases, the outermost fibres become plastified and the diagram is that of Fig. 5.3b where h_0 is the thickness of the zone which is still elastic. Finally, when the bending moment reaches the *fully plastic moment* $\mathcal{M}_\mathrm{P}$, the diagram becomes that of Fig. 5.3b'.

The bending moment is defined from the stresses in a section of width b by

$$\mathcal{M} = -\int_{-h/2}^{h/2} b\sigma_x(y)y\,\mathrm{d}y . \tag{5.16}$$

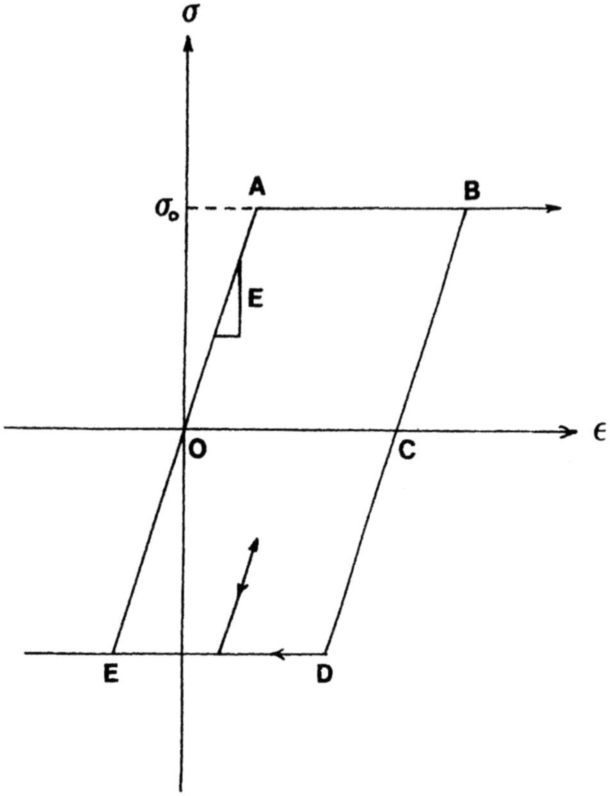

Fig. 5.2. Stress–strain diagram of a perfect elastoplastic material

Let $\mathcal{M}_e$ be the elastic moment such that the shape remains triangular, the outer fibre just reaching the elastic limit σ_0, and let K_e be the corresponding curvature. From (5.4) we have

$$\mathcal{M}_e = \frac{I\sigma(y)}{y} = \frac{bh^2}{6}\sigma_0 , \tag{5.17}$$

$$K_e = \frac{1}{R_e} = \frac{\sigma(y)}{yE} = \frac{2\sigma_0}{Eh} . \tag{5.18}$$

Let $\mathcal{M}_p$ be the absolute value of the fully plastic moment (see Fig. 5.3b'), which is, from (5.16)

$$\boxed{\mathcal{M}_p = \frac{bh^2}{4}\sigma_0} . \tag{5.19}$$

We have thus

$$\mathcal{M}_e = \frac{2}{3}\mathcal{M}_p . \tag{5.20}$$

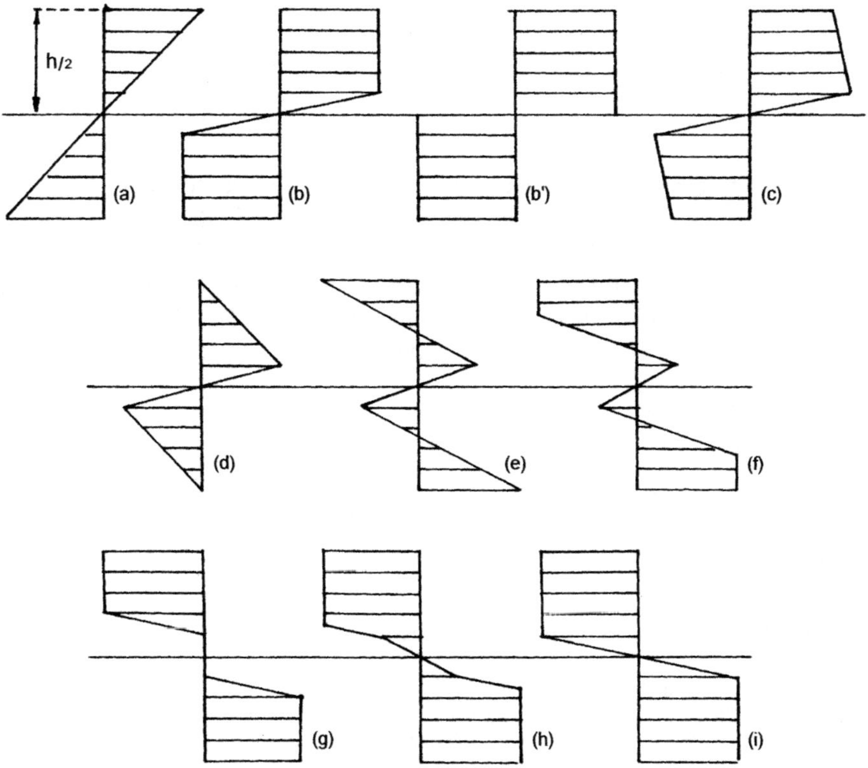

Fig. 5.3. Diagram of stress distribution in a section during elastoplastic bending.

The curvature–moment relations and the dissipated energy have been studied by Hodge [1] and by Kim and Kim [2] in the framework of the elastoplastic peeling. We will reduce the bending moments to this fully plastic moment, and the curvatures to the curvature K_e letting

$$m = \frac{\mathcal{M}}{\mathcal{M}_p} \leq 1 , \tag{5.21}$$

$$k = \frac{K}{K_e} . \tag{5.22}$$

Then there are four stages to the development of our discussion.

First, in the elastic domain (branch OA of Fig. 5.2), (5.3) can be written as

$$K = \frac{12\mathcal{M}}{Ebh^3} , \qquad \mathcal{M} \leq \mathcal{M}_e .$$

Hence

$$\frac{K}{K_e} = \frac{3}{2}\frac{\mathcal{M}}{\mathcal{M}_p} ,$$

or else

$$\boxed{m = \frac{2}{3}k\,, \qquad 0 \le k \le 1}\,. \tag{5.23}$$

Second, let us compute the bending moment and the curvature in the elasto-plastic loading of Fig. 5.3b. Let the interval $(-h_0/2,\, h_0/2)$ be the thickness on which the deformation is purely elastic. We have seen that if R is the radius of curvature of the neutral axis, the strain at height y is y/R (elastic strain or not). At height $h_0/2$ the strain is thus $\varepsilon_\mathrm{p} = h_0/2R$, and hence (by Hooke's law) $\sigma_0 = h_0 E/2R$. Thus we have

$$K = \frac{1}{R} = \frac{2\sigma_0}{h_0 E} \tag{5.24a}$$

and hence

$$k = \frac{h}{h_0} \ge 1\,. \tag{5.24b}$$

From (5.16), the bending moment is given by

$$\mathcal{M} = 2b \int_0^{h_0/2} \sigma_x(y) y \,\mathrm{d}y + 2b \int_{h_0/2}^{h/2} \sigma_0 y \,\mathrm{d}y\,,$$

with

$$\frac{\sigma_x}{\sigma_0} = \frac{y}{h_0/2}\,,$$

i.e.

$$\mathcal{M} = \frac{b\sigma_0 h^2}{4}\left(1 - \frac{h_0^2}{3h^2}\right)$$

or else, using (5.19) and (5.24b)

$$\boxed{m = 1 - \frac{1}{3k^2}\,, \qquad k \ge 1}\,. \tag{5.25}$$

These relations between curvature and bending moment are displayed in Fig. 5.4. When $m \to 1$, $h_0 \to 0$ and the curvature increases indefinitely. It is said that a plastic knee-joint is formed when the curvature increases at constant bending moment.

Third, at point B (moment m_B, curvature k_B, thickness of the elastic zone h_B, distribution $\sigma_\mathrm{B}(y)$), let the moment decrease, which amounts to adding a negative moment to m_B. All fibres unload elastically. The stress state is that of Fig. 5.3b, to which is added a multiple of the stress distribution of Fig. 5.3a with the sign changed. We obtain the diagram of Fig. 5.3c. From this stress distribution we can compute the bending moment and curvature, since the latter, from (5.4), is proportional to the slope of the distribution at $y = 0$. We can thus construct the diagram for moment–curvature, as shown in Fig. 5.4. At point B any quantity can be expressed as a function of k_B:

$$m_B = 1 - \frac{1}{3k_B^2} ,$$

$$\frac{h_B}{h} = \frac{1}{k_B} ,$$

$$\frac{\sigma_B(y)}{\sigma_0} = \frac{2y}{h_B} = \frac{2y}{h} k_B , \qquad 0 \le y \le \frac{h_B}{2} , \qquad 0 \le \frac{2y}{h} \le \frac{1}{k_B} ,$$

$$\frac{\sigma_B(y)}{\sigma_0} = 1 , \qquad h_B \le 2y \le h , \qquad \frac{1}{k_B} \le \frac{2y}{h} \le 1 .$$

Let us superimpose onto this distribution a distribution such that

$$m' = \frac{2}{3} k' .$$

From (5.16) the bending moment is the superimposition of two moments, thus:

$$m = m_B + m'$$

and the curvature is given by the new slope at the origin:

$$k = k_B + k'$$

hence

$$\boxed{m = 1 - \frac{1}{3k_B^2} + \frac{2}{3}(k - k_B)} . \tag{5.26}$$

At point D where the bending moment cancels, the beam acts under the residual stresses and there is a spontaneous curvature given by

$$k = k_B + \frac{1}{2k_B^2} - \frac{3}{2} . \tag{5.27}$$

The straight lines OA and BE are parallel, with $\overline{BE} = 2\overline{AO}$. Hence $m_B - m_E = 2m_e = 4/3$, and

$$k_B - k_E = 2 ,$$

with the curvature k_E positive only if $k_B > 2$.

Fourth, beyond the point E, corresponding to Fig. 5.3e, with coordinates

$$k_E = k_B - 2 ,$$

$$m_E = 1 - \frac{1}{3k_B^2} - \frac{4}{3} ,$$

a second plastification appears, and the stress distribution is limited to $-\sigma_0$ at the top level (Fig. 5.3f). This amounts to a superimposition onto the distribution $(m_{\rm B}, k_{\rm B})$ of a distribution limited to $-2\sigma_0$, or (which is equivalent) to superimposition of twice a distribution (m'', k'') limited to $-\sigma_0$:

$$\begin{aligned} m &= m_{\rm B} + 2m'' , \\ k &= k_{\rm B} + 2k'' , \end{aligned}$$

with

$$m'' = -\left(1 - \frac{1}{3k''^2}\right) .$$

Hence

$$m = \left(1 - \frac{1}{3k_{\rm B}^2}\right) - 2\left(1 - \frac{1}{3\big((k - k_{\rm B})/2\big)^2}\right)$$

or, for the inverse plastic loading,

$$\boxed{m = -1 - \frac{1}{3k_{\rm B}^2} + \frac{8}{3(k_{\rm B} - k)^2}} . \tag{5.28}$$

In particular, to obtain a null curvature (Fig. 5.3g) one must apply a moment

$$m_{\rm F} = -1 + \frac{7}{3k_{\rm B}^2} .$$

The energy dissipated per unit length when a beam is bent up to the curvature $k_{\rm B}$ and then straightened is in the general case represented by the area OABDEFO of Fig. 5.4, i.e. the area under the curves defined by (5.23, 25, 26 and 28). Thus

$$W_{\rm P} = \oint \mathcal{M}(K)\, {\rm d}K = \mathcal{M}_{\rm p} K_{\rm e} \oint m(k)\, {\rm d}k .$$

However, when the imposed curvature is less than unity, events remain in the elastic regime and no energy is dissipated, and when $k_{\rm B} \leq 2$ the branch EF is not involved. Let us compute these various integrals:

$$1: \quad \int_{\rm O}^{\rm A} m(k)\, {\rm d}k = \int_0^1 \frac{2k}{3}\, {\rm d}k = \frac{1}{3} , \tag{5.29}$$

$$2: \quad \int_A^B m(k)\, {\rm d}k = \int_1^{k_{\rm B}} \left(1 - \frac{1}{3k^2}\right) {\rm d}k = k_{\rm B} - \frac{4}{3} + \frac{1}{3k_{\rm B}} , \tag{5.30}$$

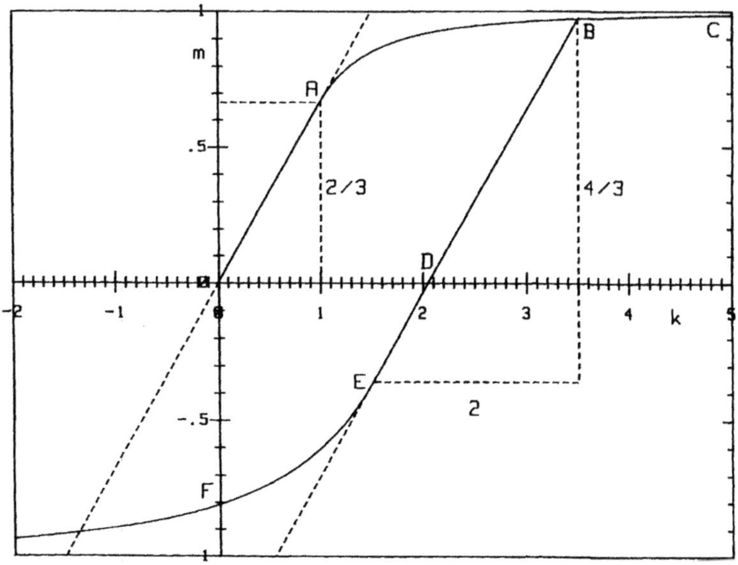

Fig. 5.4. Elastoplastic bending of a beam, showing moment vs. curvature

$$3: \quad \int_{\mathrm{B}}^{\mathrm{E}} m(k)\,\mathrm{d}k = \int_{k_{\mathrm{B}}}^{k_{\mathrm{B}}-2} \left(1 - \frac{1}{3k_{\mathrm{B}}^2} - \frac{2}{3}(k_{\mathrm{B}} - k)\right) \mathrm{d}k$$

$$= \frac{2}{3k_{\mathrm{B}}^2} - \frac{2}{3}\,, \tag{5.31}$$

$$3': \quad \int_{\mathrm{B}}^{\mathrm{E'}} m(k)\,\mathrm{d}k = \int_{k_{\mathrm{B}}}^{0} \left(1 - \frac{1}{3k_{\mathrm{B}}^2} - \frac{2}{3}(k_{\mathrm{B}} - k)\right) \mathrm{d}k$$

$$= \frac{k_{\mathrm{B}}^2}{3} - k_{\mathrm{B}} + \frac{1}{3k_{\mathrm{B}}}\,, \tag{5.32}$$

$$4: \quad \int_{\mathrm{E}}^{\mathrm{F}} m(k)\,\mathrm{d}k = \int_{k_{\mathrm{B}}-2}^{0} \left(-1 - \frac{1}{3k_{\mathrm{B}}^2} + \frac{8}{3(k - k_{\mathrm{B}})^2}\right) \mathrm{d}k$$

$$= k_{\mathrm{B}} - \frac{10}{3} + \frac{9}{3k_{\mathrm{B}}} - \frac{2}{3k_{\mathrm{B}}^2}\,. \tag{5.33}$$

The energy dissipated corresponds to $(1+2+3')$ if $1 \le k_{\mathrm{B}} \le 2$, and to $(1+2+3+4)$ if $k_{\mathrm{B}} \ge 2$. We have thus

$$W_{\mathrm{p}} = \mathcal{M}_{\mathrm{p}} K_{\mathrm{e}} \left(\frac{k_{\mathrm{B}}^2}{3} - 1 + \frac{2}{3k_{\mathrm{B}}}\right), \qquad 1 \le k_{\mathrm{B}} \le 2\,, \tag{5.34}$$

$$W_{\mathrm{p}} = \mathcal{M}_{\mathrm{p}} K_{\mathrm{e}} \left(2k_{\mathrm{B}} - 5 + \frac{10}{3k_{\mathrm{B}}}\right), \qquad k_{\mathrm{B}} \ge 2\,. \tag{5.35}$$

For a given beam, the dissipated energy therefore depends only on the maximum curvature reached. Note that

$$\mathcal{M}_{\mathrm{p}} K_{\mathrm{e}} = \frac{\sigma_0^2}{2E} bh$$

is the maximum elastic energy which can be stored per unit length of beam. Equations (5.34) and (5.35) were given by Kim and Kim [2] in 1988.

5.2 The Double Cantilever Beam

This method of the *double cantilever beam (DCB)*, also named the *cleavage technique* was used for the first time by Obreimov [3] in 1930 to determine the surface energy of mica, and then by Benbow and Roesler [4] in 1957.

5.2.1 Elementary Theory

Let us consider (see Fig. 5.5) two beams of width b, thickness h in adhesive contact with one another. Let L be the length of the arms of the double cantilever, and δ the opening under a force P or a wedge (it is the sum of the elastic displacements of the two arms). The theory of beams tells that, neglecting shear forces and rotation of the built-in section, we have

$$\delta = 2 \frac{4PL^3}{bEh^3} , \tag{5.36}$$

which is the first equation of state of the system. Elastic and potential energies (at constant P) being $\frac{1}{2}P\delta$ and $-P\delta$, we have (with $\mathrm{d}A = -b\,\mathrm{d}L$) the second equation of state of the system

$$\boxed{G = \frac{12P^2L^2}{b^2Eh^3} = \frac{3Eh^3\delta^2}{16L^4} = \frac{12\mathcal{M}_f^2}{b^2Eh^3}} , \tag{5.37a}$$

where $\mathcal{M}_f = PL$ is the bending moment at the built-in section. Correct formulae for G or γ seem to have been given for the first time by Gilman [5] in 1960.

Note that it is possible to eliminate Young's modulus by measuring simultaneously force and displacement, whereby

$$G = \frac{3P\delta}{2bL} . \tag{5.37b}$$

Note also that if the critical load P_{c} and the critical displacement δ_{c} depend on the length L of the arm (the length of the crack), the critical bending moment does not depend on it (at least in this elementary theory).

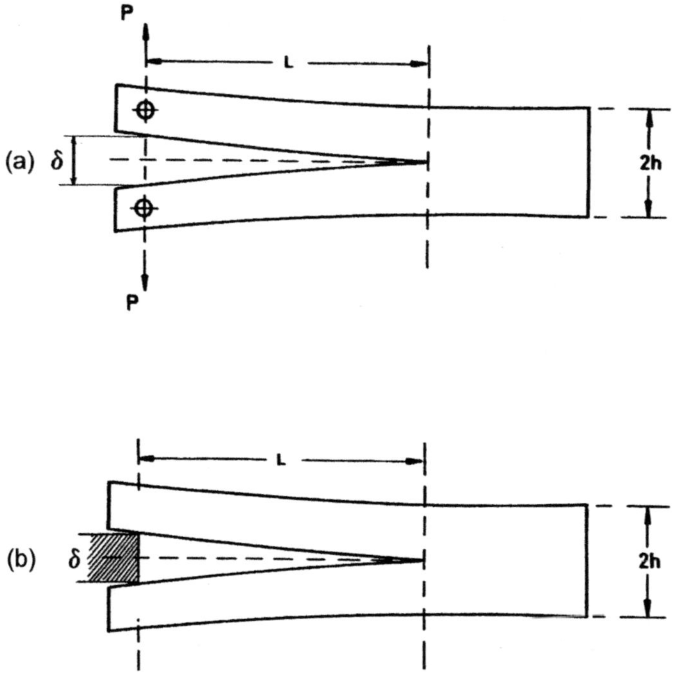

Fig. 5.5. Schema of a double cantilever beam, at fixed load (**a**) or at fixed grips (**b**)

If a testing machine of stiffness k_m is used, and if a displacement

$$\Delta = \delta + \frac{P}{k_\mathrm{m}} \tag{5.38}$$

is imposed, we have, using (3.12),

$$\boxed{G = \frac{3Eh^3}{16L^4} \frac{\Delta^2}{\left(1 + (Ebh^3)/(8k_\mathrm{m}L^3)\right)^2}} \; . \tag{5.37c}$$

Equilibrium defined by $G = w$ is unstable at fixed load, stable at fixed grips δ, and neutral at fixed bending moment. For an imposed displacement Δ, the stability is given by

$$\left(\frac{\partial G}{\partial A}\right)_\Delta = \frac{3Eh^3\delta^2}{4bL^5} \left[1 - \frac{3Ebh^3}{16L^3} \frac{1}{k_\mathrm{m} + (Ebh^3)/(8L^3)}\right] > 0 \, , \tag{5.39}$$

i.e.

$$k_\mathrm{m} > \frac{Ebh^3}{16L^3} \, , \tag{5.40}$$

which is, indeed, the condition of (3.19). In effect, at equilibrium we have

$$P^2 = \frac{wEb^2h^3}{12L^2} = \frac{wEb^2h^3}{12}\sqrt{\frac{w}{3Eh^3}\frac{4}{\delta}} ,$$

and hence

$$-\frac{\mathrm{d}P}{\mathrm{d}\delta} = \frac{2}{P\delta^2}\frac{wEb^2h^3}{12}\sqrt{\frac{w}{3Eh^3}} = \frac{Ebh^3}{16L^3} .$$

Let L_0 be the initial length of the arm and Δ_0 the displacement corresponding to $G = w$. With

$$k_0 = \left(\frac{\partial P}{\partial \delta}\right)_A = \frac{Ebh^3}{8L_0^3} , \tag{5.41}$$

(5.37c) becomes

$$\frac{G}{w} = \left(\frac{\Delta}{\Delta_0}\right)^2 \left(\frac{L_0}{L}\right)^4 \left[\frac{1 + k_0/k_\mathrm{m}}{1 + (k_0/k_\mathrm{m})(L_0/L)^3}\right]^2 \tag{5.42}$$

and is displayed in Fig. 5.6 for $k_\mathrm{m}/k_0 = 0.25$, 0.5 and 10, and for various imposed displacements Δ/Δ_0. For $k_\mathrm{m} < k_0/2$, G begins to increase and then decreases (with a maximum for a length L given by equality in (5.40)), until $G = w$ once more, this equilibrium therefore being stable.

5.2.2 Influence of the Shear Force

From the stored elastic energy, given by (5.10), or from the deflection, given by (5.11), we can deduce immediately that

$$G = \frac{12P^2L^2}{Eb^2h^3}\left(1 + \frac{1+\nu}{5}\frac{h^2}{L^2}\right) , \tag{5.43a}$$

$$G = \frac{3Eh^3\delta^2}{16L^4}\frac{1 + (1+\nu)h^2/5L^2}{[1 + 3(1+\nu)h^2/5L^2]^2} . \tag{5.43b}$$

For $L = 2h$ and $\nu = 0.25$, the corrections necessary to bring these expressions in line with the elementary theory (5.37a) are about 6% for that in P^2, but they reach 25% for that in δ^2. In the cleavage technique, a wedge of thickness δ is driven in, and the length L of the arms are measured after a return to equilibrium. This length is the shorter, the higher the energy of adhesion. Application of the elementary formula can lead to considerable errors if h/L is not sufficiently small.

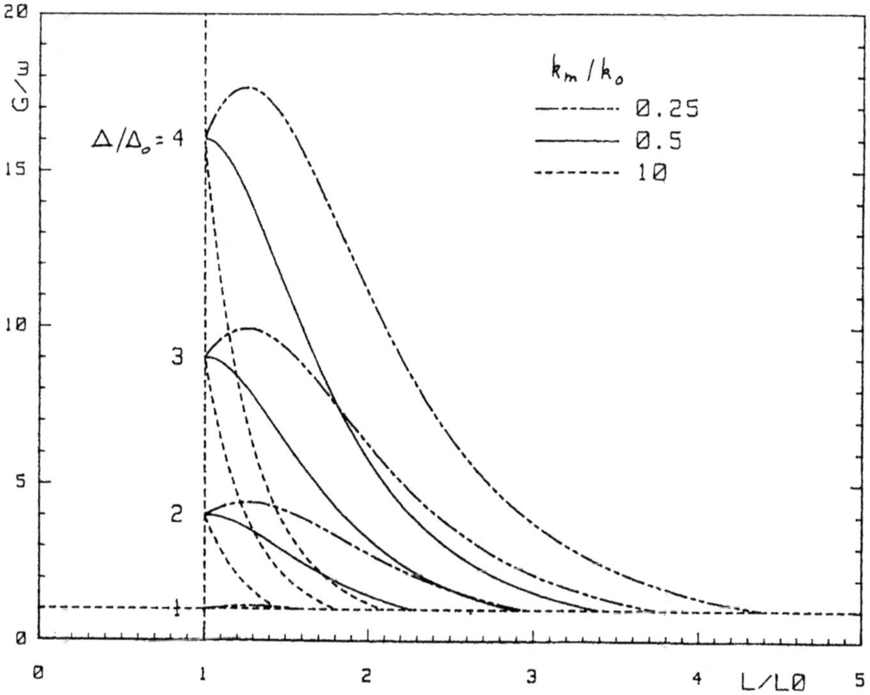

Fig. 5.6. Variation of the energy release rate G with the length L of the crack for various imposed displacements Δ. There is influence from the stiffness of the measuring apparatus on the return to equilibrium

5.2.3 Rotation of the Built-in Section

Hitherto, we have assumed that elastic energy was only stored in the bent arms of the DCB. This is not exact since, as noted by Berry [6] in 1963, the uncleaved part is not rigid, but is deformed in a manner which depends on local stresses at the crack tip, and hence on the length L of the crack. This problem was further examined by Gillis and Gilman [7] in 1964. They assumed that, instead of a perfect built-in section at $x = L$ (a tangential connection), there is a small rotation α of the sections. A term αL must thus be added to δ in (5.11), where α is proportional to P and a power function (n) of L:

$$\alpha = \frac{kL^n}{EI} P\,,$$

or else

$$\alpha = \frac{c}{EI} \frac{L^n}{h^{n-2}} P\,,$$

to make the constant adimensional (Wiederhorn *et al.* [8]). We can see furthermore that the rotation, and thus the strain, in the uncleaved part is proportional to the bending moment PL at $x = L$, i.e. that $n = 1$. This can be expressed as

$$\alpha L = \frac{4PL^3}{bEh^3} 3c\frac{h}{L} . \tag{5.44}$$

We thus have for the DCB

$$\delta = \frac{8PL^3}{Ebh^3}\left[1 + 3c\frac{h}{L} + \frac{3}{5}(1+\nu)\frac{h^2}{L^2}\right] , \tag{5.45}$$

and hence

$$G = \frac{12P^2L^2}{Eb^2h^3}\left[1 + 2c\frac{h}{L} + \frac{1+\nu}{5}\frac{h^2}{L^2}\right] , \tag{5.46a}$$

with the value $c = 0.63$ obtained by experimental adjustment.

Mostovoy *et al.* [9] in 1967 took into account the rotation of the built-in section by empirically augmenting by a length the length L_0 of the arms of the DCB, and obtained, for $\nu = 1/3$,

$$G = \frac{12P^2L^2}{Eb^2h^3}\left[\left(1 + \frac{L_0}{L}\right)^2 + \frac{h^2}{3L^2}\right] , \tag{5.46b}$$

the experimental adjustment giving $L_0 = 0.6\, h$.

These arguments were improved by Kanninen [10] in 1973[1] by modelling the lower half of the uncleaved part (of length L^*) as an elastic foundation of modulus k (Fig. 5.7). In the chosen axes, P, T, and y'' are positive and $\mathcal{M}_f < 0$. Compared with Sect. 5.1 the x are changed into $-(L+x)$. For $x \leq 0$ we must solve (5.14) with $\lambda = 0$ and with, for boundary conditions,

$$y''(-L) = 0 ,$$
$$y'''(-L) = \frac{P}{EI} .$$

The solution is

$$y = \frac{P}{6EI}(x^3 + 3Lx^2) + D_3x + D_4 , \qquad x \leq 0 . \tag{5.47}$$

For $x \geq 0$, the solution given by (5.15) must take the same values y, y', y'' and y''' at $x = 0$ as the previous equation, and satisfy to the boundary conditions

$$y''(L^*) = 0 ,$$
$$y'''(L^*) = 0 .$$

[1] In fact these calculations are the same as those done well before for peeling by Spies [11] in 1953, by Inoue and Kobatake [12] in 1959, and by Kaelble [13] in 1960.

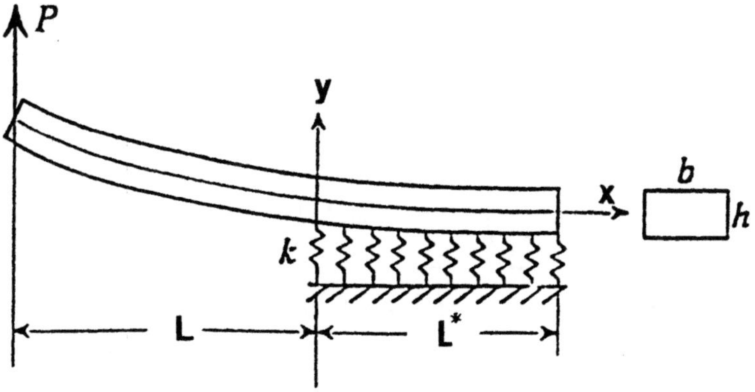

Fig. 5.7. Upper half of a DCB. The reaction forces of the lower part are modelled by an elastic foundation of modulus k

We will considerably simplify the analysis of Kanninen by cancelling the terms exponentially increasing with x, and keeping only the solution

$$y = \mathrm{e}^{-\lambda x}(A \sin \lambda x + B \cos \lambda x) , \qquad x \geq 0 . \tag{5.48}$$

Hence

$$\begin{aligned}
y' &= -\lambda \mathrm{e}^{-\lambda x}[(A+B)\sin \lambda x + (B-A)\cos \lambda x] ,\\
y'' &= 2\lambda^2 \mathrm{e}^{-\lambda x}[B \sin \lambda x - A \cos \lambda x] ,\\
y''' &= 2\lambda^3 \mathrm{e}^{-\lambda x}[(A-B)\sin \lambda x + (A+B)\cos \lambda x] .
\end{aligned}$$

The identity of the values of y and its derivatives at $x = 0$ leads to the following relations:

$$\begin{aligned}
y(0) &= \mathrm{D}_4 = B ,\\
y'(0) &= \theta_0 = \mathrm{D}_3 = \lambda(A-B) ,\\
y''(0) &= \frac{1}{R_0} = \frac{PL}{EI} = -2\lambda^2 A ,\\
y'''(0) &= \frac{P}{EI} = 2\lambda^3(A+B) ,
\end{aligned}$$

i.e.

$$A = -\frac{P}{2EI\lambda^3}\lambda L , \tag{5.49}$$

$$B = \frac{P}{2EI\lambda^3}(1+\lambda L) . \tag{5.50}$$

Equations (5.47) and (5.48) become

$$y = \frac{P}{2EI\lambda^3}\left[\frac{\lambda^3 x^3}{3} + L\lambda^3 x^2 - \lambda(1+2\lambda L)x + (1+\lambda L)\right], \qquad x \le 0, \tag{5.51}$$

$$\boxed{y = \frac{PL}{2EI\lambda^2}e^{-\lambda x}\left[\left(1+\frac{1}{\lambda L}\right)\cos\lambda x - \sin\lambda x\right], \qquad x \ge 0}. \tag{5.52}$$

The deflection of the beam at $x = -L$ is given by

$$\delta = \frac{P}{6EI\lambda^3}(2\lambda^3 L^3 + 6\lambda^2 L^2 + 6\lambda L + 3), \tag{5.53}$$

and the rotation of the built-in section at $x = 0$ by

$$y'(0) = -\frac{P}{2EI\lambda^2}(1+2\lambda L), \tag{5.54a}$$

i.e.

$$\theta_0 = -\frac{1}{\lambda R_0}\left(1+\frac{1}{2\lambda L}\right). \tag{5.54b}$$

For a DCB the opening is twice the deflection:

$$\delta = \frac{8PL^3}{Ebh^3}\left[\left(1+\frac{1}{\lambda L}\right)^3 + \frac{1}{2\lambda^3 L^3}\right], \tag{5.55a}$$

and from the potential and elastic energies we can obtain

$$\boxed{G = \frac{12P^2L^2}{b^2Eh^3}\left(1+\frac{1}{\lambda L}\right)^2}, \tag{5.56a}$$

$$G = \frac{3Eh^3\delta^2}{16L^4}\frac{(1+1/\lambda L)^2}{\left[(1+1/\lambda L)^3 + 1/2\lambda^3L^3\right]^2}, \tag{5.57a}$$

$$G = \frac{3P\delta}{2bL}\frac{(1+1/\lambda L)^2}{(1+1/\lambda L)^3 + 1/2\lambda^3L^3}.$$

Neglecting the correction term in the denominator, the last two equations can be simplified as

$$\boxed{G = \frac{3Eh^3\delta^2}{16L^4}\frac{1}{(1+1/\lambda L)^4}}, \tag{5.58a}$$

$$G = \frac{3P\delta}{2bL}\frac{1}{1+1/\lambda L}.$$

It remains to determine the foundation modulus k appearing in λ, as per (5.13). Concerning this point, Fig. 5.7 is misleading because it suggests that k represents the stiffness of the lower arm of the double cantilever. Nothing of the kind applies. If the DCB is symmetrical, or if the lower beam has a high modulus, the line situated at $y = -h/2$ is not deformed. It is the neutral fibre, by its displacement of the quantity y, which compresses or extends the lower half of the upper beam. The mean strain of this part is $\varepsilon_y = y/(h/2)$, and its mean stress is $\sigma_y = ky/b$; hence by writing the relation $\sigma = \varepsilon E$ between mean stress and mean strain, we get

$$\boxed{k = \frac{2Eb}{h}} \tag{5.59a}$$

and

$$\boxed{\lambda = \frac{6^{1/4}}{h}} \, . \tag{5.59b}$$

These two last formulae were given by Inoue and Kobatake [12] in 1959. The deflection due to the rotation of the built-in section, given in (5.54), is

$$\theta_0 L = \frac{12PL^3}{Ebh^3} 6^{-1/4} \frac{h}{L} \left(1 + \frac{1}{2\lambda L}\right) , \tag{5.60}$$

which must be compared with (5.44). We can see that the constant c of Wiederhorn *et al.* [8] is in fact $c = 6^{-1/4} = 0.64$.

Using (5.59b), the above equations can be written as

$$\delta = \frac{8PL^3}{Ebh^3} \left(1 + 1.92\frac{h}{L} + 1.22\frac{h^2}{L^2} + 0.39\frac{h^3}{L^3}\right) , \tag{5.55b}$$

$$G = \frac{12P^2L^2}{Eb^2h^3} \left(1 + 0.64\frac{h}{L}\right)^2 , \tag{5.56b}$$

$$G = \frac{3Eh^3\delta^2}{16L^4} \frac{(1 + 0.64h/L)^2}{\left[(1 + 0.64h/L)^3 + 0.13h^3/L^3\right]^2} , \tag{5.57b}$$

$$G = \frac{3Eh^3\delta^2}{16L^4} \frac{1}{(1 + 0.64h/L)^4} , \tag{5.58b}$$

$$G = \frac{3P\delta}{2bL} \frac{1}{1 + 0.64h/L} .$$

Equation (5.58b) differs from (5.57b) by less than 6% for $L > h$. Equation (5.56b) and (5.58b) show that the correction amounts to replacing L by $L + 0.64h$; hence the expression *augmented double cantilever beam* used by Kanninen [10].

Equation (5.56b) is a result close to that obtained by a collocation method[2] by Strawley and Gross [14] in 1967:

$$G = \frac{12P^2L^2}{Eb^2h^3}\left(1 + 0.687\frac{h}{L}\right)^2 , \tag{5.61}$$

which is valid for $L > h/2$.

The method of the beam on elastic foundation was used independently by Freiman *et al.* [15] in 1973 to obtain

$$G = \frac{12P^2L^2}{Eb^2h^3}\left[\left(1 + 0.64\frac{h}{L}\right)^2 + \frac{1+\nu}{6}\frac{h^2}{L^2}\right] . \tag{5.62}$$

Computations of elasticity by Fichter [16] in 1983, using Fourier transforms and the Wiener–Hopf technique, gave as an approximate solution

$$G = \frac{12P^2L^2}{Eb^2h^3}\left(1 + 0.6728\frac{h}{L} + 0.0377\frac{h^3}{L^3}\right)^2 , \tag{5.63}$$

valid until $h/L = 1$. Finally, Foote and Buchwald [17], using the same technique, obtained in 1985 an exact solution valid for any h/L. An approximation valid for $L^* > 2h$, differing from the exact solution by less than 2% is

$$G = \frac{12P^2L^2}{Eb^2h^3}\left[1 + 0.673\frac{h}{L} + \frac{1}{\sqrt{6\pi}}\left(\frac{h}{L}\right)^{3/2} - \frac{h/L}{2.823\,(L/h)^{0.619} + 1.486}\right]^2 . \tag{5.64}$$

For $L \ll h$ we can obtain once more the formula

$$G = \frac{2P^2}{\pi Eb^2L}$$

given by Irwin [18] for a lateral crack in a plate, valid for $L < h/100$.

These various formulae are compared in Fig. 5.8 with the solution by Foote and Buchwald [17] given in (5.64). For $h/L < 0.3$, the more precise formulae are those of Freiman *et al.* [15] and of Fichter [16]. For short arms the best approximation is that of Freiman *et al.* [15]. The formula of Strawley and Gross [14], shown in (5.61), remains the simplest and is a good approximation for $h/L < 2$.

[2] Research into a stress function solution of the biharmonic equation satisfying the boundary conditions at some points on the boundary.

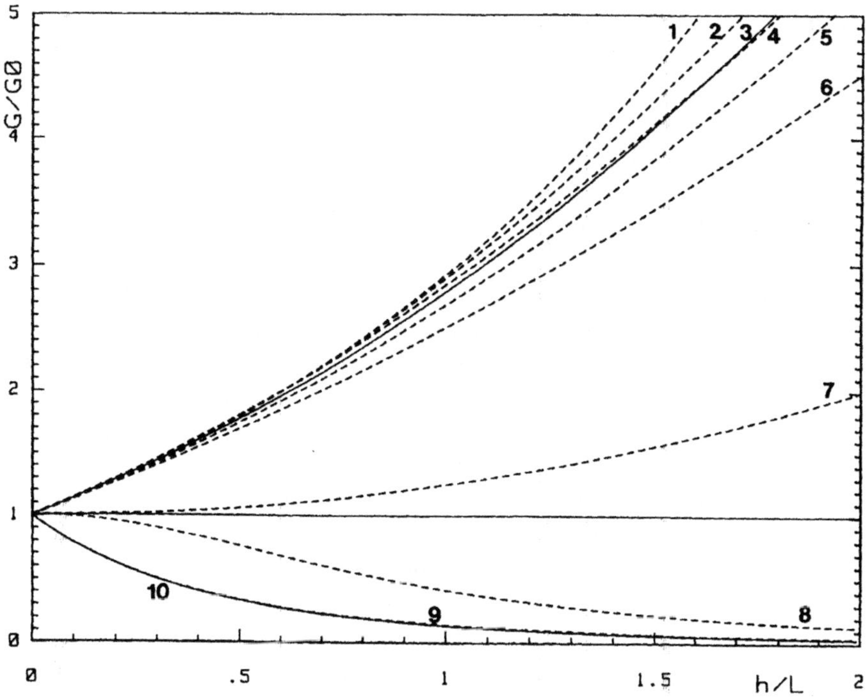

Fig. 5.8. Comparison of various correction terms proposed for a DCB with short arms: (1): Fichter, (5.63); (2): Freiman *et al.*, (5.62); (3): Foote *et al.*, (5.64); (4): Strawley and Gross, (5.61); (5): Kanninen, (5.56b). (6): Wiederhorn *et al.*, (5.46a); (7): shear force alone, (5.43a); (8): shear force alone, (5.43b); (9): Kanninen, (5.58b); (10): Kanninen, (5.57b).

Figure 5.8 shows that the expression for G as a function of P given by the elementary theory must be multiplied by 3 when $L = h$, and the Kanninen formula, (5.57b), shows that the expression of G as a function of δ given by the elementary theory must be divided by 7.6 when $L = h$. Elementary theory therefore underestimates the energy release rate at fixed load, and it overestimates it at fixed grips. When we use the cleavage technique to measure high surface energies, errors can be considerable if h/L is not small and if correction terms are neglected.

5.2.4 Cleavage Stresses

We have seen that the neutral fibre in its vertical displacement compresses or dilates the lower half of the upper beam, whose mean strain is $\varepsilon_y = 2y/h$. The shape of this deformed neutral fibre is given by (5.52). Hence, by Hooke's law, the mean cleavage stress is

$$\boxed{\sigma_y(x) = \sigma_{\mathrm{m}} \mathrm{e}^{-\lambda x} \left[\left(1 + \frac{1}{\lambda L} \right) \cos \lambda x - \sin \lambda x \right]}\,, \tag{5.65}$$

with

$$\sigma_{\rm m} = \frac{12PL}{bh^4\lambda^2} = \frac{12\,|\mathcal{M}|}{bh^4\lambda^2}\,, \tag{5.66a}$$

which, using (5.59b), becomes

$$\sigma_{\rm m} = 2\sqrt{6}\frac{PL}{bh^2} = 2\sqrt{6}\frac{|\mathcal{M}|}{bh^2}\,. \tag{5.66b}$$

Note that this equation was already given by Spies [11] in 1953 for $\pi/2$ peeling. It is a damped sinusoid (Fig. 5.9). The stress is maximum at $x = 0$, where it is tensile, increasing linearly with the bending moment $|\mathcal{M}| = PL$ at $x = 0$. It becomes a compression having its maximum in the vicinity of $x = h$. This compressive stress ahead of the crack tip was clearly revealed by Wan *et al.* [19] in cleaving mica: two coplanar cracks do not attract each other as normal, but at long range they actually repel due to a compressive stress acting between them.

These cleavage stresses, lying essentially from $x = 0$ to $x = 3h$, which equilibrate with the applied force P. We have, in effect,

$$\int_0^\infty \sigma_y b\,\mathrm{d}x = P\,. \tag{5.67}$$

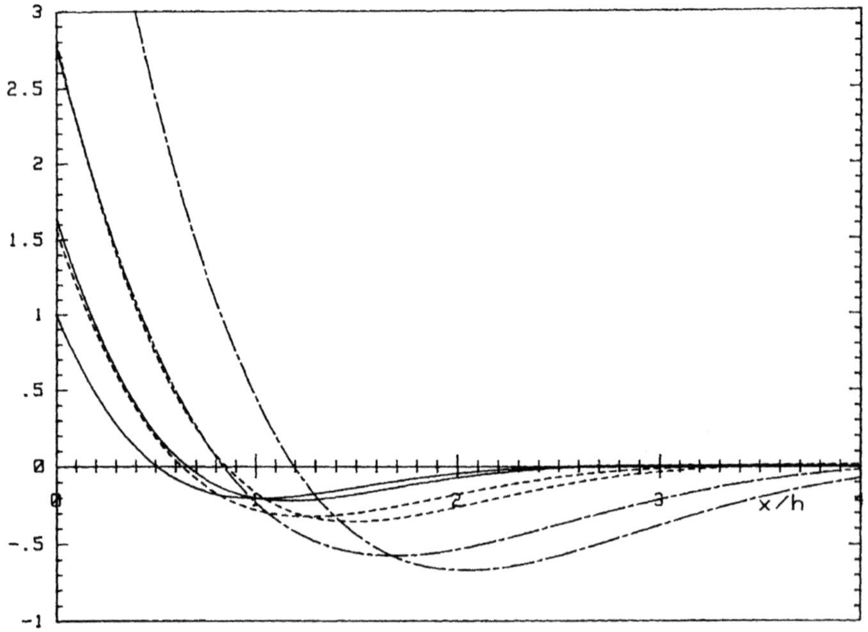

Fig. 5.9. Cleavage stress $\sigma/\sigma_{\rm m}$ ahead of the crack tip for two values of L: $L \to \infty$ and $L = h$. Values are $\lambda/\lambda_0 = 1$ (*solid line*); $\lambda/\lambda_0 = 0.8$ (*dashed line*); $\lambda/\lambda_0 = 0.6$ (*dotted-dashed line*). For λ_0, see (5.75)

If we recall that the energy of adhesion w represents the sum of the work done first against the elastic restoring forces, then against the adhesion forces in the strict sense, we can see that these elastic restoring forces equilibrate with the force P, the adhesion forces being negligible. The discussion on JKR and DMT contacts is also relevant here.

5.2.5 Asymmetric Double Cantilever Beam

If the double cantilever beam is asymmetrical (E_1 and h_1 for one arm, E_2 and h_2 for the other), an asymmetry is introduced into the elastic field at the crack tip, characterized by stress intensity factors K_{I} and K_{II}. This mode mix is characterized by a phase angle

$$\psi = \tan^{-1}\left(\frac{K_{\mathrm{II}}}{K_{\mathrm{I}}}\right) . \tag{5.68}$$

The question, still being debated, is to know whether or not the resistance to rupture depends on this phase angle. In the thermodynamic description, equilibrium is given by $G = w$ and w does *not* depend on the phase angle if G is correctly computed.

There is no difficulty in extending the elementary theory to an asymmetrical DCB, the crack being assumed to propagate in its initial plane. The opening is

$$\delta = \frac{4PL^3}{b}\left(\frac{1}{E_1h_1^3} + \frac{1}{E_2h_2^3}\right) , \tag{5.69}$$

and the energy release rate is given by

$$G = \frac{6P^2L^2}{b^2}\left(\frac{1}{E_1h_1^3} + \frac{1}{E_2h_2^3}\right) , \tag{5.70a}$$

$$G = \frac{3\delta^2}{8L^4}\frac{E_1h_1^3E_2h_2^3}{E_1h_1^3 + E_2h_2^3} . \tag{5.70b}$$

In particular, if Young's moduli are identical and if we let $h_1 = h$ and $h_2 = nh$ ($n > 1$), we get

$$G = \frac{12P^2L^2}{Eb^2h^3}\frac{1+n^3}{2n^3} , \tag{5.71a}$$

$$G = \frac{3Eh^3\delta^2}{16L^4}\frac{2n^3}{1+n^3} . \tag{5.71b}$$

Modifying similarly the theory of Kanninen, (5.56b) and (5.58b) become

$$G = \frac{6P^2L^2}{b^2}\left[\frac{F_1^2}{E_1h_1^3} + \frac{F_2^2}{E_2h_2^3}\right] , \tag{5.72}$$

$$G = \frac{3\delta^2}{8L^4}\frac{F_1^2/E_1h_1^3 + F_2^2/E_2h_2^3}{\left(F_1^3/E_1h_1^3 + F_2^3/E_2h_2^3\right)^2} , \tag{5.73}$$

with

$$F_i = 1 + 0.64\frac{h_i}{L} .$$

Equation (5.73) has been used by Creton *et al.* [20] in 1992 with good results.

5.2.6 Influence of an Intercalary Film

An original idea due to Burns and Lawn [21] in 1968 was to introduce a wetting liquid between two long strips of glass, which are thus held together by capillary pressure, and to introduce a wedge of thickness δ at an extremity. The two arms are thus forced apart, causing the liquid meniscus to move along the arms down to an equilibrium point, giving two bent arms of length L. For viscous liquids this equilibrium can take a long time to reach. Application of (5.37a) allows us to compute G which, at equilibrium, is equal to twice the surface energy γ of the liquid. Note that the elastic energy is only stored in the glass, and that there are nowhere infinite stresses, and so no stress intensity factor. Recently Thouless [22] has extended this experiment by using two strips of different thickness h and nh. Whatever value n takes, the equilibrium

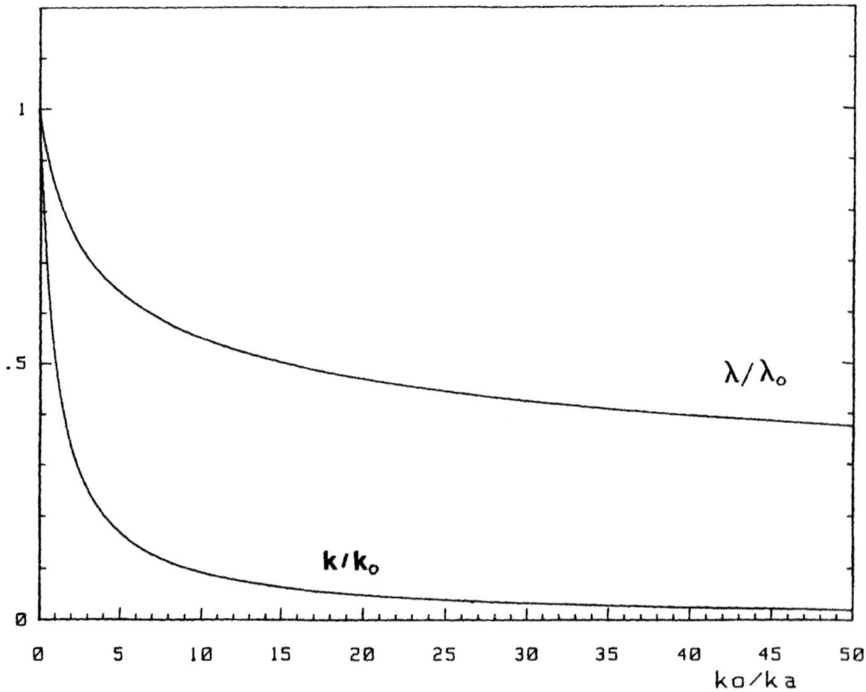

Fig. 5.10. Relative variation of the foundation modulus k and of λ as a function of the ratio k_0/k_a

position corresponds to $G = w = 2\gamma_L \cos\theta$, (where θ is the contact angle), with G given by (5.71b).

Consider a symmetrical DCB (E, h) glued by a thin adhesive film of thickness $2h_a$ and modulus E_a. If the film is thin and the modulus of the adhesive low, we can in a first approximation compute G as for a homogeneous DCB to determine the energy of adhesion w of the adhesive. (We can assume that the crack propagates in the adhesive and neglect the modification of rigidity due to the adhesive remaining on each arm after rupture.) In fact, this film modifies essentially the foundation modulus k of the Kanninen theory. This problem was studied by Penado [23] in 1993. We can consider that adherent and adhesive act as two springs in series, so that the foundation modulus k of the sandwich is given by

$$\frac{1}{k} = \frac{1}{k_0} + \frac{1}{k_a}, \tag{5.74}$$

where k_0 is the foundation modulus of the homogeneous DCB ($k_0 = 2Eb/h$ in Kanninen's theory), and k_a is that of the adhesive. Assuming the adhesive is in plane stress, Hooke's law gives $\sigma_y = \varepsilon_y E_a$, where $\varepsilon_y = y/h_a$ is the mean strain. The foundation modulus being defined by $\sigma_y = k_a y/b$, we get

$$k_a = \frac{bE_a}{h_a}.$$

Equation (5.56a) and (5.58a) give the energy release rates,[3] λ and λ_0 being defined by

$$\lambda = \left(\frac{3k}{Ebh^3}\right)^{1/4}, \tag{5.75a}$$

$$\lambda_0 = \left(\frac{3k_0}{Ebh^3}\right)^{1/4} = \frac{6^{1/4}}{h}. \tag{5.75b}$$

Figure 5.10 displays the variations of k/k_0 and λ/λ_0 as a function of the ratio

$$\frac{k_0}{k_a} = 2\frac{Eh_a}{E_a h}. \tag{5.76}$$

The modulus and the thickness play a symmetrical role: doubling the film thickness or halving, its Young's modulus gives the same result.

[3] In fact Penado [23] claims to have a better result by letting $k_0 = 4Eb/h$ and adding a term of shear force:

$$G = \frac{12P^2L^2}{b^2Eh^3}\left[\left(1 + \frac{1}{\lambda L}\right)^2 + \frac{1+\nu}{4}\frac{h^2}{L^2}\right].$$

On the other hand, he assumes that the adhesive is in plain strain ($\varepsilon_z = 0$), and takes $k = E_a b/(1-\nu^2)h_a$.

Equations (5.55a), (5.56a) and (5.58a) can also be written

$$\delta = \frac{8PL^3}{Ebh^3}\left[\left(1+\frac{\lambda_0}{\lambda}0.64\frac{h}{L}\right)^3+\frac{1}{2}\left(\frac{\lambda_0}{\lambda}0.64\frac{h}{L}\right)^3\right], \tag{5.77}$$

$$G = \frac{12P^2L^2}{Eb^2h^3}\left(1+\frac{\lambda_0}{\lambda}0.64\frac{h}{L}\right)^2, \tag{5.78}$$

$$G = \frac{3Eh^3\delta^2}{16L^4}\frac{1}{\left(1+\frac{\lambda_0}{\lambda}0.64\frac{h}{L}\right)^{-4}}. \tag{5.79}$$

These corrective terms are displayed in Fig. 5.11. Knowing the ratios h_a/h and E_a/E, we can deduce k_0/k_a and then λ/λ_0 and the corrective term.

In the same manner, the cleavage stresses, given by (5.65), can be written as

$$\sigma_y(x) = \sigma_m \frac{e^{-(\lambda/\lambda_0)\,6^{1/4}(x/h)}}{(\lambda/\lambda_0)^2} \times\left[\left(1+\frac{0.64h/L}{\lambda/\lambda_0}\right)\cos\left(\frac{\lambda}{\lambda_0}6^{1/4}\frac{x}{h}\right)-\sin\left(\frac{\lambda}{\lambda_0}6^{1/4}\frac{x}{h}\right)\right],$$

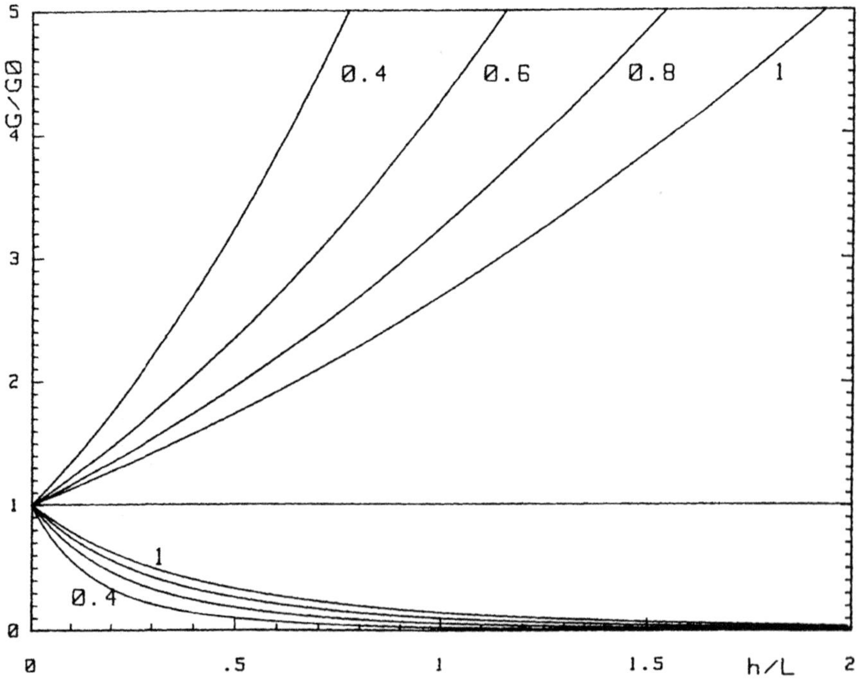

Fig. 5.11. Corrective terms as a function of h/L for various values of λ/λ_0 (various combinations of thickness and Young's modulus of the adhesive)

with

$$\sigma_{\mathrm{m}} = \frac{12PL}{bh^4\lambda_0^2} \,.$$

This form assumes that we can continue to write $\varepsilon_y = 2y/h$, i.e. that h_{a} remains small compared with h. These stresses are displayed in Fig. 5.9 for various values of the ratio λ/λ_0. We should note that they are larger and spread longer behind the crack front for small λ.

Other things being equal, to have a critical load P_{c} of the higher value, one must have a small energy release rate, and hence a thin adhesive film with a high Young's modulus.

5.3 Peeling

5.3.1 Elementary Theory

Consider, in relation to Fig. 5.12, the peeling of a tape of width b, thickness h and Young's modulus E adhering to a rigid substrate. Let P be the applied load, θ the peeling angle and L the peeled length (assumed very long). Under the action of the load P, the tape extends by

$$\delta = \frac{PL}{Ebh} \,,$$

and the stored elastic energy is thus

$$U_{\mathrm{E}} = \frac{1}{2}P\delta = \frac{P^2L}{2Ebh} \,.$$

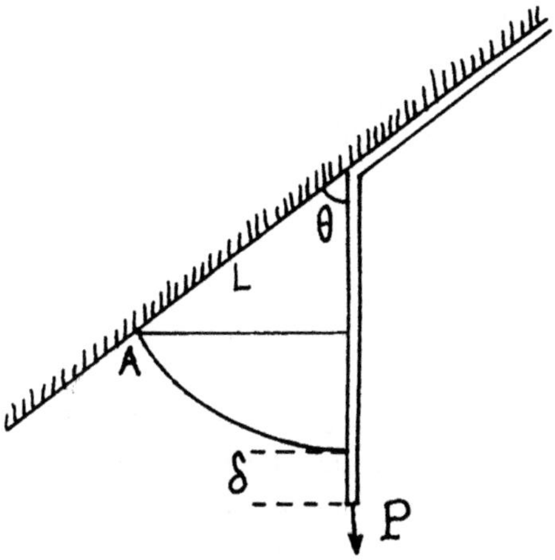

Fig. 5.12. Peeling of an elastic tape under an angle θ

(Elastic energy is also stored in the bending of the tape, but if the peeling angle and the thickness of the tape do not vary, this energy remains constant during peeling, and its contribution to G is zero.)

Compared with its original point A on the substrate, the load P has fallen by $L + \delta - L\cos\theta$ and its potential energy is:

$$U_{\mathrm{P}} = -PL\left(1 - \cos\theta + \frac{P}{Ebh}\right) .$$

The mechanical energy is

$$U_{\mathrm{E}} + U_{\mathrm{P}} = -PL(1 - \cos\theta) - \frac{P^2 L}{2Ebh} ,$$

and hence the energy release rate is given by

$$\boxed{G = \frac{P}{b}(1 - \cos\theta) + \frac{P^2}{2b^2 Eh}} . \tag{5.80}$$

The term in P was given initially by Rivlin [24] in 1944, and by Deryagin and Krotova [25], and the corrective term by Lindley [26] in 1971 and by Kendall [27] and Anderson *et al.* [28]. The energy release rate being independent of the area A, we have $(\partial G/\partial A) = 0$ and the equilibrium defined by $G = w$ is a *neutral equilibrium*. As soon as the load P exceeds the value P_{c} given by this equation of the second degree, the crack propagates with a constant extension force $G - w$.

Figure 5.13 displays the contribution of the two terms to the reduced adherence force P/bw. We can see that the classical approximation

$$\frac{P}{b} = \frac{w}{1 - \cos\theta}$$

is valid only when $\sqrt{w/2Eh} \ll 1 - \cos\theta$. To reduce the elastic deformation of the peeled tape, one often glues an inextensible fabric on the back of the tape, which is equivalent to increasing its Young's modulus and rendering the elastic term negligible. When this term is negligible, (5.80) shows that we must have

$$P_{90^\circ} = 2P_{180^\circ} . \tag{5.81}$$

5.3.2 Shape of a Peeled Band

The calculation for the shape of a peeled band was given by Kendall [29] in 1973. Consider Fig. 5.14: the peel force P is directed along the y-axis, and the peel point is at $x = d$ and $y = 0$. The rigid substrate makes an angle φ_0 with the x-axis, and the peel angle is $\theta = \pi/2 - \varphi_0$. At a point (x, y) the

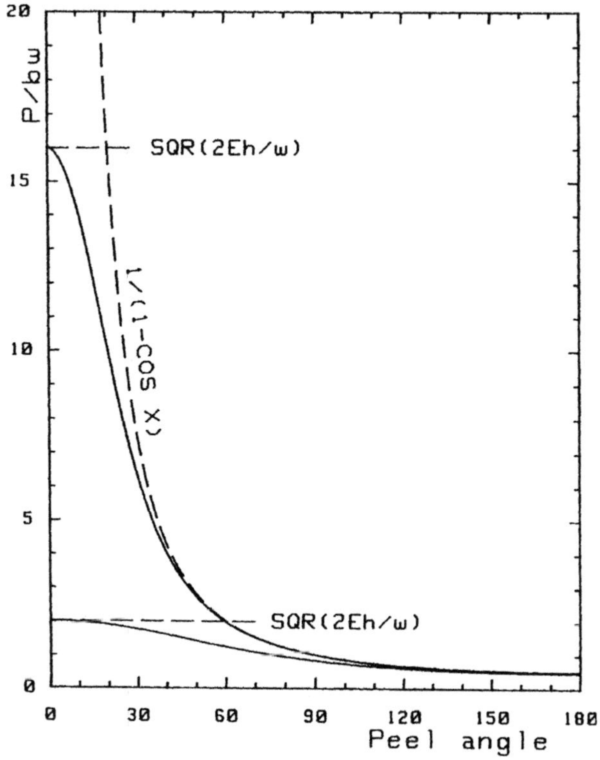

Fig. 5.13. Peel force as a function of the peel angle for two values of $\sqrt{2Eh/w}$

bending moment is $\mathcal{M}_f = -Px$ and the radius of curvature is R. As seen in Sect. 5.1.1, (5.6), the equation of elastica, neglecting the shear force, is given by the differential equation

$$\frac{y''}{(1+y'^2)^{3/2}} = \frac{Px}{EI} , \tag{5.82}$$

i.e.

$$\frac{\mathrm{d}y'}{(1+y'^2)^{3/2}} = \frac{Px}{EI}\,\mathrm{d}x ,$$

which integrates to

$$\frac{y'}{(1+y'^2)^{1/2}} = \frac{P}{2EI}x^2 + C ,$$

or else, with $y' = -\tan\varphi$,

$$-\sin\varphi = \frac{P}{2EI}x^2 + C .$$

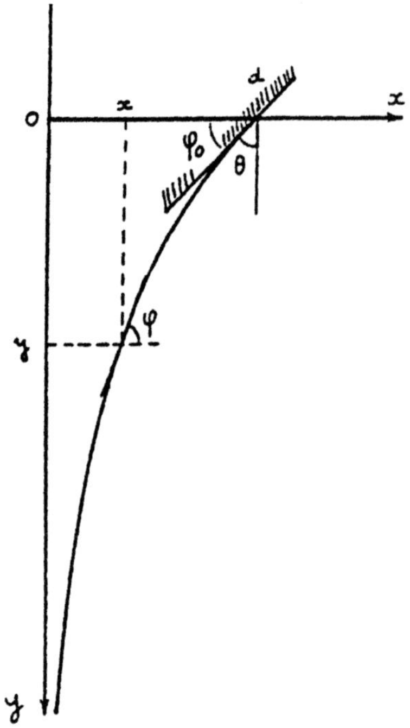

Fig. 5.14. A peeled band

The constant C is obtained by making $\varphi = \pi/2$ at $x = 0$, so that

$$x^2 = \frac{2EI}{P}(1 - \sin\varphi) . \tag{5.83}$$

In particular, the distance d between the peeling point and the axis of the force P (the length of the lever arm) is obtained by making $x = d$ and $\varphi = \varphi_0$. Then

$$d^2 = \frac{2EI}{P}(1 - \sin\varphi_0) , \tag{5.84}$$

i.e.

$$\boxed{d^2 = \frac{2EI}{P}(1 - \cos\theta)} , \tag{5.85}$$

a formula given in 1959 by Inoue and Kobatake [12] and by Kaelble [30]. Eliminating $P/2EI$ between (5.83) and (5.84) we get

$$\sin\varphi = 1 - \frac{x^2}{d^2}(1 - \sin\varphi_0) . \tag{5.86}$$

Let

$$r = \frac{x}{d}\sqrt{1 - \sin\varphi_0} .$$

Then (5.86) becomes

$$\tan\varphi = -\frac{\mathrm{d}y}{\mathrm{d}x} = \frac{1-r^2}{\sqrt{1-(1-r^2)^2}} = \frac{1-r^2}{r\sqrt{2-r^2}} ,$$

i.e.

$$\frac{\sqrt{1-\sin\varphi_0}}{d}\,\mathrm{d}y = -\frac{1-r^2}{r\sqrt{2-r^2}}\,\mathrm{d}r ,$$

which is easily integrated to give, with $\sin\varphi_0 = \cos\theta$:

$$\boxed{\begin{aligned}\frac{y}{d}\sqrt{1-\cos\theta} = \frac{1}{\sqrt{2}}\ln&\left[\frac{\sqrt{2}+\sqrt{2-\frac{x^2}{d^2}(1-\cos\theta)}}{\frac{x}{d}\left(\sqrt{2}+\sqrt{1+\cos\theta}\right)}\right]\\ &+\sqrt{1+\cos\theta}-\sqrt{2-\frac{x^2}{d^2}(1-\cos\theta)}\end{aligned}} \quad . \tag{5.87}$$

Figure 5.15 displays, in reduced coordinates x/d and y/d, the shape of the band for various peel angles. Comparison with experiments [29] is excellent.

From (5.3), we can note that the radius of curvature at the peeling point is given by

$$R_0 = \frac{EI}{Pd}$$

or, by (5.85),

$$R_0 = \frac{d}{2(1-\cos\theta)} = \frac{1}{2}\sqrt{\frac{Ebh^3}{6P(1-\cos\theta)}} , \tag{5.88}$$

a result given by Kaelble [30]. These radii of curvature vary as $1/\sqrt{P}$. They are drawn in Fig. 5.15 for $\theta = 90°$ ($R_0 = d/2$) and $\theta = 180°$ ($R_0 = d/4$).

Extracting $(1-\cos\theta)$ from (5.85) and introducing the bending moment at the crack tip, $\mathcal{M}_f = -Pd$, the energy release rate can be written as

$$\boxed{G = \frac{6\mathcal{M}_f^2}{Eb^2h^3} + \frac{P^2}{2b^2Eh}} \quad . \tag{5.89}$$

We obtain the same term as for the DCB, within a factor of two.

5.3.3 Rotation of the Built-in Section

As for the DCB (Sect. 5.2.3 and 5.2.6), we can consider the unpeeled part as a beam on an elastic foundation with a foundation modulus k, and we will let

$$\lambda = \left(\frac{3k}{Ebh^3}\right)^{1/4} \quad . \tag{5.90}$$

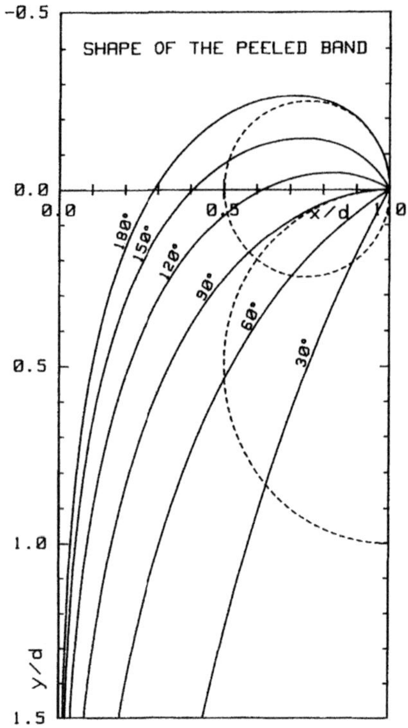

Fig. 5.15. Shape of the peeled band for various peel angles

We can again use Fig. 5.5, but this time with P tangent to the neutral axis in $x = -L$. The shear force is no longer $T = P$ but $T = P \sin\theta$. The bending moment at the crack tip is no longer $\mathcal{M}_f = -PL$ but $\mathcal{M}_f = -Pd$. (Strictly speaking, d is no longer given by (5.85), which assumes a perfect build-in.) The shape of the neutral fibre is thus given by (5.48) as

$$y = \mathrm{e}^{-\lambda x}(A \sin \lambda x + B \cos \lambda x) , \qquad x \geq 0 ,$$

with more general boundary conditions at $x = 0$ of

$$y''(0) = -\frac{\mathcal{M}_f}{EI} = -2\lambda^2 A ,$$
$$y'''(0) = \frac{T}{EI} = 2\lambda^3(A + B) .$$

We then get

$$y = \frac{\mathrm{e}^{-\lambda y}}{2EI\lambda^3}\left[\lambda \mathcal{M}_f \sin \lambda x + (T - \lambda \mathcal{M}_f) \cos \lambda x\right] , \qquad x \geq 0 , \tag{5.91a}$$

or else

$$y = \frac{Pd}{2EI\lambda^2}\mathrm{e}^{-\lambda x}\left[\left(1 + \frac{\sin\theta}{\lambda d}\right) \cos \lambda x - \sin \lambda x\right] , \qquad x \geq 0 . \tag{5.91b}$$

The rotation of the built-in section at $x = 0$ is thus

$$y'(0) = -\frac{Pd}{2EI\lambda}\left(2 + \frac{\sin\theta}{\lambda d}\right) . \tag{5.92}$$

Note that these equations assume small curvatures in the unpeeled section.

5.3.4 Cleavage Stresses

Taking $\varepsilon_y = 2y/h$ for the mean strain, Hooke's law gives

$$\sigma_y(x) = \frac{12}{bh^4}\frac{\mathrm{e}^{-\lambda x}}{\lambda^3}\left[\lambda \mathcal{M}_f \sin\lambda x + (T - \lambda\mathcal{M}_f)\cos\lambda x\right] , \qquad x \geq 0 , \tag{5.93a}$$

or else

$$\sigma_y(x) = \frac{12Pd}{bh^4\lambda^2}\mathrm{e}^{-\lambda x}\left[\left(1 + \frac{\sin\theta}{\lambda d}\right)\cos\lambda x - \sin\lambda x\right] , \qquad x \geq 0 . \tag{5.93b}$$

If there is no intercalary film, λ is given by (5.59b) and we get

$$\sigma_y(x) = \frac{2}{b}\lambda\mathrm{e}^{-\lambda x}\left[\lambda \mathcal{M}_f \sin\lambda x + (T - \lambda\mathcal{M}_f)\cos\lambda x\right] , \qquad x \geq 0 , \tag{5.94a}$$

or else

$$\boxed{\sigma_y(x) = \sigma_{\mathrm{m}}\mathrm{e}^{-\lambda x}\left[\left(1 + \frac{\sin\theta}{\lambda d}\right)\cos\lambda x - \sin\lambda x\right] , \qquad x \geq 0} , \tag{5.94b}$$

with

$$\sigma_{\mathrm{m}} = 2\sqrt{6}\frac{Pd}{bh^2} = 2\sqrt{6}\frac{|\mathcal{M}_f|}{bh^2} , \tag{5.95}$$

$$\frac{\sin\theta}{\lambda d} = 6^{1/4}\cos\frac{\theta}{2}\sqrt{\frac{2P}{Ebh}} . \tag{5.96}$$

These curves have the same shape as for a DCB, (damped sinusoids, as per Fig. 5.9), provided that λL is replaced by $\lambda d/\sin\theta$ in (5.65). This compressive zone ahead of the crack can be seen easily on adhesive tapes. Placing two tapes in contact on a small area, and then trying to peel, one observes that the contact area moves without any complete separation of the tapes occuring. It is the compressive zone which establishes a new area of contact when peeling is done. This phenomenon was commented on by de Bruyne [31] in 1957. We can verify that

$$\int_0^\infty b\sigma_y \,\mathrm{d}x = P\sin\theta .$$

There are these elastic stresses in the unpeeled part which sustain the normal applied load $P \sin\theta$. Note that (5.91a) was given for the first time by Spies [11] in 1953, and (5.93a) by Kaelble [13] in 1960.

Consider now the case where an intercalary film (glue) of thickness h_a and modulus E_a exists. As seen for the DCB, adherent and adhesive constitute two springs in series:

$$k_0 = \frac{2bE}{h} , \tag{5.97}$$

$$k_a = \frac{bE_a}{h_a} , \tag{5.98}$$

so that the foundation modulus of the sandwich is given by

$$\frac{1}{k} = \frac{1}{k_0} + \frac{1}{k_a} . \tag{5.99}$$

We will assume $h_a \ll h$, so that we can write $\varepsilon_y = 2y/h$. If, morever, E_a is sufficiently small to have $k_a \ll k_0$, we can write

$$\boxed{\lambda = \left(\frac{3E_a}{Eh^3h_a}\right)^{1/4}} . \tag{5.100}$$

The expression for mean cleavage stresses in the adhesive has been perfectly verified by finite-element calculations by Crocombe and Adams [32] in 1981. These numerical calculations have also shown that the proportion of mode II at the crack tip was independent of the peel angle and depended only on the Poisson ratio.

Besides this stress $\sigma_y(x)$ normal to the interface, we can also compute, for $x = 0$, the traction stress on the more extended fibre (along the interface). This stress is the sum of the flexure stress $\mathcal{M}_f h/2I$ from (5.4), and of the tensile stress due to the component $P \cos\theta$:

$$\sigma_x(0) = \left[\frac{6PE}{bh}(1 - \cos\theta)\right]^{1/2} + \frac{P\cos\theta}{bh} . \tag{5.101}$$

5.3.5 Shear Stresses

The problem of shear stresses at the interface was studied by Kaelble [13] in 1960. Consider a ribbon of height h, width b, modulus E lying on an elastic layer of thickness h_a and for shear modulus μ_a as shown in Fig. 5.16. Examine the equilibrium of an element of length $\mathrm{d}x$ of the band on its elastic substrate. It is submitted on its left to a normal force $N = \sigma_x bh$, at its right to a force $N + \mathrm{d}N$, and at its interface to a tangential force $Q = \tau_{xy} b\,\mathrm{d}x$. The horizontal equilibrium requires

$$\mathrm{d}N + Q = 0$$

i.e.

$$d\sigma_x h = \tau_{xy}\, dx \,.$$

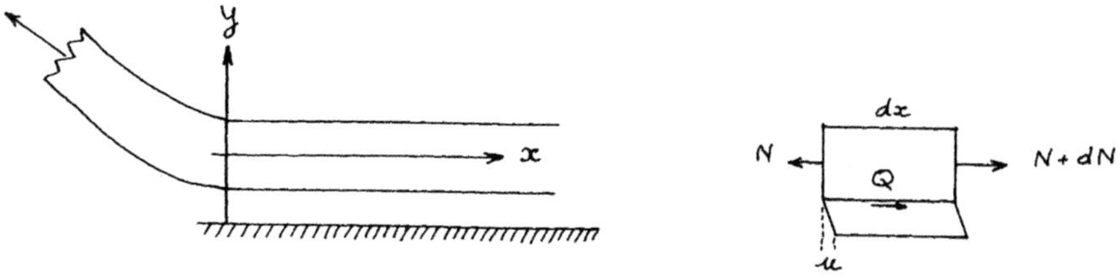

Fig. 5.16. Ribbon on an elastic layer

If u is the horizontal displacement of the interface, we have

$$\tau_{xy} = \mu_a \frac{u}{h_a}$$

and hence

$$\frac{d\sigma_x}{dx} = \frac{\tau_{xy}}{h} = \mu_a \frac{u}{hh_a} \,. \tag{5.102}$$

By differentiation of this expression, we obtain

$$\frac{d^2\sigma_x}{dx^2} = \frac{\mu_a}{hh_a}\frac{du}{dx} \,.$$

On the other hand, we have

$$\frac{du}{dx} = \varepsilon_x = \frac{\sigma_x}{E} \,.$$

Letting

$$\boxed{\lambda_T = \left(\frac{\mu_a}{Ehh_a}\right)^{1/2}} \,, \tag{5.103}$$

we thus have

$$\frac{d^2\sigma_x}{dx^2} - \lambda_T^2 \sigma_x = 0 \,, \tag{5.104}$$

of which the general solution is

$$\sigma_x = C_1 e^{-\lambda_T x} + C_2 e^{+\lambda_T x} \,.$$

As the stress must approach zero at large x, we keep only the solution

$$\sigma_x = C\mathrm{e}^{-\lambda_\mathrm{T} x} . \tag{5.105}$$

At $x = 0$ we have

$$\sigma_x(0) = \frac{P\cos\theta}{bh} ,$$

and hence

$$\sigma_x = \frac{P\cos\theta}{bh}\mathrm{e}^{-\lambda_\mathrm{T} x}$$

and by differentiation

$$\frac{\mathrm{d}\sigma_x}{\mathrm{d}x} = -\frac{P\cos\theta}{bh}\lambda_\mathrm{T}\mathrm{e}^{-\lambda_\mathrm{T} x} .$$

Using (5.102), this equation can be written as

$$\boxed{\tau_{xy} = -\frac{P\cos\theta}{b}\lambda_\mathrm{T}\mathrm{e}^{-\lambda_\mathrm{T} x}} . \tag{5.106}$$

We can verify that this shear stress transmits the force $P\cos\theta$ to the elastic foundation, since we have

$$\int_0^\infty \tau_{xy} b\,\mathrm{d}x = -P\lambda_\mathrm{T}\cos\theta \int_0^\infty \mathrm{e}^{-\lambda_\mathrm{T} x}\,\mathrm{d}x = -P\cos\theta .$$

This exponential decrease of the shear stress and cleavage stresses explains the independence of the peel force of the length remaining to peel.

5.3.6 Influence of Prestresses

Now consider a ribbon stretched with a force F on a rigid substrate. The stress in the ribbon can schematize the stresses introduced by differential thermal expansion, or the residual stresses existing after warming and then cooling the ribbon. Then peel this ribbon under a load P and an angle θ. We will assume that the other extremity is clamped to the substrate to avoid any spontaneous peeling.

Let $\varepsilon = F/Ebh$ be the strain imposed to the stretched ribbon and L_0 the real length of contact (corresponding to an initial length, not stretched, $L_0(1-\varepsilon)$). The elastic energy stored by the pre-stress is $\frac{1}{2}\varepsilon^2 E$ per unit volume, i.e. $F^2 L_0/2bEh$. When the area of contact varies by $\mathrm{d}A = -b\,|\,\mathrm{d}L_0|$ this energy decreases by $-F^2\,\mathrm{d}A/2b^2Eh$, but the element of length released undergoes the action of the load P and stores the energy $-P^2\,\mathrm{d}A/2b^2Eh$. The change in elastic energy is then given by

$$\mathrm{d}U_\mathrm{E} = -\frac{P^2 - F^2}{2b^2Eh}\,\mathrm{d}A .$$

When the length $|\,\mathrm{d}L_0|$ is released, the load drops by $|\,\mathrm{d}L_0|\,(1-\cos\theta)$, contracting by $|\,\mathrm{d}L_0|\,\varepsilon$ due to the removal of the force F, but lenthening by $|\,\mathrm{d}L_0|\,P/bEh$ due to the load P. Finally, the load drops by

$$\mathrm{d}\delta = |\,\mathrm{d}L_0|\,(1-\cos\theta) - \frac{F-P}{bEh}\,|\,\mathrm{d}L_0|\ .$$

The variation of potential energy is

$$\mathrm{d}U_{\mathrm{P}} = -P\,\mathrm{d}\delta = \left[\frac{P}{b}(1-\cos\theta) + \frac{P(P-F)}{b^2Eh}\right]\mathrm{d}A\ ,$$

and hence the energy release rate [33, 34] becomes

$$\boxed{G = \frac{P}{b}(1-\cos\theta) + \frac{(P-F)^2}{2b^2Eh}} \tag{5.107}$$

and the adherence force becomes

$$\frac{P}{bEh} = \varepsilon - (1-\cos\theta) \pm \sqrt{(1-\cos\theta)^2 + \frac{2w}{Eh} - 2\varepsilon(1-\cos\theta)}\ . \tag{5.108}$$

The influence of pre-stress is bad (it increases G) when F is negative, i.e. when the tape is placed after compression. When the tape is placed after stretching, ($F > 0$), it could be beneficial since $(P-F)^2$ can be higher than P^2. Figures 5.17 and 5.18 display the case for $\theta = 90°$ and $\theta = 20°$. For $\theta = 90°$, dashed lines represent the Kendall [35] approximation

$$\frac{P}{b} = w - \frac{1}{2}\varepsilon^2 Eh\ ,$$

which has been rather well verified experimentally.

At large angles a pre-tension is generally bad, but at small angles its influence is consiberable. For a given w, we can see from (5.108) that the maximum value of P is equal to the pre-tension F. In this case, the strain in the peeled band is the same as ε in the unpeeled part, and there is only a variation of the potential energy of the load.

5.3.7 Peeling at Small Angles: Dislocations

In 1978, Kendall [36] made the following curious observation. After applying a sufficient load, he steadily decreased the peel angle. For an angle about 5°, peeling stopped even although the computed G was higher than w: the peeled band adhered again to the surface (probably by an electrostatic forces). To restart the peeling at the same velocity as before (i.e. have the same G) it was necessary to double the applied load.

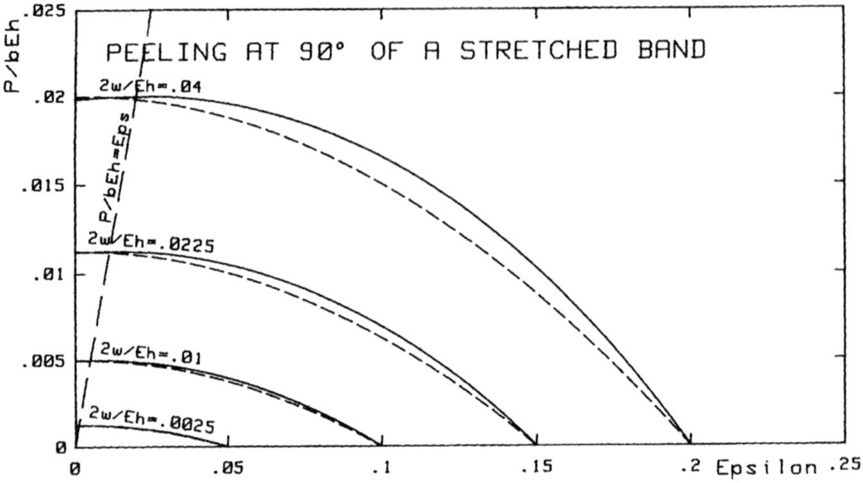

Fig. 5.17. Peeling at 90° of a tape attached after stretching

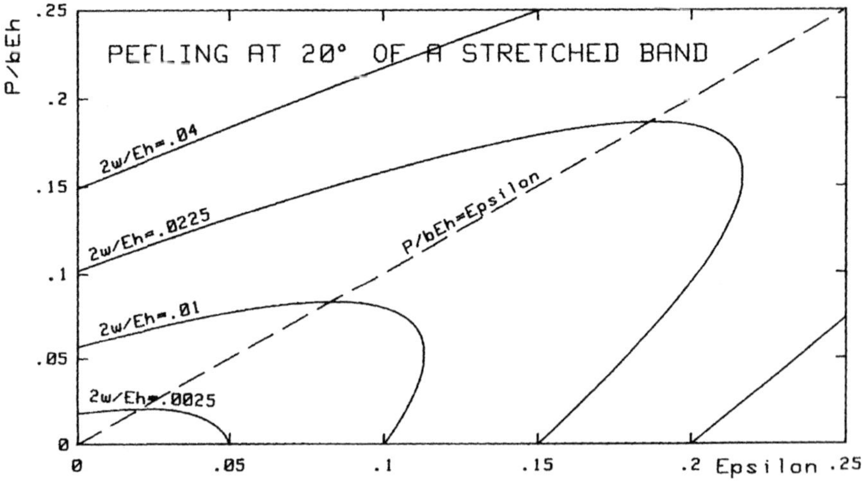

Fig. 5.18. Peeling at 20° of a tape attached after stretching

The explanation provided is as follows. The crack is at equilibrium for $G = w$, with G given by (5.80). For $G > w$ the crack advances. We might think that the condition for resticking would be $G < w$ with the same expression for G, but this is not correct for the band is stretched under the action of the load P and resticks in a stretched state. There is no variation of elastic energy when the crack recedes, but only a variation of potential energy. The condition for receding is thus $P(1 - \cos\theta)/b < w$.

If we return to Fig. 5.13, this illustrates that peeling occurs above the full line and the resticking below the dashed line. There thus exists a zone between the two curves, with small angles, where peeling and sticking are both possible. Once the stuck band has stretched under the beneficial strain $\varepsilon = P_1/bEh$, it is necessary to apply a force $P_2 > P_1$ in order to peel it. If we neglect, as Kendell did, the terms in $1 - \cos\theta$, the initial peeling force

$$P_1 = \sqrt{2b^2 Ehw}$$

has become the pre-stress force F, and (5.107) shows that we must excert a force

$$P_2 = F + \sqrt{2b^2 Ehw} = 2P_1$$

to peel it.

Peeling, followed by strain and resticking is equivalent to a *dislocation.* (It suffices to replace the word "peeling" by the word "cut" to recover the definition of a dislocation.)

Further discussion is provided by Lake and Stevenson [37] in 1981. When the peel force at small angles is important and the elastic deformation is high, the lateral contraction due to the Poisson effect is no longer negligible, and the peel front is no longer linear: it becomes V-shaped, with the tip of the V at the rear. The separation then occurs by propagation of detachment and attachment waves parallel to the crack front, and the peel force can increase by an order of magnitude.

5.4 The Lap Joint

5.4.1 Elementary Theory

The simplest example of a lap joint is that of a band of width b, thickness h and Young's modulus E attached to a rigid half-space and peeled at 0°. Equation (5.80) gives, for $\cos\theta = 1$,

$$G = \frac{P^2}{2b^2 Eh}, \tag{5.109}$$

i.e. an adherence force of

$$\frac{P}{b} = \sqrt{2wEh} \tag{5.110}$$

independent of the contact area.

Another simple example is that of two long ribbons of width b, thickness h_1 and h_2, Young's moduli E_1 an E_2, placed on one another, as shown in Fig. 5.19. If each ribbon is pulled with force P, the displacement δ of the load P corresponds to the sum of the elongations of the free segments of length l_1 and l_2, and of the segment l of thickness $h_1 + h_2$, i.e.

$$\delta = \frac{P}{b}\left(\frac{l_1}{E_1 h_1} + \frac{l_2}{E_2 h_2} + \frac{l}{E_1 h_1 + E_2 h_2}\right).$$

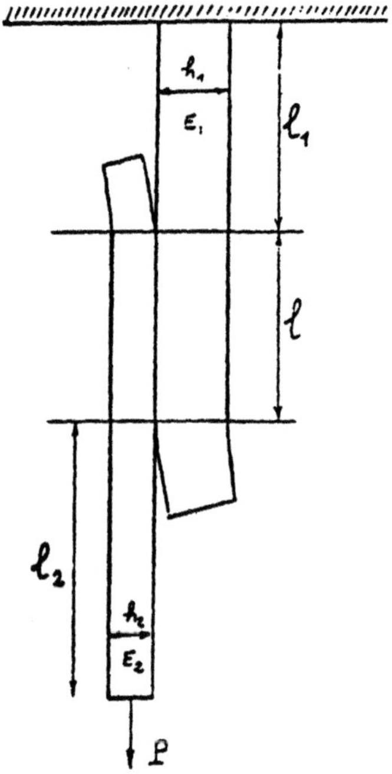

Fig. 5.19. Lap joint

The mechanical energy $U_{\mathrm{T}} = U_{\mathrm{E}} + U_{\mathrm{P}}$ of the system is thus:

$$U_{\mathrm{T}} = -\frac{P^2}{2b}\left(\frac{l_1}{E_1 h_1} + \frac{l_2}{E_2 h_2} + \frac{l}{E_1 h_1 + E_2 h_2}\right) .$$

As the problem is not symmetrical, there are two energy release rates to compute according to whether $\mathrm{d}A = -b\,\mathrm{d}l_1$ or $\mathrm{d}A = -b\,\mathrm{d}l_2$. Suppose that l_2 remains constant. We will have then $\mathrm{d}l = -\,\mathrm{d}l_1$, and hence

$$\boxed{G_1 = \frac{P^2}{2b^2}\left(\frac{1}{E_1 h_1} - \frac{1}{E_1 h_1 + E_2 h_2}\right)} . \tag{5.111}$$

So (5.109) is obtained again when $E_2 h_2$ is infinite. For G_2, it suffices to interchange $E_1 h_1$ and $E_2 h_2$. The crack will start on the side where G is the largest, i.e. on the side of the band which has the product Eh the smallest. The energy release rate G being independent of the contact area, the equilibrium is neutral at fixed load, and the adherence force, given by $G = w$, if $E_1 h_1 < E_2 h_2$, is

$$\boxed{\frac{P}{b} = \sqrt{2w\frac{E_1 h_1}{E_2 h_2}(E_1 h_1 + E_2 h_2)}} . \tag{5.112}$$

This formula was given and experimentally verified by Kendall [38] in 1975.

In reality, the problem is more complicated for, as shown by Goland and Reissner [39] in 1944, a bending moment appears in the joint and the peeled bands, due to the fact that forces are not aligned. We will restrict ourselves to the simple symmetrical case where $E_1 = E_2$, $h_1 = h_2$ and $l_1 = l_2$, and to the approximation of the short joint studied by Kendall [40]. In this latter case, the joint of length l turns as a whole by a small angle θ (see Fig. 5.20). We thus have a peeling of that joint under the angle 0, and neglecting the elastic term in (5.80) the energy release rate is

$$G = \frac{P}{b}(1 - \cos\theta) .$$

Extracting $(1 - \cos\theta)$ from (5.85), with $I = bh^3/12$, we get

$$G = \frac{6P^2d^2}{Eb^2h^3} ,$$

where d is the distance between the axis of the force P and the neutral fibre at the peeling point. Further examination of Fig. 5.20 shows that

$$\frac{h}{2} = \frac{l}{2}\tan\theta + \frac{d}{\cos\theta} . \tag{5.113}$$

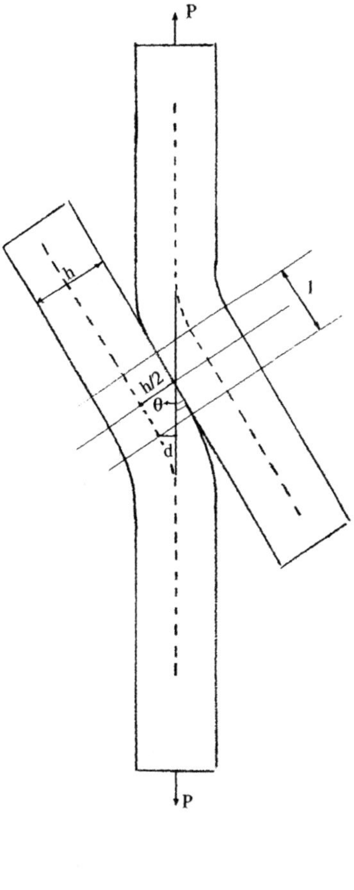

Fig. 5.20. Bending of a lap joint when two ribbons are placed one on another

The angle of rotation θ being small, we have as a first approximation

$$d \cong h/2\,,$$

and we get

$$G = \frac{3P^2}{2b^2Eh}\,, \tag{5.114}$$

an approximation given by Kendall [40]. There must be a transition between $G = P^2/4b^2Eh$, from (5.111), for a long joint and $G = 3P^2/2b^2Eh$ for a short joint. To see that, it is interesting to develop (5.113) to first order in θ, and (5.85) to second order. We then get

$$h = l\theta + 2d\,,$$

$$d = \theta\sqrt{\frac{EI}{P}}\,.$$

Hence

$$\theta = \sqrt{\frac{3P}{Ebh}}\left(1 + \frac{ul}{2}\right)^{-1}\,, \tag{5.115}$$

with

$$u = \sqrt{\frac{P}{EI}} = \sqrt{\frac{12P}{Ebh^3}}\,, \tag{5.116}$$

a parameter used by Goland and Reissner [39]. The energy release rate

$$G = \frac{P}{b}\frac{\theta^2}{2}$$

is written as

$$G = \frac{3P^2}{2b^2Eh}\left(1 - \frac{ul}{2}\right)^2 \quad \text{for a short joint.} \tag{5.117a}$$

A short joint is thus defined by $ul \ll 1$, i.e. $l \ll \sqrt{EI/P}$. Experimentally, Kendall [40] found that the adherence of a very short joint is 0.39 times that for a long joint, a value near $1/\sqrt{6} = 0.408$. Equation (5.117a), with $G = w$ and $y = P/b$, can still be written

$$\frac{l}{h} = \sqrt{\frac{Eh}{3y}} - \frac{Eh}{3y}\sqrt{\frac{2w}{y}}\,,$$

an in this form it can be compared with the experimental results of Kendall. Figure 5.21 shows, however, that this approximation is not very good (see the dashed line).

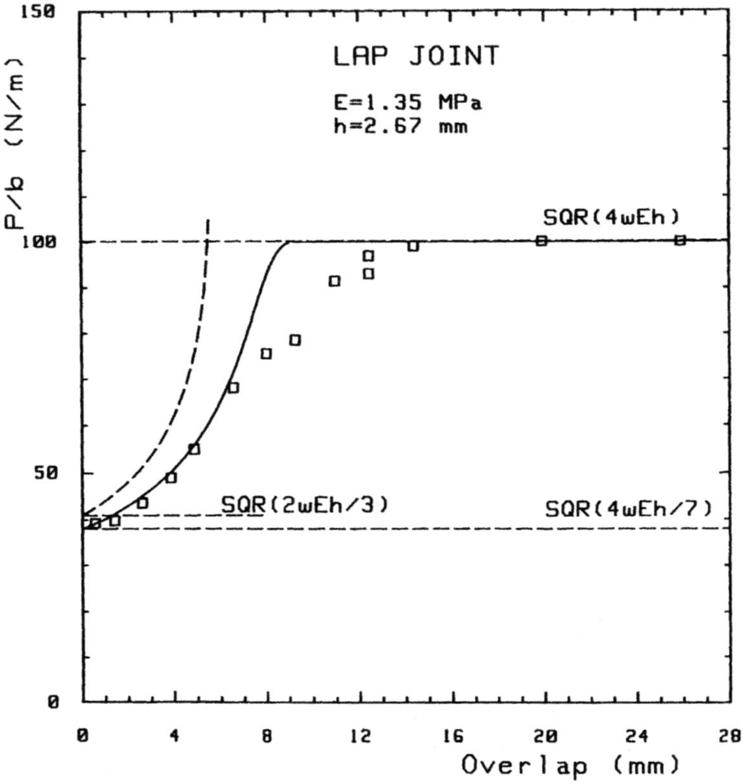

Fig. 5.21. Peel force of a lap joint as a function of the overlap length, showing the force necessary to propagste a crack at 10 µm/s. Experimental results of Kendall [40]

A better approximation consists in taking into account the elastic deformation of the band. We shall restrict ourselves, by symmetry, to the half-system of length $l_1 + l/2$. The elongation under the load P is then given by

$$\delta = \frac{P}{b}\left(\frac{l_1}{Eh} + \frac{l/2}{2Eh}\right) ,$$

and the elastic energy is given by

$$U_{\mathrm{E}} = \frac{P^2}{2bEH}\left(l_1 + \frac{l}{4}\right) .$$

For a length variation $\mathrm{d}l_1$, the load falls by

$$\begin{aligned} \mathrm{d}\Delta &= \mathrm{d}l_1(1-\cos\theta) + \mathrm{d}\delta \\ &\cong \mathrm{d}l_1\frac{\theta^2}{2} + \mathrm{d}\delta \\ &\cong \mathrm{d}l_1\frac{3P}{2bEh}\left(1-\frac{ul}{2}\right)^2 + \frac{P}{bEh}\left(\mathrm{d}l_1 + \frac{\mathrm{d}l}{4}\right) . \end{aligned}$$

The potential energy thus varies by

$$dU_P = -\frac{P^2}{2bEh}\left[3\,dl_1\left(1-\frac{ul}{2}\right)^2 + 2\,dl_1 + \frac{dl}{2}\right] .$$

Finally, with $dl_1 = -\,dl/2$, the mechanical energy varies by

$$dU_T = -\frac{P^2}{4bEh}\left[1+6\left(1-\frac{ul}{2}\right)^2\right] dl_1 ,$$

and hence

$$\boxed{G = \frac{P^2}{4b^2Eh}\left[1+6\left(1-\frac{ul}{2}\right)^2\right] \quad \text{for a short joint}} , \tag{5.117b}$$

instead of

$$\boxed{G = \frac{P^2}{4b^2Eh} \quad \text{for a long joint}} . \tag{5.118}$$

This correction adds an extra term to (5.117a) for the long joint. These results show that G increases when the area of contact decreases, and that the equilibrium $G = w$ becomes unstable for short joints. Figure 5.21 shows that the approximation is more in line with the experiment now and that the ratio of adherence for (short joint/long joint) is equal to $1/\sqrt{7} = 0.378$.

5.4.2 "Dislocations" in Lap Joints

When a lap joint is pulled, the free end of each band lift up and turn by an angle θ. That angle has been computed by Kendall [41] from (5.85), giving the distance d between the axis of the force (on the neutral fibre) and the peeling point (taken on the neutral fibre) (Fig. 5.22). For a joint of thickness $2h$ and a band of thickness h, we have

$$d_1 = \mathrm{AM} = \sqrt{\frac{2EI_1}{P}(1-\cos\theta)} ,$$

$$d_2 = \mathrm{BN} = \sqrt{\frac{2EI}{P}(1-\cos\theta)} ,$$

and the sum is equal to $\mathrm{AB}\cos\theta = \frac{1}{2}h\cos\theta$, i.e.

$$(1+\sqrt{8})\sqrt{\frac{Ebh^3}{6P}(1-\cos\theta)} = \frac{1}{2}h\cos\theta .$$

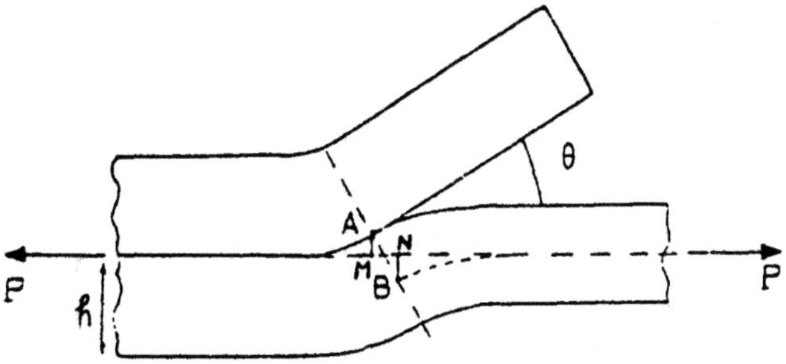

Fig. 5.22. A lap joint being pulled

By expanding $\cos\theta$, Kendall obtains:

$$\theta = \frac{1}{1+\sqrt{8}}\sqrt{\frac{3P}{bEh}}\,, \tag{5.119}$$

a result rather well verified experimentally.

Applying now a small transverse force T on the extremity of the freed band of length c, we can see in Fig. 5.23 after Kendall [41] that it takes the shape defined by

$$\frac{\mathrm{d}^2 y}{\mathrm{d}x^2} = -\frac{\mathcal{M}_f}{EI} = -\frac{T(c-x)}{EI}\,.$$

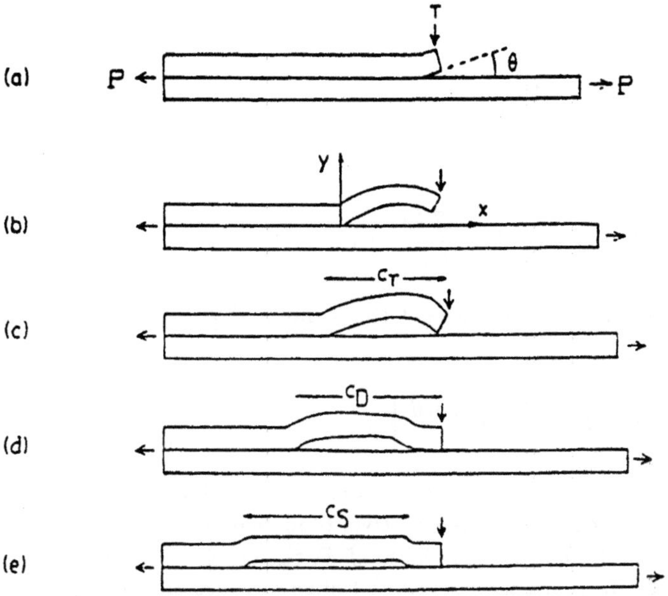

Fig. 5.23. A freed band subject to a transverse force T

Integrating twice, with the conditions $\mathrm{d}y/\,\mathrm{d}x = \tan\theta$ and $y = 0$ at $x = 0$, we get:

$$y = \frac{T}{6EI}(x^3 - 3cx^2) + x\tan\theta\,.$$

The extremity $x = c$ touches the stretched band for a length $c = c_T$ given by $y = 0$, i.e.

$$c_T = \sqrt{\frac{3EI\tan\theta}{T}}\,,$$

or, with $\tan\theta \cong \theta$,

$$c_T^2 = \frac{Ebh^3}{4T}\frac{1}{1+\sqrt{8}}\sqrt{\frac{3P}{bEh}}\,. \tag{5.120}$$

This introduction of a transverse force T modifies the value of G (which was $G = P^2/4b^2Eh$, from (5.111)) due to the storage of elastic energy in the bent band. To the total energy, which was

$$U_{\mathrm{T}} = -\frac{P^2}{2b}\left(\frac{c}{Eh} + \frac{l_2}{Eh} + \frac{l}{2Eh}\right)\,,$$

we must add the potential energy of the load T, namely

$$U'_{\mathrm{P}} = -\frac{T^2c^3}{3EI} + cT\tan\theta\,,$$

and the elastic energy, namely

$$U'_{\mathrm{E}} = \int_0^c \frac{\mathcal{M}_f^2}{2EI}\,\mathrm{d}x = \frac{T^2c^3}{6EI}\,.$$

Hence

$$U_{\mathrm{T}} = -\frac{P^2}{2bEh}\left(c + l_2 + \frac{l}{2}\right) - \frac{2T^2c^3}{Ebh^3} + cT\tan\theta\,.$$

With $\mathrm{d}l = -\,\mathrm{d}c$ and l_2 constant, we get the formula given by Kendall [41]:

$$\boxed{G = \frac{P^2}{4b^2Eh} + \frac{6T^2}{b^2Eh}\frac{c^2}{h^2} - \frac{T}{b}\tan\theta}\,. \tag{5.121}$$

For small values of c/h, application of a force $T > 0$ reduces G, and $T < 0$ increases it. But, as $(\partial G/\partial A)_{P,T}$ is negative, G increases constantly and the equilibrium is unstable. For a given value of P, the crack will move initially faster or slower according to the sign of T, but in any case it will accelerate, as observed by Kendall.

When the peeled band reaches the length c_T an touches the stretched band, things change. A dislocation is formed, and the cracking slows down and then stops. In fact the contact is well established only for the length $c_D \cong 2c_T$. From this point the two bands act in parallel subjected to the force P. Calculation shows that

$$G = \frac{P^2}{4b^2Eh}\frac{c_D^2}{c^2} .$$

As $(\partial G/\partial A)_P$ is now positive, the cracking slows down and stops when the equilibrium $G = w$ is reached. Experimental results are in good qualitative agreement with this theory. If one continues to pull, the band which was in contact lifts up once more and then touches the stretched band again, and a new dislocation is formed. Thus a succession of crack propagation and resticking phases take place.

We can observe these complex phenomena each time a crack, after opening, is forced to close under the action of external constraints.

5.4.3 Other Examples of Spontaneous Peeling

We have discussed above the spontaneous peeling of a stretched band adhering to a rigid substrate. Another example is the following [42]. Consider, as in Fig. 5.24, a band (1) (E_1, h_1) of length $2l$, width b placed on a band (2) (E_2, h_2) of the same width and of length $2(l+u)$. When the second band is pulled, band (1) peels. As the system is symmetrical, let us examine the half to the right in Fig. 5.24 and neglect the bending due to the excentred load in the part of thickness $h_1 + h_2$.

Let x be the peeled length. Hooke's law gives the displacement δ of the force P such that

$$\delta = \frac{P}{b}\left(\frac{u+x}{E_2h_2} + \frac{l-x}{E_1h_1 + E_2h_2}\right) ,$$

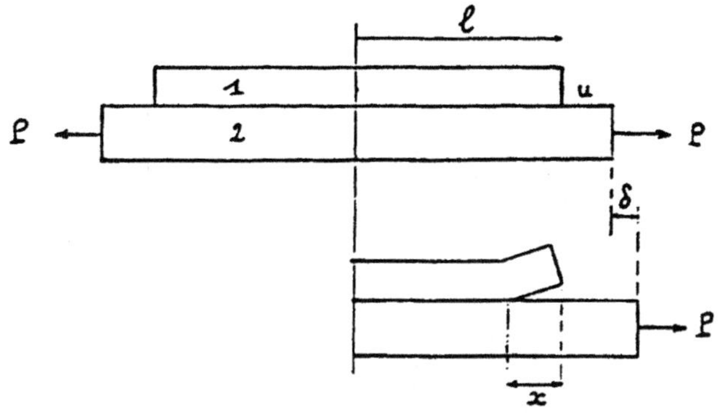

Fig. 5.24. Two bands, one peeled away from the other

and hence the mechanical energy (elastic plus potential) is given by

$$U_{\mathrm{T}} = -\frac{1}{2}P\delta = -\frac{P^2}{2b}\left(\frac{u+x}{E_2h_2} + \frac{l-x}{E_1h_1+E_2h_2}\right) .$$

By a variation of contact area $\mathrm{d}A = -b\,\mathrm{d}x$ at constant load P, we get

$$\boxed{G = \frac{P^2}{2b^2}\left(\frac{1}{E_2h_2} - \frac{1}{E_1h_1+E_2h_2}\right)} . \qquad (5.122)$$

This equation is the same as for the lap joint, as per (5.111), although the loading system is slightly different. The reason is that in both cases the bending due to the excentric load has been neglected.

Replacing P by its value in function of δ and x, we get

$$G = \frac{\delta^2}{2}\frac{E_1h_1\left(1+\frac{E_1h_1}{E_2h_2}\right)}{\left[l + x\frac{E_1h_1}{E_2h_2} + u\left(1+\frac{E_1h_1}{E_2h_2}\right)\right]^2} . \qquad (5.123)$$

We can see that, at fixed load P, equilibrium is neutral and the condition $G = w$ defines the force under which the crack will propagate until complete separation. In contrast, at fixed grips δ equilibrium is always stable. Increasing δ leads to reducing the contact area down to a new equilibrium. We are thus dealing with a *controlled rupture* of the joint. We can of course compute the critical elongation at which the area of contact reduces to a line. Making $x = l$ in $G = w$, we get

$$\varepsilon_{\mathrm{c}} = \frac{\delta_{\mathrm{c}}}{l+u} = \sqrt{2w\left(\frac{1}{E_1h_1} + \frac{1}{E_2h_2}\right)} .$$

These various expressions have been well verified experimentally in the case of rubber strips [42].

If we cut in its middle the band (2) which is pulled (see Fig. 5.25), we have the image of a joint with a butt strap. Compared with the previous example, four starting points for the crack are possible instead of two. If we note that the load is transmitted to the two halves of band (2) through the cover-plate, we see that we are dealing with two single laps, a problem treated above. The crack will start from the side which has the lower product Eh, i.e. at the level of the cut if the E_1h_1 of the cover plate is lower than E_2h_2; otherwise it starts at the extremities of the cover-plate, as in the previous example.

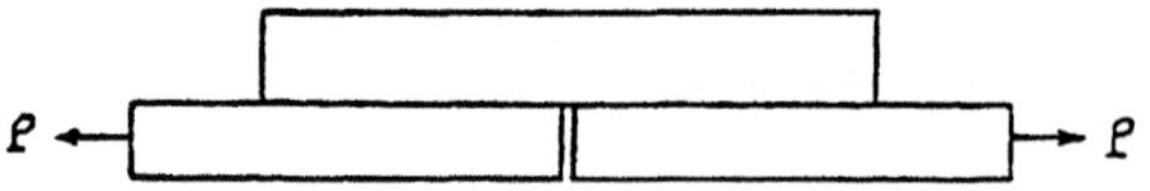

Fig. 5.25. Schema of a joint with a butt strap

Placing two identical bands (E_1, h_1) on both sides of a band (2) (E_2, h_2), we have the image of a double-butt strap joint. Calculation is as above, considering only the right-hand side with a thickness $h_2 + 2h_1$ of the central part. The mechanical energy is:

$$U_{\mathrm{T}} = -\frac{P^2}{2b}\left(\frac{u+x}{E_2 h_2} + \frac{l-x}{E_2 h_2 + 2E_1 h_1}\right).$$

However, we must note that the created area is no longer $b\,\mathrm{d}x$ but $2b\,\mathrm{d}x$, and hence the value of G is given by

$$G = -\frac{1}{2b}\frac{\partial U_{\mathrm{T}}}{\partial x},$$

$$\boxed{G = \frac{P^2}{4b^2}\left(\frac{1}{E_2 h_2} - \frac{1}{E_2 h_2 + 2E_1 h_1}\right)}. \tag{5.124}$$

In this case, the cracks at the extremities of the cover-plates begin to propagate as soon as $G > w$.

5.5 Double Torsion

The test for double torsion was introduced by Outwater and Gerry [43] in 1969. It is displayed in Fig. 5.26. The sample of length L, width $2b$ and thickness h is supported by two rollers. A load P is applied by two hemispheres on the upper side, and each arm of thickness b of the pre-cracked plate (the crack length c is generally larger than $2b$) is subjected to a torsional moment

$$M_{\mathrm{t}} = \frac{P}{2}d. \tag{5.125}$$

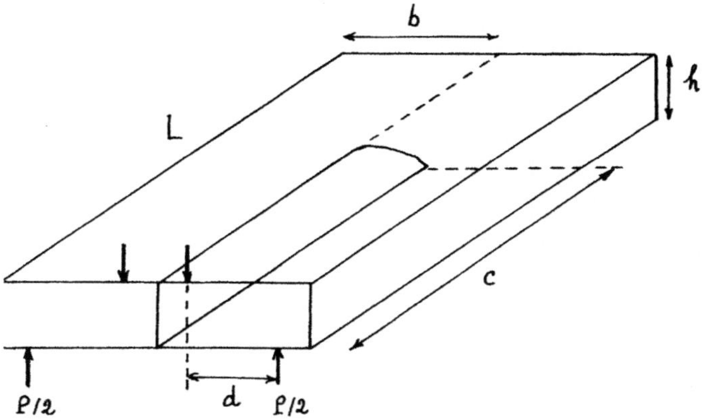

Fig. 5.26. Double torsion test

The crack propagates in Mode I by symmetry, but the crack front is observed not to be perpendicular to the plate: it is not rectilinear but curved, with a crack longer by the value Δc on the lower face (the stress intensity factor must be constant along the crack front). The crack velocity being perpendicular to the crack front, the velocity dc/dt is only apparent and higher than the actual one.

5.5.1 Elementary Theory

Let us apply the theory of torsion of thin beams ($h \ll b$). The rotation of the section is proportional to the length c of the arm and is:

$$\theta = \frac{M_t c}{\mu I_0}, \tag{5.126}$$

where μ is the shear modulus and

$$I_0 = \frac{bh^3}{3}.$$

This angle is small an can be replaced by its tangent

$$\tan\theta = \frac{\delta}{d},$$

where δ is the displacement of the point of application of the upper load, as drawn in Fig. 5.27, and which is thus

$$\delta = \frac{3d^2}{2\mu bh^3} Pc. \tag{5.127}$$

The potential energy is then

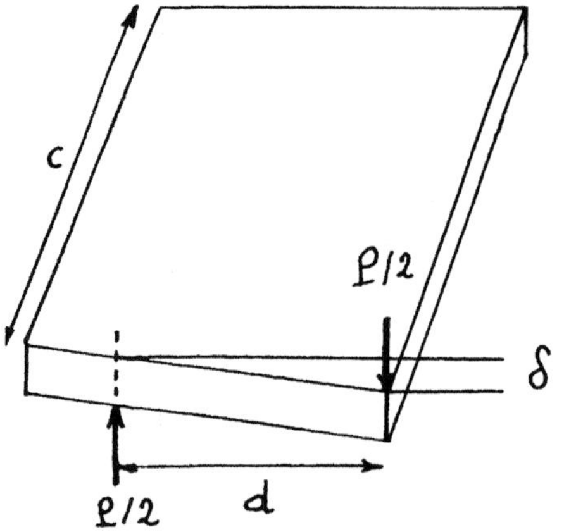

Fig. 5.27. Theory of torsion of thin beams: displacement under load

$$U_{\mathrm{P}} = -2\frac{P}{2}\delta = -\frac{3P^2cd^2}{2\mu bh^3} \ . \tag{5.128}$$

The elastic energy stored in the two arms is

$$U_{\mathrm{E}} = 2\int_0^c \frac{M_{\mathrm{t}}^2}{2\mu I_0}\,\mathrm{d}c = \frac{M_{\mathrm{t}}^2 c}{\mu I_0} \ . \tag{5.129}$$

Differentiating at constant P the total mechanical energy $U_{\mathrm{E}} + U_{\mathrm{P}}$ with respect to A, with $\mathrm{d}A = -h\,\mathrm{d}c$ (assuming that the crack front moves without deformation), we get

$$\boxed{G = \frac{3d^2}{4\mu bh^4}P^2 = \frac{\mu bh^2}{3d^2c^2}\delta^2 = \frac{P\delta}{2hc}} \ , \tag{5.130}$$

a formula given by Outwater and Gerry [43]. The equilibrium defined by $G = w$ is thus neutral at fixed load P and stable at fixed grips δ.

As for any geometry which is stable at fixed grips, the apparent velocity of the crack $\mathrm{d}c/\,\mathrm{d}t$ can be obtained directly from the rate of load relaxation $\mathrm{d}P/\,\mathrm{d}t$ at fixed displacements δ and from the final length c_0 of the crack. Differentiating (5.127) at constant δ, we get:

$$\frac{\mathrm{d}c}{\mathrm{d}t} = -\frac{c}{P}\frac{\mathrm{d}P}{\mathrm{d}t} \ ,$$

and as at constant δ we have $P_0c_0 = Pc$, we get

$$\frac{\mathrm{d}c}{\mathrm{d}t} = -\frac{P_0c_0}{P^2}\frac{\mathrm{d}P}{\mathrm{d}t} \ . \tag{5.131}$$

With the help of a tensile test machine, a force P can be rapidly applied and then the cross-head locked. With the crack advance at fixed displacements, the gradual relaxation of the load is then recorded. After a time, when the relaxation is no longer measurable, the experiment is stopped and the final length of the crack c_0 (corresponding to the load P_0) is measured. For any value of P, the apparent crack velocity is given by (5.131) and G by (5.130). With this method, crack velocities of the order 10^{-8} m/s can be coped with. In fact, it is the velocity normal to the crack front, which must be related to the energy release rate G, and this demands a careful study of the shape of the crack front. Approximating the crack front to an inclined straight line, we have

$$v = \frac{h}{\sqrt{h^2 + (\Delta c)^2}}\frac{\mathrm{d}c}{\mathrm{d}t} \ .$$

This method of relaxation was proposed by Evans [44] in 1972.

5.5.2 Influence of the Thickness

This problem was examined by Fuller [45] in 1979. It is known since Barré de Saint Venant in 1855 that for beams of any section the sections do not remain planes during torsion but show a warping. The exact calculation for a rectangular section (cf. Timoshenko and Goodier [46]) gives

$$\theta = \frac{M_t c}{\mu I_0} \frac{1}{\psi(h/b)} ,$$

where

$$\psi(h/b) = 1 - \frac{192}{\pi^5} \frac{h}{b} \sum_{n=1,3,5,\ldots}^{\infty} \frac{1}{n^5} \tanh \frac{\pi n b}{2h} .$$

Equations (5.127) and (5.130) become

$$\delta = \frac{3d^2}{2\mu b h^3 \psi(h/b)} Pc , \tag{5.127a}$$

$$\boxed{G = \frac{3d^2 P^2}{4\mu b h^4 \psi(h/b)} = \frac{\mu b h^2 \psi(h/b) \delta^2}{3d^2 c^2} = \frac{P\delta}{2hc}} . \tag{5.130a}$$

An approximation better than 0.1% for $0 < h/b < 1$ was given by Fuller [45]:

$$\psi(h/b) = 1 - 0.6302 \frac{h}{b} + 1.20 \frac{h}{b} \mathrm{e}^{-\pi b/h} .$$

This corrective term varies from 1 to 0.42 when h/b varies from 0 to 1.

5.6 The Blister Test

The blister test was introduced by Dannenberg [47] in 1961. The idea is to inject a fluid under a pressure p (corresponding to a load P) between a substrate and a coating of thickness h and modulus E so that the coating peels, forming a blister (Fig. 5.28). An initial decohesion zone of radius a_0 is generally introduced at the start. If thickness h is small compared with the radius a, the system can be considered as a membrane subjected to tension only (no flexion).

As a first approximation, the shape of the blister is a portion of a sphere of radius R, with height δ (with $\delta \ll R$). Figure 5.28 leads to geometrical relations

$$\delta = R(1 - \cos\alpha) \approx R\frac{\alpha^2}{2} ,$$

$$a = R \sin\alpha \approx R\alpha ,$$

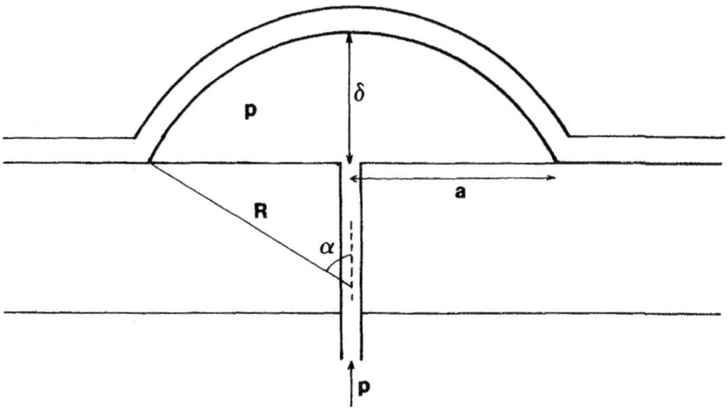

Fig. 5.28. Injection of fluid to form a blister under a thin coating

and hence

$$\alpha \approx \frac{2\delta}{a} ,$$
$$R \approx \frac{a^2}{2\delta} .$$

The volume of the portion of sphere is given by

$$V = \int_{R-\delta}^{R} A(z)\,\mathrm{d}z = \int_{R-\delta}^{R} \pi(R^2 - z^2)\,\mathrm{d}z$$
$$= \frac{\pi\delta^2}{3}(3R - \delta) .$$

For $\delta \ll R$ and with $\delta = a^2/2R$, the volume is given by

$$V = \frac{1}{2}\pi a^2 \delta . \tag{5.132}$$

Let us compute the deflection δ of the blister under pressure p. Let σ_r be the stress exerted by the spherical membrane on its substrate. Vertical equilibrium requires

$$\pi a^2 p = 2\pi a h \sigma_r \sin\alpha , \tag{5.133}$$

or

$$p = \frac{2\sigma_r h}{R} \approx \frac{4\sigma_r h \delta}{a^2} .$$

The pressure p is thus small compared with the stress σ_r. Writing now Hooke's law, noting that the extension is zero along the circle of radius a, we obtain

$$\varepsilon_r \approx \frac{1}{E}[\sigma_r - \nu\sigma_\theta] ,$$
$$\varepsilon_\theta \approx \frac{1}{E}[\sigma_\theta - \nu\sigma_r] = 0 ,$$

and hence

$$\varepsilon_r \approx \frac{1-\nu^2}{E}\sigma_r \, .$$

Evaluating the extension ε_r from the arc of circle $R\alpha$, we get

$$\varepsilon_r = \frac{R\alpha - R\sin\alpha}{R\sin\alpha} \approx \frac{\alpha^2}{6} \approx \frac{2}{3}\frac{\delta^2}{a^2} \, .$$

We thus have

$$\sigma_r \approx \frac{2E\delta^2}{3(1-\nu^2)a^2} \, ,$$

and (5.133) gives

$$\delta^3 = \frac{3(1-\nu^2)}{8}\frac{pa^4}{Eh} \, . \tag{5.134}$$

This equation for $\delta(p, a)$ is the first equation of state of the system.

In reality, the shape is more complex than a portion of a sphere, and more precise calculations by Hencky [48] give for the volume

$$V = C_1 \pi a^2 \delta \, , \tag{5.135}$$

with C_1 dependent on ν ($C_1 = 0.518$ for $\nu = 0.3$ and $C_1 = 0.519$ for $\nu = 0.5$). For the deflection Hencky established

$$\boxed{\delta = C_2 \left(\frac{pa^4}{Eh}\right)^{1/3}} \, , \tag{5.136}$$

with $C_2 = 0.662$ for $\nu = 0.3$ and $C_2 = 0.595$ for $\nu = 0.5$. The difference from the simple theory is only a few percentage points.

To compute the elastic energy U_{E} of the membrane of radius a, we assume it has been deformed at constant radius a by an incompressible liquid subjected to a force F acting on a piston of section S. If x is the displacement of the piston, the elastic energy of the membrane is given by

$$U_{\mathrm{E}} = \int F(x)\, \mathrm{d}x = \int \frac{F(x)}{S} S\, \mathrm{d}x = \int p\, \mathrm{d}V \, .$$

Using (5.135) and (5.136), which show that p is proportional to V^3, we get

$$U_{\mathrm{E}} = \int p\, \mathrm{d}V = \frac{pV}{4} \, . \tag{5.137}$$

Let us examine now the above system with a compressible fluid under a load P, and study the energy of the system when a varies at constant P

(p and V thus being constant). The elastic energy of the membrane is given by (5.137), that of the fluid is constant since there is constant p, and the total energy of the load P is given by

$$U_{\mathrm{P}} = -Px = -pV \; . \tag{5.138}$$

The total energy of the system $U_{\mathrm{T}} = U_{\mathrm{E}} + U_{\mathrm{P}}$, is thus given as

$$U_{\mathrm{T}} = -\frac{3pV}{4} = -\frac{3p}{4} C_1 \pi a^2 C_2 \left(\frac{pa^4}{Eh}\right)^{1/3} \; . \tag{5.139}$$

The energy release rate given by $G = (\partial U_{\mathrm{T}} / \mathrm{d}A)_p$ with $\mathrm{d}A = -2\pi a \, \mathrm{d}a$ is given by

$$\boxed{G = \frac{5C_1}{4} C_2 \left(\frac{p^4 a^4}{Eh}\right)^{1/3} = \frac{5C_1}{4} p\delta} \; . \tag{5.140}$$

This formula was given by Gent and Lewandowski [49] in 1987. At equilibrium defined by $G = w$, the radius a of the blister and its deflection are inversely proportional to the pressure. As for the flat punch, this equilibrium is unstable. When the pressure p increases, the deflection of the membrane begins to increase at constant radius $a = a_0$ according to (5.136). Then, when the pressure reaches the critical value

$$p_{\mathrm{c}} = \left(\frac{4w}{5C_1 C_2}\right)^{3/4} \frac{(Eh)^{1/4}}{a_0} \tag{5.141}$$

and the deflection reaches the critical deflection

$$\delta_{\mathrm{c}} = \left(\frac{4wC_2^3}{5C_1 Eh}\right)^{1/4} a_0 \tag{5.142}$$

defined by $G = w$, the radius of the blister increases catastrophically at constant pressure p_{c} until there is complete decohesion of the coating. In reduced coordinates, (5.136) can be written as

$$\frac{p}{p_{\mathrm{c}}} = \frac{a_0^4}{a^4} \left(\frac{\delta}{\delta_{\mathrm{c}}}\right)^3 \; .$$

To have a stable propagation, it is necessary to have a constant mass of fluid in the blister [50].

References

1. P.H. Hodge: *Plastic analysis of structures* (McGraw-Hill, 1959)
2. K.S. Kim, J. Kim: Trans. ASME, J. Eng. Mat. Technol. **110**, 266 (1988)
3. J.W. Obreimov: Proc. Roy. Soc. A **127**, 290 (1930)
4. J.J. Benbow, F.C. Roesler: Proc. Phys. Soc. **13**, 201 (1957)
5. J.J. Gilman: J. Appl. Phys. **31**, 2208 (1960)
6. J.P. Berry: J. Appl. Phys. **34**, 62 (1963)
7. P.P. Gillis, J.J. Gilman: J. Appl. Phys. **35**, 647 (1964)
8. S.M. Wiederhorn, A.M. Shorb, R.L. Moses: J. Appl. Phys. **39**, 1569 (1968)
9. S. Mostovoy, P.B. Crosley, E.J. Ripling: J. Materials **2**, 661 (1967)
10. M.F. Kanninen: Int. Journ. of Fracture **9**, 83 (1973)
11. G.J. Spies: Aircraft Eng. **25**, 64 (1953)
12. Y. Inoue, Y. Kobatake: Appl. Sci. Res. **A8**, 321 (1959)
13. H. Kaelble: Trans. Soc. Rheol. **4**, 45 (1960)
14. J.E. Strawley, B. Gross: Mater. Res. Std **7**, 155 (1967)
15. S.W. Freiman, D.R. Mulville, P.W. Mast: J. Mater. Sci. **8**, 1527 (1973)
16. W.B. Fichter: Int. J. Fracture **22**, 133 (1983)
17. R.M.L. Foote, V.T. Buchwald: Int. J. Fracture **29**, 125 (1985)
18. G.R. Irwin: J. Appl. Mech. **24**, 361 (1957)
19. K.T. Wan, B.R. Lawn, R.G. Horn: J. Mater. Res. **7**, 1584 (1992)
20. C. Creton, E.J. Kramer, C.Y. Hui, H.R. Brown: Macromolecules **25**, 3075 (1992)
21. S.J. Burns, B.R. Lawn: Int. J. Fracture **4**, 339 (1968)
22. M.D. Thouless: Acta metall. mater. **40**, 1281 (1992)
23. F.E. Penado: J. Composite Mater. **27**, 383 (1993)
24. R.S. Rivlin: Paint Technol. **9**, 215 (1944)
25. B.V. Deryagin, N.A. Krotova: Dokl. Akad. Nauk SSR **61**, 849 (1948)
26. P.B. Lindley: J. Inst. Rubber Ind. **5**, 243 (1971)
27. K. Kendall: J. Phys. D: Appl. Phys. **8**, 1449 (1975)
28. G.P. Anderson, K.L. Devries, M.L. Williams: Exp. Mech. **24**, 11 (1976)
29. K. Kendall: J. Adhesion **5**, 105 (1973)
30. D.H. Kaelble: Trans. Soc. Rheol. **3**, 161 (1959)
31. N.A. de Bruyne: Nature **180**, 262 (1957)
32. A.D. Crocombe, R.D. Adams: J. Adhesion **12**, 127 (1981)
33. D. Maugis: Le Vide **186**, 1 (1977)
34. J.G. Williams: J. Strain Anal. **28**, 237 (1993)
35. K. Kendall: J. Phys. D: Appl. Phys. **6**, 1782 (1973)
36. K. Kendall: J. Phys. D: Appl. Phys. **11**, 1519 (1978)
37. G.J. Lake, A. Stevenson: J. Adhesion **12**, 13 (1981)
38. K. Kendall: J. Phys. D: Appl. Phys. **8**, 512 (1975)
39. M. Goland, E. Reissner: J. Appl. Mech. Trans. ASME, A17 (1944)
40. K. Kendall: J. Adhesion **6**, 137 (1975)
41. K. Kendall: Phil. Mag. A **43**, 713 (1981)
42. M. Barquins: J. Appl. Polym. Sci. **29**, 3269 (1984)
43. J.O. Outwater, D.J. Gerry: J. Adhesion **1**, 290 (1969)
44. A.G. Evans: J. Mater. Sci. **7**, 1137 (1972)
45. E.R. Fuller: 'An evaluation of Double-Torsion testing-Analysis.' In: *Fracture mechanics applied to brittle materials*, ASTM STP 678, ed. by S.W. Freiman (1979) pp. 3–18

46. S. Timoshenko, J.N. Goodier: *Théorie de l'élasticité.* (Béranger, Paris 1961)
47. H. Dannenberg: J. Appl. Polym. Sci. **5**, 125 (1961)
48. H. Hencky: Z. Math. Phys. **63**, 311 (1915)
49. A.N. Gent, L.H. Lewandowski: J. Appl. Polym. Sci. **33**, 1567 (1987)
50. K.T. Wan, Y.W. Mai: Acta Metall. Mater. **43**, 4109 (1995)

Appendix A

With regard to the Bessel functions, let us recall the classical formulae:

$$2nJ_n(x) = xJ_{n-1}(x) + xJ_{n+1}(x) , \tag{A1}$$

$$\frac{\mathrm{d}}{\mathrm{d}x}(x^n J_n) = x^n J_{n-1} , \tag{A2}$$

$$\frac{\mathrm{d}}{\mathrm{d}x}(x^{-n} J_n) = -x^n J_{n+1} . \tag{A3}$$

In particular:

$$J_0' = -J_1 . \tag{A4}$$

We will need also the cosine and sine Fourier transforms of the Bessel functions [1]:

$$\int_0^\infty \cos(\xi t) J_0(\xi r)\,\mathrm{d}\xi = \frac{H(r-t)}{\sqrt{r^2 - t^2}} , \tag{A5}$$

$$\int_0^\infty \sin(\xi t) J_0(\xi r)\,\mathrm{d}\xi = \frac{H(t-r)}{\sqrt{t^2 - r^2}} , \tag{A6}$$

$$\int_0^\infty \cos(\xi t) J_1(\xi r)\,\mathrm{d}\xi = \frac{1}{r}\left[1 - \frac{t}{\sqrt{t^2 - r^2}} H(t-r)\right] , \tag{A7}$$

$$\int_0^\infty \sin(\xi t) J_1(\xi r)\,\mathrm{d}\xi = \frac{t}{r\sqrt{r^2 - t^2}} H(r-t) , \tag{A8}$$

where $H(x)$ is the step function of value zero if $x < 0$ and value 1 if $x > 0$. We will use also the relationships

$$\int_0^\infty \frac{1}{\xi} \sin(\xi t) J_0(\xi r)\,\mathrm{d}\xi = \begin{cases} \dfrac{\pi}{2} , & r \le t , \\ \sin^{-1}\dfrac{t}{r} , & r \ge t , \end{cases} \tag{A9}$$

$$\int_0^\infty \frac{1}{\xi} \sin(\xi t) J_1(\xi r)\,\mathrm{d}\xi = \frac{1}{r}\left[t - H(t-r)\sqrt{t^2 - r^2}\right] , \tag{A10}$$

$$\int_0^\infty \frac{1}{\xi} \cos(\xi t) J_1(\xi r)\,\mathrm{d}\xi = H(r-t) \cos\left(\sin^{-1}\frac{t}{r}\right) , \tag{A11}$$

$$\int_0^\infty \xi^{-1} \sin \xi \mathrm{e}^{-\xi z} \, \mathrm{d}\xi = \tan^{-1} \frac{1}{z} \, , \tag{A12}$$

$$\int_0^\infty \sin \xi \mathrm{e}^{-\xi z} \, \mathrm{d}\xi = \frac{1}{1+z^2} \, , \tag{A13}$$

$$\int_0^\infty \xi \sin \xi \mathrm{e}^{-\xi z} \, \mathrm{d}\xi = \frac{2z}{(1+z^2)^2} \, . \tag{A14}$$

Sneddon [2] gives the following results for $n > -1$:

$$\begin{aligned} H_n[\xi^{-1}\mathrm{e}^{-\xi z}; r] &= \int_0^\infty \mathrm{e}^{-\xi z} J_n(\xi r) \, \mathrm{d}\xi \\ &= \frac{1}{\sqrt{r^2+z^2}} \left(\frac{r}{z+\sqrt{r^2+z^2}} \right)^n \, , \end{aligned} \tag{A15}$$

$$\begin{aligned} H_n[\mathrm{e}^{-\xi z}; r] &= \int_0^\infty \xi \mathrm{e}^{-\xi z} J_n(\xi r) \, \mathrm{d}\xi \\ &= \frac{z+n\sqrt{r^2+z^2}}{(r^2+z^2)^{3/2}} \left(\frac{r}{z+\sqrt{r^2+z^2}} \right)^n \, . \end{aligned} \tag{A16}$$

Introducing the notation

$$Z_n^m = \int_0^\infty \xi^{n-1} \mathrm{e}^{-\xi z} J_m(\xi r) \, \mathrm{d}\xi \, , \tag{A17}$$

we deduce that

$$Z_1^0 = \int_0^\infty \mathrm{e}^{-\xi z} J_0(\xi r) \, \mathrm{d}\xi = \frac{1}{\sqrt{r^2+z^2}} \, , \tag{A18}$$

$$\begin{aligned} Z_1^1 &= \int_0^\infty \mathrm{e}^{-\xi z} J_1(\xi r) \, \mathrm{d}\xi = \frac{r}{r^2+z^2+z\sqrt{r^2+z^2}} \\ &= \frac{1}{r} \left(1 - \frac{z}{\sqrt{r^2+z^2}} \right) \, , \end{aligned} \tag{A19}$$

$$Z_2^0 = \int_0^\infty \xi \mathrm{e}^{-\xi z} J_0(\xi r) \, \mathrm{d}\xi = \frac{z}{(r^2+z^2)^{3/2}} \, , \tag{A20}$$

$$Z_2^1 = \int_0^\infty \xi \mathrm{e}^{-\xi z} J_1(\xi r) \, \mathrm{d}\xi = \frac{r}{(r^2+z^2)^{3/2}} \, . \tag{A21}$$

Differentiating (A20) with respect to z, we get

$$Z_3^0 = \int_0^\infty \xi^2 \mathrm{e}^{-\xi z} J_0(\xi r) \, \mathrm{d}\xi = \frac{2z^2 - r^2}{(r^2+z^2)^{5/2}} \, , \tag{A22}$$

whereas differentiating with respect to r, we get

$$Z_3^1 = \int_0^\infty \xi^2 \mathrm{e}^{-\xi z} J_1(\xi r) \, \mathrm{d}\xi = \frac{3rz}{(r^2+z^2)^{5/2}} \, . \tag{A23}$$

Integrating (A18) and (A19) with respect to z, we get

$$Z_0^0 = \int_0^\infty \frac{1}{\xi} \mathrm{e}^{-\xi z} J_0(\xi r)\,\mathrm{d}\xi = -\ln\left(z + \sqrt{r^2+z^2}\right)$$
$$= -\left(\sinh^{-1}\frac{z}{r} + \ln r\right) , \tag{A24}$$

$$Z_0^1 = \int_0^\infty \frac{1}{\xi} \mathrm{e}^{-\xi z} J_1(\xi r)\,\mathrm{d}\xi = \frac{\sqrt{r^2+z^2}-z}{r} . \tag{A25}$$

Continuing the integrations, we find that

$$Z_{-1}^0 = \int_0^\infty \frac{1}{\xi^2} \mathrm{e}^{-\xi z} J_0(\xi r)\,\mathrm{d}\xi = -zZ_0^0 - \sqrt{r^2+z^2} , \tag{A26}$$

$$Z_{-1}^1 = \int_0^\infty \frac{1}{\xi^2} \mathrm{e}^{-\xi z} J_1(\xi r)\,\mathrm{d}\xi = \frac{1}{2}(rZ_0^0 - zZ_0^1) , \tag{A27}$$

$$Z_{-2}^0 = \int_0^\infty \frac{1}{\xi^3} \mathrm{e}^{-\xi z} J_0(\xi r)\,\mathrm{d}\xi = \frac{1}{4}\left[3z\sqrt{r^2+z^2} + (2z^2-r^2)Z_0^0\right] , \tag{A28}$$

$$Z_{-2}^1 = \int_0^\infty \frac{1}{\xi^3} \mathrm{e}^{-\xi z} J_1(\xi r)\,\mathrm{d}\xi = \frac{1}{6r}\left[(r^2+z^2)^{3/2} - z^3\right] + \frac{r}{2}Z_{-1}^0$$
$$= \frac{1}{6}Z_0^1\left[r^2 + 2z^2 + z\sqrt{r^2+z^2}\right] + \frac{r}{2}Z_{-1}^0 . \tag{A29}$$

Note that using the recurrence relationship (A1), we get

$$Z_{n-1}^m = \frac{r}{2m}[Z_n^{m-1} + Z_n^{m+1}] . \tag{A30}$$

Computations of Integrals C_n^m and S_n^m

The integrals C_n^m and S_n^m, which occur in the calculation of the circular flat punch, can be written as

$$\begin{aligned} C_n^m &= \int_0^\infty p^{n-1} \cos p\,\mathrm{e}^{-p\zeta} J_m(p\varrho)\,\mathrm{d}p , \\ S_n^m &= \int_0^\infty p^{n-1} \sin p\,\mathrm{e}^{-p\zeta} J_m(p\varrho)\,\mathrm{d}p . \end{aligned} \tag{A31}$$

They can be deduced from integrals Z_n^m by replacing z by $\zeta+\mathrm{i}$, and identifying real and imaginary parts so that

$$Z_n^m = C_n^m - \mathrm{i}S_n^m . \tag{A32}$$

Let us introduce the following notations

$$\begin{aligned} s &= \sqrt{(\varrho^2+\zeta^2-1)^2 + 4\zeta^2} \\ &= \sqrt{(\varrho^2+\zeta^2+1)^2 - 4\varrho^2} , \end{aligned} \tag{A33}$$

$$\sin\varphi = \frac{2\zeta}{s}\,, \tag{A34}$$

$$\cos\varphi = \frac{\varrho^2+\zeta^2-1}{s}\,, \qquad 0<\varphi<\pi\,, \tag{A35}$$

$$\tan\theta = \frac{1}{\zeta} \tag{A36}$$

(where the contours of φ are displayed in Fig. 4.8), and compute for example C_0^0 and S_0^0 from Z_0^0:

$$\begin{aligned} C_0^0 - \mathrm{i}S_0^0 &= \int_0^\infty \frac{1}{p}\mathrm{e}^{-p(\zeta+\mathrm{i})} J_0(p\varrho)\,\mathrm{d}p \\ &= -\ln\left[\zeta+\mathrm{i}+\sqrt{(\zeta+\mathrm{i})^2+\varrho^2}\right] \\ &= -\ln\left[\zeta+\mathrm{i}+s^{1/2}\left(\cos\frac{\varphi}{2}+\mathrm{i}\sin\frac{\varphi}{2}\right)\right]\,. \end{aligned}$$

Recalling that, as a generalization,

$$\ln(x+\mathrm{i}y) = \ln\sqrt{x^2+y^2} + \mathrm{i}\tan^{-1}\frac{y}{x}\,,$$

we have

$$C_0^0 = -\ln\sqrt{\left(\zeta+s^{1/2}\cos\frac{\varphi}{2}\right)^2+\left(1+s^{1/2}\sin\frac{\varphi}{2}\right)^2} \tag{A37}$$

and

$$\begin{aligned} S_0^0 &= \tan^{-1}\frac{1+s^{1/2}\sin(\varphi/2)}{\zeta+s^{1/2}\cos(\varphi/2)} \\ &= \tan^{-1}\left(\frac{1}{s^{1/2}\cos(\varphi/2)}\right)\,. \end{aligned} \tag{A38}$$

Similarly, we get

$$C_0^1 = -\frac{1}{\varrho}\left(\zeta - s^{1/2}\cos\frac{\varphi}{2}\right)\,, \tag{A39}$$

$$S_0^1 = \frac{1}{\varrho}\left(1 - s^{1/2}\sin\frac{\varphi}{2}\right)\,, \tag{A40}$$

$$C_1^0 = \frac{1}{s^{1/2}}\cos\frac{\varphi}{2}\,, \tag{A41}$$

$$S_1^0 = \frac{1}{s^{1/2}}\sin\frac{\varphi}{2}\,, \tag{A42}$$

$$C_1^1 = \frac{1}{\varrho} - \frac{\sqrt{1+\zeta^2}}{\varrho s^{1/2}}\cos\left(\theta-\frac{\varphi}{2}\right)\,, \tag{A43}$$

$$S_1^1 = \frac{\sqrt{1+\zeta^2}}{\varrho s^{1/2}}\sin\left(\theta-\frac{\varphi}{2}\right)\,, \tag{A44}$$

$$C_2^0 = \frac{\sqrt{1+\zeta^2}}{s^{3/2}} \cos\left(\frac{3\varphi}{2} - \theta\right) , \tag{A45}$$

$$S_2^0 = \frac{\sqrt{1+\zeta^2}}{s^{3/2}} \sin\left(\frac{3\varphi}{2} - \theta\right) , \tag{A46}$$

$$C_2^1 = \frac{\varrho}{s^{3/2}} \cos\frac{3\varphi}{2} , \tag{A47}$$

$$S_2^1 = \frac{\varrho}{s^{3/2}} \sin\frac{3\varphi}{2} , \tag{A48}$$

results given by Sneddon [3, 4].

These formulae can be simplified by introducing the parameter u, the positive root of the equation

$$\frac{\varrho^2}{1+u} + \frac{\zeta^2}{u} = 1 , \tag{A49}$$

i.e.

$$u = \frac{1}{2}(\varrho^2 + \zeta^2 - 1 + s) \tag{A50}$$

and

$$\frac{\zeta}{\sqrt{u}} = \sqrt{1 - \frac{\varrho^2}{u+1}} . \tag{A51}$$

We then have

$$s^{1/2} \cos\frac{\varphi}{2} = \sqrt{u} , \tag{A52}$$

$$s^{1/2} \sin\frac{\varphi}{2} = \frac{\zeta}{\sqrt{u}} , \tag{A53}$$

$$us = u^2 + \zeta^2 . \tag{A54}$$

Note that if s_1 and s_2 are the distances of a point $\mathrm{M}(\varrho, \zeta)$ to the points $(-1, 0)$ and $(1, 0)$, we have

$$s = \sqrt{s_1 s_2} ,$$

$$\sqrt{u+1} = \frac{1}{2}(s_1 + s_2) .$$

The C_n^m and S_n^m are thus written as

$$C_0^0 = \int_0^\infty \frac{1}{p} \cos p\,\mathrm{e}^{-p\zeta} J_0(p\varrho)\,\mathrm{d}p = -\ln\sqrt{(1+u)\left(1 + \frac{\zeta}{\sqrt{u}}\right)^2} , \tag{A55}$$

$$S_0^0 = \int_0^\infty \frac{1}{p} \sin p\,\mathrm{e}^{-p\zeta} J_0(p\varrho)\,\mathrm{d}p = \tan^{-1}\frac{1}{\sqrt{u}} , \tag{A56}$$

$$C_0^1 = \int_0^\infty \frac{1}{p} \cos p \, \mathrm{e}^{-p\zeta} J_1(p\varrho) \, \mathrm{d}p = \frac{\sqrt{u}}{\varrho}\left(1 - \frac{\zeta}{\sqrt{u}}\right) , \tag{A57}$$

$$S_0^1 = \int_0^\infty \frac{1}{p} \sin p \, \mathrm{e}^{-p\zeta} J_1(p\varrho) \, \mathrm{d}p = \frac{1}{\varrho}\left(1 - \frac{\zeta}{\sqrt{u}}\right) , \tag{A58}$$

$$C_1^0 = \int_0^\infty \cos p \, \mathrm{e}^{-p\zeta} J_0(p\varrho) \, \mathrm{d}p = \frac{\sqrt{u}}{s} , \tag{A59}$$

$$S_1^0 = \int_0^\infty \sin p \, \mathrm{e}^{-p\zeta} J_0(p\varrho) \, \mathrm{d}p = \frac{\zeta}{s\sqrt{u}} , \tag{A60}$$

$$C_1^1 = \int_0^\infty \cos p \, \mathrm{e}^{-p\zeta} J_1(p\varrho) \, \mathrm{d}p = \frac{1}{\varrho}\left(1 - \frac{\zeta}{\sqrt{u}} \frac{1+u}{s}\right) , \tag{A61}$$

$$S_1^1 = \int_0^\infty \sin p \, \mathrm{e}^{-p\zeta} J_1(p\varrho) \, \mathrm{d}p = \frac{\varrho\sqrt{u}}{s(1+u)} , \tag{A62}$$

$$\begin{aligned} S_2^0 &= \int_0^\infty p \sin p \, \mathrm{e}^{-p\zeta} J_0(p\varrho) \, \mathrm{d}p \\ &= \frac{1}{s^3\sqrt{u}}\left[- s(\zeta^2 + u) + 4\zeta^2(1+u)\right] , \end{aligned} \tag{A63}$$

$$S_2^1 = \int_0^\infty p \sin p \, \mathrm{e}^{-p\zeta} J_1(p\varrho) \, \mathrm{d}p = \frac{\varrho}{s^2} \frac{\zeta}{\sqrt{u}}\left(3 - \frac{4\zeta^2}{us}\right) . \tag{A64}$$

By the same method, we have

$$C_{-1}^0 = \int_0^\infty \frac{1}{p^2} \cos p \, \mathrm{e}^{-p\zeta} J_0(p\varrho) \, \mathrm{d}p = -(\zeta C_0^0 + S_0^0 + \sqrt{u}) , \tag{A65}$$

$$S_{-1}^0 = \int_0^\infty \frac{1}{p^2} \sin p \, \mathrm{e}^{-p\zeta} J_0(p\varrho) \, \mathrm{d}p = C_0^0 - \zeta S_0^0 + \frac{\zeta}{\sqrt{u}} , \tag{A66}$$

$$C_{-1}^1 = \int_0^\infty \frac{1}{p^2} \cos p \, \mathrm{e}^{-p\zeta} J_1(p\varrho) \, \mathrm{d}p = \frac{1}{2}(\varrho C_0^0 - \zeta C_0^1 - S_0^1) , \tag{A67}$$

$$S_{-1}^1 = \int_0^\infty \frac{1}{p^2} \sin p \, \mathrm{e}^{-p\zeta} J_1(p\varrho) \, \mathrm{d}p = \frac{1}{2}(\varrho S_0^0 - \zeta S_0^1 + C_0^1) , \tag{A68}$$

$$\begin{aligned} C_{-2}^0 &= \int_0^\infty \frac{1}{p^3} \cos p \, \mathrm{e}^{-p\zeta} J_0(p\varrho) \, \mathrm{d}p \\ &= \frac{1}{4}\left[\frac{3\zeta}{\sqrt{u}}(u-1) + (2\zeta^2 - \varrho^2 - 2)C_0^0\right] + \zeta S_0^0 , \end{aligned} \tag{A69}$$

$$\begin{aligned} S_{-2}^0 &= \int_0^\infty \frac{1}{p^3} \sin p \, \mathrm{e}^{-p\zeta} J_0(p\varrho) \, \mathrm{d}p \\ &= \frac{1}{4}\left[-3\frac{u+\zeta^2}{\sqrt{u}} + (2\zeta^2 - \varrho^2 - 2)S_0^0\right] - \zeta C_0^0 , \end{aligned} \tag{A70}$$

$$C_{-2}^{1} = \int_0^\infty \frac{1}{p^3} \cos p \, \mathrm{e}^{-p\zeta} J_1(p\varrho) \, \mathrm{d}p$$

$$= \frac{u\sqrt{u}}{6\varrho}\left[1 - \left(\frac{\zeta}{\sqrt{u}}\right)^3\right] + \frac{\zeta}{2} S_0^1 + \frac{\varrho}{2} C_0^0 \,, \tag{A71}$$

$$S_{-2}^{1} = \int_0^\infty \frac{1}{p^3} \sin p \, \mathrm{e}^{-p\zeta} J_1(p\varrho) \, \mathrm{d}p$$

$$= -\frac{1}{6\varrho}\left[1 - \left(\frac{\zeta}{\sqrt{u}}\right)^3\right] - \frac{\zeta}{2} C_0^1 + \frac{\varrho}{2} S_{-1}^0 \,. \tag{A72}$$

In-surface integrals S_0^0, S_0^1 and S_1^0 take the simple values given by (A9), (A10) and (A6), so that

$$S_0^0(\varrho, 0) = \int_0^\infty \frac{1}{p} \sin p \, J_0(p\varrho) \, \mathrm{d}p = \begin{cases} \pi/2 \,, & \varrho < 1 \,, \\ \sin^{-1}(1/\varrho) \,, & \varrho > 1 \,, \end{cases} \tag{A73}$$

$$S_0^1(\varrho, 0) = \int_0^\infty \frac{1}{p} \sin p \, J_1(p\varrho) \, \mathrm{d}p = \frac{1}{\varrho}\left[1 - H(1-\varrho)\sqrt{1-\varrho^2}\right] \,, \tag{A74}$$

$$S_1^0(\varrho, 0) = \int_0^\infty \sin p \, J_0(p\varrho) \, \mathrm{d}p = \frac{H(1-\varrho)}{\sqrt{1-\varrho^2}} \,. \tag{A75}$$

When $\varrho \to 0$, we have $J_0(p\varrho) \to 1$ and $J_1(p\varrho) \approx p\varrho/2$, so that *on the z-axis* the S_n^m are written as

$$S_0^0(0, \zeta) = \int_0^\infty \frac{1}{p} \sin p \, \mathrm{e}^{-p\zeta} \, \mathrm{d}p = \tan^{-1} \frac{1}{\zeta} \,, \tag{A76}$$

$$S_1^0(0, \zeta) = \frac{2}{\varrho} S_0^1(0, \zeta) = \int_0^\infty \sin p \, \mathrm{e}^{-p\zeta} \, \mathrm{d}p = \frac{1}{1+\zeta^2} \,, \tag{A77}$$

$$S_2^0(0, \zeta) = \frac{2}{\varrho} S_1^1(0, \zeta) = \int_0^\infty p \sin p \, \mathrm{e}^{-p\zeta} \, \mathrm{d}p = \frac{2\zeta}{(1+\zeta^2)^2} \,. \tag{A78}$$

References

1. I.N. Sneddon: Proc. Glasgow Math. Assoc. **4**, 108 (1960)
2. I.N. Sneddon: *The use of integral transforms* (McGraw Hill 1972)
3. I.N. Sneddon: Proc. Camb. Phil. Soc. **42**, 29 (1946)
4. I.N. Sneddon: Proc. Roy. Soc. A **187**, 229 (1946)

Appendix B

Inversion of an Abel Integral

Let us consider the integral

$$\boxed{f(y) = \int_a^y \frac{g(x)\,\mathrm{d}x}{\sqrt{y^2 - x^2}}} \,. \tag{B1}$$

The function $f(y)$ is assumed to be known, and one must deduce $g(x)$. Multiply both sides by $y/\sqrt{t^2 - y^2}$ and integrate from $y = a$ to $y = t$. We then have

$$\begin{aligned}\int_a^t f(y)\frac{y\,\mathrm{d}y}{\sqrt{t^2 - y^2}} &= \int_a^t \frac{y}{\sqrt{t^2 - y^2}}\left[\int_a^y \frac{g(x)\,\mathrm{d}x}{\sqrt{y^2 - x^2}}\right]\mathrm{d}y \\ &= \int_{y=a}^{y=t}\int_{x=a}^{x=y} \frac{y g(x)\,\mathrm{d}x\,\mathrm{d}y}{\sqrt{t^2 - y^2}\sqrt{y^2 - x^2}} \,.\end{aligned} \tag{B2}$$

The field of integration of the double integral is displayed in Fig. B1. Interchange the order of integration. We get:

$$\int_a^t f(y)\frac{y\,\mathrm{d}y}{\sqrt{t^2 - y^2}} = \int_{x=a}^{x=t} \frac{1}{2}g(x)\left[\int_{y=x}^{y=t} \frac{2y\,\mathrm{d}y}{\sqrt{t^2 - y^2}\sqrt{y^2 - x^2}}\right]\mathrm{d}x \,.$$

The inner integral is tabulated and is

$$\int_x^t \frac{\mathrm{d}(y^2)}{\sqrt{t^2 - y^2}\sqrt{y^2 - x^2}} = \left[2\tan^{-1}\sqrt{\frac{y^2 - x^2}{t^2 - y^2}}\right]_{y=x}^{y=t} = \pi \,.$$

We thus have

$$\int_a^t f(y)\frac{y\,\mathrm{d}y}{\sqrt{t^2 - y^2}} = \frac{\pi}{2}\int_a^t g(x)\,\mathrm{d}x \,.$$

We deduce, by differentiation

$$\boxed{g(t) = \frac{2}{\pi}\frac{\mathrm{d}}{\mathrm{d}t}\int_a^t \frac{yf(y)\,\mathrm{d}y}{\sqrt{t^2-y^2}}}\,. \tag{B3}$$

This function can also be written under another form. After integration by parts, we have

$$g(t) = \frac{2}{\pi}\frac{\mathrm{d}}{\mathrm{d}t}\left[f(a)\sqrt{t^2-a^2} + \int_a^t f'(y)\sqrt{t^2-y^2}\,\mathrm{d}y\right]\,.$$

If we then differentiate with respect to t, we get

$$\boxed{g(t) = \frac{2}{\pi}\left[f(a)\frac{t}{\sqrt{t^2-a^2}} + t\int_a^t \frac{f'(y)\,\mathrm{d}y}{\sqrt{t^2-y^2}}\right]}\,. \tag{B4}$$

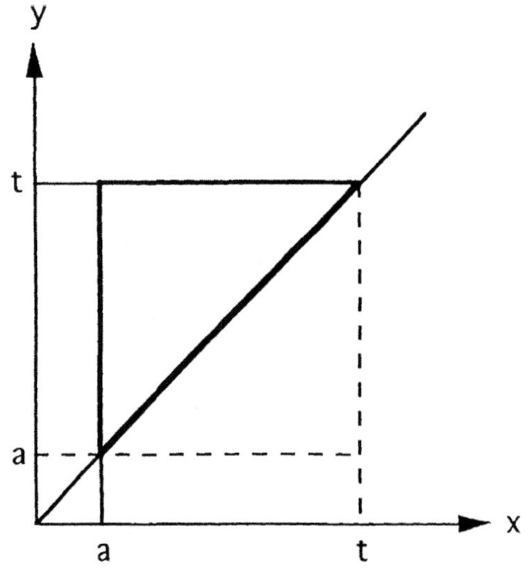

Fig. B1. Field of integration for (B2) double integral

Index

Springer Series in Solid-State Sciences

Editors: M. Cardona P. Fulde K. von Klitzing H.-J. Queisser

1 **Principles of Magnetic Resonance**
3rd Edition By C. P. Slichter

2 **Introduction to Solid-State Theory**
By O. Madelung

3 **Dynamical Scattering of X-Rays in Crystals** By Z. G. Pinsker

4 **Inelastic Electron Tunneling Spectroscopy**
Editor: T. Wolfram

5 **Fundamentals of Crystal Growth I**
Macroscopic Equilibrium and Transport Concepts
By F. E. Rosenberger

6 **Magnetic Flux Structures in Superconductors** By R. P. Huebener

7 **Green's Functions in Quantum Physics**
2nd Edition
By E. N. Economou

8 **Solitons and Condensed Matter Physics**
Editors: A. R. Bishop and T. Schneider

9 **Photoferroelectrics** By V. M. Fridkin

10 **Phonon Dispersion Relations in Insulators** By H. Bilz and W. Kress

11 **Electron Transport in Compound Semiconductors** By B. R. Nag

12 **The Physics of Elementary Excitations**
By S. Nakajima, Y. Toyozawa, and R. Abe

13 **The Physics of Selenium and Tellurium**
Editors: E. Gerlach and P. Grosse

14 **Magnetic Bubble Technology** 2nd Edition
By A. H. Eschenfelder

15 **Modern Crystallography I**
Fundamentals of Crystals
Symmetry, and Methods of Structural Crystallography
2nd Edition
By B. K. Vainshtein

16 **Organic Molecular Crystals**
Their Electronic States By E. A. Silinsh

17 **The Theory of Magnetism I**
Statics and Dynamics
By D. C. Mattis

18 **Relaxation of Elementary Excitations**
Editors: R. Kubo and E. Hanamura

19 **Solitons** Mathematical Methods for Physicists
By. G. Eilenberger

20 **Theory of Nonlinear Lattices**
2nd Edition By M. Toda

21 **Modern Crystallography II**
Structure of Crystals 2nd Edition
By B. K. Vainshtein, V. L. Indenbom, and V. M. Fridkin

22 **Point Defects in Semiconductors I**
Theoretical Aspects
By M. Lannoo and J. Bourgoin

23 **Physics in One Dimension**
Editors: J. Bernasconi and T. Schneider

24 **Physics in High Magnetics Fields**
Editors: S. Chikazumi and N. Miura

25 **Fundamental Physics of Amorphous Semiconductors** Editor: F. Yonezawa

26 **Elastic Media with Microstructure I**
One-Dimensional Models By I. A. Kunin

27 **Superconductivity of Transition Metals**
Their Alloys and Compounds
By S. V. Vonsovsky, Yu. A. Izyumov, and E. Z. Kurmaev

28 **The Structure and Properties of Matter**
Editor: T. Matsubara

29 **Electron Correlation and Magnetism in Narrow-Band Systems** Editor: T. Moriya

30 **Statistical Physics I** Equilibrium Statistical Mechanics 2nd Edition
By M. Toda, R. Kubo, N. Saito

31 **Statistical Physics II** Nonequilibrium Statistical Mechanics 2nd Edition
By R. Kubo, M. Toda, N. Hashitsume

32 **Quantum Theory of Magnetism**
2nd Edition By R. M. White

33 **Mixed Crystals** By A. I. Kitaigorodsky

34 **Phonons: Theory and Experiments I**
Lattice Dynamics and Models of Interatomic Forces By P. Brüesch

35 **Point Defects in Semiconductors II**
Experimental Aspects
By J. Bourgoin and M. Lannoo

36 **Modern Crystallography III**
Crystal Growth
By A. A. Chernov

37 **Modern Chrystallography IV**
Physical Properties of Crystals
Editor: L. A. Shuvalov

38 **Physics of Intercalation Compounds**
Editors: L. Pietronero and E. Tosatti

39 **Anderson Localization**
Editors: Y. Nagaoka and H. Fukuyama

40 **Semiconductor Physics** An Introduction
6th Edition By K. Seeger

41 **The LMTO Method**
Muffin-Tin Orbitals and Electronic Structure
By H. L. Skriver

42 **Crystal Optics with Spatial Dispersion, and Excitons** 2nd Edition
By V. M. Agranovich and V. L. Ginzburg

43 **Structure Analysis of Point Defects in Solids**
An Introduction to Multiple Magnetic Resonance Spectroscopy
By J.-M. Spaeth, J. R. Niklas, and R. H. Bartram

44 **Elastic Media with Microstructure II**
Three-Dimensional Models By I. A. Kunin

45 **Electronic Properties of Doped Semiconductors**
By B. I. Shklovskii and A. L. Efros

46 **Topological Disorder in Condensed Matter**
Editors: F. Yonezawa and T. Ninomiya

Springer Series in Solid-State Sciences

Editors: M. Cardona P. Fulde K. von Klitzing H.-J. Queisser

47 **Statics and Dynamics of Nonlinear Systems**
Editors: G. Benedek, H. Bilz, and R. Zeyher

48 **Magnetic Phase Transitions**
Editors: M. Ausloos and R. J. Elliott

49 **Organic Molecular Aggregates**
Electronic Excitation and Interaction Processes
Editors: P. Reineker, H. Haken, and H. C. Wolf

50 **Multiple Diffraction of X-Rays in Crystals**
By Shih-Lin Chang

51 **Phonon Scattering in Condensed Matter**
Editors: W. Eisenmenger, K. Laßmann, and S. Döttinger

52 **Superconductivity in Magnetic and Exotic Materials** Editors: T. Matsubara and A. Kotani

53 **Two-Dimensional Systems, Heterostructures, and Superlattices**
Editors: G. Bauer, F. Kuchar, and H. Heinrich

54 **Magnetic Excitations and Fluctuations**
Editors: S. W. Lovesey, U. Balucani, F. Borsa, and V. Tognetti

55 **The Theory of Magnetism II** Thermodynamics and Statistical Mechanics By D. C. Mattis

56 **Spin Fluctuations in Itinerant Electron Magnetism** By T. Moriya

57 **Polycrystalline Semiconductors**
Physical Properties and Applications
Editor: G. Harbeke

58 **The Recursion Method and Its Applications**
Editors: D. G. Pettifor and D. L. Weaire

59 **Dynamical Processes and Ordering on Solid Surfaces** Editors: A. Yoshimori and M. Tsukada

60 **Excitonic Processes in Solids**
By M. Ueta, H. Kanzaki, K. Kobayashi, Y. Toyozawa, and E. Hanamura

61 **Localization, Interaction, and Transport Phenomena** Editors: B. Kramer, G. Bergmann, and Y. Bruynseraede

62 **Theory of Heavy Fermions and Valence Fluctuations** Editors: T. Kasuya and T. Saso

63 **Electronic Properties of Polymers and Related Compounds**
Editors: H. Kuzmany, M. Mehring, and S. Roth

64 **Symmetries in Physics** Group Theory Applied to Physical Problems 2nd Edition
By W. Ludwig and C. Falter

65 **Phonons: Theory and Experiments II**
Experiments and Interpretation of Experimental Results By P. Brüesch

66 **Phonons: Theory and Experiments III**
Phenomena Related to Phonons
By P. Brüesch

67 **Two-Dimensional Systems: Physics and New Devices**
Editors: G. Bauer, F. Kuchar, and H. Heinrich

68 **Phonon Scattering in Condensed Matter V**
Editors: A. C. Anderson and J. P. Wolfe

69 **Nonlinearity in Condensed Matter**
Editors: A. R. Bishop, D. K. Campbell, P. Kumar, and S. E. Trullinger

70 **From Hamiltonians to Phase Diagrams**
The Electronic and Statistical-Mechanical Theory of sp-Bonded Metals and Alloys By J. Hafner

71 **High Magnetic Fields in Semiconductor Physics**
Editor: G. Landwehr

72 **One-Dimensional Conductors**
By S. Kagoshima, H. Nagasawa, and T. Sambongi

73 **Quantum Solid-State Physics**
Editors: S. V. Vonsovsky and M. I. Katsnelson

74 **Quantum Monte Carlo Methods in Equilibrium and Nonequilibrium Systems** Editor: M. Suzuki

75 **Electronic Structure and Optical Properties of Semiconductors** 2nd Edition
By M. L. Cohen and J. R. Chelikowsky

76 **Electronic Properties of Conjugated Polymers**
Editors: H. Kuzmany, M. Mehring, and S. Roth

77 **Fermi Surface Effects**
Editors: J. Kondo and A. Yoshimori

78 **Group Theory and Its Applications in Physics**
2nd Edition
By T. Inui, Y. Tanabe, and Y. Onodera

79 **Elementary Excitations in Quantum Fluids**
Editors: K. Ohbayashi and M. Watabe

80 **Monte Carlo Simulation in Statistical Physics**
An Introduction 3rd Edition
By K. Binder and D. W. Heermann

81 **Core-Level Spectroscopy in Condensed Systems**
Editors: J. Kanamori and A. Kotani

82 **Photoelectron Spectroscopy**
Principle and Applications 2nd Edition
By S. Hüfner

83 **Physics and Technology of Submicron Structures**
Editors: H. Heinrich, G. Bauer, and F. Kuchar

84 **Beyond the Crystalline State** An Emerging Perspective By G. Venkataraman, D. Sahoo, and V. Balakrishnan

85 **The Quantum Hall Effects**
Fractional and Integral 2nd Edition
By T. Chakraborty and P. Pietiläinen

86 **The Quantum Statistics of Dynamic Processes**
By E. Fick and G. Sauermann

87 **High Magnetic Fields in Semiconductor Physics II**
Transport and Optics Editor: G. Landwehr

88 **Organic Superconductors** 2nd Edition
By T. Ishiguro, K. Yamaji, and G. Saito

89 **Strong Correlation and Superconductivity**
Editors: H. Fukuyama, S. Maekawa, and A. P. Malozemoff

Springer Series in Solid-State Sciences

Editors: M. Cardona P. Fulde K. von Klitzing H.-J. Queisser

Managing Editor: H. K.V. Lotsch

90 **Earlier and Recent Aspects of Superconductivity**
Editors: J. G. Bednorz and K. A. Müller

91 **Electronic Properties of Conjugated Polymers III** Basic Models and Applications
Editors: H. Kuzmany, M. Mehring, and S. Roth

92 **Physics and Engineering Applications of Magnetism** Editors: Y. Ishikawa and N. Miura

93 **Quasicrystals** Editors: T. Fujiwara and T. Ogawa

94 **Electronic Conduction in Oxides**
By N. Tsuda, K. Nasu, A. Yanase, and K. Siratori

95 **Electronic Materials**
A New Era in Materials Science
Editors: J. R. Chelikowsky and A. Franciosi

96 **Electron Liquids** 2nd Edition By A. Isihara

97 **Localization and Confinement of Electrons in Semiconductors**
Editors: F. Kuchar, H. Heinrich, and G. Bauer

98 **Magnetism and the Electronic Structure of Crystals** By V.A. Gubanov, A.I. Liechtenstein, and A.V. Postnikov

99 **Electronic Properties of High-T_c Superconductors and Related Compounds**
Editors: H. Kuzmany, M. Mehring, and J. Fink

100 **Electron Correlations in Molecules and Solids** 3rd Edition By P. Fulde

101 **High Magnetic Fields in Semiconductor Physics III** Quantum Hall Effect, Transport and Optics By G. Landwehr

102 **Conjugated Conducting Polymers**
Editor: H. Kiess

103 **Molecular Dynamics Simulations**
Editor: F. Yonezawa

104 **Products of Random Matrices** in Statistical Physics By A. Crisanti, G. Paladin, and A. Vulpiani

105 **Self-Trapped Excitons**
2nd Edition By K. S. Song and R. T. Williams

106 **Physics of High-Temperature Superconductors**
Editors: S. Maekawa and M. Sato

107 **Electronic Properties of Polymers**
Orientation and Dimensionality of Conjugated Systems Editors: H. Kuzmany, M. Mehring, and S. Roth

108 **Site Symmetry in Crystals**
Theory and Applications 2nd Edition
By R. A. Evarestov and V. P. Smirnov

109 **Transport Phenomena in Mesoscopic Systems** Editors: H. Fukuyama and T. Ando

110 **Superlattices and Other Heterostructures**
Symmetry and Optical Phenomena 2nd Edition
By E. L. Ivchenko and G. E. Pikus

111 **Low-Dimensional Electronic Systems**
New Concepts
Editors: G. Bauer, F. Kuchar, and H. Heinrich

112 **Phonon Scattering in Condensed Matter VII**
Editors: M. Meissner and R. O. Pohl

113 **Electronic Properties of High-T_c Superconductors**
Editors: H. Kuzmany, M. Mehring, and J. Fink

114 **Interatomic Potential and Structural Stability**
Editors: K. Terakura and H. Akai

115 **Ultrafast Spectroscopy of Semiconductors and Semiconductor Nanostructures**
2nd Edition By J. Shah

116 **Electron Spectrum of Gapless Semiconductors**
By J. M. Tsidilkovski

117 **Electronic Properties of Fullerenes**
Editors: H. Kuzmany, J. Fink, M. Mehring, and S. Roth

118 **Correlation Effects in Low-Dimensional Electron Systems**
Editors: A. Okiji and N. Kawakami

119 **Spectroscopy of Mott Insulators and Correlated Metals**
Editors: A. Fujimori and Y. Tokura

120 **Optical Properties of III–V Semiconductors**
The Influence of Multi-Valley Band Structures
By H. Kalt

121 **Elementary Processes in Excitations and Reactions on Solid Surfaces**
Editors: A. Okiji, H. Kasai, and K. Makoshi

122 **Theory of Magnetism**
By K. Yosida

123 **Quantum Kinetics in Transport and Optics of Semiconductors**
By H. Haug and A.-P. Jauho

124 **Relaxations of Excited States and Photo-Induced Structural Phase Transitions**
Editor: K. Nasu

125 **Physics and Chemistry of Transition-Metal Oxides**
Editors: H. Fukuyama and N. Nagaosa

Printing (Computer to Film): Saladruck, Berlin
Binding: Lüderitz & Bauer, Berlin

GPSR Compliance

The European Union's (EU) General Product Safety Regulation (GPSR) is a set of rules that requires consumer products to be safe and our obligations to ensure this.

If you have any concerns about our products, you can contact us on ProductSafety@springernature.com

In case Publisher is established outside the EU, the EU authorized representative is:

Springer Nature Customer Service Center GmbH
Europaplatz 3
69115 Heidelberg, Germany

Batch number: 10755284

Printed by Printforce, the Netherlands